ENGINEERING ELECTROMAGNETICS

McGraw-Hill Series in Electrical Engineering

Consulting Editor

Stephen W. Director, *Carnegie-Mellon University*

Circuits and Systems
Communications and Signal Processing
Control Theory
Electronics and Electronic Circuits
Power and Energy
Electromagnetics
Computer Engineering
Introductory
Radar and Antennas
VLSI

Previous Consulting Editors

Ronald N. Bracewell, Colin Cherry, James F. Gibbons, Willis W. Harman, Hubert Heffner, Edward W. Herold, John G. Linvill, Simon Ramo, Ronald A. Rohrer, Anthony E. Siegman, Charles Susskind, Frederick E. Terman, John G. Truxal, Ernst Weber, and John R. Whinnery

Electromagnetics

Consulting Editor

Stephen W. Director, *Carnegie-Mellon University*

Chen: *Theory of Electromagnetic Waves: A Coordinate-Free Approach*
Dearhold and McSpadden: *Electromagnetic Wave Propagation*
Goodman: *Introduction to Fourier Optics*
Hayt: *Engineering Electromagnetics*
Kraus: *Electromagnetics*
Paul and Nasar: *Introduction to Electromagnetic Fields*
Plonus: *Applied Electromagnetics*

ENGINEERING ELECTROMAGNETICS

Fifth Edition

William H. Hayt, Jr.

Professor of Electrical Engineering,
Emeritus
Purdue University

McGraw-Hill, Inc.

New York St. Louis San Francisco Auckland Bogotá
Caracas Lisbon London Madrid Mexico Milan
Montreal New Delhi Paris San Juan Singapore
Sydney Tokyo Toronto

This book was set in Times Roman by Syntax International.
The editors were Alar E. Elken and Steven Tenney;
the designer was Amy Becker;
the production supervisor was Diane Renda.
New drawings were done by J & R Services, Inc.
Arcata Graphics/Martinsburg was printer and binder.

ENGINEERING ELECTROMAGNETICS

6 7 8 9 0 AGM AGM 9 0 9 8 7 6 5 4

ISBN 0-07-027406-1

Library of Congress Cataloging-in-Publication Data

Hayt, William Hart, (date).
Engineering electromagnetics/William H. Hayt, Jr.—5th ed.
p. cm.
Includes bibliographies and index.
ISBN 0-07-027406-1
1. Electromagnetic theory. I. Title.
QC670.H39 1989
530.1'41—dc19 88–22030

This book is printed on acid-free paper.

CONTENTS

Preface xi

Chapter 1 Vector Analysis 1

1.1 Scalars and Vectors 2
1.2 Vector Algebra 3
1.3 The Cartesian Coordinate System 4
1.4 Vector Components and Unit Vectors 6
1.5 The Vector Field 9
1.6 The Dot Product 10
1.7 The Cross Product 12
1.8 Other Coordinate Systems: Circular Cylindrical Coordinates 14
1.9 The Spherical Coordinate System 19

Chapter 2 Coulomb's Law and Electric Field Intensity 26

2.1 The Experimental Law of Coulomb 27
2.2 Electric Field Intensity 30
2.3 Field Due to a Continuous Volume Charge Distribution 34
2.4 Field of a Line Charge 37
2.5 Field of a Sheet of Charge 43
2.6 Streamlines and Sketches of Fields 45

Chapter 3 Electric Flux Density, Gauss's Law, and Divergence 52

3.1 Electric Flux Density 52
3.2 Gauss's Law 56
3.3 Application of Gauss's Law: Some Symmetrical Charge Distributions 60
3.4 Application of Gauss's Law: Differential Volume Element 65
3.5 Divergence 68
3.6 Maxwell's First Equation (Electrostatics) 71
3.7 The Vector Operator ∇ and the Divergence Theorem 72

Chapter 4 Energy and Potential 80

4.1 Energy Expended in Moving a Point Charge in an Electric Field 81
4.2 The Line Integral 82
4.3 Definition of Potential Difference and Potential 87
4.4 The Potential Field of a Point Charge 89
4.5 The Potential Field of a System of Charges: Conservative Property 91
4.6 Potential Gradient 95
4.7 The Dipole 102
4.8 Energy Density in the Electrostatic Field 106

Chapter 5 Conductors, Dielectrics, and Capacitance 114

5.1 Current and Current Density 115
5.2 Continuity of Current 117
5.3 Metallic Conductors 119
5.4 Conductor Properties and Boundary Conditions 124
5.5 The Method of Images 129
5.6 Semiconductors 131
5.7 The Nature of Dielectric Materials 132
5.8 Boundary Conditions for Perfect Dielectric Materials 138
5.9 Capacitance 144
5.10 Several Capacitance Examples 147
5.11 Capacitance of a Two-Wire Line 150

Chapter 6 Experimental Mapping Methods 162

6.1 Curvilinear Squares 163
6.2 The Iteration Method 169
6.3 Current Analogies 175
6.4 Physical Models 178

Chapter 7 Poisson's and Laplace's Equations 188

7.1 Poisson's and Laplace's Equations 189
7.2 Uniqueness Theorem 191
7.3 Examples of the Solution of Laplace's Equation 193
7.4 Example of the Solution of Poisson's Equation 200
7.5 Product Solution of Laplace's Equation 204

Chapter 8 The Steady Magnetic Field 216

8.1 Biot-Savart Law 217
8.2 Ampère's Circuital Law 224
8.3 Curl 231
8.4 Stokes' Theorem 238
8.5 Magnetic Flux and Magnetic Flux Density 243
8.6 The Scalar and Vector Magnetic Potentials 246
8.7 Derivation of Steady-Magnetic-Field Laws 253

Chapter 9 Magnetic Forces, Materials, and Inductance 265

9.1 Force on a Moving Charge 266
9.2 Force on a Differential Current Element 267
9.3 Force between Differential Current Elements 271
9.4 Force and Torque on a Closed Circuit 273
9.5 The Nature of Magnetic Materials 278
9.6 Magnetization and Permeability 282
9.7 Magnetic Boundary Conditions 287
9.8 The Magnetic Circuit 289
9.9 Potential Energy and Forces on Magnetic Materials 296
9.10 Inductance and Mutual Inductance 298

Chapter 10 Time-Varying Fields and Maxwell's Equations 311

10.1 Faraday's Law 312
10.2 Displacement Current 318
10.3 Maxwell's Equations in Point Form 322
10.4 Maxwell's Equations in Integral Form 324
10.5 The Retarded Potentials 326

Chapter 11 The Uniform Plane Wave 336

11.1 Wave Motion in Free Space 336
11.2 Wave Motion in Perfect Dielectrics 345
11.3 Plane Waves in Lossy Dielectrics 347
11.4 The Poynting Vector and Power Considerations 353
11.5 Propagation in Good Conductors: Skin Effect 357
11.6 Reflection of Uniform Plane Waves 364
11.7 Standing-Wave Ratio 370

Chapter 12 Transmission Lines 381

12.1 The Transmission-Line Equations 382
12.2 Transmission-Line Parameters 387
12.3 Some Transmission-Line Examples 393
12.4 Graphical Methods 397
12.5 Several Practical Problems 404

Chapter 13 Several Other Applications of Maxwell's Equations 414

13.1 The Laws of Circuit Theory 414
13.2 The Resonant Coaxial Cavity 417
13.3 Radiation 427

Appendix A Vector Analysis 440

A.1 General Curvilinear Coordinates 440
A.2 Divergence, Gradient, and Curl in General Curvilinear Coordinates 441
A.3 Vector Identities 443

Appendix B Units 445

Appendix C Material Constants 450

Appendix D Answers to Odd-Numbered Problems 454

Index 461

PREFACE

Although most electrical engineering curricula begin with a study of electric circuits, it is now recognized that the more basic theory of the electric and magnetic fields deserves subsequent attention in the curriculum. Some familiarity with circuit concepts as well as a knowledge of calculus allows a treatment of field theory in the junior year that proceeds through Maxwell's equations and justifies the approximations leading to circuit theory.

This textbook uses Maxwell's equations as the central theme. These equations are developed from the historical approach in which the relevant experimental laws are gradually introduced and manipulated with the help of a steadily increasing knowledge of vector calculus. Maxwell's equations are identified as they occur, even as they apply to static fields, and a certain sense of accomplishment and perhaps even familiarity should be felt when the theory is finally completed. Several applications of these equations are described in the final chapters, including wave motion, skin effect, transmission-line phenomena, circuit theory, and the resonant cavity. A first look at radiation and antennas is also included.

The material is more than adequate for a one-semester course. Depending on the instructional level, it may be desirable to omit portions of the chapters on experimental mapping methods, the solutions of Laplace's equation, transmission lines, or the application of Maxwell's equations.

This book has been written with the goal of making it as easy as possible for the student to learn independently. This has been done by applying a carefully graduated scale of difficulty within each chapter and among the chapters themselves, by providing numerical examples wherever possible and numerous examples interpreting and applying every basic result, and by including a large number of drill problems with answers. Excessive reliance upon analytic and vector geometry and their involvement with interpreting fields has been avoided.

The more difficult material has been placed near the ends of the chapters or at the end of the study of some definite phase of the subject. The slower student, who cannot assimilate the same amount of material as a faster one, will thus be attracted to the more basic material at the beginning of each chapter. Since the subject matter of the following chapter is not usually built on the most advanced material of the previous chapter, the student can proceed without mastering that material. The more advanced material does, however, offer a needed challenge to the better students.

Drill problems are placed at the ends of most sections in which a formula or law is introduced that is capable of expression in problem form. The drill problems usually have several parts, and as a self-study aid, all the answers are given immediately below the problem in the same order as the parts of the problem. The problems appearing at the ends of the chapters are a little more difficult and perhaps a little more interesting; the answers to the odd-numbered ones appear in Appendix D, and the solutions to all the problems are given in the booklet that instructors may request from the publisher. The order of the problems corresponds to that of the text material.

The problems in this edition are all new, except for four or five well-used problems that supply missing steps in the theory. A few other problems that are favorites of the author have been thinly disguised and used again. There are also a few more problems in this edition than there are in any previous edition. As a parenthetical remark to the instructor, most of the problems in the other editions are still quite applicable, although an occasional symbol change is needed. The proliferation of copying machines and printing methods also makes it easy to use these older problems, and the publisher freely gives permission to cannibalize the earlier editions. And as a parenthetical remark to the student, note that anyone who works all the problems in all the editions is very hard to surprise on the next test.

Most of the goals of the first four editions continue to be valid here. However, over 30 years of usage have prompted many students, colleagues, and correspondents to suggest improvements which are now part of the text.

A majority of the changes involve additional numerical examples. Thus, in Chap. 1, an example has been added to illustrate the computation of scalar and vector projections. Sections 2.2 and 2.3 of the fourth edition have been combined in a new Sec. 2.2 in the fifth edition. This section also includes a new example showing the calculation of the electric field caused by several point charges. The nomenclature used in the uniform line charge examples has been improved, and a numerical exercise involving a line charge that does not lie along the z axis has been added. A new example in Sec. 2.6 illustrates a method of finding the equation of a specific streamline passing through a specific point.

Four new numerical examples have been inserted in Chap. 3, and Chap. 4 includes new examples illustrating the application of a closed line integral in a nonconservative field as well as several applications of the gradient.

Some improved numerical values are used in Sec. 5.6, and five additional new examples appear in Chap. 5.

Chapter 6 now includes the concept of sheet resistance, and a new example showing how the iteration method may be applied to a configuration having irregular or curved boundaries.

A numerical example showing the procedure for calculating the magnetic field of a finite length current element appears in Sec. 8.1.

Chapter 9 has an improved description of the Hall-effect voltage and the calculation of force and torque between current-carrying conductors. A new table has been included to provide a direct comparison of the six different types of magnetic materials. An example is also included that illustrates the use of the boundary conditions in calculating values of the magnetic field.

A new numerical example in Chap. 11 helps illustrate the conversion between the frequency and time domains for both vector and scalar fields. The relationship between velocity, wavelength, and the phase constant is explained and illustrated more carefully. Also, the numerical accuracy has been improved in the example illustrating the calculation of the parameters for a uniform plane wave propagating in a lossy dielectric. The same parameters are then calculated by approximate methods and the results are compared with the more accurate values. Finally, numerical examples have been included that should help clarify the Poynting-vector calculations for a uniform plane wave traveling through a lossy dielectric as well as in a good conductor.

Section 12.2 on transmission-line parameters has been largely rewritten to present the formulas for calculating C, L, G, R, and Z_0 for the coaxial, two-wire, and strip transmission lines at both low and high frequencies. A new example shows the calculation of the reflection coefficient, standing-wave ratio, and input impedance for a line terminated in a general load impedance. Also, additional explanatory material has been inserted to illustrate the interpretation and use of the Smith chart.

Appendix B has been rewritten to conform with the recent definition of the velocity of light as 299 792 458 meters per second. Table C.4 in Appendix C now gives improved values for the charge and mass of an electron, the permittivity of free space, and the velocity of light.

A systematic change found in the fifth edition is the use of ρ_v for volume charge density.

The bibliographies appearing at the ends of all the chapters include several new books, are missing a few older references, and cite the latest edition of each work.

The author again gratefully acknowledges the help of Professor W. L. Weeks of Purdue University in spotting errata in the manuscript. Many additional reviewers have also suggested a number of the changes that have been made in this edition. These include John A. Buck, Georgia Institute of Technology; Robert Chute, Lawrence Institute of Technology; Walter Kurt Kahn, George Washington University; Sigrid McAfce, Rutgers University; Leonard S. Taylor, University of Maryland; E. M. Thomson, University of Florida; Martin Uman, University of Florida; and Walter Weeks, Purdue University. Students have also written to point out errors and to identify confusing paragraphs. All this help is greatly appreciated, and the author sincerely hopes this group effort will continue.

William H. Hayt, Jr.

ENGINEERING ELECTROMAGNETICS

CHAPTER 1

VECTOR ANALYSIS

Vector analysis is a mathematical subject which is much better taught by mathematicians than by engineers. Most junior and senior engineering students, however, have not had the time (or perhaps the inclination) to take a course in vector analysis, although it is likely that many elementary vector concepts and operations were introduced in the calculus sequence. These fundamental concepts and operations are covered in this chapter, and the time devoted to them now should depend on past exposure.

The viewpoint here is also that of the engineer or physicist and not that of the mathematician in that proofs are indicated rather than rigorously expounded and the physical interpretation is stressed. It is easier for engineers to take a more rigorous and complete course in the mathematics department after they have been presented with a few physical pictures and applications.

It is possible to study electricity and magnetism without the use of vector analysis, and some engineering students may have done so in a previous electrical engineering or basic physics course. Carrying this elementary work a bit further, however, soon leads to line-filling equations often composed of terms which all look about the same. A quick glance at one of these long equations discloses little of the physical nature of the equation and may even lead to slighting an old friend.

Vector analysis is a mathematical shorthand. It has some new symbols, some new rules, and a pitfall here and there like most new fields, and it demands concentration, attention, and practice. The drill problems, first met at the end of Sec. 1.4, should be considered an integral part of the text and should all be worked. They should not prove to be difficult if the material in the accompanying section of the text has been thoroughly understood. It takes a little longer to "read" the chapter this way, but the investment in time will produce a surprising interest.

1.1 SCALARS AND VECTORS

The term *scalar* refers to a quantity whose value may be represented by a single (positive or negative) real number. The x, y, and z we used in basic algebra are scalars, and the quantities they represent are scalars. If we speak of a body falling a distance L in a time t, or the temperature T at any point in a bowl of soup whose coordinates are x, y, and z, then L, t, T, x, y, and z are all scalars. Other scalar quantities are mass, density, pressure (but not force), volume, and volume resistivity. Voltage is also a scalar quantity, although the complex representation of a sinusoidal voltage, an artificial procedure, produces a *complex scalar*, or *phasor*, which requires two real numbers for its representation, such as amplitude and phase angle, or real part and imaginary part.

A *vector* quantity has both a magnitude[1] and a direction in space. We shall be concerned with two- and three-dimensional spaces only, but vectors may be defined in n-dimensional space in more advanced applications. Force, velocity, acceleration, and a straight line from the positive to the negative terminal of a storage battery are examples of vectors. Each quantity is characterized by both a magnitude and a direction.

We shall be mostly concerned with scalar and vector *fields*. A field (scalar or vector) may be defined mathematically as some function of that vector which connects an arbitrary origin to a general point in space. We usually find it possible to associate some physical effect with a field, such as the force on a compass needle in the earth's magnetic field, or the movement of smoke particles in the field defined by the vector velocity of air in some region of space. Note that the field concept invariably is related to a region. Some quantity is defined at every point in a region. Both *scalar fields* and *vector fields* exist. The temperature throughout the bowl of soup and the density at any point in the earth are examples of scalar fields. The gravitational and magnetic fields of the earth, the voltage gradient in a cable, and the temperature gradient in a soldering-iron tip are examples of vector fields. The value of a field varies in general with both position and time.

In this book, as in most others using vector notation, vectors will be indicated by boldface type, for example, **A**. Scalars are printed in italic type, for example, A. When writing longhand or using a typewriter, it is customary to draw a line or an arrow over a vector quantity to show its vector character. (CAUTION: This is the first pitfall. Sloppy notation, such as the omission of the line or arrow symbol for a vector, is the major cause of errors in vector analysis.)

---

[1] We adopt the convention that "magnitude" infers "absolute value"; the magnitude of any quantity is therefore always positive.

1.2 VECTOR ALGEBRA

With the definitions of vectors and vec[illegible] ceed to define the rules of vector arithme[illegible] calculus. Some of the rules will be similar to [illegible] differ slightly, and some will be entirely new and [illegible] for a vector represents more information than does [illegible]lished, we may pro- tion of two vectors, for example, will be more involved [illegible](later) of vector two scalars. [illegible]ra, some will

The rules are those of a branch of mathematics which is expected, Everyone "plays by the same rules," and we, of course, are merely [illegible]plica- at and interpret these rules. However, it is enlightening to conside[illegible]of pioneers in the field. We are making our own rules, and we can make any [illegible] wish. The only requirement is that the rules be self-consistent. Of course, it w[illegible] be nice if the rules agreed with those of scalar algebra where possible, and it woul[illegible] be even nicer if the rules enabled us to solve a few practical problems.

One should not fall into the trap of "algebra worship" and believe that the rules of college algebra were delivered unto man at the Creation. These rules are merely self-consistent and extremely useful. There are other less familiar algebras, however, with very different rules. In Boolean algebra the product AB can be only unity or zero. Vector algebra has its own set of rules, and we must be constantly on guard against the mental forces exerted by the more familiar rules of scalar algebra.

Vectorial addition follows the parallelogram law, and this is easily, if inaccurately, accomplished graphically. Figure 1.1 shows the sum of two vectors, $\mathbf{A}$ and $\mathbf{B}$. It is easily seen that $\mathbf{A} + \mathbf{B} = \mathbf{B} + \mathbf{A}$, or that vector addition obeys the commutative law. Vector addition also obeys the associative law,

$$\mathbf{A} + (\mathbf{B} + \mathbf{C}) = (\mathbf{A} + \mathbf{B}) + \mathbf{C}$$

Note that when a vector is drawn as an arrow of finite length, its location is defined to be at the tail end of the arrow.

Coplanar vectors, or vectors lying in a common plane, such as those shown in Fig. 1.1, which both lie in the plane of the paper, may also be added by expressing each vector in terms of "horizontal" and "vertical" components and adding the corresponding components.

FIGURE 1.1
Two vectors may be added graphically either by drawing both vectors from a common origin and completing the parallelogram or by beginning the second vector from the head of the first and completing the triangle; either method is easily extended to three or more vectors.

...y likewise be added by expressing the vectors ...nd adding the corresponding components. Ex-...ation will be given after vector components are dis-

Vectors ...subtraction of vectors follows easily from that for addition, in terms ...express **A** − **B** as **A** + (−**B**); the sign and direction of the second ample...ssed, and this vector is then added to the first by the rule for vector cus...

...ctors may be multiplied by scalars. The magnitude of the vector changes, ...ts direction does not when the scalar is positive, although it reverses direction when multiplied by a negative scalar. Multiplication of a vector by a scalar also obeys the associative and distributive laws of algebra, leading to

$$(r + s)(\mathbf{A} + \mathbf{B}) = r(\mathbf{A} + \mathbf{B}) + s(\mathbf{A} + \mathbf{B}) = r\mathbf{A} + r\mathbf{B} + s\mathbf{A} + s\mathbf{B}$$

Division of a vector by a scalar is merely multiplication by the reciprocal of that scalar.

The multiplication of a vector by a vector is discussed in Secs. 1.6 and 1.7 below.

Two vectors are said to be equal if their difference is zero, or $\mathbf{A} = \mathbf{B}$ if $\mathbf{A} - \mathbf{B} = 0$.

In our use of vector fields we shall always add and subtract vectors which are defined at the same point. For example, the *total* magnetic field about a small horseshoe magnet will be shown to be the sum of the fields produced by the earth and the permanent magnet; the total field at any point is the sum of the individual fields at that point.

If we are not considering a vector *field*, however, we may add or subtract vectors which are not defined at the same point. For example, the sum of the gravitational force acting on a 150-lb_f (pound-force) man at the North Pole and that acting on a 175-lb_f man at the South Pole may be obtained by shifting each force vector to the South Pole before addition. The resultant is a force of 25 lb_f directed toward the center of the earth at the South Pole; if we wanted to be difficult, we could just as well describe the force as 25 lb_f directed *away* from the center of the earth (or "upward") at the North Pole.[2]

1.3 THE CARTESIAN COORDINATE SYSTEM

In order to describe a vector accurately, some specific lengths, directions, angles, projections, or components must be given. There are three simple methods of doing this, and about eight or ten other methods which are useful in very special cases. We are going to use only the three simple methods, and the simplest of these is the *cartesian*, or *rectangular*, *coordinate system.*

---

[2] Someone has also pointed out that the force might be described at the equator as being in a "northerly" direction. He is right, but enough is enough.

In the cartesian coordinate system we set up three coordinate axes mutually at right angles to each other, and call them the x, y, and z axes. It is customary to choose a *right-handed* coordinate system, in which a rotation (through the smaller angle) of the x axis into the y axis would cause a right-handed screw to progress in the direction of the z axis. Using the right hand, the thumb, forefinger, and middle finger may then be identified, respectively, as the x, y, and z axes. Figure 1.2*a* shows a right-handed cartesian coordinate system.

A point is located by giving its x, y, and z coordinates. These are, respectively, the distances from the origin to the intersection of a perpendicular dropped

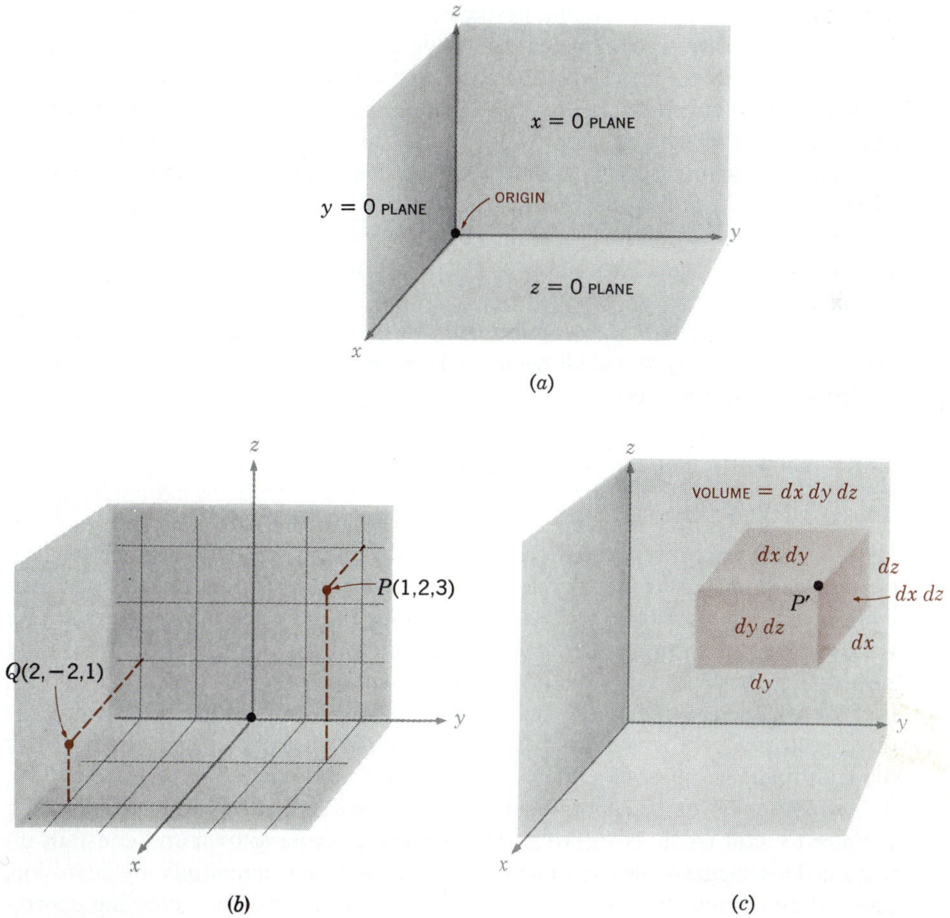

FIGURE 1.2
(*a*) A right-handed cartesian coordinate system. If the curved fingers of the right hand indicate the direction through which the x axis is turned into coincidence with the y axis, the thumb shows the direction of the z axis. (*b*) The location of points $P(1,2,3)$ and $Q(2,-2,1)$. (*c*) The differential volume element in cartesian coordinates; dx, dy, and dz are, in general, independent differentials.

from the point to the x, y, and z axes. An alternative method of interpreting coordinate values, and a method corresponding to that which *must* be used in all other coordinate systems, is to consider the point as being at the common intersection of three surfaces, the planes $x =$ constant, $y =$ constant, and $z =$ constant, the constants being the coordinate values of the point.

Figure 1.2*b* shows the points P and Q whose coordinates are (1,2,3) and (2, −2,1), respectively. Point P is therefore located at the common point of intersection of the planes $x = 1$, $y = 2$, and $z = 3$, while point Q is located at the intersection of the planes $x = 2$, $y = -2$, $z = 1$.

As we encounter other coordinate systems in Secs. 1.8 and 1.9, we should expect points to be located by the common intersection of three surfaces, not necessarily planes, but still mutually perpendicular at the point of intersection.

If we visualize three planes intersecting at the general point P, whose coordinates are x, y, and z, we may increase each coordinate value by a differential amount and obtain three slightly displaced planes intersecting at point P', whose coordinates are $x + dx$, $y + dy$, and $z + dz$. The six planes define a rectangular parallelepiped whose volume is $dv = dx\,dy\,dz$; the surfaces have differential areas dS of $dx\,dy$, $dy\,dz$, and $dz\,dx$. Finally, the distance dL from P to P' is the diagonal of the parallelepiped and has a length of $\sqrt{(dx)^2 + (dy)^2 + (dz)^2}$. The volume element is shown in Fig. 1.2*c*; point P' is indicated, but point P is located at the only invisible corner.

All this is familiar from trigonometry or solid geometry and as yet involves only scalar quantities. We shall begin to describe vectors in terms of a coordinate system in the next section.

1.4 VECTOR COMPONENTS AND UNIT VECTORS

To describe a vector in the cartesian coordinate system, let us first consider a vector $\mathbf{r}$ extending outward from the origin. A logical way to identify this vector is by giving the three *component vectors*, lying along the three coordinate axes, whose vector sum must be the given vector. If the component vectors of the vector $\mathbf{r}$ are $\mathbf{x}$, $\mathbf{y}$, and $\mathbf{z}$, then $\mathbf{r} = \mathbf{x} + \mathbf{y} + \mathbf{z}$. The component vectors are shown in Fig. 1.3*a*. Instead of one vector, we now have three, but this is a step forward, because the three vectors are of a very simple nature; each is always directed along one of the coordinate axes.

In other words, the component vectors have a magnitude which depends on the given vector (such as $\mathbf{r}$ above), but they each have a known and constant direction. This suggests the use of *unit vectors* having unit magnitude, by definition, and directed along the coordinate axes in the direction of the increasing coordinate values. We shall reserve the symbol $\mathbf{a}$ for a unit vector and identify the direction of the unit vector by an appropriate subscript. Thus $\mathbf{a}_x$, $\mathbf{a}_y$, and $\mathbf{a}_z$ are the unit vectors in the cartesian coordinate system.[3] They are directed along the x, y, and z axes, respectively, as shown in Fig. 1.3*b*.

---

[3] The symbols **i**, **j**, and **k** are also commonly used for the unit vectors in cartesian coordinates.

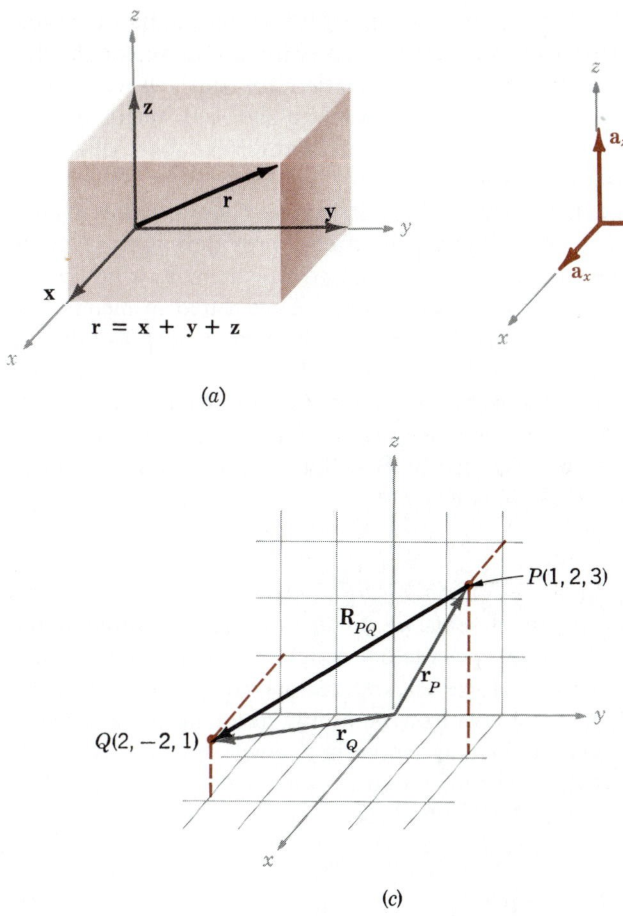

FIGURE 1.3
(*a*) The component vectors **x**, **y**, and **z** of vector **r**. (*b*) The unit vectors of the cartesian coordinate system have unit magnitude and are directed toward increasing values of their respective variables. (*c*) The vector $\mathbf{R}_{PQ}$ is equal to the vector difference $\mathbf{r}_Q - \mathbf{r}_P$.

If the component vector **y** happens to be two units in magnitude and directed toward increasing values of *y*, we should then write $\mathbf{y} = 2\mathbf{a}_y$. A vector $\mathbf{r}_P$ pointing from the origin to point $P(1,2,3)$ is written $\mathbf{r}_P = \mathbf{a}_x + 2\mathbf{a}_y + 3\mathbf{a}_z$. The vector from *P* to *Q* may be obtained by applying the rule of vector addition. This rule shows that the vector from the origin to *P* plus the vector from *P* to *Q* is equal to the vector from the origin to *Q*. The desired vector from $P(1,2,3)$ to $Q(2,-2,1)$ is therefore

$$\mathbf{R}_{PQ} = \mathbf{r}_Q - \mathbf{r}_P = (2-1)\mathbf{a}_x + (-2-2)\mathbf{a}_y + (1-3)\mathbf{a}_z$$
$$= \mathbf{a}_x - 4\mathbf{a}_y - 2\mathbf{a}_z$$

The vectors $\mathbf{r}_P$, $\mathbf{r}_Q$, and $\mathbf{R}_{PQ}$ are shown in Fig. 1.3*c*.

This last vector does not extend outward from the origin, as did the vector **r** we initially considered. However, we have already learned that vectors having the same magnitude and pointing in the same direction are equal, so we see that to help our visualization processes we are at liberty to slide any vector over to the origin before determining its component vectors. Parallelism must, of course, be maintained during the sliding process.

If we are discussing a force vector **F**, or indeed any vector other than a displacement-type vector such as **r**, the problem arises of providing suitable letters for the three component vectors. It would not do to call them **x**, **y**, and **z**, for these are displacements, or directed distances, and are measured in meters (abbreviated m) or some other unit of length. The problem is most often avoided by using *component scalars*, simply called *components*, F_x, F_y, and F_z. The components are the signed magnitudes of the component vectors. We may then write $\mathbf{F} = F_x\mathbf{a}_x + F_y\mathbf{a}_y + F_z\mathbf{a}_z$. The component vectors are $F_x\mathbf{a}_x$, $F_y\mathbf{a}_y$, and $F_z\mathbf{a}_z$.

Any vector **B** then may be described by $\mathbf{B} = B_x\mathbf{a}_x + B_y\mathbf{a}_y + B_z\mathbf{a}_z$. The magnitude of **B**, written $|\mathbf{B}|$, or simply B, is given by

$$|\mathbf{B}| = \sqrt{B_x^2 + B_y^2 + B_z^2} \tag{1}$$

Each of the three coordinate systems we discuss will have its three fundamental and mutually perpendicular unit vectors which are used to resolve any vector into its component vectors. However, unit vectors are not limited to this application. It is often helpful to be able to write a unit vector having a specified direction. This is simply done, for a unit vector in a given direction is merely a vector in that direction divided by its magnitude. A unit vector in the **r** direction is $\mathbf{r}/\sqrt{x^2 + y^2 + z^2}$, and a unit vector in the direction of the vector **B** is

$$\mathbf{a}_B = \frac{\mathbf{B}}{\sqrt{B_x^2 + B_y^2 + B_z^2}} = \frac{\mathbf{B}}{|\mathbf{B}|} \tag{2}$$

For example, the unit vector directed from the origin toward the point $G(2, -2, -1)$ would be obtained by first designating a vector **G** that extends from the origin to $G(2, -2, -1)$,

$$\mathbf{G} = 2\mathbf{a}_x - 2\mathbf{a}_y - \mathbf{a}_z$$

then finding the magnitude of **G**,

$$|\mathbf{G}| = \sqrt{(2)^2 + (-2)^2 + (-1)^2} = 3$$

and finally expressing the desired unit vector as the quotient,

$$\mathbf{a}_G = \frac{\mathbf{G}}{|\mathbf{G}|} = \tfrac{2}{3}\mathbf{a}_x - \tfrac{2}{3}\mathbf{a}_y - \tfrac{1}{3}\mathbf{a}_z$$

A special identifying symbol is desirable for a unit vector so that its character is immediately apparent. Symbols which have been used are $\mathbf{u}_B$, $\mathbf{a}_B$, $\mathbf{1}_B$, or even $\mathbf{b}$. We shall consistently use the lowercase **a** with an appropriate subscript.

[NOTE: Throughout the text, drill problems appear following sections in which a new principle is introduced in order to allow the student to test his or her understanding of the basic fact itself. The problems are useful in gaining familiarization with new terms and ideas and should all be worked. More general problems appear at the ends of the chapters. The answers to the drill problems are given in the same order as the parts of the problem.]

D1.1. Given the two vectors, $\mathbf{r}_A = -\mathbf{a}_x - 3\mathbf{a}_y - 4\mathbf{a}_z$ and $\mathbf{r}_B = 2\mathbf{a}_x + 2\mathbf{a}_y + 2\mathbf{a}_z$, and point $C(1,3,4)$, find: *(a)* $\mathbf{R}_{AB}$; *(b)* $|\mathbf{r}_A|$; *(c)* $\mathbf{a}_A$; *(d)* $\mathbf{a}_{AB}$; *(e)* a unit vector directed from C toward A.

Ans. $3\mathbf{a}_x + 5\mathbf{a}_y + 6\mathbf{a}_z$; 5.10; $-0.1961\mathbf{a}_x - 0.588\mathbf{a}_y - 0.784\mathbf{a}_z$; $0.359\mathbf{a}_x + 0.598\mathbf{a}_y + 0.717\mathbf{a}_z$; $-0.1961\mathbf{a}_x - 0.588\mathbf{a}_y - 0.784\mathbf{a}_z$

1.5 THE VECTOR FIELD

We have already defined a vector field as a vector function of a position vector. In general, the magnitude and direction of the function will change as we move throughout the region, and the value of the vector function must be determined from the coordinate values of the point in question. Since we have considered only the cartesian coordinate system, we should expect the vector to be a function of the variables x, y, and z.

If we again represent the position vector as $\mathbf{r}$, then a vector field $\mathbf{G}$ can be expressed in functional notation as $\mathbf{G}(\mathbf{r})$; a scalar field T is written as $T(\mathbf{r})$.

If we inspect the velocity of the water in the ocean in some region near the surface where tides and currents are important, we might decide to represent it by a velocity vector which is in any direction, even up or down. If the z axis is taken as upward, the x axis in a northerly direction, the y axis to the west, and the origin at the surface, we have a right-handed coordinate system and may write the velocity vector as $\mathbf{v} = v_x\mathbf{a}_x + v_y\mathbf{a}_y + v_z\mathbf{a}_z$, or $\mathbf{v}(\mathbf{r}) = v_x(\mathbf{r})\mathbf{a}_x + v_y(\mathbf{r})\mathbf{a}_y + v_z(\mathbf{r})\mathbf{a}_z$; each of the components v_x, v_y, and v_z may be a function of the three variables x, y, and z. If the problem is simplified by assuming that we are in some portion of the Gulf Stream where the water is moving only to the north, then v_y and v_z are zero. Further simplifying assumptions might be made if the velocity falls off with depth and changes very slowly as we move north, south, east, or west. A suitable expression could be $\mathbf{v} = 2e^{z/100}\mathbf{a}_x$. We have a velocity of 2 m/s (meters per second) at the surface and a velocity of 0.368×2, or 0.736 m/s, at a depth of 100 m ($z = -100$), and the velocity continues to decrease with depth; in this example the vector velocity has a constant direction.

While the example given above is fairly simple and only a rough approximation to a physical situation, a more exact expression would be correspondingly more complex and difficult to interpret. We shall come across many fields in our study of electricity and magnetism which are simpler than the velocity example, an example in which only one component and one variable were involved (the x component and the variable z). We shall also study more complicated fields, and methods of interpreting these expressions physically will be discussed then.

D1.2. Given the vector field, $\mathbf{F} = 0.4(y - 2x)\mathbf{a}_x - [200/(x^2 + y^2 + z^2)]\mathbf{a}_z$: (*a*) evaluate $|\mathbf{F}|$ at $P(-4,3,5)$; (*b*) find a unit vector specifying the direction of $\mathbf{F}$ at P. Describe the locus of all points for which: (*c*) $F_x = 1$; (*d*) $|F_z| = 2$.

Ans. 5.95; $0.740\mathbf{a}_x - 0.673\mathbf{a}_z$; plane: $y = 2x + 2.5$; sphere: $x^2 + y^2 + z^2 = 100$

1.6 THE DOT PRODUCT

We now consider the first of two types of vector multiplication. The second type will be discussed in the following section.

Given two vectors **A** and **B**, the *dot product*, or *scalar product*, is defined as the product of the magnitude of **A**, the magnitude of **B**, and the cosine of the smaller angle between them,

$$\mathbf{A} \cdot \mathbf{B} = |\mathbf{A}|\,|\mathbf{B}| \cos \theta_{AB} \tag{3}$$

The dot appears between the two vectors and should be made heavy for emphasis. The dot, or scalar, product is a scalar, as one of the names implies, and it obeys the commutative law,

$$\mathbf{A} \cdot \mathbf{B} = \mathbf{B} \cdot \mathbf{A} \tag{4}$$

for the sign of the angle does not affect the cosine term. The expression $\mathbf{A} \cdot \mathbf{B}$ is read "**A** dot **B**."

Perhaps the most common application of the dot product is in mechanics, where a constant force **F** applied over a straight displacement **L** does an amount of work $FL \cos \theta$, which is more easily written $\mathbf{F} \cdot \mathbf{L}$. We might anticipate one of the results of Chap. 4 by pointing out that if the force varies along the path, integration is necessary to find the total work, and the result becomes

$$\text{Work} = \int \mathbf{F} \cdot d\mathbf{L}$$

Another example might be taken from magnetic fields, a subject about which we shall have a lot more to say later. The total flux Φ crossing a surface of area S is given by BS if the magnetic flux density B is perpendicular to the surface and uniform over it. We define a *vector surface* **S** as having the usual area for its magnitude and having a direction *normal* to the surface (avoiding for the moment the problem of which of the two possible normals to take). The flux crossing the surface is then $\mathbf{B} \cdot \mathbf{S}$. This expression is valid for any direction of the uniform magnetic flux density. However, if the flux density is not constant over the surface, the total flux is $\Phi = \int \mathbf{B} \cdot d\mathbf{S}$. Integrals of this general form appear in Chap. 3.

Finding the angle between two vectors in three-dimensional space is often a job we would prefer to avoid, and for that reason the definition of the dot product is usually not used in its basic form. A more helpful result is obtained by considering two vectors whose cartesian components are given, such as $\mathbf{A} = A_x\mathbf{a}_x + A_y\mathbf{a}_y + A_z\mathbf{a}_z$ and $\mathbf{B} = B_x\mathbf{a}_x + B_y\mathbf{a}_y + B_z\mathbf{a}_z$. The dot product also obeys the distributive law, and, therefore, $\mathbf{A} \cdot \mathbf{B}$ yields the sum of nine scalar terms, each involving the dot product of two unit vectors. Since the angle between two different unit vectors of the cartesian coordinate system is 90°, we then have

$$\mathbf{a}_x \cdot \mathbf{a}_y = \mathbf{a}_y \cdot \mathbf{a}_x = \mathbf{a}_x \cdot \mathbf{a}_z = \mathbf{a}_z \cdot \mathbf{a}_x = \mathbf{a}_y \cdot \mathbf{a}_z = \mathbf{a}_z \cdot \mathbf{a}_y = 0$$

The remaining three terms involve the dot product of a unit vector with itself, which is unity, giving finally

$$\boxed{\mathbf{A} \cdot \mathbf{B} = A_x B_x + A_y B_y + A_z B_z} \tag{5}$$

which is an expression involving no angles.

A vector dotted with itself yields the magnitude squared, or

$$\boxed{\mathbf{A} \cdot \mathbf{A} = A^2 = |\mathbf{A}|^2} \tag{6}$$

and any unit vector dotted with itself is unity,

$$\mathbf{a}_A \cdot \mathbf{a}_A = 1$$

One of the most important applications of the dot product is that of finding the component of a vector in a given direction. Referring to Fig. 1.4*a*, we can obtain the component (scalar) of $\mathbf{B}$ in the direction specified by the unit vector $\mathbf{a}$ as

$$\mathbf{B} \cdot \mathbf{a} = |\mathbf{B}|\,|\mathbf{a}| \cos \theta_{Ba} = |\mathbf{B}| \cos \theta_{Ba}$$

The sign of the component is positive if $0 \le \theta_{Ba} < 90°$ and negative whenever $90° < \theta_{Ba} \le 180°$.

In order to obtain the component *vector* of $\mathbf{B}$ in the direction of $\mathbf{a}$, we simply multiply the component (scalar) by $\mathbf{a}$, as illustrated by Fig. 1.4*b*. For example, the component of $\mathbf{B}$ in the direction of $\mathbf{a}_x$ is $\mathbf{B} \cdot \mathbf{a}_x = B_x$, and the component vector is $B_x \mathbf{a}_x$, or $(\mathbf{B} \cdot \mathbf{a}_x)\mathbf{a}_x$. Hence, the problem of finding the component of a vector in any desired direction becomes the problem of finding a unit vector in that direction, and that we can do.

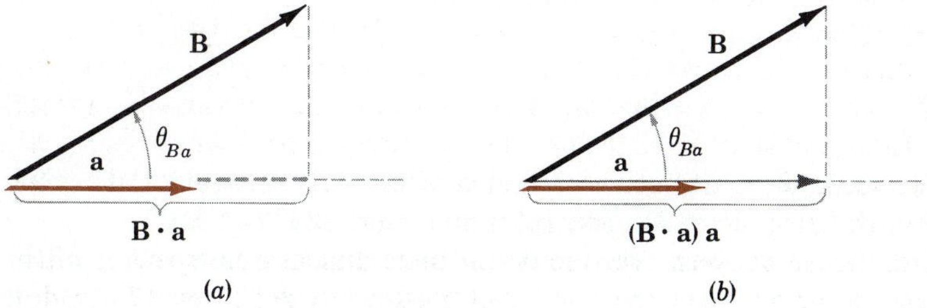

FIGURE 1.4
(*a*) The component (scalar) of **B** in the direction of the unit vector **a** is **B · a**. (*b*) The component vector of **B** in the direction of the unit vector **a** is (**B · a**)**a**.

The geometrical term *projection* is also used with the dot product. Thus, $\mathbf{B}\cdot\mathbf{a}$ is the projection of $\mathbf{B}$ in the $\mathbf{a}$ direction.

As an example of these definitions and operations, let us consider a vector field $\mathbf{G} = y\mathbf{a}_x - 2.5x\mathbf{a}_y + 3\mathbf{a}_z$ at a point $Q(4,5,2)$. Thus, $\mathbf{G}(\mathbf{r}_Q) = 5\mathbf{a}_x - 10\mathbf{a}_y + 3\mathbf{a}_z$. We now let a direction be defined by $\mathbf{a}_N = (2\mathbf{a}_x + \mathbf{a}_y - 2\mathbf{a}_z)/3$. Thus, $\mathbf{G}\cdot\mathbf{a}_N = (5\mathbf{a}_x - 10\mathbf{a}_y + 3\mathbf{a}_z)\cdot(2\mathbf{a}_x + \mathbf{a}_y - 2\mathbf{a}_z)/3 = (10 - 10 - 6)/3 = -2$, the scalar component of $\mathbf{G}$ at Q in the direction of $\mathbf{a}_N$, and the vector component is $-2\mathbf{a}_N = (-1.333\mathbf{a}_x - 0.667\mathbf{a}_y + 1.333\mathbf{a}_z)$. The angle between $\mathbf{G}(\mathbf{r}_Q)$ and $\mathbf{a}_N$ is found from

$$\mathbf{G}\cdot\mathbf{a}_N = |\mathbf{G}|\cos\theta_{Ga}$$

$$-2 = \sqrt{25 + 100 + 9}\cos\theta_{Ga}$$

and

$$\theta_{Ga} = \cos^{-1}\frac{-2}{\sqrt{134}} = 99.9^\circ$$

D1.3. Given points $A(2,5,-1)$, $B(3,-2,4)$, and $C(-2,3,1)$, find: (*a*) $\mathbf{R}_{AB}\cdot\mathbf{R}_{AC}$; (*b*) the angle between $\mathbf{R}_{AB}$ and $\mathbf{R}_{AC}$; (*c*) the length of the projection of $\mathbf{R}_{AB}$ on $\mathbf{R}_{AC}$; (*d*) the vector projection of $\mathbf{R}_{AB}$ on $\mathbf{R}_{AC}$.

Ans. 20; 61.9°; 4.08; $-3.33\mathbf{a}_x - 1.667\mathbf{a}_y + 1.667\mathbf{a}_z$

1.7 THE CROSS PRODUCT

Given two vectors $\mathbf{A}$ and $\mathbf{B}$, we shall now define the *cross product*, or *vector product*, of $\mathbf{A}$ and $\mathbf{B}$, written with a cross between the two vectors as $\mathbf{A}\times\mathbf{B}$ and read "**A** cross **B**." The cross product $\mathbf{A}\times\mathbf{B}$ is a vector; the magnitude of $\mathbf{A}\times\mathbf{B}$ is equal to the product of the magnitudes of **A**, **B**, and the sine of the smaller angle between **A** and **B**; the direction of $\mathbf{A}\times\mathbf{B}$ is perpendicular to the plane containing **A** and **B** and is along that one of the two possible perpendiculars which is in the direction of advance of a right-handed screw as **A** is turned into **B**. This direction is illustrated in Fig. 1.5. Remember that either vector may be moved about at will, maintaining its direction constant, until the two vectors have a "common origin." This determines the plane containing both. However, in most of our applications we shall be concerned with vectors defined at the same point.

As an equation we can write

$$\boxed{\mathbf{A}\times\mathbf{B} = \mathbf{a}_N|\mathbf{A}||\mathbf{B}|\sin\theta_{AB}} \tag{7}$$

where an additional statement, such as that given above, is still required to explain the direction of the unit vector $\mathbf{a}_N$. The subscript stands for "normal."

Reversing the order of the vectors **A** and **B** results in a unit vector in the opposite direction, and we see that the cross product is not commutative, for $\mathbf{B}\times\mathbf{A} = -(\mathbf{A}\times\mathbf{B})$.

If the definition of the cross product is applied to the unit vectors $\mathbf{a}_x$ and $\mathbf{a}_y$, we find $\mathbf{a}_x\times\mathbf{a}_y = \mathbf{a}_z$, for each vector has unit magnitude, the two vectors are perpendicular, and the rotation of $\mathbf{a}_x$ into $\mathbf{a}_y$ indicates the positive z direction by the

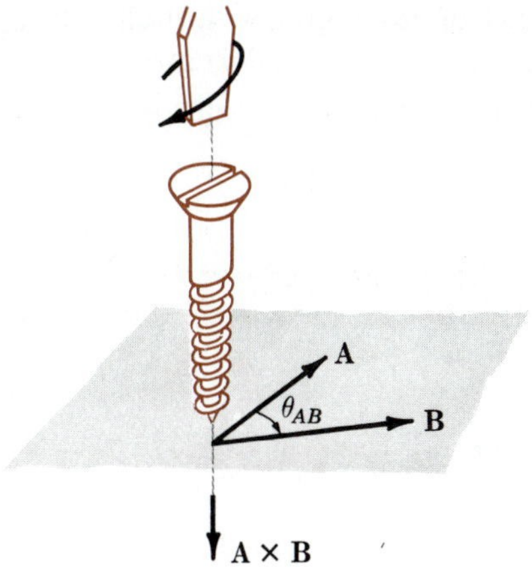

FIGURE 1.5
The direction of $\mathbf{A} \times \mathbf{B}$ is in the direction of advance of a right-handed screw as $\mathbf{A}$ is turned into $\mathbf{B}$.

definition of a right-handed coordinate system. In a similar way $\mathbf{a}_y \times \mathbf{a}_z = \mathbf{a}_x$, and $\mathbf{a}_z \times \mathbf{a}_x = \mathbf{a}_y$. Note the alphabetic symmetry. As long as the three vectors $\mathbf{a}_x$, $\mathbf{a}_y$, and $\mathbf{a}_z$ are written in order (and assuming that $\mathbf{a}_x$ follows $\mathbf{a}_z$, like three elephants in a circle holding tails, so that we could also write $\mathbf{a}_y$, $\mathbf{a}_z$, $\mathbf{a}_x$ or $\mathbf{a}_z$, $\mathbf{a}_x$, $\mathbf{a}_y$), then the cross and equal sign may be placed in either of the two vacant spaces. As a matter of fact, it is now simpler to define a right-handed cartesian coordinate system by saying that $\mathbf{a}_x \times \mathbf{a}_y = \mathbf{a}_z$.

A simple example of the use of the cross product may be taken from geometry or trigonometry. To find the area of a parallelogram, the product of the lengths of two adjacent sides is multiplied by the sine of the angle between them. Using vector notation for the two sides, we then may express the (scalar) area as the *magnitude* of $\mathbf{A} \times \mathbf{B}$, or $|\mathbf{A} \times \mathbf{B}|$.

The cross product may be used to replace the right-hand rule familiar to all electrical engineers. Consider the force on a straight conductor of length $\mathbf{L}$, where the direction assigned to $\mathbf{L}$ corresponds to the direction of the steady current I, and a uniform magnetic field of flux density $\mathbf{B}$ is present. Using vector notation, we may write the result neatly as $\mathbf{F} = I\mathbf{L} \times \mathbf{B}$. This relationship will be obtained later in Chap. 9.

The evaluation of a cross product by means of its definition turns out to be more work than the evaluation of the dot product from its definition, for not only must we find the angle between the vectors, but we must find an expression for the unit vector $\mathbf{a}_N$. This work may be avoided by using cartesian components for the two vectors $\mathbf{A}$ and $\mathbf{B}$ and expanding the cross product as a sum of nine simpler cross products, each involving two unit vectors,

$$\begin{aligned}\mathbf{A} \times \mathbf{B} = & A_xB_x\mathbf{a}_x \times \mathbf{a}_x + A_xB_y\mathbf{a}_x \times \mathbf{a}_y + A_xB_z\mathbf{a}_x \times \mathbf{a}_z \\ & + A_yB_x\mathbf{a}_y \times \mathbf{a}_x + A_yB_y\mathbf{a}_y \times \mathbf{a}_y + A_yB_z\mathbf{a}_y \times \mathbf{a}_z \\ & + A_zB_x\mathbf{a}_z \times \mathbf{a}_x + A_zB_y\mathbf{a}_z \times \mathbf{a}_y + A_zB_z\mathbf{a}_z \times \mathbf{a}_z\end{aligned}$$

We have already found that $\mathbf{a}_x \times \mathbf{a}_y = \mathbf{a}_z$, $\mathbf{a}_y \times \mathbf{a}_z = \mathbf{a}_x$, and $\mathbf{a}_z \times \mathbf{a}_x = \mathbf{a}_y$, and it follows that $\mathbf{a}_y \times \mathbf{a}_x = -\mathbf{a}_z$, $\mathbf{a}_z \times \mathbf{a}_y = -\mathbf{a}_x$, and $\mathbf{a}_x \times \mathbf{a}_z = -\mathbf{a}_y$. The three remaining terms are zero, for the cross product of any vector with itself is zero, since the included angle is zero. These results may be combined to give

$$\mathbf{A} \times \mathbf{B} = (A_y B_z - A_z B_y)\mathbf{a}_x + (A_z B_x - A_x B_z)\mathbf{a}_y + (A_x B_y - A_y B_x)\mathbf{a}_z \tag{8}$$

or written as a determinant in a more easily remembered form,

$$\mathbf{A} \times \mathbf{B} = \begin{vmatrix} \mathbf{a}_x & \mathbf{a}_y & \mathbf{a}_z \\ A_x & A_y & A_z \\ B_x & B_y & B_z \end{vmatrix} \tag{9}$$

Thus, if $\mathbf{A} = 2\mathbf{a}_x - 3\mathbf{a}_y + \mathbf{a}_z$ and $\mathbf{B} = -4\mathbf{a}_x - 2\mathbf{a}_y + 5\mathbf{a}_z$, we have

$$\mathbf{A} \times \mathbf{B} = \begin{vmatrix} \mathbf{a}_x & \mathbf{a}_y & \mathbf{a}_z \\ 2 & -3 & 1 \\ -4 & -2 & 5 \end{vmatrix}$$

$$= [(-3)(5) - (1)(-2)]\mathbf{a}_x - [(2)(5) - (1)(-4)]\mathbf{a}_y + [(2)(-2) - (-3)(-4)]\mathbf{a}_z$$

$$= -13\mathbf{a}_x - 14\mathbf{a}_y - 16\mathbf{a}_z$$

D1.4. A triangle is defined by the three points $A(2,-5,1)$, $B(-3,2,4)$, and $C(0,3,1)$. Find: (*a*) $\mathbf{R}_{BC} \times \mathbf{R}_{BA}$; (*b*) the area of the triangle; (*c*) a unit vector perpendicular to the plane in which the triangle is located.

Ans. $-24\mathbf{a}_x - 6\mathbf{a}_y - 26\mathbf{a}_z$; 17.94; $\pm(0.669\mathbf{a}_x + 0.1672\mathbf{a}_y + 0.724\mathbf{a}_z)$

1.8 OTHER COORDINATE SYSTEMS: CIRCULAR CYLINDRICAL COORDINATES

The cartesian coordinate system is in general the one in which students prefer to work every problem. This often means a lot more work for the student, because many problems possess a type of symmetry which pleads for a more logical treatment. It is easier to do now, once and for all, the work required to become familiar with cylindrical and spherical coordinates, instead of applying an equal or greater effort to every problem involving cylindrical or spherical symmetry later. With this future saving of labor in mind, we shall take a careful and unhurried look at cylindrical and spherical coordinates.

The circular cylindrical coordinate system is the three-dimensional version of the polar coordinates of analytic geometry. In the two-dimensional polar coordinates, a point was located in a plane by giving its distance ρ from the origin, and the angle ϕ between the line from the point to the origin and an arbitrary radial line, taken as $\phi = 0$.[4] A three-dimensional coordinate system, circular cylin-

[4] The two variables of polar coordinates are commonly called r and θ. With three coordinates, however, it is more common to use ρ for the radius variable of cylindrical coordinates and r for the (different) radius variable of spherical coordinates. Also, the angle variable of cylindrical coordinates is customarily called ϕ because everyone uses θ for a different angle in spherical coordinates. The angle ϕ is common to both cylindrical and spherical coordinates. See?

drical coordinates, is obtained by also specifying the distance z of the point from an arbitrary $z = 0$ reference plane which is perpendicular to the line $\rho = 0$. For simplicity, we usually refer to circular cylindrical coordinates simply as cylindrical coordinates. This will not cause any confusion in reading this book, but it is only fair to point out that there are such systems as elliptic cylindrical coordinates, hyperbolic cylindrical coordinates, parabolic cylindrical coordinates, and others.

We no longer set up three axes as in cartesian coordinates, but must instead consider any point as the intersection of three mutually perpendicular surfaces. These surfaces are a circular cylinder (ρ = constant), a plane (ϕ = constant), and another plane (z = constant). This corresponds to the location of a point in a cartesian coordinate system by the intersection of three planes (x = constant, y = constant, and z = constant). The three surfaces of circular cylindrical coordinates are shown in Fig. 1.6*a*. Note that three such surfaces may be passed through any point, unless it lies on the z axis, in which case one plane suffices.

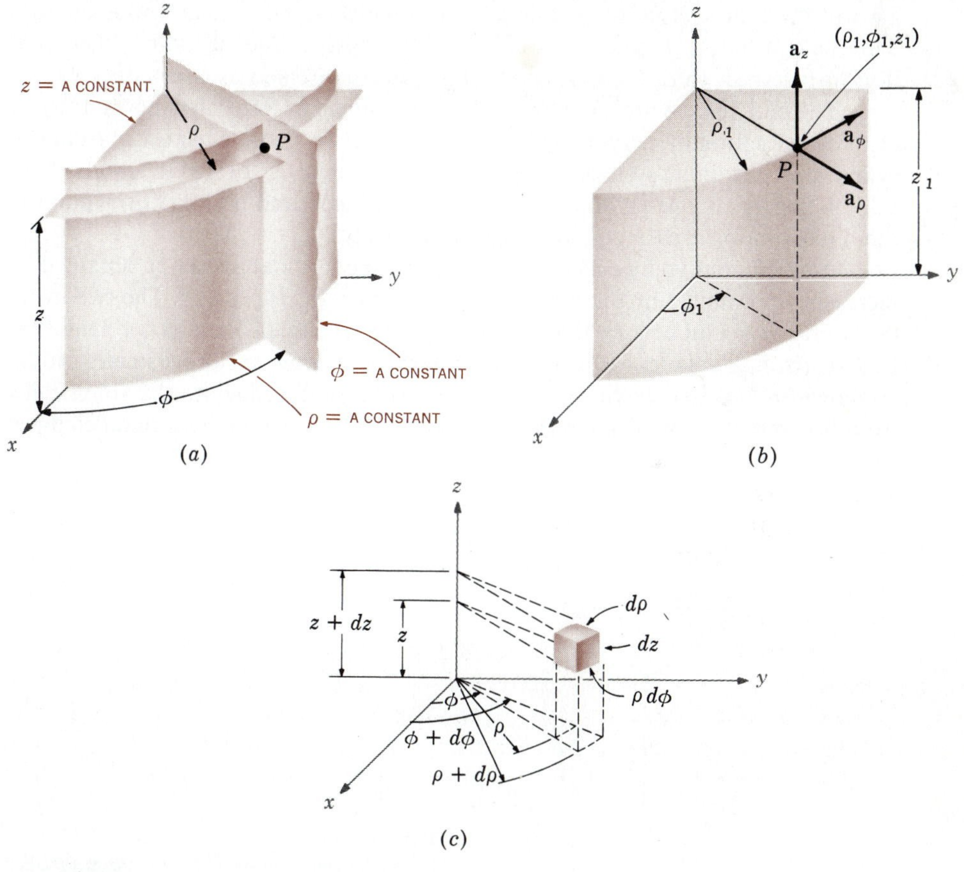

FIGURE 1.6
(*a*) The three mutually perpendicular surfaces of the circular cylindrical coordinate system. (*b*) The three unit vectors of the circular cylindrical coordinate system. (*c*) The differential volume unit in the circular cylindrical coordinate system; $d\rho$, $\rho\, d\phi$, and dz are all elements of length.

Three unit vectors must also be defined, but we may no longer direct them along the "coordinate axes," for such axes exist only in cartesian coordinates. Instead, we take a broader view of the unit vectors in cartesian coordinates and realize that they are directed toward increasing coordinate values and are perpendicular to the surface on which that coordinate value is constant; i.e., the unit vector $\mathbf{a}_x$ is normal to the plane $x =$ constant and points toward larger values of x. In a corresponding way we may now define three unit vectors in cylindrical coordinates, $\mathbf{a}_\rho$, $\mathbf{a}_\phi$, and $\mathbf{a}_z$.

The unit vector $\mathbf{a}_\rho$ at a point $P(\rho_1,\phi_1,z_1)$ is directed radially outward, normal to the cylindrical surface $\rho = \rho_1$. It lies in the planes $\phi = \phi_1$ and $z = z_1$. The unit vector $\mathbf{a}_\phi$ is normal to the plane $\phi = \phi_1$, points in the direction of increasing ϕ, lies in the plane $z = z_1$, and is tangent to the cylindrical surface $\rho = \rho_1$. The unit vector $\mathbf{a}_z$ is the same as the unit vector $\mathbf{a}_z$ of the cartesian coordinate system. Figure 1.6*b* shows the three unit vectors in cylindrical coordinates.

In cartesian coordinates, the unit vectors are not functions of the coordinates. Two of the unit vectors in cylindrical coordinates, $\mathbf{a}_\rho$ and $\mathbf{a}_\phi$, however, do vary with the coordinate ϕ, since their directions change. In integration or differentiation with respect to ϕ, then, $\mathbf{a}_\rho$ and $\mathbf{a}_\phi$ must not be treated as constants.

The unit vectors are again mutually perpendicular, for each is normal to one of the three mutually perpendicular surfaces, and we may define a right-handed cylindrical coordinate system as one in which $\mathbf{a}_\rho \times \mathbf{a}_\phi = \mathbf{a}_z$, or (for those who have flexible fingers) as one in which the thumb, forefinger, and middle finger point in the direction of increasing ρ, ϕ, and z, respectively.

A differential volume element in cylindrical coordinates may be obtained by increasing ρ, ϕ, and z by the differential increments $d\rho$, $d\phi$, and dz. The two cylinders of radius ρ and $\rho + d\rho$, the two radial planes at angles ϕ and $\phi + d\phi$, and the two "horizontal" planes at "elevations" z and $z + dz$ now enclose a small volume, as shown in Fig. 1.6*c*, having the shape of a truncated wedge. As the volume element becomes very small, its shape approaches that of a rectangular parallelepiped

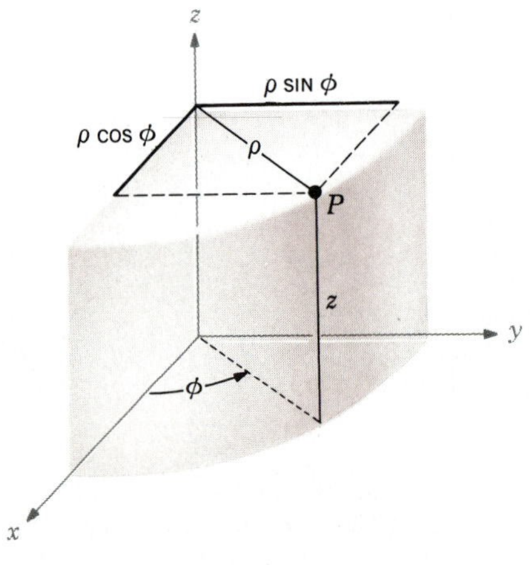

FIGURE 1.7
The relationship between the cartesian variables x, y, z and the cylindrical coordinate variables ρ, ϕ, z. There is no change in the variable z between the two systems.

having sides of length $d\rho$, $\rho\, d\phi$, and dz. Note that $d\rho$ and dz are dimensionally lengths, but $d\phi$ is not; $\rho\, d\phi$ is the length. The surfaces have areas of $\rho\, d\rho\, d\phi$, $d\rho\, dz$, and $\rho\, d\phi\, dz$, and the volume becomes $\rho\, d\rho\, d\phi\, dz$.

The variables of the rectangular and cylindrical coordinate systems are easily related to each other. With reference to Fig. 1.7, we see that

$$\begin{aligned} x &= \rho \cos \phi \\ y &= \rho \sin \phi \\ z &= z \end{aligned} \tag{10}$$

From the other viewpoint, we may express the cylindrical variables in terms of x, y, and z:

$$\begin{aligned} \rho &= \sqrt{x^2 + y^2} \qquad (\rho \geq 0) \\ \phi &= \tan^{-1} \frac{y}{x} \\ z &= z \end{aligned} \tag{11}$$

We shall consider the variable ρ to be positive or zero, thus using only the positive sign for the radical in (11). The proper value of the angle ϕ is determined by inspecting the signs of x and y. Thus, if $x = -3$ and $y = 4$, we find that the point lies in the second quadrant so that $\rho = 5$ and $\phi = 126.9°$. For $x = 3$ and $y = -4$, we have $\phi = -53.1°$ or $306.9°$, whichever is more convenient.

Using (10) or (11), scalar functions given in one coordinate system are easily transformed into the other system.

A vector function in one coordinate system, however, requires two steps in order to transform it to another coordinate system, because a different set of component vectors is generally required. That is, we may be given a cartesian vector

$$\mathbf{A} = A_x \mathbf{a}_x + A_y \mathbf{a}_y + A_z \mathbf{a}_z$$

where each component is given as a function of x, y, and z, and we need a vector in cylindrical coordinates

$$\mathbf{A} = A_\rho \mathbf{a}_\rho + A_\phi \mathbf{a}_\phi + A_z \mathbf{a}_z$$

where each component is given as a function of ρ, ϕ, and z.

To find any desired component of a vector, we recall from the discussion of the dot product that a component in a desired direction may be obtained by taking the dot product of the vector and a unit vector in the desired direction. Hence,

$$A_\rho = \mathbf{A} \cdot \mathbf{a}_\rho \qquad \text{and} \qquad A_\phi = \mathbf{A} \cdot \mathbf{a}_\phi$$

Expanding these dot products, we have

$$A_\rho = (A_x \mathbf{a}_x + A_y \mathbf{a}_y + A_z \mathbf{a}_z) \cdot \mathbf{a}_\rho = A_x \mathbf{a}_x \cdot \mathbf{a}_\rho + A_y \mathbf{a}_y \cdot \mathbf{a}_\rho \tag{12}$$

$$A_\phi = (A_x \mathbf{a}_x + A_y \mathbf{a}_y + A_z \mathbf{a}_z) \cdot \mathbf{a}_\phi = A_x \mathbf{a}_x \cdot \mathbf{a}_\phi + A_y \mathbf{a}_y \cdot \mathbf{a}_\phi \tag{13}$$

TABLE 1.1
Dot products of unit vectors in cylindrical and cartesian coordinate systems

	$\mathbf{a}_\rho$	$\mathbf{a}_\phi$	$\mathbf{a}_z$
$\mathbf{a}_x \cdot$	$\cos\phi$	$-\sin\phi$	0
$\mathbf{a}_y \cdot$	$\sin\phi$	$\cos\phi$	0
$\mathbf{a}_z \cdot$	0	0	1

and

$$A_z = (A_x\mathbf{a}_x + A_y\mathbf{a}_y + A_z\mathbf{a}_z) \cdot \mathbf{a}_z = A_z\mathbf{a}_z \cdot \mathbf{a}_z = A_z \tag{14}$$

since $\mathbf{a}_z \cdot \mathbf{a}_\rho$ and $\mathbf{a}_z \cdot \mathbf{a}_\phi$ are zero.

In order to complete the transformation of the components, it is necessary to know the dot products $\mathbf{a}_x \cdot \mathbf{a}_\rho$, $\mathbf{a}_y \cdot \mathbf{a}_\rho$, $\mathbf{a}_x \cdot \mathbf{a}_\phi$, and $\mathbf{a}_y \cdot \mathbf{a}_\phi$. Applying the definition of the dot product, we see that since we are concerned with unit vectors, the result is merely the cosine of the angle between the two unit vectors in question. Referring to Fig. 1.7 and thinking mightily, we identify the angle between $\mathbf{a}_x$ and $\mathbf{a}_\rho$ as ϕ, and thus $\mathbf{a}_x \cdot \mathbf{a}_\rho = \cos\phi$, but the angle between $\mathbf{a}_y$ and $\mathbf{a}_\rho$ is $90° - \phi$, and $\mathbf{a}_y \cdot \mathbf{a}_\rho = \cos(90° - \phi) = \sin\phi$. The remaining dot products of the unit vectors are found in a similar manner, and the results are tabulated as functions of ϕ in Table 1.1.

Transforming vectors from cartesian to cylindrical coordinates or vice versa is therefore accomplished by using (10) or (11) to change variables, and by using the dot products of the unit vectors given in Table 1.1 to change components. The two steps may be taken in either order.

As an example, let us transform the vector $\mathbf{B} = y\mathbf{a}_x - x\mathbf{a}_y + z\mathbf{a}_z$ into cylindrical coordinates. The new components are

$$B_\rho = \mathbf{B} \cdot \mathbf{a}_\rho = y(\mathbf{a}_x \cdot \mathbf{a}_\rho) - x(\mathbf{a}_y \cdot \mathbf{a}_\rho)$$
$$= y\cos\phi - x\sin\phi = \rho\sin\phi\cos\phi - \rho\cos\phi\sin\phi = 0$$
$$B_\phi = \mathbf{B} \cdot \mathbf{a}_\phi = y(\mathbf{a}_x \cdot \mathbf{a}_\phi) - x(\mathbf{a}_y \cdot \mathbf{a}_\phi)$$
$$= -y\sin\phi - x\cos\phi = -\rho\sin^2\phi - \rho\cos^2\phi = -\rho$$

Thus,

$$\mathbf{B} = -\rho\mathbf{a}_\phi + z\mathbf{a}_z$$

D1.5. Given points $A(x = 2, y = 3, z = -1)$ and $B(\rho = 4, \phi = -50°, z = 2)$, find the distance from: (*a*) A to the origin; (*b*) B to the origin; (*c*) A to B.

Ans. 3.74; 4.47; 6.79

D1.6. Transform each of the following vectors to cylindrical coordinates at the point specified: (*a*) $5\mathbf{a}_x$ at $P(\rho = 4, \phi = 120°, z = 2)$; (*b*) $5\mathbf{a}_x$ at $Q(x = 3, y = 4, z = -1)$; (*c*) $4\mathbf{a}_x - 2\mathbf{a}_y - 4\mathbf{a}_z$ at $A(x = 2, y = 3, z = 5)$.

Ans. $-2.50\mathbf{a}_\rho - 4.33\mathbf{a}_\phi$; $3\mathbf{a}_\rho - 4\mathbf{a}_\phi$; $0.555\mathbf{a}_\rho - 4.44\mathbf{a}_\phi - 4\mathbf{a}_z$

1.9 THE SPHERICAL COORDINATE SYSTEM

We have no two-dimensional coordinate system to help us understand the three-dimensional spherical coordinate system, as we have for the circular cylindrical coordinate system. In certain respects we can draw on our knowledge of the latitude-and-longitude system of locating a place on the surface of the earth, but usually we consider only points on the surface and not those below or above ground.

Let us start by building a spherical coordinate system on the three cartesian axes (Fig. 1.8*a*). We first define the distance from the origin to any point as r. The surface $r =$ constant is a sphere.

The second coordinate is an angle θ between the z axis and the line drawn from the origin to the point in question. The surface $\theta =$ constant is a cone, and the two surfaces, cone and sphere, are everywhere perpendicular along their intersection, which is a circle of radius $r \sin \theta$. The coordinate θ corresponds to latitude, except that latitude is measured from the equator and θ is measured from the "North Pole."

The third coordinate ϕ is also an angle and is exactly the same as the angle ϕ of cylindrical coordinates. It is the angle between the x axis and the projection in the $z = 0$ plane of the line drawn from the origin to the point. It corresponds to the angle of longitude, but the angle ϕ increases to the "east." The surface $\phi =$ constant is a plane passing through the $\theta = 0$ line (or the z axis).

We should again consider any point as the intersection of three mutually perpendicular surfaces—a sphere, a cone, and a plane—each oriented in the manner described above. The three surfaces are shown in Fig. 1.8*b*.

Three unit vectors may again be defined at any point. Each unit vector is perpendicular to one of the three mutually perpendicular surfaces and oriented in that direction in which the coordinate increases. The unit vector $\mathbf{a}_r$ is directed radially outward, normal to the sphere $r =$ constant, and lies in the cone $\theta =$ constant and the plane $\phi =$ constant. The unit vector $\mathbf{a}_\theta$ is normal to the conical surface, lies in the plane, and is tangent to the sphere. It is directed along a line of "longitude" and points "south." The third unit vector $\mathbf{a}_\phi$ is the same as in cylindrical coordinates, being normal to the plane and tangent to both the cone and sphere. It is directed to the "east."

The three unit vectors are shown in Fig. 1.8*c*. They are, of course, mutually perpendicular, and a right-handed coordinate system is defined by causing $\mathbf{a}_r \times \mathbf{a}_\theta = \mathbf{a}_\phi$. Our system is right-handed, as an inspection of Fig. 1.8*c* will show, on application of the definition of the cross product. The right-hand rule serves

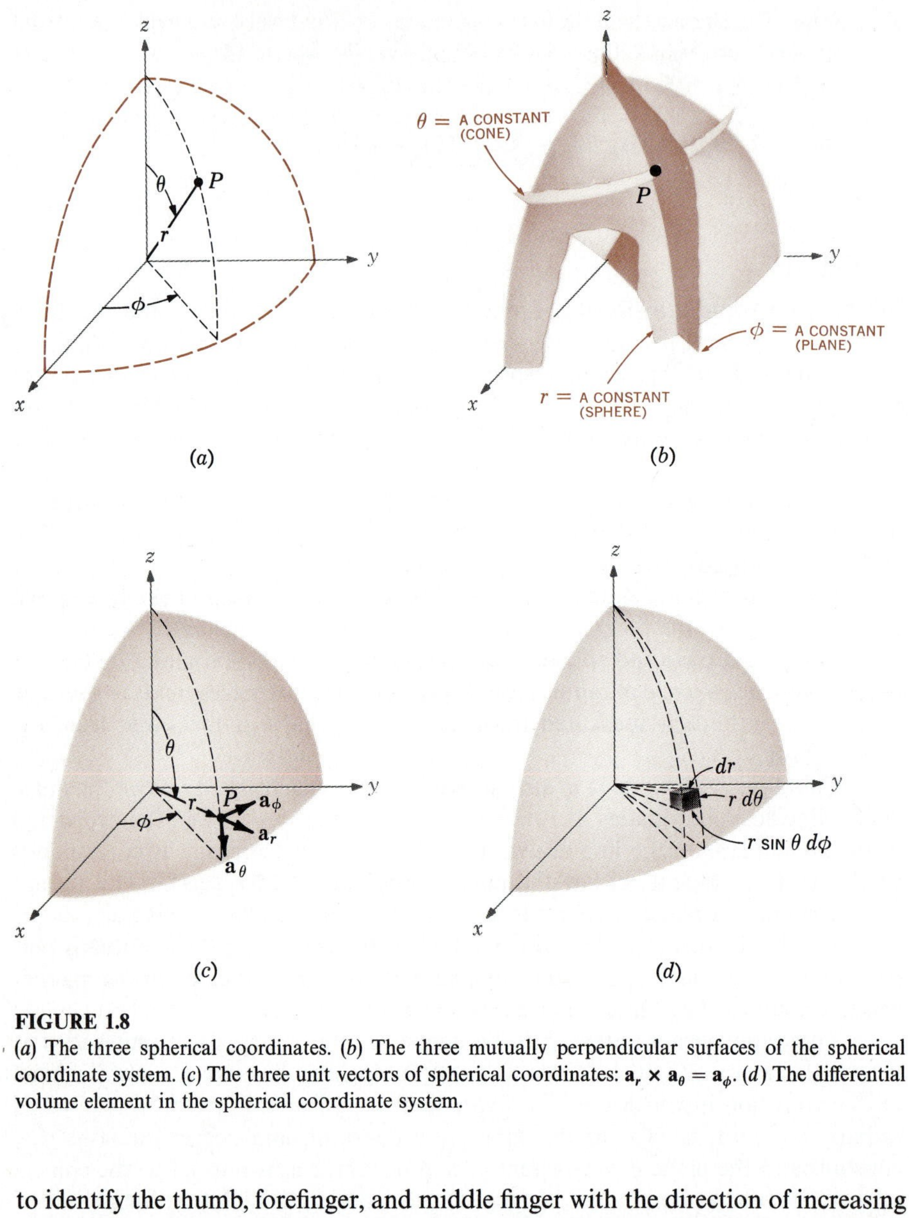

FIGURE 1.8
(*a*) The three spherical coordinates. (*b*) The three mutually perpendicular surfaces of the spherical coordinate system. (*c*) The three unit vectors of spherical coordinates: $\mathbf{a}_r \times \mathbf{a}_\theta = \mathbf{a}_\phi$. (*d*) The differential volume element in the spherical coordinate system.

to identify the thumb, forefinger, and middle finger with the direction of increasing r, θ, and ϕ, respectively. (Note that the identification in cylindrical coordinates was with ρ, ϕ, and z, and in cartesian coordinates with x, y, and z.)

A differential volume element may be constructed in spherical coordinates by increasing r, θ, and ϕ by dr, $d\theta$, and $d\phi$, as shown in Fig. 1.8*d*. The distance between the two spherical surfaces of radius r and $r + dr$ is dr; the distance between the two cones having generating angles of θ and $\theta + d\theta$ is $r\,d\theta$; and the distance between the two radial planes at angles ϕ and $\phi + d\phi$ is found to be $r \sin\theta\, d\phi$, after a few moments of trigonometric thought. The surfaces have areas of $r\,dr\,d\theta$, $r\sin\theta\,dr\,d\phi$, and $r^2\sin\theta\,d\theta\,d\phi$, and the volume is $r^2\sin\theta\,dr\,d\theta\,d\phi$.

The transformation of scalars from the cartesian to the spherical coordinate system is easily made by using Fig. 1.8*a* to relate the two sets of variables:

$$\begin{aligned} x &= r\sin\theta\cos\phi \\ y &= r\sin\theta\sin\phi \\ z &= r\cos\theta \end{aligned} \tag{15}$$

The transformation in the reverse direction is achieved with the help of

$$\begin{aligned} r &= \sqrt{x^2+y^2+z^2} \qquad && (r \geq 0) \\ \theta &= \cos^{-1}\frac{z}{\sqrt{x^2+y^2+z^2}} \qquad && (0^\circ \leq \theta \leq 180^\circ) \\ \phi &= \tan^{-1}\frac{y}{x} \end{aligned} \tag{16}$$

The radius variable r is nonnegative, and θ is restricted to the range from 0° to 180°, inclusive. The angles are placed in the proper quadrants by inspecting the signs of x, y, and z.

The transformation of vectors requires the determination of the products of the unit vectors in cartesian and spherical coordinates. We work out these products from Fig. 1.8*c* and a pinch of trigonometry. Since the dot product of any spherical unit vector with any cartesian unit vector is the component of the spherical vector in the direction of the cartesian vector, the dot products with $\mathbf{a}_z$ are found to be

$$\begin{aligned} \mathbf{a}_z\cdot\mathbf{a}_r &= \cos\theta \\ \mathbf{a}_z\cdot\mathbf{a}_\theta &= -\sin\theta \\ \mathbf{a}_z\cdot\mathbf{a}_\phi &= 0 \end{aligned}$$

The dot products involving $\mathbf{a}_x$ and $\mathbf{a}_y$ require first the projection of the spherical unit vector on the xy plane and then the projection onto the desired axis. For example, $\mathbf{a}_r\cdot\mathbf{a}_x$ is obtained by projecting $\mathbf{a}_r$ onto the xy plane, giving $\sin\theta$, and then projecting $\sin\theta$ on the x axis, which yields $\sin\theta\cos\phi$. The other dot products are found in a like manner, and all are shown in Table 1.2.

TABLE 1.2
Dot products of unit vectors in spherical and cartesian coordinate systems

	$\mathbf{a}_r$	$\mathbf{a}_\theta$	$\mathbf{a}_\phi$
$\mathbf{a}_x\cdot$	$\sin\theta\cos\phi$	$\cos\theta\cos\phi$	$-\sin\phi$
$\mathbf{a}_y\cdot$	$\sin\theta\sin\phi$	$\cos\theta\sin\phi$	$\cos\phi$
$\mathbf{a}_z\cdot$	$\cos\theta$	$-\sin\theta$	0

The transformation procedure may be illustrated by considering the vector $\mathbf{G} = (xz/y)\mathbf{a}_x$. We find the three spherical components by dotting $\mathbf{G}$ with the appropriate unit vectors, and we change variables during the procedure:

$$G_r = \mathbf{G} \cdot \mathbf{a}_r = \frac{xz}{y} \mathbf{a}_x \cdot \mathbf{a}_r = \frac{xz}{y} \sin\theta \cos\phi$$

$$= r \sin\theta \cos\theta \frac{\cos^2\phi}{\sin\phi}$$

$$G_\theta = \mathbf{G} \cdot \mathbf{a}_\theta = \frac{xz}{y} \mathbf{a}_x \cdot \mathbf{a}_\theta = \frac{xz}{y} \cos\theta \cos\phi$$

$$= r \cos^2\theta \frac{\cos^2\phi}{\sin\phi}$$

$$G_\phi = \mathbf{G} \cdot \mathbf{a}_\phi = \frac{xz}{y} \mathbf{a}_x \cdot \mathbf{a}_\phi = \frac{xz}{y}(-\sin\phi)$$

$$= -r \cos\theta \cos\phi$$

Collecting these results, we have

$$\mathbf{G} = r\cos\theta\cos\phi(\sin\theta\cot\phi\,\mathbf{a}_r + \cos\theta\cot\phi\,\mathbf{a}_\theta - \mathbf{a}_\phi)$$

Appendix A describes the general curvilinear coordinate system of which the cartesian, circular cylindrical, and spherical coordinate systems are special cases. The first section of this appendix could well be scanned now.

D1.7. Given the points $A(x = 2, y = 3, z = -1)$ and $B(r = 4, \theta = 25°, \phi = 120°)$, find: (*a*) the spherical coordinates of A; (*b*) the cartesian coordinates of B; (*c*) the distance from A to B.

Ans. $r = 3.74$, $\theta = 105.5°$, $\phi = 56.3°$; $x = -0.845$, $y = 1.464$, $z = 3.63$; 5.64

D1.8. Transform each of the following vectors to spherical coordinates at the point specified: (*a*) $5\mathbf{a}_x$ at $B(r = 4, \theta = 25°, \phi = 120°)$; (*b*) $5\mathbf{a}_x$ at $A(x = 2, y = 3, z = -1)$; (*c*) $4\mathbf{a}_x - 2\mathbf{a}_y - 4\mathbf{a}_z$ at $P(x = -2, y = -3, z = 4)$.

Ans. $-1.057\mathbf{a}_r - 2.27\mathbf{a}_\theta - 4.33\mathbf{a}_\phi$; $2.67\mathbf{a}_r - 0.741\mathbf{a}_\theta - 4.16\mathbf{a}_\phi$; $-3.34\mathbf{a}_r + 2.27\mathbf{a}_\theta + 4.44\mathbf{a}_\phi$

SUGGESTED REFERENCES

1. Grossman, S. I.: "Calculus," 3d ed., Academic Press and Harcourt Brace Jovanovich, Publishers, Orlando, 1984. Vector algebra and cylindrical and spherical coordinates appear in chap. 17, and vector calculus is introduced in chap. 20.

2. Spiegel, M. R.: "Vector Analysis," Schaum Outline Series, McGraw-Hill Book Company, New York, 1959. A large number of examples and problems with answers are provided in this concise, inexpensive member of an outline series.

3. Swokowski, E. W.: "Calculus with Analytic Geometry," 3d ed., Prindle, Weber, & Schmidt, Boston, 1984. Vector algebra and the cylindrical and spherical coordinate systems are discussed in chap. 14, and vector calculus appears in chap. 18.

4. Thomas, G. B., Jr., and R. L. Finney: "Calculus and Analytic Geometry," 6th ed., Addison-Wesley Publishing Company, Reading, Mass., 1984. Vector algebra and the three coordinate systems we use are discussed in chap. 13. Other vector operations are discussed in chaps. 15 and 17.

PROBLEMS

1. Let each of the vectors $\mathbf{A} = 5\mathbf{a}_x - \mathbf{a}_y + 3\mathbf{a}_z$, $\mathbf{B} = -2\mathbf{a}_x + 2\mathbf{a}_y + 4\mathbf{a}_z$, and $\mathbf{C} = 3\mathbf{a}_y - 4\mathbf{a}_z$ extend outward from the origin of a cartesian coordinate system to points A, B, and C, respectively. Find a unit vector directed from point A toward: (*a*) the origin; (*b*) point B; (*c*) a point equidistant from B and C on the line BC. (*d*) Find the length of the perimeter of the triangle ABC.

2. If $\mathbf{A} = 2\mathbf{a}_x + 6\mathbf{a}_y - 3\mathbf{a}_z$ and $\mathbf{B} = -3\mathbf{a}_x - 4\mathbf{a}_y - 5\mathbf{a}_z$, find: (*a*) a unit vector in the direction of $\mathbf{A} - \mathbf{B}$; (*b*) the magnitude of $\mathbf{A} + \mathbf{B}$; (*c*) $|[(\mathbf{A} + 3\mathbf{B})/|\mathbf{A} + 3\mathbf{B}|]|\mathbf{B}$.

3. The vector $\mathbf{R}_{AB}$ extends from $A(1,2,3)$ to B. If the length of $\mathbf{R}_{AB}$ is 10 units and its direction is given by $\mathbf{a} = 0.6\mathbf{a}_x + 0.64\mathbf{a}_y + 0.48\mathbf{a}_z$, find the coordinates of B.

4. Given two points, $M(2,5,-3)$ and $N(-3,1,4)$: (*a*) find their separation; (*b*) find the distance from the origin to the midpoint of the line MN; (*c*) find a unit vector in the direction of $\mathbf{R}_{MN}$; (*d*) find the point of intersection of the line MN and the $x = 0$ plane.

5. Given the vector field, $\mathbf{G} = 2x^2y\mathbf{a}_x - 2(z - x)\mathbf{a}_y + 3xyz\mathbf{a}_z$, find: (*a*) $\mathbf{G}$ at $P(2,-3,4)$; (*b*) a unit vector in the direction of $\mathbf{G}$ at P; (*c*) the (scalar) equation of the surface on which $|\mathbf{G}| = 100$; (*d*) the y coordinate of $Q(-3,y,5)$ if $|\mathbf{G}_Q| = 100$ and $y > 0$; (*e*) the distance between P and Q.

6. A vector field is given as $\mathbf{G} = 2x^2y\mathbf{a}_x - 2(z - x)\mathbf{a}_y + 3xyz\mathbf{a}_z$. Prepare sketches of G_x, G_y, G_z, and $|\mathbf{G}|$, all evaluated along the line $x = 2$, $y = -1$, for $0 \le z \le 10$.

7. A vector field is specified as $\mathbf{F} = 2(x + y)\sin \pi z\mathbf{a}_x - (x^2 + y)\mathbf{a}_y + [10/(x^2 + y^2)]\mathbf{a}_z$. Specify the locus of all points at which: (*a*) $F_x = 0$; (*b*) $F_y = 0$; (*c*) $|F_z| = 1$.

8. Given fields $\mathbf{G}_1 = 5(x + y)\mathbf{a}_x + 10\mathbf{a}_y$ and $\mathbf{G}_2 = 5\mathbf{a}_x + 2xy\mathbf{a}_y$, find at $P(3,2,0)$: (*a*) $|\mathbf{G}_1|$; (*b*) $|\mathbf{G}_2|$; (*c*) $\mathbf{a}_1$, a unit vector in the direction of $\mathbf{G}_1$; (*d*) $\mathbf{a}_2$, a unit vector in the direction of $\mathbf{G}_2$; (*e*) $\mathbf{a}_{12}$, a unit vector in the direction of $(\mathbf{G}_1 + \mathbf{G}_2)$.

9. Given the field, $\mathbf{G} = [2x/(1 + y^2)]\mathbf{a}_x + (y + z + 1)\mathbf{a}_y + (5x - z^2)\mathbf{a}_z$, find: (*a*) a unit vector in the direction of $\mathbf{G}$ at $P(1,2,-3)$; (*b*) the angle between $\mathbf{G}$ and the $y = 0$ plane at $Q(2,0,4)$; (*c*) the value of $\int_{y=0}^{2}\int_{x=1}^{3} \mathbf{G} \cdot dx\,dy\,\mathbf{a}_z$ at the plane $z = 1$.

10. Given points $P(2,5,-1)$, $Q(-1,-4,1)$, and $T(5,0,2)$, find: (*a*) the vector $\mathbf{R}_{PQ}$; (*b*) a unit vector in the direction of $\mathbf{R}_{PQ}$; (*c*) the length of the perimeter of the triangle PQT; (*d*) the interior angle at Q; (*e*) the vector projection of $\mathbf{R}_{PQ}$ on $\mathbf{R}_{PT}$; (*f*) the length of the altitude of the triangle that extends from Q perpendicularly to the opposite side or its extension.

11. Given points $E(2,5,1)$, $F(-1,4,-2)$, and $G(3,-2,4)$, find: (*a*) a unit vector directed from E towards F; (*b*) the angle between $\mathbf{R}_{EF}$ and $\mathbf{R}_{EG}$; (*c*) the length of the perimeter of triangle EFG; (*d*) the scalar projection of $\mathbf{R}_{EF}$ on $\mathbf{R}_{EG}$.

12. Express in cartesian components: (*a*) the vector **G** extending from the origin to the midpoint of the line joining $A(2,-3,5)$ to $B(6,-5,5)$; (*b*) the vector **D** extending from $C(-2,7,3)$ to the midpoint of the line joining A to B; (*c*) the component of $\mathbf{R}_{AB}$ that is in the direction of $\mathbf{R}_{AC}$; (*d*) a unit vector in the direction of $\mathbf{R}_{BC}$.

13. Given $\mathbf{A} = -4\mathbf{a}_x + 2\mathbf{a}_y + 3\mathbf{a}_z$ and $\mathbf{B} = 3\mathbf{a}_x + 4\mathbf{a}_y - \mathbf{a}_z$, find: (*a*) the magnitude of $5\mathbf{A} - 2\mathbf{B}$; (*b*) a unit vector in the direction of $(5\mathbf{A} - 2\mathbf{B})/|\mathbf{A}|$; (*c*) the vector component of **A** that is parallel to **B**; (*d*) the vector component of **A** that is perpendicular to **B**.

14. Given the three points $A(2,-1,2)$, $B(-1,1,4)$, and $C(4,3,-1)$, find: (*a*) the angle between $\mathbf{R}_{AB}$ and $\mathbf{R}_{AC}$; (*b*) the (scalar) area of triangle ABC; (*c*) a unit vector perpendicular to ABC.

15. Find in rectangular components the unit vector which is: (*a*) in the direction of **E** at $P(2,3,-4)$ if $\mathbf{E} = (x^2 + y^2 + z^2)(x\mathbf{a}_x/\sqrt{y^2 + z^2} + y\mathbf{a}_y/\sqrt{x^2 + z^2} + z\mathbf{a}_z/\sqrt{x^2 + y^2})$; (*b*) perpendicular to the plane passing through $M(1,-5,5)$, $N(-2,4,0)$, and $Q(2,3,4)$, and having a positive x component. (*c*) Find the angle between the vectors $\mathbf{R}_{MN}$ and $\mathbf{R}_{MQ}$.

16. Given vectors from the origin $\mathbf{r}_A = 3\mathbf{a}_x + 8\mathbf{a}_y + \mathbf{a}_z$, $\mathbf{r}_B = \mathbf{a}_x - 6\mathbf{a}_y + 5\mathbf{a}_z$, and $\mathbf{r}_C = 7\mathbf{a}_x - 4\mathbf{a}_z$, find: (*a*) a unit vector perpendicular to both $\mathbf{r}_A$ and $\mathbf{r}_B$; (*b*) a unit vector perpendicular to all the vectors, $\mathbf{r}_A - \mathbf{r}_B$, $\mathbf{r}_B - \mathbf{r}_C$, and $\mathbf{r}_C - \mathbf{r}_A$; (*c*) the area of the triangle defined by the heads of the vectors $\mathbf{r}_A$, $\mathbf{r}_B$, and $\mathbf{r}_C$.

17. Given the three points $M(6,2,-3)$, $N(-2,3,0)$, and $P(-4,6,5)$, find: (*a*) the area of the triangle they define; (*b*) a unit vector perpendicular to this triangular surface; (*c*) a unit vector bisecting the interior angle of the triangle at M.

18. (*a*) Express the field $\mathbf{F} = 2xyz\mathbf{a}_x - 5(x + y + z)\mathbf{a}_z$ in cylindrical coordinates (variables and components). (*b*) Find $|\mathbf{F}|$ at $P(\rho = 2, \phi = 60°, z = 3)$.

19. Given the points $P(\rho = 5, \phi = 60°, z = 2)$ and $Q(\rho = 2, \phi = 110°, z = -1)$: (*a*) find the distance $|\mathbf{R}_{PQ}|$; (*b*) give a unit vector in cartesian coordinates at P that is directed towards Q; (*c*) give a unit vector in cylindrical coordinates at P that is directed towards Q.

20. Find in cylindrical components: (*a*) a unit vector at $P(\rho = 5, \phi = 53.13°, z = -2)$ in the direction of $\mathbf{F} = \rho z \cos\phi\, \mathbf{a}_\rho - \rho z \sin\phi\, \mathbf{a}_\phi + \rho z \mathbf{a}_z$; (*b*) a unit vector at P parallel to $\mathbf{a}_x$; (*c*) a unit vector at $Q(\rho = 5, \phi = -36.87°, z = -2)$ parallel to $\mathbf{a}_x$; (*d*) $\mathbf{G} = 2\mathbf{a}_x - 4\mathbf{a}_y + 4\mathbf{a}_z$ at P.

21. (*a*) Give the vector in cartesian coordinates that extends from $P(\rho = 4, \phi = 10°, z = 1)$ to $Q(\rho = 7, \phi = 75°, z = 4)$. (*b*) Give the vector in cylindrical coordinates at $M(x = 5, y = 1, z = 2)$ that extends to $N(2,4,6)$. (*c*) How far is it from $A(110,60°,-20)$ to $B(30,125°,10)$?

22. Let $\mathbf{E} = [(10/\rho) + 8 \sin\phi]\mathbf{a}_\rho + 8 \cos\phi\, \mathbf{a}_\phi - 5\mathbf{a}_z$. Sketch $|\mathbf{E}|$: (*a*) vs. ϕ for $\rho = 2$, $z = 3$; (*b*) vs. ρ for $\phi = 0°$, $z = 3$; (*c*) vs. ρ for $\phi = 90°$, $z = 3$.

23. Given points $A(x = 2, y = 3, z = -1)$ and $B(\rho = 4, \phi = -50°, z = 2)$, find a unit vector in cylindrical coordinates: (*a*) at point B directed toward point A; (*b*) at point A directed toward point B.

24. (*a*) Find the volume defined by $4 < \rho < 6$, $30° < \phi < 60°$, $2 < z < 5$. (*b*) What is the length of the longest straight line that lies entirely within the volume? (*c*) Find the total area of the surface.

25. Using the coordinate system named, give the vector at point $A(2, -1, -3)$ that extends to $B(1,3,4)$: (*a*) cartesian; (*b*) cylindrical; (*c*) spherical.

26. An electric field intensity is given as $\mathbf{E} = (100 \cos \theta)\mathbf{a}_r/r^3 + (50 \sin \theta)\mathbf{a}_\theta/r^3$. At the point whose spherical coordinates are $r = 2$, $\theta = 60°$, $\phi = 20°$, find: (*a*) $|\mathbf{E}|$; (*b*) a unit vector (in cartesian coordinates) in the direction of $\mathbf{E}$.

27. (*a*) Find $\mathbf{a}_x$ in spherical components at $P(3, -4, 5)$. (*b*) Find $\mathbf{a}_\theta$ in cartesian components at P.

28. Given the points $M(r = 5, \theta = 20°, \phi = 120°)$ and $N(r = 2, \theta = 80°, \phi = 30°)$: (*a*) find the distance from M to N; (*b*) give a unit vector in cartesian coordinates at M that is directed toward N; (*c*) give a unit vector in spherical coordinates at M that is directed toward N.

29. A closed surface is defined in spherical coordinates by $3 < r < 5$, $0.1\pi < \theta < 0.3\pi$, $1.2\pi < \phi < 1.6\pi$. (*a*) Find the volume enclosed. (*b*) Find the distance from $P_1(r = 3, \theta = 0.1\pi, \phi = 1.2\pi)$ to $P_2(r = 5, \theta = 0.3\pi, \phi = 1.6\pi)$. (*c*) Find the total surface area.

30. (*a*) Give the vector in cartesian coordinates that extends from $P(r = 4, \theta = 20°, \phi = 10°)$ to $Q(r = 7, \theta = 120°, \phi = 75°)$. (*b*) Give the vector in spherical coordinates at $M(x = 5, y = 1, z = 2)$ that extends to $N(2,4,6)$. (*c*) How far is it from $A(r = 110, \theta = 30°, \phi = 60°)$ to $B(r = 30, \theta = 75°, \phi = 125°)$?

31. Points $A(r = 100, \theta = 90°, \phi = 0°)$ and $B(r = 100, \theta = 90°, \phi = 5°)$ are located on the surface of a 100-m-radius sphere. (*a*) What is their separation, using a path on the spherical surface? (*b*) What is their separation, using a straight-line path?

32. (*a*) Express $\mathbf{a}_\rho$ in spherical components and variables. (*b*) Express $\mathbf{a}_r$ in cylindrical components and variables.

CHAPTER 2

COULOMB'S LAW AND ELECTRIC FIELD INTENSITY

Now that we have formulated a new language in the first chapter, we shall establish a few basic principles of electricity and attempt to describe them in terms of it. If we had used vector calculus for several years and already had a few correct ideas about electricity and magnetism, we might jump in now with both feet and present a handful of equations, including Maxwell's equations and a few other auxiliary equations, and proceed to describe them physically by virtue of our knowledge of vector analysis. This is perhaps the ideal way, starting with the most general results and then showing that Ohm's, Gauss's, Coulomb's, Faraday's, Ampère's, Biot-Savart's, Kirchhoff's, and a few less familiar laws are all special cases of these equations. It is philosophically satisfying to have the most general result and to feel that we are able to obtain the results for any special case at will. However, such a jump would lead to many frantic cries of "Help" and not a few drowned students.

Instead we shall present at decent intervals the experimental laws mentioned above, expressing each in vector notation, and use these laws to solve a number of simple problems. In this way our familiarity with both vector analysis and electric and magnetic fields will gradually increase, and by the time we have finally reached our handful of general equations, little additional explanation will be required. The entire field of electromagnetic theory is then open to us, and we may use Maxwell's equations to describe wave propagation, radiation from antennas, skin effect, waveguides and transmission lines, and traveling-wave tubes, and even to obtain a new insight into the ordinary power transformer.

In this chapter we shall restrict our attention to *static* electric fields in *vacuum* or *free space*. Such fields, for example, are found in the focusing and deflection systems of electrostatic cathode-ray tubes. For all practical purposes, our results will also be applicable to air and other gases. Other materials will be introduced in Chap. 5, and time-varying fields will be introduced in Chap. 10.

We shall begin by describing a quantitative experiment performed in the seventeenth century.

2.1 THE EXPERIMENTAL LAW OF COULOMB

Records from at least 600 B.C. show evidence of the knowledge of static electricity. The Greeks were responsible for the term "electricity," derived from their word for amber, and they spent many a leisure hour rubbing a small piece of amber on their sleeves and observing how it would then attract pieces of fluff and stuff. However, their main interest lay in philosophy and logic, not in experimental science, and it was many centuries before the attracting effect was considered to be anything other than magic or a "life force."

Dr. Gilbert, physician to Her Majesty the Queen of England, was the first to do any true experimental work with this effect and in 1600 stated that glass, sulfur, amber, and other materials which he named would "not only draw to themselves straws and chaff, but all metals, wood, leaves, stone, earths, even water and oil."

Shortly thereafter a colonel in the French Army Engineers, Col. Charles Coulomb, a precise and orderly minded officer, performed an elaborate series of experiments using a delicate torsion balance, invented by himself, to determine quantitatively the force exerted between two objects each having a static charge of electricity. His published result is now known to many high school students and bears a great similarity to Newton's gravitational law (discovered about a hundred years earlier). Coulomb stated that the force between two very small objects separated in a vacuum or free space by a distance which is large compared to their size is proportional to the charge on each and inversely proportional to the square of the distance between them, or

$$F = k\frac{Q_1 Q_2}{R^2}$$

where Q_1 and Q_2 are the positive or negative quantities of charge, R is the separation, and k is a proportionality constant. If the International System of Units[1]

---

[1] The International System of Units (an mks system) is described in Appendix B. Abbreviations for the units are given in Table B.1. Conversions to other systems of units are given in Table B.2, while the prefixes designating powers of ten in SI appear in Table B.3.

(SI) is used, Q is measured in coulombs (C), R is in meters, and the force should be in newtons (N). This will be achieved if the constant of proportionality k is written as

$$k = \frac{1}{4\pi\epsilon_0}$$

The factor 4π will appear in the denominator of Coulomb's law but will not appear in the more useful equations (including Maxwell's equations) which we shall obtain with the help of Coulomb's law. The new constant ϵ_0 is called the *permittivity of free space* and has the magnitude, measured in farads per meter (F/m),

$$\epsilon_0 = 8.854 \times 10^{-12} \doteq \frac{1}{36\pi} 10^{-9} \qquad \text{F/m} \tag{1}$$

The quantity ϵ_0 is not dimensionless, for Coulomb's law shows that it has the label $C^2/N \cdot m^2$. We shall later define the farad and show that it has the dimensions $C^2/N \cdot m$; we have anticipated this definition by using the unit F/m in (1) above.

Coulomb's law is now

$$F = \frac{Q_1 Q_2}{4\pi\epsilon_0 R^2} \tag{2}$$

Not all SI units are as familiar as the English units we use daily, but they are now standard in electrical engineering and physics. The newton is a unit of force that is equal to 0.2248 lb_f, and is the force required to give a 1-kilogram (kg) mass an acceleration of 1 meter per second per second (m/s^2). The coulomb is an extremely large unit of charge, for the smallest known quantity of charge is that of the electron (negative) or proton (positive), given in mks units as $1.602\,18 \times 10^{-19}$ C, and hence a negative charge of one coulomb represents about 6×10^{18} electrons.[2] Coulomb's law shows that the force between two charges of one coulomb each, separated by one meter, is 9×10^9 N, or about one million tons. The electron has a rest mass of $9.109\,39 \times 10^{-31}$ kg and has a radius of the order of magnitude of 3.8×10^{-15} m. This does not mean that the electron is spherical in shape, but merely serves to describe the size of the region in which a slowly moving electron has the greatest probability of being found. All other known charged particles, including the proton, have larger masses, and larger radii, and occupy a probabilistic volume larger than does the electron.

In order to write the vector form of (2), we need the additional fact (furnished also by Colonel Coulomb) that the force acts along the line joining the two charges and is repulsive if the charges are alike in sign and attractive if they are of opposite sign. Let the vector $\mathbf{r}_1$ locate Q_1 while $\mathbf{r}_2$ locates Q_2. Then the vector $\mathbf{R}_{12} = \mathbf{r}_2 - \mathbf{r}_1$

[2] The charge and mass of an electron and other physical constants are tabulated in Table C.4 of Appendix C.

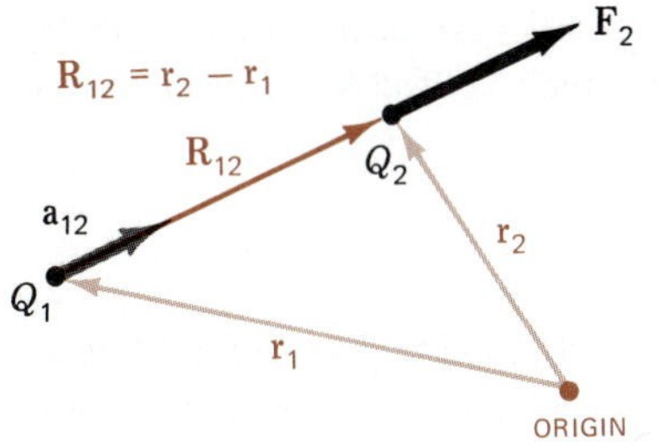

FIGURE 2.1
If Q_1 and Q_2 have like signs, the vector force $\mathbf{F}_2$ on Q_2 is in the same direction as the vector $\mathbf{R}_{12}$.

represents the directed line segment from Q_1 to Q_2, as shown in Fig. 2.1. The vector $\mathbf{F}_2$ is the force on Q_2 and is shown for the case where Q_1 and Q_2 have the same sign. The vector form of Coulomb's law is

$$\mathbf{F}_2 = \frac{Q_1 Q_2}{4\pi\epsilon_0 R_{12}^2}\mathbf{a}_{12} \tag{3}$$

where $\mathbf{a}_{12}$ = a unit vector in the direction of $\mathbf{R}_{12}$, or

$$\mathbf{a}_{12} = \frac{\mathbf{R}_{12}}{|\mathbf{R}_{12}|} = \frac{\mathbf{R}_{12}}{R_{12}} = \frac{\mathbf{r}_2 - \mathbf{r}_1}{|\mathbf{r}_2 - \mathbf{r}_1|} \tag{4}$$

As an example of the use of the vector form of Coulomb's law, consider a charge of 3×10^{-4} C at $P(1,2,3)$ and a charge of -10^{-4} C at $Q(2,0,5)$ in a vacuum. Then

$$Q_1 = 3 \times 10^{-4} \qquad Q_2 = -10^{-4}$$

$$\mathbf{R}_{12} = \mathbf{r}_2 - \mathbf{r}_1 = (2-1)\mathbf{a}_x + (0-2)\mathbf{a}_y + (5-3)\mathbf{a}_z = \mathbf{a}_x - 2\mathbf{a}_y + 2\mathbf{a}_z$$

$$\mathbf{a}_{12} = \frac{\mathbf{a}_x - 2\mathbf{a}_y + 2\mathbf{a}_z}{3}$$

$$\mathbf{F}_2 = \frac{3 \times 10^{-4}(-10^{-4})}{4\pi(1/36\pi)10^{-9} \times 9}\left(\frac{\mathbf{a}_x - 2\mathbf{a}_y + 2\mathbf{a}_z}{3}\right)$$

$$= -30\left(\frac{\mathbf{a}_x - 2\mathbf{a}_y + 2\mathbf{a}_z}{3}\right) \quad \text{N}$$

The magnitude of the force is 30 N (or about 7 lb_f), and the direction is specified by the unit vector, which has been left in parentheses to display the magnitude of the force. The force on Q_2 may also be considered as three component forces,

$$\mathbf{F}_2 = -10\mathbf{a}_x + 20\mathbf{a}_y - 20\mathbf{a}_z$$

The force expressed by Coulomb's law is a mutual force, for each of the two charges experiences a force of the same magnitude, although of opposite direction. We might equally well have written

$$\mathbf{F}_1 = -\mathbf{F}_2 = \frac{Q_1 Q_2}{4\pi\epsilon_0 R_{12}^2}\mathbf{a}_{21} = -\frac{Q_1 Q_2}{4\pi\epsilon_0 R_{12}^2}\mathbf{a}_{12} \tag{5}$$

Coulomb's law is linear, for if we multiply Q_1 by a factor n, the force on Q_2 is also multiplied by the same factor n. It is also true that the force on a charge in the presence of several other charges is the sum of the forces on that charge due to each of the other charges acting alone.

D2.1. A 2-mC (millicoulomb) positive charge is located in vacuum at $P_1(3,-2,-4)$, and a 5-μC (microcoulomb) negative charge is at $P_2(1,-4,2)$. (*a*) Find the vector force on the negative charge. (*b*) What is the magnitude of the force on the charge at P_1?

Ans. $0.616\mathbf{a}_x + 0.616\mathbf{a}_y - 1.848\mathbf{a}_z$ N; 2.04 N

2.2 ELECTRIC FIELD INTENSITY

If we now consider one charge fixed in position, say Q_1, and move a second charge slowly around, we note that there exists everywhere a force on this second charge; in other words, this second charge is displaying the existence of a force *field*. Call this second charge a test charge Q_t. The force on it is given by Coulomb's law,

$$\mathbf{F}_t = \frac{Q_1 Q_t}{4\pi\epsilon_0 R_{1t}^2}\mathbf{a}_{1t}$$

Writing this force as a force per unit charge gives

$$\boxed{\frac{\mathbf{F}_t}{Q_t} = \frac{Q_1}{4\pi\epsilon_0 R_{1t}^2}\mathbf{a}_{1t}} \tag{6}$$

The quantity on the right side of (6) is a function only of Q_1 and the directed line segment from Q_1 to the position of the test charge. This describes a vector field and is called the *electric field intensity*.

We define the electric field intensity as the vector force on a unit positive test charge. We would not *measure* it experimentally by finding the force on a 1-C test charge, however, for this would probably cause such a force on Q_1 as to change the position of that charge.

Electric field intensity must be measured by the unit newtons per coulomb—the force per unit charge. Again anticipating a new dimensional quantity, the *volt* (V), to be presented in Chap. 4 and having the label of joules per coulomb (J/C) or newton-meters per coulomb (N·m/C), we shall at once measure electric field intensity in the practical units of volts per meter (V/m). Using a capital letter **E** for electric field intensity, we have finally

$$\boxed{\mathbf{E} = \frac{\mathbf{F}_t}{Q_t}} \tag{7}$$

$$\mathbf{E} = \frac{Q_1}{4\pi\epsilon_0 R_{1t}^2}\mathbf{a}_{1t} \tag{8}$$

Equation (7) is the defining expression for electric field intensity, and (8) is the expression for the electric field intensity due to a single point charge Q_1 in a vacuum. In the succeeding sections we shall obtain and interpret expressions for the electric field intensity due to more complicated arrangements of charge, but now let us see what information we can obtain from (8), the field of a single point charge.

First, let us dispense with most of the subscripts in (8), reserving the right to use them again any time there is a possibility of misunderstanding:

$$\boxed{\mathbf{E} = \frac{Q}{4\pi\epsilon_0 R^2}\,\mathbf{a}_R} \tag{9}$$

We should remember that R is the magnitude of the vector $\mathbf{R}$, the directed line segment from the point at which the point charge Q is located to the point at which $\mathbf{E}$ is desired, and $\mathbf{a}_R$ is a unit vector in the $\mathbf{R}$ direction.[3]

Let us arbitrarily locate Q_1 at the center of a spherical coordinate system. The unit vector $\mathbf{a}_R$ then becomes the radial unit vector $\mathbf{a}_r$, and R is r. Hence

$$\mathbf{E} = \frac{Q_1}{4\pi\epsilon_0 r^2}\,\mathbf{a}_r \tag{10}$$

or

$$E_r = \frac{Q_1}{4\pi\epsilon_0 r^2}$$

The field has a single radial component, and its inverse-square-law relationship is quite obvious.

Writing these expressions in cartesian coordinates for a charge Q at the origin, we have $\mathbf{R} = \mathbf{r} = x\mathbf{a}_x + y\mathbf{a}_y + z\mathbf{a}_z$, $\mathbf{a}_R = \mathbf{a}_r = (x\mathbf{a}_x + y\mathbf{a}_y + z\mathbf{a}_z)/\sqrt{x^2 + y^2 + z^2}$, and then

$$\mathbf{E} = \frac{Q}{4\pi\epsilon_0(x^2 + y^2 + z^2)}\left(\frac{x}{\sqrt{x^2 + y^2 + z^2}}\,\mathbf{a}_x + \frac{y}{\sqrt{x^2 + y^2 + z^2}}\,\mathbf{a}_y + \frac{z}{\sqrt{x^2 + y^2 + z^2}}\,\mathbf{a}_z\right) \tag{11}$$

This expression no longer shows immediately the simple nature of the field, and its complexity is the price we pay for solving a problem having spherical symmetry in a coordinate system with which we may (temporarily) have more familiarity.

Without using vector analysis, the information contained in (11) would have to be expressed in three equations, one for each component, and in order to obtain

[3] We firmly intend to avoid confusing r and $\mathbf{a}_r$ with R and $\mathbf{a}_R$. The first two refer specifically to the spherical coordinate system, whereas R and $\mathbf{a}_R$ do not refer to any coordinate system—the choice is still available to us.

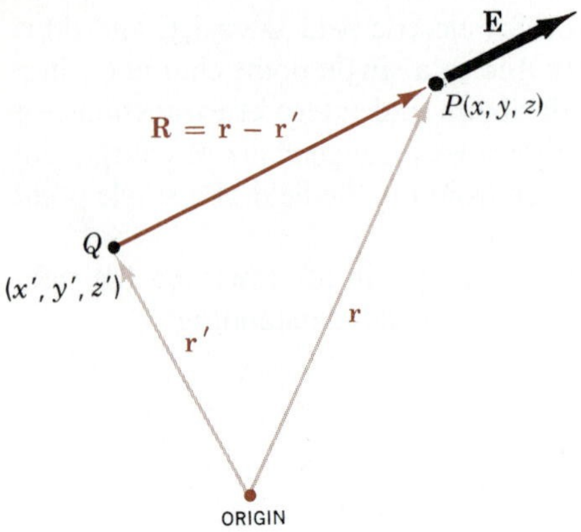

FIGURE 2.2
The vector $\mathbf{r}'$ locates the point charge Q, the vector $\mathbf{r}$ identifies the general point in space $P(x,y,z)$, and the vector $\mathbf{R}$ from Q to $P(x,y,z)$ is then $\mathbf{R} = \mathbf{r} - \mathbf{r}'$.

the equations we should have to break up the magnitude of the electric field intensity into the three components by finding the projection on each coordinate axis. Using vector notation, this is done automatically when we write the unit vector.

If we consider a charge which is *not* at the origin of our coordinate system, the field no longer possesses spherical symmetry (nor cylindrical symmetry, unless the charge lies on the z axis), and we might as well use cartesian coordinates. For a charge Q located at the source point $\mathbf{r}' = x'\mathbf{a}_x + y'\mathbf{a}_y + z'\mathbf{a}_z$, as illustrated in Fig. 2.2, we find the field at a general field point $\mathbf{r} = x\mathbf{a}_x + y\mathbf{a}_y + z\mathbf{a}_z$ by expressing $\mathbf{R}$ as $\mathbf{r} - \mathbf{r}'$, and then

$$\mathbf{E}(\mathbf{r}) = \frac{Q}{4\pi\epsilon_0|\mathbf{r}-\mathbf{r}'|^2}\frac{\mathbf{r}-\mathbf{r}'}{|\mathbf{r}-\mathbf{r}'|} = \frac{Q(\mathbf{r}-\mathbf{r}')}{4\pi\epsilon_0|\mathbf{r}-\mathbf{r}'|^3}$$

$$= \frac{Q[(x-x')\mathbf{a}_x + (y-y')\mathbf{a}_y + (z-z')\mathbf{a}_z]}{4\pi\epsilon_0[(x-x')^2 + (y-y')^2 + (z-z')^2]^{3/2}} \qquad (12)$$

Earlier, we defined a vector field as a vector function of a position vector, and this is emphasized by letting $\mathbf{E}$ be symbolized in functional notation by $\mathbf{E}(\mathbf{r})$.

Equation (11) is merely a special case of (12), where $x' = y' = z' = 0$.

Since the coulomb forces are linear, the electric field intensity due to two point charges, Q_1 at $\mathbf{r}_1$ and Q_2 at $\mathbf{r}_2$, is the sum of the forces on Q_t caused by Q_1 and Q_2 acting alone, or

$$\mathbf{E}(\mathbf{r}) = \frac{Q_1}{4\pi\epsilon_0|\mathbf{r}-\mathbf{r}_1|^2}\mathbf{a}_1 + \frac{Q_2}{4\pi\epsilon_0|\mathbf{r}-\mathbf{r}_2|^2}\mathbf{a}_2$$

where $\mathbf{a}_1$ and $\mathbf{a}_2$ are unit vectors in the direction of $(\mathbf{r} - \mathbf{r}_1)$ and $(\mathbf{r} - \mathbf{r}_2)$, respectively. The vectors $\mathbf{r}, \mathbf{r}_1, \mathbf{r}_2, \mathbf{r} - \mathbf{r}_1, \mathbf{r} - \mathbf{r}_2, \mathbf{a}_1$, and $\mathbf{a}_2$ are shown in Fig. 2.3.

If we add more charges at other positions, the field due to n point charges is

$$\mathbf{E}(\mathbf{r}) = \frac{Q_1}{4\pi\epsilon_0|\mathbf{r}-\mathbf{r}_1|^2}\mathbf{a}_1 + \frac{Q_2}{4\pi\epsilon_0|\mathbf{r}-\mathbf{r}_2|^2}\mathbf{a}_2 + \cdots + \frac{Q_n}{4\pi\epsilon_0|\mathbf{r}-\mathbf{r}_n|^2}\mathbf{a}_n \qquad (13)$$

This expression takes up less space when we use a summation sign $\sum$ and a summing integer m which takes on all integral values between 1 and n,

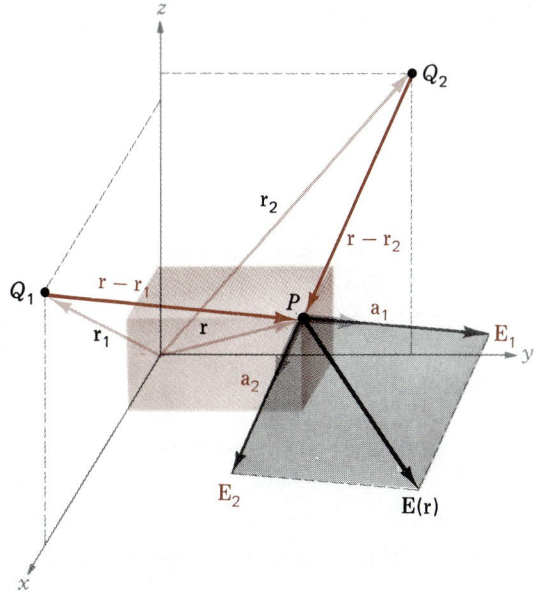

FIGURE 2.3
The vector addition of the total electric field intensity at P due to Q_1 and Q_2 is made possible by the linearity of Coulomb's law.

$$\mathbf{E}(\mathbf{r}) = \sum_{m=1}^{n} \frac{Q_m}{4\pi\epsilon_0|\mathbf{r} - \mathbf{r}_m|^2} \mathbf{a}_m \tag{14}$$

When expanded, (14) is identical with (13), and students unfamiliar with summation signs should check that result.

As an example of the determination of the electric field intensity produced by a point-charge distribution, let us find $\mathbf{E}$ at $P(1,1,1)$ caused by four identical 3-nC (nanocoulomb) point charges located at $P_1(1,1,0)$, $P_2(-1,1,0)$, $P_3(-1,-1,0)$, and $P_4(1,-1,0)$, as shown in Fig. 2.4. We find that $\mathbf{r} = \mathbf{a}_x + \mathbf{a}_y + \mathbf{a}_z$, $\mathbf{r}_1 = \mathbf{a}_x + \mathbf{a}_y$,

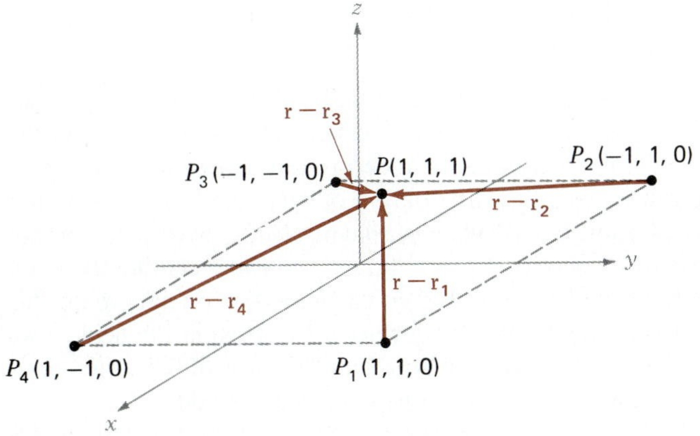

FIGURE 2.4
A symmetrical distribution of four identical 3-nC point charges produces a field at P, $\mathbf{E} = 6.82\mathbf{a}_x + 6.82\mathbf{a}_y + 32.8\mathbf{a}_z$ V/m.

and thus $\mathbf{r} - \mathbf{r}_1 = \mathbf{a}_z$. Similarly, $\mathbf{r} - \mathbf{r}_2 = 2\mathbf{a}_x + \mathbf{a}_z$, $\mathbf{r} - \mathbf{r}_3 = 2\mathbf{a}_x + 2\mathbf{a}_y + \mathbf{a}_z$, and $\mathbf{r} - \mathbf{r}_4 = 2\mathbf{a}_y + \mathbf{a}_z$. We have, therefore,

$$\frac{Q}{4\pi\epsilon_0} = \frac{3 \times 10^{-9}}{4\pi(8.854 \times 10^{-12})} = 26.96 \qquad \text{V}\cdot\text{m}$$

and

$$\mathbf{E} = 26.96\left[\frac{\mathbf{a}_z}{1^3} + \frac{2\mathbf{a}_x + \mathbf{a}_z}{(\sqrt{5})^3} + \frac{2\mathbf{a}_x + 2\mathbf{a}_y + \mathbf{a}_z}{(\sqrt{9})^3} + \frac{2\mathbf{a}_y + \mathbf{a}_z}{(\sqrt{5})^3}\right]$$

or

$$\mathbf{E} = 6.82\mathbf{a}_x + 6.82\mathbf{a}_y + 32.8\mathbf{a}_z \qquad \text{V/m}$$

D2.2. Calculate **E** at $M(3,-4,2)$ in free space caused by: (*a*) a charge $Q_1 = 2\ \mu\text{C}$ at $P_1(0,0,0)$; (*b*) a charge $Q_2 = 3\ \mu\text{C}$ at $P_2(-1,2,3)$; (*c*) a charge $Q_1 = 2\ \mu\text{C}$ at $P_1(0,0,0)$, and a charge $Q_2 = 3\ \mu\text{C}$ at $P_2(-1,2,3)$.

Ans. $345\mathbf{a}_x - 460\mathbf{a}_y + 230\mathbf{a}_z$ V/m; $280\mathbf{a}_x - 419\mathbf{a}_y - 69.9\mathbf{a}_z$ V/m; $625\mathbf{a}_x - 880\mathbf{a}_y + 160.3\mathbf{a}_z$ V/m

D2.3. Evaluate the sums:

$$(a)\quad \sum_{m=1}^{5} \frac{2m}{(m^2 + 3^2)^{1.5}}$$

$$(b)\quad \sum_{m=1}^{10} \frac{(-1)^{m+1}}{m} \sin\frac{\pi}{2m}$$

Ans. 0.342; 0.747

2.3 FIELD DUE TO A CONTINUOUS VOLUME CHARGE DISTRIBUTION

If we now visualize a region of space filled with a tremendous number of charges separated by minute distances, such as the space between the control grid and the cathode in the electron-gun assembly of a cathode-ray tube operating with space charge, we see that we can replace this distribution of very small particles with a smooth continuous distribution described by a *volume charge density*, just as we describe water as having a density of 1 g/cm^3 (gram per cubic centimeter) even though it consists of atomic- and molecular-sized particles. We are able to do this only if we are uninterested in the small irregularities (or ripples) in the field as we move from electron to electron or if we care little that the mass of the water actually increases in small but finite steps as each new molecule is added.

This is really no limitation at all, for our end results as electrical engineers are almost always in terms of a current in a receiving antenna, a voltage in an electronic circuit, or a charge on a capacitor, or in general in terms of some large-scale *macroscopic* phenomenon. It is very seldom that we must know a current electron by electron.[4]

We denote volume charge density by ρ_v, having the units of coulombs per cubic meter (C/m^3).

The small amount of charge ΔQ in a small volume Δv is

$$\Delta Q = \rho_v \Delta v \tag{15}$$

and we may define ρ_v mathematically by using a limiting process on (15),

$$\boxed{\rho_v = \lim_{\Delta v \to 0} \frac{\Delta Q}{\Delta v}} \tag{16}$$

The total charge within some finite volume is obtained by integrating throughout that volume,

$$\boxed{Q = \int_{\text{vol}} dQ = \int_{\text{vol}} \rho_v \, dv} \tag{17}$$

Only one integral sign is customarily indicated, but the differential dv signifies integration throughout a volume, and hence a triple integration. Fortunately, we may content ourselves for the most part with no more than the indicated integration, for multiple integrals are very difficult to evaluate in all but the most symmetrical problems.

As an example of the evaluation of a volume integral, let us find the total charge contained in the 2-cm length of the electron beam shown in Fig. 2.5. At the instant shown, we assume the volume charge density to be

$$\rho_v = -5 \times 10^{-6} e^{-10^5 \rho z} \qquad \text{C/m}^3$$

The volume differential in cylindrical coordinates is given in Sec. 1.8; therefore,

$$Q = \int_{0.02}^{0.04} \int_0^{2\pi} \int_0^{0.01} -5 \times 10^{-6} e^{-10^5 \rho z} \rho \, d\rho \, d\phi \, dz$$

We integrate first with respect to ϕ, since it is so easy,

$$Q = \int_{0.02}^{0.04} \int_0^{0.01} -10^{-5} \pi e^{-10^5 \rho z} \rho \, d\rho \, dz$$

and then with respect to z, because this will simplify the last integration with respect to ρ,

$$Q = \int_0^{0.01} \left(\frac{-10^{-5}\pi}{-10^5 \rho} e^{-10^5 \rho z} \rho \, d\rho \right)_{z=0.02}^{z=0.04}$$

$$= \int_0^{0.01} -10^{-10} \pi (e^{-2000\rho} - e^{-4000\rho}) \, d\rho$$

---

[4] A study of the noise generated by electrons or ions in transistors, vacuum tubes, and resistors, however, requires just such an examination of the charge.

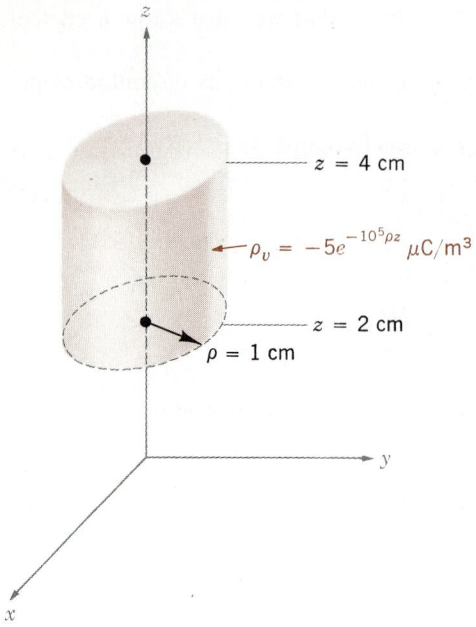

FIGURE 2.5
The total charge contained within the right circular cylinder may be obtained by evaluating $Q = \int_{\text{vol}} \rho_v \, dv$.

Finally,

$$Q = -10^{-10}\pi \left(\frac{e^{-2000\rho}}{-2000} - \frac{e^{-4000\rho}}{-4000} \right)_0^{0.01}$$

or

$$Q = -10^{-10}\pi \left(\frac{1}{2000} - \frac{1}{4000} \right) = \frac{-\pi}{40} \qquad \text{pC}$$

where pC indicates picocoulombs.

As a rough estimate, if we assume these electrons are moving at a constant velocity of 10 percent of the velocity of light, this 2-cm-long packet will have moved 2 cm in $\frac{2}{3}$ ns, and the current is about equal to the quotient,

$$\frac{\Delta Q}{\Delta t} = \frac{-(\pi/40)10^{-12}}{(\frac{2}{3})10^{-9}}$$

or approximately 118 μA.

The incremental contribution to the electric field intensity at $\mathbf{r}$ produced by an incremental charge ΔQ at $\mathbf{r}'$ is

$$\Delta \mathbf{E}(\mathbf{r}) = \frac{\Delta Q}{4\pi\epsilon_0 |\mathbf{r} - \mathbf{r}'|^2} \frac{\mathbf{r} - \mathbf{r}'}{|\mathbf{r} - \mathbf{r}'|} = \frac{\rho_v \Delta v}{4\pi\epsilon_0 |\mathbf{r} - \mathbf{r}'|^2} \frac{\mathbf{r} - \mathbf{r}'}{|\mathbf{r} - \mathbf{r}'|}$$

If we sum the contributions of all the volume charge in a given region and let the volume element Δv approach zero as the number of these elements becomes infinite, the summation becomes an integral,

$$\mathbf{E}(\mathbf{r}) = \int_{\mathrm{vol}} \frac{\rho_v(\mathbf{r}')\,dv'}{4\pi\epsilon_0|\mathbf{r}-\mathbf{r}'|^2} \frac{\mathbf{r}-\mathbf{r}'}{|\mathbf{r}-\mathbf{r}'|} \tag{18}$$

This is again a triple integral, and (except in the drill problem that follows) we shall do our best to avoid actually performing the integration.

The significance of the various quantities under the integral sign of (18) might stand a little review. The vector $\mathbf{r}$ from the origin locates the field point where $\mathbf{E}$ is being determined, while the vector $\mathbf{r}'$ extends from the origin to the source point where $\rho_v(\mathbf{r}')\,dv'$ is located. The scalar distance between the source point and the field point is $|\mathbf{r}-\mathbf{r}'|$, and the fraction $(\mathbf{r}-\mathbf{r}')/|\mathbf{r}-\mathbf{r}'|$ is a unit vector directed from source point to field point. The variables of integration are x', y', and z' in cartesian coordinates.

D2.4. Find the total charge inside each of the volumes indicated: (*a*) $\rho_v = 10z^2e^{-0.1x}\sin \pi y$, $-1 \le x \le 2$, $0 \le y \le 1$, $3 \le z \le 3.6$; (*b*) $\rho_v = 4xyz^2$, $0 \le \rho \le 2$, $0 \le \phi \le \pi/2$, $0 \le z \le 3$; (*c*) $\rho_v = 3\pi \cos^2\theta \cos^2\phi/[2r^2(r^2+1)]$; universe.

Ans. 119.5 C; 72.0 C; 15.50 C

2.4 FIELD OF A LINE CHARGE

Up to this point we have considered two types of charge distribution, the point charge and charge distributed throughout a volume with a density of ρ_v C/m^3. If we now consider a filamentlike distribution of volume charge density, such as a very fine, sharp beam in a cathode-ray tube or a charged conductor of very small radius, we find it convenient to treat the charge as a line charge of density ρ_L C/m. In the case of the electron beam the charges are in motion and it is true that we do not have an electrostatic problem. However, if the electron motion is steady and uniform (a dc beam) and if we ignore for the moment the magnetic field which is produced, the electron beam may be considered as composed of stationary electrons, for snapshots taken at any time will show the same charge distribution.

Let us assume a straight line charge extending along the z axis in a cylindrical coordinate system from $-\infty$ to ∞, as shown in Fig. 2.6. We desire the electric field intensity $\mathbf{E}$ at any and every point resulting from a *uniform* line charge density ρ_L.

Symmetry should always be considered first in order to determine two specific factors: (1) with which coordinates the field does *not* vary, and (2) which components of the field are *not* present. The answers to these questions then tell us which components are present and with which coordinates they do vary.

Referring to Fig. 2.6, we realize that as we move around the line charge, varying ϕ while keeping ρ and z constant, the line charge appears the same from every angle. In other words, azimuthal symmetry is present, and no field component may vary with ϕ.

Again, if we maintain ρ and ϕ constant while moving up and down the line charge by changing z, the line charge still recedes into infinite distance in both

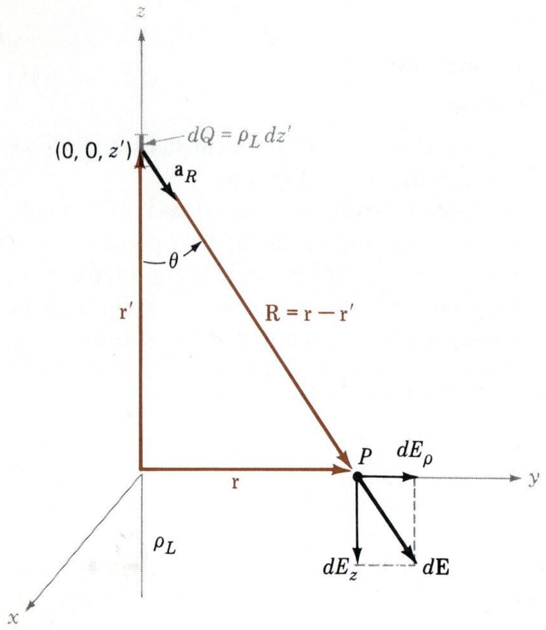

FIGURE 2.6
The contribution $d\mathbf{E} = dE_\rho \mathbf{a}_\rho + dE_z \mathbf{a}_z$ to the electric field intensity produced by an element of charge $dQ = \rho_L \, dz'$ located a distance z' from the origin. The linear charge density is uniform and extends along the entire z axis.

directions and the problem is unchanged. This is axial symmetry and leads to fields which are not functions of z.

If we maintain ϕ and z constant and vary ρ, the problem changes, and Coulomb's law leads us to expect the field to become weaker as ρ increases. Hence, by a process of elimination we are led to the fact that the field varies only with ρ.

Now, which components are present? Each incremental length of line charge acts as a point charge and produces an incremental contribution to the electric field intensity which is directed away from the bit of charge (assuming a positive line charge). No element of charge produces a ϕ component of electric intensity; E_ϕ is zero. However, each element does produce an E_ρ and an E_z component, but the contribution to E_z by elements of charge which are equal distances above and below the point at which we are determining the field will cancel.

We therefore have found that we have only an E_ρ component and it varies only with ρ. Now to find this component.

We choose a point $P(0,y,0)$ on the y axis at which to determine the field. This is a perfectly general point in view of the lack of variation of the field with ϕ and z. Applying (12) to find the incremental field at P due to the incremental charge $dQ = \rho_L \, dz'$, we have

$$d\mathbf{E} = \frac{\rho_L \, dz' (\mathbf{r} - \mathbf{r}')}{4\pi\epsilon_0 |\mathbf{r} - \mathbf{r}'|^3}$$

where

$$\mathbf{r} = y\mathbf{a}_y = \rho\mathbf{a}_\rho$$

$$\mathbf{r}' = z'\mathbf{a}_z$$

and

$$\mathbf{r} - \mathbf{r}' = \rho\mathbf{a}_\rho - z'\mathbf{a}_z$$

Therefore,

$$d\mathbf{E} = \frac{\rho_L\, dz'\,(\rho\mathbf{a}_\rho - z'\mathbf{a}_z)}{4\pi\epsilon_0(\rho^2 + z'^2)^{3/2}}$$

Since only the E_ρ component is present, we may simplify:

$$dE_\rho = \frac{\rho_L \rho\, dz'}{4\pi\epsilon_0(\rho^2 + z'^2)^{3/2}}$$

and

$$E_\rho = \int_{-\infty}^{\infty} \frac{\rho_L \rho\, dz'}{4\pi\epsilon_0(\rho^2 + z'^2)^{3/2}}$$

Integrating by integral tables or change of variable, $z' = \rho \cot \theta$, we have

$$E_\rho = \frac{\rho_L}{4\pi\epsilon_0}\,\rho\left(\frac{1}{\rho^2}\frac{z'}{\sqrt{\rho^2 + z'^2}}\right)_{-\infty}^{\infty}$$

and

$$\boxed{E_\rho = \frac{\rho_L}{2\pi\epsilon_0 \rho}} \qquad (19)$$

This is the desired answer, but there are many other ways of obtaining it. We might have used the angle θ as our variable of integration, for $z' = \rho \cot \theta$ from Fig. 2.6 and $dz' = -\rho \csc^2 \theta\, d\theta$. Since $R = \rho \csc \theta$, our integral becomes, simply,

$$dE_\rho = \frac{\rho_L\, dz'}{4\pi\epsilon_0 R^2} \sin\theta = -\frac{\rho_L \sin\theta\, d\theta}{4\pi\epsilon_0 \rho}$$

$$E_\rho = -\frac{\rho_L}{4\pi\epsilon_0 \rho}\int_{\pi}^{0} \sin\theta\, d\theta = \frac{\rho_L}{4\pi\epsilon_0 \rho} \cos\theta\Big]_{\pi}^{0}$$

$$= \frac{\rho_L}{2\pi\epsilon_0 \rho}$$

Here the integration was simpler, but some experience with problems of this type is necessary before we can unerringly choose the simplest variable of integration at the beginning of the problem.

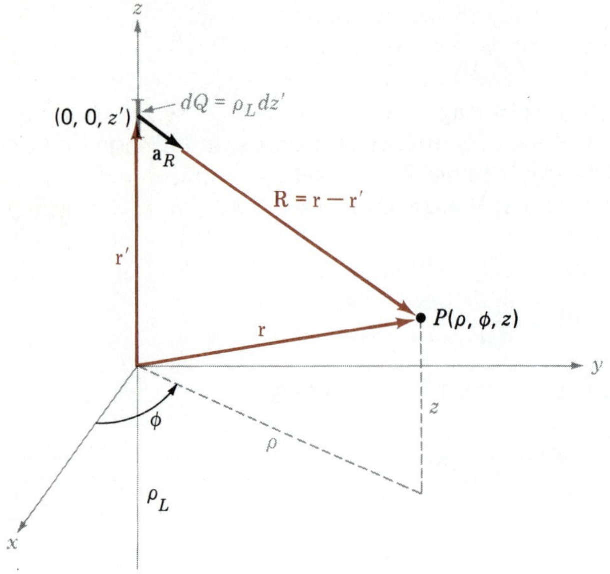

FIGURE 2.7
The geometry of the problem for the field about an infinite line charge leads to more difficult integrations when symmetry is ignored.

We might also have considered (18) as our starting point,

$$\mathbf{E} = \int_{\text{vol}} \frac{\rho_v\, dv'\, (\mathbf{r} - \mathbf{r}')}{4\pi\epsilon_0 |\mathbf{r} - \mathbf{r}'|^3}$$

letting $\rho_v\, dv' = \rho_L\, dz'$ and integrating along the line which is now our "volume" containing all the charge. Suppose we do this and forget everything we have learned from the symmetry of the problem. Choose point P now at a general location (ρ,ϕ,z) (Fig. 2.7) and write

$$\mathbf{r} = \rho\mathbf{a}_\rho + z\mathbf{a}_z$$

$$\mathbf{r}' = z'\mathbf{a}_z$$

$$\mathbf{R} = \mathbf{r} - \mathbf{r}' = \rho\mathbf{a}_\rho + (z - z')\mathbf{a}_z$$

$$R = \sqrt{\rho^2 + (z - z')^2}$$

$$\mathbf{a}_R = \frac{\rho\mathbf{a}_\rho + (z - z')\mathbf{a}_z}{\sqrt{\rho^2 + (z - z')^2}}$$

$$\mathbf{E} = \int_{-\infty}^{\infty} \frac{\rho_L\, dz' \left[\rho\mathbf{a}_\rho + (z - z')\mathbf{a}_z\right]}{4\pi\epsilon_0\left[\rho^2 + (z - z')^2\right]^{3/2}}$$

$$= \frac{\rho_L}{4\pi\epsilon_0}\left\{\int_{-\infty}^{\infty} \frac{\rho\, dz'\, \mathbf{a}_\rho}{\left[\rho^2 + (z - z')^2\right]^{3/2}} + \int_{-\infty}^{\infty} \frac{(z - z')\, dz'\, \mathbf{a}_z}{\left[\rho^2 + (z - z')^2\right]^{3/2}}\right\}$$

Before integrating a vector expression, we must know whether or not the vector under the integral sign (here the unit vectors $\mathbf{a}_\rho$ and $\mathbf{a}_z$) varies with the variable of integration (here dz'). If it does not, then it is a constant and may be removed from within the integral, leaving a scalar which may be integrated by normal methods. Our unit vectors, of course, cannot change in magnitude, but a change in direction is just as troublesome. Fortunately, the direction of $\mathbf{a}_\rho$ does not change with z' (nor with ρ, but it does change with ϕ), and $\mathbf{a}_z$ is constant always.

Hence we remove the unit vectors from the integrals and again integrate with tables or by changing variables,

$$\mathbf{E} = \frac{\rho_L}{4\pi\epsilon_0}\left\{\mathbf{a}_\rho \int_{-\infty}^{\infty} \frac{\rho\, dz'}{[\rho^2 + (z - z')^2]^{3/2}} + \mathbf{a}_z \int_{-\infty}^{\infty} \frac{(z - z')\, dz'}{[\rho^2 + (z - z')^2]^{3/2}}\right\}$$

$$= \frac{\rho_L}{4\pi\epsilon_0}\left\{\left[\mathbf{a}_\rho \rho \frac{1}{\rho^2} \frac{-(z - z')}{\sqrt{\rho^2 + (z - z')^2}}\right]_{-\infty}^{\infty} + \left[\mathbf{a}_z \frac{1}{\sqrt{\rho^2 + (z - z')^2}}\right]_{-\infty}^{\infty}\right\}$$

$$= \frac{\rho_L}{4\pi\epsilon_0}\left[\mathbf{a}_\rho \frac{2}{\rho} + \mathbf{a}_z(0)\right] = \frac{\rho_L}{2\pi\epsilon_0 \rho}\,\mathbf{a}_\rho$$

Again we obtain the same answer, as we should, for there is nothing wrong with the method except that the integration was harder and there were two integrations to perform. This is the price we pay for neglecting the consideration of symmetry and plunging doggedly ahead with mathematics. Look before you integrate.

Other methods for solving this basic problem will be discussed later after we introduce Gauss's law and the concept of potential.

Now let us consider the answer itself,

$$\boxed{\mathbf{E} = \frac{\rho_L}{2\pi\epsilon_0 \rho}\,\mathbf{a}_\rho} \tag{20}$$

We note that the field falls off inversely with the distance to the charged line, as compared with the point charge, where the field decreased with the square of the distance. Moving ten times as far from a point charge leads to a field only 1 percent the previous strength, but moving ten times as far from a line charge only reduces the field to 10 percent of its former value. An analogy can be drawn with a source of illumination, for the light intensity from a point source of light also falls off inversely as the square of the distance to the source. The field of an infinitely long fluorescent tube thus decays inversely as the first power of the radial distance to the tube, and we should expect the light intensity about a finite-length tube to obey this law near the tube. As our point recedes farther and farther from a finite-length tube, however, it eventually looks like a point source and the field obeys the inverse-square relationship.

Before leaving this introductory look at the field of the infinite line charge, we should recognize the fact that not all line charges are located along the z axis.

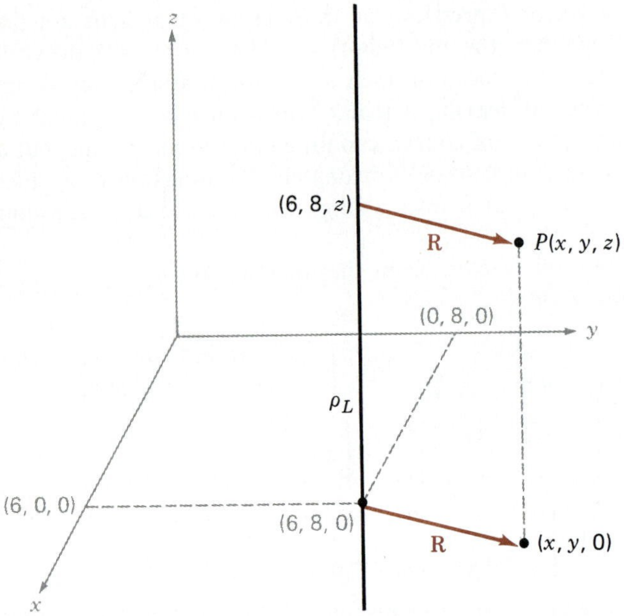

FIGURE 2.8
A point $P(x,y,z)$ is identified near an infinite uniform line charge located at $x = 6$, $y = 8$.

As an example, let us consider an infinite line charge parallel to the z axis at $x = 6$, $y = 8$, Fig. 2.8. We wish to find $\mathbf{E}$ at the general field point $P(x,y,z)$.

We replace ρ in (20) by the radial distance between the line charge and point P, $R = \sqrt{(x-6)^2 + (y-8)^2}$, and let $\mathbf{a}_\rho$ be $\mathbf{a}_R$. Thus,

$$\mathbf{E} = \frac{\rho_L}{2\pi\epsilon_0\sqrt{(x-6)^2 + (y-8)^2}}\mathbf{a}_R$$

where

$$\mathbf{a}_R = \frac{\mathbf{R}}{|\mathbf{R}|} = \frac{(x-6)\mathbf{a}_x + (y-8)\mathbf{a}_y}{\sqrt{(x-6)^2 + (y-8)^2}}$$

Therefore,

$$\mathbf{E} = \frac{\rho_L}{2\pi\epsilon_0}\frac{(x-6)\mathbf{a}_x + (y-8)\mathbf{a}_y}{(x-6)^2 + (y-8)^2}$$

We again note that the field is not a function of z.

In Sec. 2.6 we shall describe how fields may be sketched and use the field of the line charge as one example.

D2.5. An infinitely long, uniform line charge is located at $y = 3$, $z = 5$. If $\rho_L = 30$ nC/m, find **E** at: (*a*) the origin; (*b*) $P_B(0,6,1)$; (*c*) $P_C(5,6,1)$.

Ans. $-47.6\mathbf{a}_y - 79.3\mathbf{a}_z$ V/m; $64.7\mathbf{a}_y - 86.3\mathbf{a}_z$ V/m; $64.7\mathbf{a}_y - 86.3\mathbf{a}_z$ V/m

2.5 FIELD OF A SHEET OF CHARGE

Another basic charge configuration is the infinite sheet of charge having a uniform density of ρ_S C/m^2. Such a charge distribution may often be used to approximate that found on the conductors of a strip transmission line or a parallel-plate capacitor. As we shall see in Chap. 5, static charge resides on conductor surfaces and not in their interiors; for this reason, ρ_S is commonly known as *surface charge density*. The charge-distribution family now is complete—point, line, surface, and volume, or Q, ρ_L, ρ_S, and ρ_v.

Let us place a sheet of charge in the *yz* plane and again consider symmetry (Fig. 2.9). We see first that the field cannot vary with *y* or with *z*, and then that the *y* and *z* components arising from differential elements of charge symmetrically located with respect to the point at which we wish the field will cancel. Hence only E_x is present, and this component is a function of *x* alone. We are again faced with a choice of many methods by which to evaluate this component, and this time we shall use but one method and leave the others as exercises for a quiet Sunday afternoon.

Let us use the field of the infinite line charge (19) by dividing the infinite sheet into differential-width strips. One such strip is shown in Fig. 2.9. The line charge density, or charge per unit length, is $\rho_L = \rho_S\,dy'$, and the distance from this line charge to our general point P on the x axis is $R = \sqrt{x^2 + y'^2}$. The contribution to E_x at P from this differential-width strip is then

$$dE_x = \frac{\rho_S\,dy'}{2\pi\epsilon_0\sqrt{x^2 + y'^2}}\cos\theta = \frac{\rho_S}{2\pi\epsilon_0}\frac{x\,dy'}{x^2 + y'^2}$$

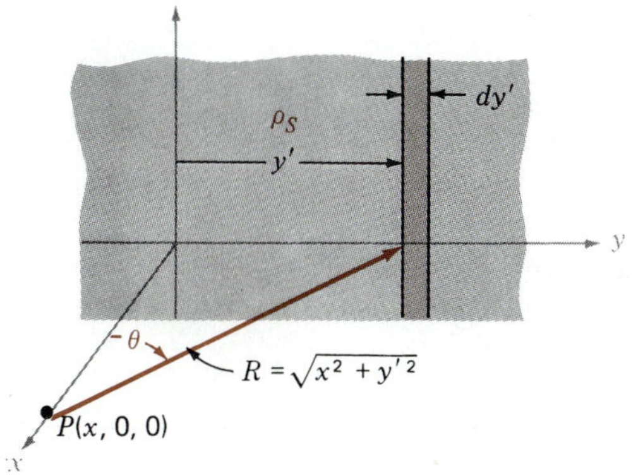

FIGURE 2.9
An infinite sheet of charge in the *yz* plane, a general point *P* on the *x* axis, and the differential-width line charge used as the element in determining the field at *P* by $d\mathbf{E} = \rho_S\,dy'\,\mathbf{a}_R/(2\pi\epsilon_0 R)$.

Adding the effects of all the strips,

$$E_x = \frac{\rho_S}{2\pi\epsilon_0}\int_{-\infty}^{\infty}\frac{x\,dy'}{x^2 + y'^2} = \frac{\rho_S}{2\pi\epsilon_0}\tan^{-1}\frac{y'}{x}\bigg]_{-\infty}^{\infty}$$

$$= \frac{\rho_S}{2\epsilon_0}$$

If the point P were chosen on the negative x axis, then

$$E_x = -\frac{\rho_S}{2\epsilon_0}$$

for the field is always directed away from the positive charge. This difficulty in sign is usually overcome by specifying a unit vector $\mathbf{a}_N$, which is normal to the sheet and directed outward, or away from it. Then

$$\boxed{\mathbf{E} = \frac{\rho_S}{2\epsilon_0}\mathbf{a}_N} \qquad (21)$$

This is a startling answer, for the field is constant in magnitude and direction. It is just as strong a million miles away from the sheet as it is right off the surface. Returning to our light analogy, we see that a uniform source of light on the ceiling of a very large room leads to just as much illumination on a square foot on the floor as it does on a square foot a few inches below the ceiling. If you desire greater illumination on this subject, it will do you no good to hold the book closer to such a light source.

If a second infinite sheet of charge, having a *negative* charge density $-\rho_S$, is located in the plane $x = a$, we may find the total field by adding the contribution of each sheet. In the region $x > a$,

$$\mathbf{E}_+ = \frac{\rho_S}{2\epsilon_0}\mathbf{a}_x \qquad \mathbf{E}_- = -\frac{\rho_S}{2\epsilon_0}\mathbf{a}_x \qquad \mathbf{E} = \mathbf{E}_+ + \mathbf{E}_- = 0$$

and for $x < 0$,

$$\mathbf{E}_+ = -\frac{\rho_S}{2\epsilon_0}\mathbf{a}_x \qquad \mathbf{E}_- = \frac{\rho_S}{2\epsilon_0}\mathbf{a}_x \qquad \mathbf{E} = \mathbf{E}_+ + \mathbf{E}_- = 0$$

and when $0 < x < a$,

$$\mathbf{E}_+ = \frac{\rho_S}{2\epsilon_0}\mathbf{a}_x \qquad \mathbf{E}_- = \frac{\rho_S}{2\epsilon_0}\mathbf{a}_x$$

and

$$\boxed{\mathbf{E} = \mathbf{E}_+ + \mathbf{E}_- = \frac{\rho_S}{\epsilon_0}\mathbf{a}_x} \qquad (22)$$

This is an important practical answer, for it is the field between the parallel plates of an air capacitor, provided the linear dimensions of the plates are very much greater than their separation and provided also that we are considering a point well removed from the edges. The field outside the capacitor, while not zero, as we found for the ideal case above, is usually negligible.

D2.6. Four infinite uniform sheets of charge are located as follows: 20 pC/m^2 at $y = 7$, -8 pC/m^2 at $y = 3$, 6 pC/m^2 at $y = -1$, and -18 pC/m^2 at $y = -4$. Find **E** at the point: (*a*) $P_A(2,6,-4)$; (*b*) $P_B(0,0,0)$; (*c*) $P_C(-1,-1.1,5)$; (*d*) $P_D(10^6,10^6,10^6)$.

Ans. $-2.26\mathbf{a}_y$ V/m; $-1.355\mathbf{a}_y$ V/m; $-2.03\mathbf{a}_y$ V/m; 0

2.6 STREAMLINES AND SKETCHES OF FIELDS

We now have vector equations for the electric field intensity resulting from several different charge configurations, and we have had little difficulty in interpreting the magnitude and direction of the field from the equations. Unfortunately, this simplicity cannot last much longer, for we have solved most of the simple cases and our new charge distributions must lead to more complicated expressions for the fields and more difficulty in visualizing the fields through the equations. However, it is true that one picture is worth about a thousand words, if we just knew what picture to draw.

Consider the field about the line charge,

$$\mathbf{E} = \frac{\rho_L}{2\pi\epsilon_0\rho}\mathbf{a}_\rho$$

Figure 2.10*a* shows a cross-sectional view of the line charge and presents what might be our first effort at picturing the field—short line segments drawn here and there having a length proportional to the magnitude of **E** and pointing in the direction of **E**. The figure fails to show the symmetry with respect to ϕ, so we try again in Fig. 2.10*b* with a symmetrical location of the line segments. The real trouble now appears—the longest lines must be drawn in the most crowded region, and this also plagues us if we use line segments of equal length but of a thickness which is proportional to E (Fig. 2.10*c*). Other schemes which have been suggested include drawing shorter lines to represent stronger fields (inherently misleading) and using intensity of color to represent stronger fields (difficult and expensive).

For the present, then, let us be content to show only the *direction* of **E** by drawing continuous lines from the charge which are everywhere tangent to **E**. Figure 2.10*d* shows this compromise. A symmetrical distribution of lines (one every 45°) indicates azimuthal symmetry, and arrowheads should be used to show direction.

These lines are usually called *streamlines*, although other terms such as flux lines and direction lines are also used. A small positive test charge placed at any

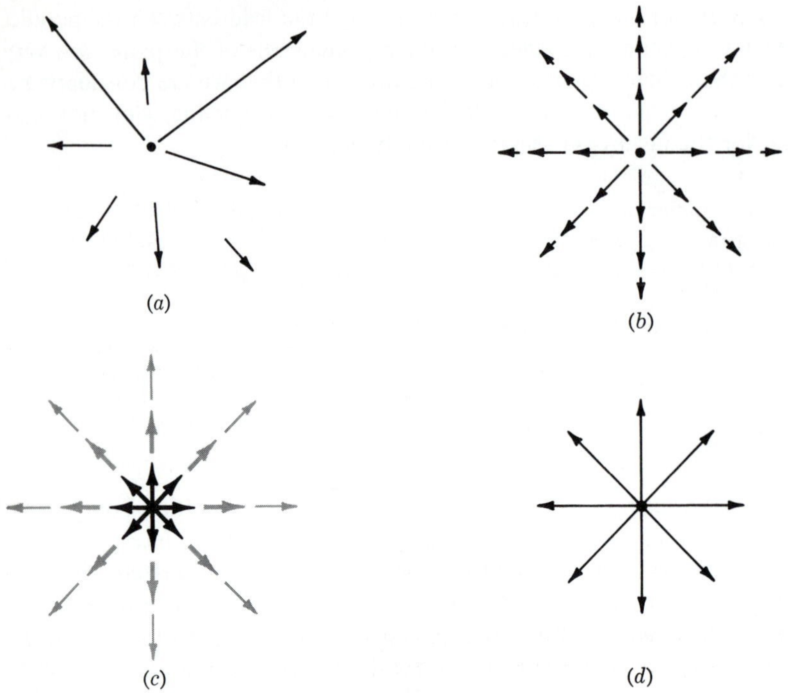

FIGURE 2.10
(*a*) One very poor sketch, (*b*) and (*c*) two fair sketches, and (*d*) the usual form of streamline sketch. In the last form, the arrows show the direction of the field at every point along the line, and the spacing of the lines is inversely proportional to the strength of the field.

point in this field and free to move would accelerate in the direction of the streamline passing through that point. If the field represented the velocity of a liquid or a gas (which, incidentally, would have to have a source at $\rho = 0$), small suspended particles in the liquid or gas would trace out the streamlines.

We shall find out later that a bonus accompanies this streamline sketch, for the magnitude of the field can be shown to be inversely proportional to the spacing of the streamlines for some important special cases. The closer they are together, the stronger is the field. At that time we shall also find an easier, more accurate method of making that type of streamline sketch.

If we attempted to sketch the field of the point charge, the variation of the field into and away from the page would cause essentially insurmountable difficulties; for this reason sketching is usually limited to two-dimensional fields.

In the case of the two-dimensional field let us arbitrarily set $E_z = 0$. The streamlines are thus confined to planes for which z is constant, and the sketch is the same for any such plane. Several streamlines are shown in Fig. 2.11, and the E_x and E_y components are indicated at a general point. Since it is apparent from

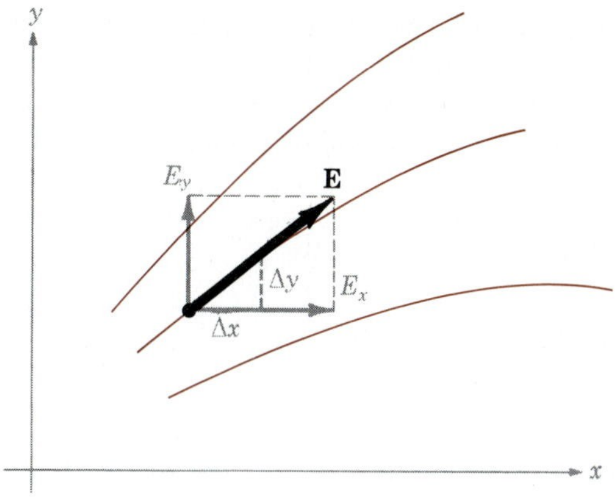

FIGURE 2.11
The equation of a streamline is obtained by solving the differential equation $E_y/E_x = dy/dx$.

the geometry that

$$\boxed{\frac{E_y}{E_x} = \frac{dy}{dx}} \tag{23}$$

a knowledge of the functional form of E_x and E_y (and the ability to solve the resultant differential equation) will enable us to obtain the equations of the streamlines.

As an illustration of this method, consider the field of the uniform line charge with $\rho_L = 2\pi\epsilon_0$,

$$\mathbf{E} = \frac{1}{\rho}\,\mathbf{a}_\rho$$

In cartesian coordinates,

$$\mathbf{E} = \frac{x}{x^2 + y^2}\,\mathbf{a}_x + \frac{y}{x^2 + y^2}\,\mathbf{a}_y$$

Thus we form the differential equation

$$\frac{dy}{dx} = \frac{E_y}{E_x} = \frac{y}{x} \qquad \text{or} \qquad \frac{dy}{y} = \frac{dx}{x}$$

Therefore,

$$\ln y = \ln x + C_1 \qquad \text{or} \qquad \ln y = \ln x + \ln C$$

from which the equations of the streamlines are obtained,

$$y = Cx$$

If we want to find the equation of one particular streamline, say that one passing through $P(-2,7,10)$, we merely substitute the coordinates of that point into our equation and evaluate C. Here, $7 = C(-2)$, and $C = -3.5$, so that $y = -3.5x$.

Each streamline is associated with a specific value of C, and the radial lines shown in Fig. 2.10d are obtained when $C = 0, 1, -1$, and $1/C = 0$.

The equations of streamlines may also be obtained directly in cylindrical or spherical coordinates. A spherical coordinate example will be examined in Sec. 4.7.

D2.7. Obtain the equation of the streamline that passes through the point $P(-2,7,10)$ in the field $\mathbf{E} =$: (*a*) $2(y-1)\mathbf{a}_x + 2x\mathbf{a}_y$; (*b*) $e^y\mathbf{a}_x + (x+1)e^y\mathbf{a}_y$.

Ans. $(y-1)^2 - x^2 = 32$; $2y - (x+1)^2 = 13$

SUGGESTED REFERENCES

1. Boast, W. B.: "Vector Fields," Harper and Row, Publishers, Incorporated, New York, 1964. This book contains numerous examples and sketches of fields.
2. Della Torre, E., and C. L. Longo: "The Electromagnetic Field," Allyn and Bacon, Inc., Boston, 1969. The authors introduce all of electromagnetic theory with a careful and rigorous development based on a single experimental law—that of Coulomb. It begins in chap. 1.
3. Schelkunoff, S. A.: "Electromagnetic Fields," Blaisdell Publishing Company, New York, 1963. Many of the physical aspects of fields are discussed early in this text without advanced mathematics.

PROBLEMS

1. The two small plastic spheres shown in Fig. 2.12 are arranged so that they can slide freely along the insulating fiber. Each has a mass of 0.1 g. The spheres are given identical positive charges. (*a*) If the lower sphere is resting against the lower knot while the upper sphere moves to a location 6 cm distant from that knot, find the charge on each. (*b*) If the charges remain the same, what angle should the fiber make with the horizontal to double the separation?

2. A charge $Q_0 = 1$ nC is located in free space at $P(a,0,0)$. Prepare a sketch of the magnitude of the force on Q_0, as a function of a, $0 \le a \le 5$ m, produced by two other charges, $Q_1 = 1$ C at (0,1,0) and $Q_2 =$: (*a*) 1 C at (0,−1,0); (*b*) −1 C at (0,−1,0).

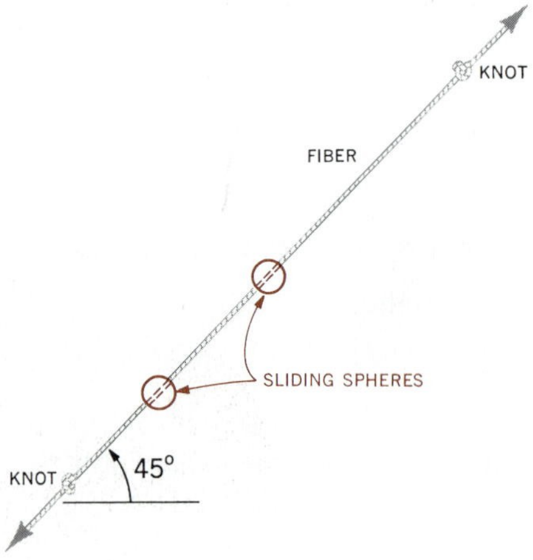

FIGURE 2.12
See Prob. 1.

3. A regular tetrahedron has vertices at $P_1(2,0,0)$, $P_2(-1,\sqrt{3},0)$, $P_3(-1,-\sqrt{3},0)$, and $P_4(0,0,2\sqrt{2})$. Charges of 1 mC are located at each of the four vertices. If the configuration is located in free space: (*a*) find the magnitude of the force on each charge, and (*b*) the angle between any two edges of the tetrahedron.

4. A point charge, $Q_1 = 10\ \mu\text{C}$, is located at $P_1(1,2,3)$ in free space, while $Q_2 = -5\ \mu\text{C}$ is at $P_2(1,2,10)$. (*a*) Find the vector force exerted on Q_2 by Q_1. (*b*) Find the coordinates of P_3 at which a point charge Q_3 experiences no force.

5. In free space, let $Q_1 = 10$ nC be at $P_1(0,-4,0)$, and $Q_2 = 20$ nC be at $P_2(0,0,4)$. (*a*) Find **E** at the origin. (*b*) Where should a 30-nC point charge be located so that $\mathbf{E} = 0$ at the origin?

6. A 5-nC point charge is located at $A(2,-1,-3)$ in free space. (*a*) Find **E** at the origin. (*b*) Plot $|\mathbf{E}(x,0,0)|$ vs. x, $-10 \le x \le 10$ m. (*c*) What is $|\mathbf{E}(x,0,0)|_{\text{max}}$?

7. A point charge, $Q_A = 1\ \mu\text{C}$, is at $A(0,0,1)$, and $Q_B = -1\ \mu\text{C}$ is at $B(0,0,-1)$. Find E_r, E_θ, and E_ϕ at $P(1,2,3)$.

8. A 20-nC point charge is located at $P(2,4,-3)$ in free space. (*a*) Find $\mathbf{E}(\mathbf{r})$. (*b*) Find **E** at $A(-3,2,0)$. (*c*) Find the locus of all points at which $E_x = 1$ V/m.

9. Eight 25-nC point charges in free space are located symmetrically on a circle of radius 0.2 m centered at the origin in the $z = 0$ plane. (*a*) At what point on the z axis is $|\mathbf{E}|$ a maximum? (*b*) What is $|\mathbf{E}|_{\text{max}}$?

10. Identical 1-μC point charges are located in free space at (0,0,1) and (0,0,−1). Prepare a plot showing $|\mathbf{E}|$ vs. z along the line $x = 0$, $y = 2$, for $|z| \le 4$.

11. Eight point charges of 1 nC each are located at the corners of a cube in free space that is 1 m on a side. Find $|\mathbf{E}|$ at the center of: (*a*) the cube; (*b*) a face; (*c*) an edge.

12. A point charge Q_1, located at the origin in free space, produces a field $\mathbf{E} = 10\mathbf{a}_r$ V/m for $r = 1$ m. (*a*) Find Q_1. (*b*) Using this value of Q_1, find **E** at $P(2,-1,5)$ in cartesian components. (*c*) Using the same value of Q_1, find **E** at $C(2,2,2)$ in cylindrical components.

13. For x, y, and z positive, let $\rho_v = 40xyz$ C/m³. Find the total charge within the region defined by: (*a*) $0 \le x,y,z \le 2$; (*b*) $x = 0$, $y = 0$, $0 \le 2x + 3y \le 10$, $0 \le z \le 2$.

14. Volume charge density within the operating region of a parallel-plate vacuum diode is given by $\rho_v = -(\frac{4}{9})\epsilon_0 V_0 d^{-4/3} x^{-2/3}$, where the cathode is located at $x = 0$ and the anode at $x = d$. The voltage between anode and cathode is V_0. If the total charge in a 1-m² cross section is −20 nC and: (*a*) $d = 8$ mm, find V_0; (*b*) $V_0 = 24$ V, find d. (*c*) If $d = 8$ mm and $V_0 = 24$ V, sketch ρ_v vs. x.

15. Volume charge density is given by $\rho_v = 10e^{-1000\rho}e^{-100z}$ C/m³. (*a*) Find the maximum value of ρ_v in the region $0 \le \rho \le 0.01$ m, $0 \le \phi < 2\pi$, $0 \le z \le 0.01$ m. (*b*) Find the total charge contained in the first octant, where x, y, and z are positive. (*c*) Find b if the total charge found in the volume $0 \le \rho \le b$, $0 \le \phi \le \pi/2$, $z \ge 0$, is half the value found in part (*b*) above.

16. Let $\rho_v = (x + 2y + 3z)$ C/m³ in the cubical region $0 \le x,y,z \le 1$ mm, and $\rho_v = 0$ outside the cube. (*a*) What is the total charge contained within the cube? (*b*) Set up the volume integral that will give $\mathbf{E}(x,0,0)$ for $x > 1$ mm. Do not integrate.

17. Volume charge density is given as $\rho_v = 10^{-5}e^{-100r} \sin\theta$ C/m^3 for $0 \le r \le 1$ cm, and $\rho_v = 0$ for $r > 1$ cm. Estimate **E** at $r = 1$ m, $\theta = 90°$, $\phi = 0$, by thinking in terms of a point charge.

18. A uniform volume charge density of 10 μC/m^3 is present in the spherical shell $0.9 < r < 1$ m, and $\rho_v = 0$ elsewhere. (*a*) Find Q_{tot}, the total charge present. (*b*) In the next chapter we will see that this symmetrical charge distribution in free space produces an electric field for $r > 1$ m that is identical to the field that would be produced by a point charge Q_{tot} at the origin. Find **E** in spherical coordinates for $r > 1$ m.

19. A line charge density of 24 nC/m is located in free space on the line $y = 1$, $z = 2$. (*a*) Find **E** at $P(6,-1,3)$. (*b*) What point charge Q_A should be located at $Q(-3,4,1)$ to cause E_y to be equal to zero at P?

20. A uniform line charge, $\rho_{L1} = 15$ nC/m, is at $y = -1$, $z = 0$, while $\rho_{L2} = -15$ nC/m is symmetrically located at $y = 1$, $z = 0$. Find **E** as a function of z at $y = 0$.

21. A uniform line charge density of 20 nC/m lies on the z axis between $z = 1$ and $z = 3$ m. No other charge is present. Find **E** at: (*a*) the origin; (*b*) $P(4,0,0)$.

22. That portion of the z axis for which $-2 < z < 2$ m carries a nonuniform charge density of $z^2 + 1$ nC/m in free space. There are no other charges anywhere. Find **E** at: (*a*) $P_A(1,0,0)$; (*b*) $P_B(2,0,0)$.

23. A sheet of charge, $\rho_S = 2$ nC/m^2, is present at the plane $x = 3$ in free space, and a line charge, $\rho_L = 20$ nC/m, is located at $x = 1$, $z = 4$. (*a*) Find the magnitude of the electric field intensity at the origin. (*b*) Find the direction of **E** at $P(4,5,6)$. (*c*) What is the force per meter length on the line charge?

24. Find **E** at $P(1,5,2)$ in free space if a point charge of 6 μC is located at $Q(0,0,1)$, a uniform line charge of 180 nC/m lies along the x axis, and a uniform sheet of charge equal to 25 nC/m^2 lies in the plane $z = -1$.

25. A line charge, $\rho_L = 50$ nC/m, is located along the line $x = 2$, $y = 5$, in free space. (*a*) Find **E** at $P(1,3,-4)$. (*b*) If the surface $x = 4$ contains a uniform surface charge density, $\rho_S = 18$ nC/m^2, at what point in the $z = 0$ plane is $\mathbf{E}_{total} = 0$?

26. The circular region, $\rho < a$, $z = 0$, carries a uniform surface charge density ρ_S. Find **E** at $P(0,0,h)$, $h > 0$.

27. In the rectangular region, $-2 < x < 2$, $-3 < y < 3$, $z = 0$, the surface charge density is given by $\rho_S = (x^2 + y^2 + 1)^{3/2}$. If no other charge is present, find **E** at $P(0,0,1)$.

28. Let $\mathbf{E} = (2x + 4y - 3)\mathbf{a}_x + (4x - 2y)\mathbf{a}_y$ and: (*a*) find the equation of the direction line passing through $P(1,2,z)$; (*b*) sketch this direction line and show **E** at P.

29. Given the field $\mathbf{E} = 15x^2y\mathbf{a}_x + 5x^3\mathbf{a}_y$, find: (*a*) the equation of the streamline that passes through $P(2,3,-4)$; (*b*) a unit vector $\mathbf{a}_E$ specifying the direction of **E** at P; (*c*) a unit vector $\mathbf{a}_\perp = (l,m,0)$ that is perpendicular to $\mathbf{a}_E$ at P and has $l > 0$.

30. If $\mathbf{E} = 200 \cosh 2x \sin 2y\, \mathbf{a}_x + 200 \sinh 2x \cos 2y\, \mathbf{a}_y$ V/m, find the equation of the direction line passing through the point $P(1,0,0)$ and sketch it in the $z = 0$ plane for $0 < x < 5$ and $0 < y < \pi/4$.

31. The electric field intensity is given as $\mathbf{E} = 5e^{-2x}(\sin 2y\ \mathbf{a}_x - \cos 2y\ \mathbf{a}_y)$ V/m. (*a*) Find the equation of the streamline passing through the point $P(0.5,\pi/10,0)$. (*b*) Find a unit vector tangent to the streamline at P.

32. Given a field $\mathbf{E}(\rho,\phi) = E_\rho(\rho,\phi)\mathbf{a}_\rho + E_\phi(\rho,\phi)\mathbf{a}_\phi$ in cylindrical coordinates, the equations of the streamlines in a plane $z =$ constant are obtained by solving the differential equation $E_\rho/E_\phi = d\rho/(\rho\, d\phi)$. Find the equation of the line passing through point $P(2,30°,0)$ for the field $\mathbf{E} = \rho \sin 2\phi\ \mathbf{a}_\rho - \rho \cos 2\phi\ \mathbf{a}_\phi$.

CHAPTER

3

ELECTRIC FLUX DENSITY, GAUSS'S LAW, AND DIVERGENCE

After drawing a few of the fields described in the previous chapter and becoming familiar with the concept of the streamlines which show the direction of the force on a test charge at every point, it is difficult to avoid giving these lines a physical significance and thinking of them as *flux* lines. No physical particle is projected radially outward from the point charge, and there are no steel tentacles reaching out to attract or repel an unwary test charge, but as soon as the streamlines are drawn on paper there seems to be a picture showing "something" is present.

It is very helpful to invent an *electric flux* which streams away symmetrically from a point charge and is coincident with the streamlines and to visualize this flux wherever an electric field is present.

This chapter introduces and uses the concept of electric flux and electric flux density to solve again several of the problems presented in the last chapter. The work here turns out to be much easier, and this is due to the extremely symmetrical problems which we are solving.

3.1 ELECTRIC FLUX DENSITY

About 1837 the Director of the Royal Society in London, Michael Faraday, became very much interested in static electric fields and the effect of various insulating materials on these fields. This problem had been bothering him during the past ten years when he was engaged experimentally in his now famous work on induced electromotive force, which we shall discuss in Chap. 10. With that subject completed, he had a pair of concentric metallic spheres constructed, the outer one consisting of two hemispheres which could be firmly clamped together. He also prepared shells of insulating material (or dielectric material, or simply *dielectric*) which would occupy the entire volume between the concentric spheres. We shall not make immediate use of his findings about dielectric materials, for we are restricting our attention to fields in free space until Chap. 5. At that time we shall see that the materials he used will be classified as ideal dielectrics.

His experiment, then, consisted essentially of the following steps:

1. With the equipment dismantled, the inner sphere was given a known positive charge.
2. The hemispheres were then clamped together around the charged sphere with about 2 cm of dielectric material between them.
3. The outer sphere was discharged by connecting it momentarily to ground.
4. The outer sphere was separated carefully, using tools made of insulating material in order not to disturb the induced charge on it, and the negative induced charge on each hemisphere was measured.

Faraday found that the total charge on the outer sphere was equal in *magnitude* to the original charge placed on the inner sphere and that this was true regardless of the dielectric material separating the two spheres. He concluded that there was some sort of "displacement" from the inner sphere to the outer which was independent of the medium, and we now refer to this flux as *displacement*, *displacement flux*, or simply *electric flux*.

Faraday's experiments also showed, of course, that a larger positive charge on the inner sphere induced a correspondingly larger negative charge on the outer sphere, leading to a direct proportionality between the electric flux and the charge on the inner sphere. The constant of proportionality is dependent on the system of units involved, and we are fortunate in our use of SI units, because the constant is unity. If electric flux is denoted by Ψ and the total charge on the inner sphere by Q, then for Faraday's experiment

$$\boxed{\Psi = Q}$$

and the electric flux Ψ is measured in coulombs.

We can obtain more quantitative information by considering an inner sphere of radius a and an outer sphere of radius b, with charges of Q and $-Q$, respectively (Fig. 3.1). The paths of electric flux Ψ extending from the inner sphere to the outer sphere are indicated by the symmetrically distributed streamlines drawn radially from one sphere to the other.

At the surface of the inner sphere, Ψ coulombs of electric flux are produced by the charge Q $(=\Psi)$ coulombs distributed uniformly over a surface having an area of $4\pi a^2$ m^2. The density of the flux at this surface is $\Psi/4\pi a^2$, or $Q/4\pi a^2$ C/m^2, and this is an important new quantity.

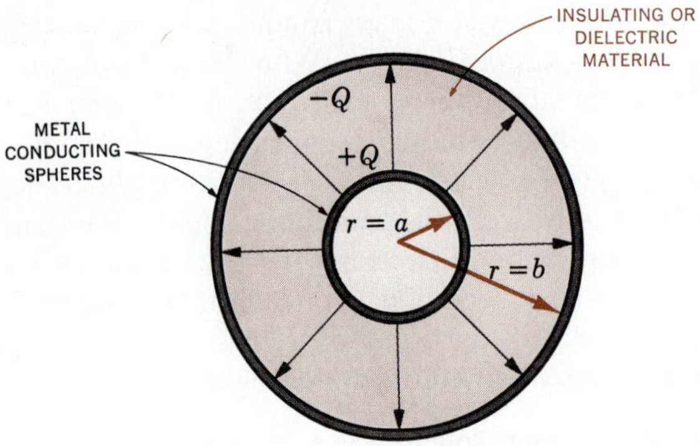

FIGURE 3.1
The electric flux in the region between a pair of charged concentric spheres. The direction and magnitude of **D** are not functions of the dielectric between the spheres.

Electric flux density, measured in coulombs per square meter (sometimes described as "lines per square meter," for each line is due to one coulomb), is given the letter **D**, which was originally chosen because of the alternate names of *displacement flux density* or *displacement density*. Electric flux density is more descriptive, however, and we shall use that term consistently.

The electric flux density **D** is a vector field and is a member of the "flux density" class of vector fields, as opposed to the "force fields" class, which includes the electric field intensity **E**. The direction of **D** at a point is the direction of the flux lines at that point, and the magnitude is given by the number of flux lines crossing a surface normal to the lines divided by the surface area.

Referring again to Fig. 3.1, the electric flux density is in the radial direction and has a value of

$$\mathbf{D}\Big|_{r=a} = \frac{Q}{4\pi a^2}\,\mathbf{a}_r \qquad \text{(inner sphere)}$$

$$\mathbf{D}\Big|_{r=b} = \frac{Q}{4\pi b^2}\,\mathbf{a}_r \qquad \text{(outer sphere)}$$

and at a radial distance r, where $a \le r \le b$,

$$\mathbf{D} = \frac{Q}{4\pi r^2}\,\mathbf{a}_r$$

If we now let the inner sphere become smaller and smaller, while still retaining a charge of Q, it becomes a point charge in the limit, but the electric flux density at a point r meters from the point charge is still given by

$$\boxed{\mathbf{D} = \frac{Q}{4\pi r^2}\,\mathbf{a}_r} \tag{1}$$

for Q lines of flux are symmetrically directed outward from the point and pass through an imaginary spherical surface of area $4\pi r^2$.

This result should be compared with Sec. 2.2, Eq. (10), the radial electric field intensity of a point charge in free space,

$$\mathbf{E} = \frac{Q}{4\pi\epsilon_0 r^2}\,\mathbf{a}_r$$

In free space, therefore,

$$\mathbf{D} = \epsilon_0 \mathbf{E} \qquad \text{(free space only)} \tag{2}$$

Although (2) is applicable only to a vacuum, it is not restricted solely to the field of a point charge. For a general volume charge distribution in free space

$$\mathbf{E} = \int_{\text{vol}} \frac{\rho_v\,dv}{4\pi\epsilon_0 R^2}\,\mathbf{a}_R \qquad \text{(free space only)} \tag{3}$$

where this relationship was developed from the field of a single point charge. In a similar manner, (1) leads to

$$\mathbf{D} = \int_{\text{vol}} \frac{\rho_v\,dv}{4\pi R^2}\,\mathbf{a}_R \tag{4}$$

and (2) is therefore true for any free-space charge configuration; we shall consider (2) as defining **D** in free space.

As a preparation for the study of dielectrics later, it might be well to point out now that, for a point charge embedded in an infinite ideal dielectric medium, Faraday's results show that (1) is still applicable, and thus so is (4). Equation (3) is not applicable, however, and so the relationship between **D** and **E** will be slightly more complicated than (2).

Since **D** is directly proportional to **E** in free space, it does not seem that it should really be necessary to introduce a new symbol. We do so for several reasons. First, **D** is associated with the flux concept, which is an important new idea. Second, the **D** fields we obtain will be a little simpler than the corresponding **E** fields, since ϵ_0 does not appear. And, finally, it helps to become a little familiar with **D** before it is applied to dielectric materials in Chap. 5.

Let us consider a simple numerical example to illustrate these new quantities and units. Suppose we have a uniform line charge of 8 nC/m lying along the z axis. The **E** field is

$$\mathbf{E} = \frac{\rho_L}{2\pi\epsilon_0 \rho}\,\mathbf{a}_\rho = \frac{8 \times 10^{-9}}{2\pi(8.854 \times 10^{-12})\rho}\,\mathbf{a}_\rho = \frac{143.8}{\rho}\,\mathbf{a}_\rho \qquad \text{V/m}$$

At $\rho = 3$ m, $\mathbf{E} = 47.9\mathbf{a}_\rho$ V/m.

Associated with the **E** field, we find

$$\mathbf{D} = \frac{\rho_L}{2\pi\rho}\mathbf{a}_\rho = \frac{8 \times 10^{-9}}{2\pi\rho}\mathbf{a}_\rho = \frac{1.273 \times 10^{-9}}{\rho}\mathbf{a}_\rho \qquad \text{C/m}^2$$

The value at $\rho = 3$ m is $\mathbf{D} = 0.424\mathbf{a}_\rho$ nC/m.

The total flux leaving a 5-m length of the line charge is equal to the total charge on that length, or $\Psi = 40$ nC.

D3.1. A 25-μC point charge is located at the origin. Calculate the electric flux passing through: (*a*) that portion of the sphere $r = 20$ cm bounded by $\theta = 0$ and π, $\phi = 0$ and $\pi/2$; (*b*) the closed surface $\rho = 0.8$ m, $z = \pm 0.5$ m; (*c*) the plane $z = 4$ m.

Ans. 6.25 μC; 25 μC; 12.5 μC

D3.2. Find **D** (in cartesian coordinates) at $P(6, 8, -10)$ caused by: (*a*) a point charge of 30 mC at the origin; (*b*) a uniform line charge $\rho_L = 40$ μC/m on the z axis; (*c*) a uniform surface charge density $\rho_S = 57.2$ $\mu\text{C/m}^2$ on the plane $x = 9$.

Ans. $5.06\mathbf{a}_x + 6.75\mathbf{a}_y - 8.44\mathbf{a}_z$ $\mu\text{C/m}^2$; $0.382\mathbf{a}_x + 0.509\mathbf{a}_y$ $\mu\text{C/m}^2$; $-28.6\mathbf{a}_x$ $\mu\text{C/m}^2$

3.2 GAUSS'S LAW

The results of Faraday's experiments with the concentric spheres could be summed up as an experimental law by stating that the electric flux passing through any imaginary spherical surface lying between the two conducting spheres is equal to the charge enclosed within that imaginary surface. This enclosed charge is distributed on the surface of the inner sphere, or it might be concentrated as a point charge at the center of the imaginary sphere. However, since one coulomb of electric flux is produced by one coulomb of charge, the inner conductor might just as well have been a cube or a brass door key and the total induced charge on the outer sphere would still be the same. Certainly the flux density would change from its previous symmetrical distribution to some unknown configuration, but $+Q$ coulombs on any inner conductor would produce an induced charge of $-Q$ coulombs on the surrounding sphere. Going one step further, we could now replace the two outer hemispheres by an empty (but completely closed) soup can. Q coulombs on the brass door key would produce $\Psi = Q$ lines of electric flux and would induce $-Q$ coulombs on the tin can.[1]

These generalizations of Faraday's experiment lead to the following statement, which is known as *Gauss's law*:

> *The electric flux passing through any closed surface is equal to the total charge enclosed by that surface.*

---

[1] If it were a perfect insulator, the soup could even be left in the can without any difference in the results.

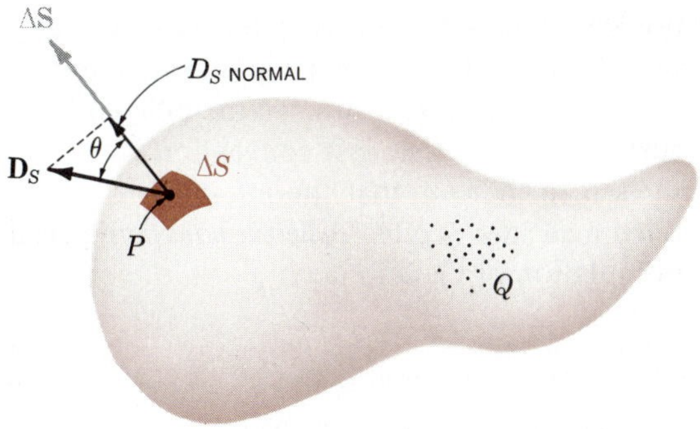

FIGURE 3.2
The electric flux density $\mathbf{D}_S$ at P due to charge Q. The total flux passing through $\Delta\mathbf{S}$ is $\mathbf{D}_S \cdot \Delta\mathbf{S}$.

The contribution of Gauss, one of the greatest mathematicians the world has ever produced, was actually not in stating the law as we have above, but in providing a mathematical form for this statement, which we shall now obtain.

Let us imagine a distribution of charge, shown as a cloud of point charges in Fig. 3.2, surrounded by a closed surface of any shape. The closed surface may be the surface of some real material, but more generally it is any closed surface we wish to visualize. If the total charge is Q, then Q coulombs of electric flux will pass through the enclosing surface. At every point on the surface the electric-flux-density vector $\mathbf{D}$ will have some value $\mathbf{D}_S$, where the subscript S merely reminds us that $\mathbf{D}$ must be evaluated at the surface, and $\mathbf{D}_S$ will in general vary in magnitude and direction from one point on the surface to another.

We must now consider the nature of an incremental element of the surface. An incremental element of area ΔS is very nearly a portion of a plane surface, and the complete description of this surface element requires not only a statement of its magnitude ΔS but also of its orientation in space. In other words, the incremental surface element is a vector quantity. The only unique direction which may be associated with $\Delta\mathbf{S}$ is the direction of the normal to that plane which is tangent to the surface at the point in question. There are, of course, two such normals, and the ambiguity is removed by specifying the outward normal whenever the surface is closed and "outward" has a specific meaning.

At any point P consider an incremental element of surface ΔS and let $\mathbf{D}_S$ make an angle θ with $\Delta\mathbf{S}$, as shown in Fig. 3.2. The flux crossing $\Delta\mathbf{S}$ is then the product of the normal component of $\mathbf{D}_S$ and $\Delta\mathbf{S}$,

$$\Delta\Psi = \text{flux crossing } \Delta\mathbf{S} = D_{S,\,\text{norm}}\,\Delta S = D_S \cos\theta\,\Delta S = \mathbf{D}_S \cdot \Delta\mathbf{S}$$

where we are able to apply the definition of the dot product developed in Chap. 1.

The *total* flux passing through the closed surface is obtained by adding the differential contributions crossing each surface element $\Delta\mathbf{S}$,

$$\Psi = \int d\Psi = \oint_{\substack{\text{closed}\\ \text{surface}}} \mathbf{D}_S \cdot d\mathbf{S}$$

The resultant integral is a *closed surface integral*, and since the surface element $d\mathbf{S}$ always involves the differentials of two coordinates, such as $dx\,dy$, $\rho\,d\phi\,d\rho$, or $r^2 \sin\theta\, d\theta\, d\phi$, the integral is a double integral. Usually only one integral sign is used for brevity, and we shall always place an S below the integral sign to indicate a surface integral, although this is not actually necessary since the differential $d\mathbf{S}$ is automatically the signal for a surface integral. One last convention is to place a small circle on the integral sign itself to indicate that the integration is to be performed over a *closed* surface. Such a surface is often called a *gaussian surface*. We then have the mathematical formulation of Gauss's law,

$$\Psi = \oint_S \mathbf{D}_S \cdot d\mathbf{S} = \text{charge enclosed} = Q \tag{5}$$

The charge enclosed might be several point charges, in which case

$$Q = \Sigma Q_n$$

or a line charge,

$$Q = \int \rho_L \, dL$$

or a surface charge,

$$Q = \int_S \rho_S \, dS \qquad \text{(not necessarily a closed surface)}$$

or a volume charge distribution,

$$Q = \int_{\text{vol}} \rho_v \, dv$$

The last form is usually used, and we should agree now that it represents any or all of the other forms. With this understanding Gauss's law may be written in terms of the charge distribution as

$$\oint_S \mathbf{D}_S \cdot d\mathbf{S} = \int_{\text{vol}} \rho_v \, dv \tag{6}$$

a mathematical statement meaning simply that the total electric flux through any closed surface is equal to the charge enclosed.

To illustrate the application of Gauss's law, let us check the results of Faraday's experiment by placing a point charge Q at the origin of a spherical coordinate system (Fig. 3.3) and by choosing our closed surface as a sphere of radius a. The electric field intensity of the point charge has been found to be

$$\mathbf{E} = \frac{Q}{4\pi\epsilon_0 r^2}\mathbf{a}_r$$

and since

$$\mathbf{D} = \epsilon_0 \mathbf{E}$$

we have, as before,

$$\mathbf{D} = \frac{Q}{4\pi r^2}\mathbf{a}_r$$

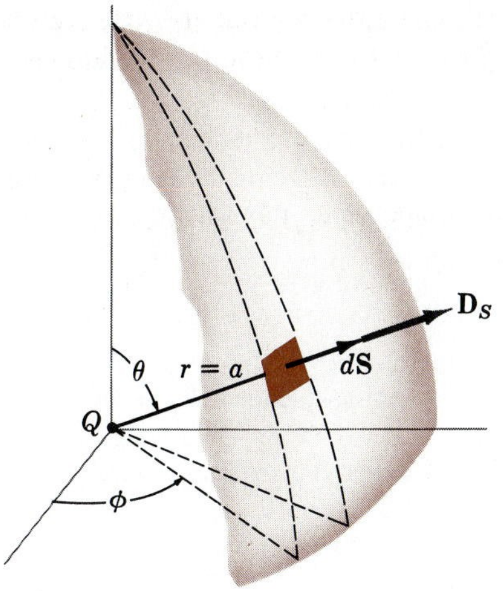

FIGURE 3.3
Application of Gauss's law to the field of a point charge Q on a spherical closed surface of radius a. The electric flux density $\mathbf{D}$ is everywhere normal to the spherical surface and has a constant magnitude at every point on it.

At the surface of the sphere,

$$\mathbf{D}_S = \frac{Q}{4\pi a^2}\mathbf{a}_r$$

The differential element of area on a spherical surface is, in spherical coordinates from Chap. 1,

$$dS = r^2 \sin\theta \, d\theta \, d\phi = a^2 \sin\theta \, d\theta \, d\phi$$

or

$$d\mathbf{S} = a^2 \sin\theta \, d\theta \, d\phi \, \mathbf{a}_r$$

The integrand is

$$\mathbf{D}_S \cdot d\mathbf{S} = \frac{Q}{4\pi a^2} a^2 \sin\theta \, d\theta \, d\phi \, \mathbf{a}_r \cdot \mathbf{a}_r = \frac{Q}{4\pi} \sin\theta \, d\theta \, d\phi$$

leading to the closed surface integral

$$\int_{\phi=0}^{\phi=2\pi} \int_{\theta=0}^{\theta=\pi} \frac{Q}{4\pi} \sin\theta \, d\theta \, d\phi$$

where the limits on the integrals have been chosen so that the integration is carried over the entire surface of the sphere once.[2] Integrating gives

$$\int_0^{2\pi} \frac{Q}{4\pi}(-\cos\theta)_0^{\pi} \, d\phi = \int_0^{2\pi} \frac{Q}{2\pi} \, d\phi = Q$$

and we obtain a result showing that Q coulombs of electric flux are crossing the surface, as we should since the enclosed charge is Q coulombs.

---

[2] Note that if θ and ϕ both cover the range from 0 to 2π, the spherical surface is covered twice.

The following section contains examples of the application of Gauss's law to problems of a simple symmetrical geometry with the object of finding the electric field intensity.

D3.3. Let $\mathbf{D} = r\mathbf{a}_r/3$ nC/m^2 in free space. (*a*) Find E at $r = 0.2$ m. (*b*) Find the total charge within the sphere $r = 0.2$ m. (*c*) Find the total electric flux leaving the sphere $r = 0.3$ m.

Ans. 7.53 V/m; 33.5 pC; 113.1 pC

D3.4. Find the total electric flux leaving the spherical surface $r = 2.5$ m given the charge configuration: (*a*) $Q = 2^{-x^2}$ nC on the x axis at $x = 0, \pm 1, \pm 2, \ldots$ m; (*b*) a line charge $\rho_L = 1/(z^2 + 1)$ nC/m on the z axis; (*c*) a surface charge $\rho_S = 1/(x^2 + y^2 + 4)$ nC/m^2 on the $z = 0$ plane.

Ans. 2.125 nC; 2.38 nC; 2.96 nC

3.3 APPLICATION OF GAUSS'S LAW: SOME SYMMETRICAL CHARGE DISTRIBUTIONS

Let us now consider how we may use Gauss's law,

$$Q = \oint_S \mathbf{D}_S \cdot d\mathbf{S}$$

to determine $\mathbf{D}_S$ if the charge distribution is known. This is an example of an integral equation in which the unknown quantity to be determined appears inside the integral.

The solution is easy if we are able to choose a closed surface which satisfies two conditions:

1. $\mathbf{D}_S$ is everywhere either normal or tangential to the closed surface, so that $\mathbf{D}_S \cdot d\mathbf{S}$ becomes either $D_S\,dS$ or zero, respectively.
2. On that portion of the closed surface for which $\mathbf{D}_S \cdot d\mathbf{S}$ is not zero, $D_S =$ constant.

This allows us to replace the dot product with the product of the scalars D_S and dS and then to bring D_S outside the integral sign. The remaining integral is then $\int_S dS$ over that portion of the closed surface which $\mathbf{D}_S$ crosses normally, and this is simply the area of this section of that surface.

Only a knowledge of the symmetry of the problem enables us to choose such a closed surface, and this knowledge is obtained easily by remembering that the electric field intensity due to a positive point charge is directed radially outward from the point charge.

Let us again consider a point charge Q at the origin of a spherical coordinate system and decide on a suitable closed surface which will meet the two requirements listed above. The surface in question is obviously a spherical surface, centered at the origin and of any radius r. $\mathbf{D}_S$ is everywhere normal to the surface; D_S has the same value at all points on the surface.

Then we have, in order,

$$Q = \oint_S \mathbf{D}_S \cdot d\mathbf{S} = \oint_{\text{sph}} D_S \, dS$$

$$= D_S \oint_{\text{sph}} dS = D_S \int_{\phi=0}^{\phi=2\pi} \int_{\theta=0}^{\theta=\pi} r^2 \sin\theta \, d\theta \, d\phi$$

$$= 4\pi r^2 D_S$$

and hence

$$D_S = \frac{Q}{4\pi r^2}$$

Since r may have any value and since $\mathbf{D}_S$ is directed radially outward,

$$\mathbf{D} = \frac{Q}{4\pi r^2}\mathbf{a}_r \qquad \mathbf{E} = \frac{Q}{4\pi\epsilon_0 r^2}\mathbf{a}_r$$

which agrees with the results of Chap. 2. The example is a trivial one, and the objection could be raised that we had to know that the field was symmetrical and directed radially outward before we could obtain an answer. This is true, and that leaves the inverse-square-law relationship as the only check obtained from Gauss's law. The example does, however, serve to illustrate a method which we may apply to other problems, including several to which Coulomb's law is almost incapable of supplying an answer.

Are there any other surfaces which would have satisfied our two conditions? The student should determine that such simple surfaces as a cube or a cylinder do not meet the requirements.

As a second example, let us reconsider the uniform line charge distribution ρ_L lying along the z axis and extending from $-\infty$ to $+\infty$. We must first obtain a knowledge of the symmetry of the field, and we may consider this knowledge complete when the answers to these two questions are known:

1. With which coordinates does the field vary (or of what variables is D à function)?
2. Which components of $\mathbf{D}$ are present?

These same questions were asked when we used Coulomb's law to solve this problem in Sec. 2.5. We found then that the knowledge obtained from answering them enabled us to make a much simpler integration. The problem could have been (and was) worked without any consideration of symmetry, but it was more difficult.

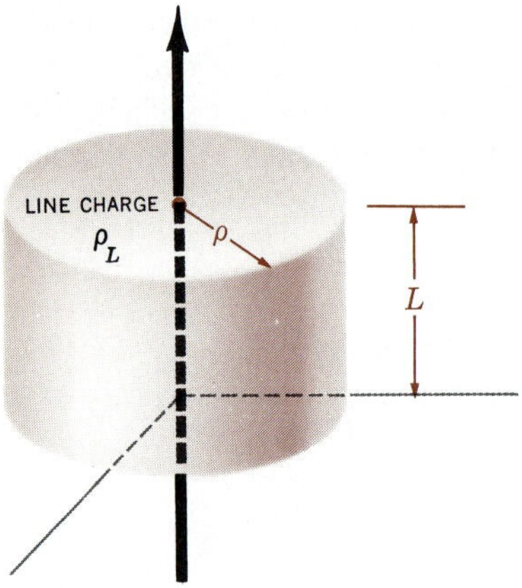

FIGURE 3.4
The gaussian surface for an infinite uniform line charge is a right circular cylinder of length L and radius ρ. **D** is constant in magnitude and everywhere perpendicular to the cylindrical surface; **D** is parallel to the end faces.

In using Gauss's law, however, it is not a question of using symmetry to simplify the solution, for the application of Gauss's law depends on symmetry, and *if we cannot show that symmetry exists then we cannot use Gauss's law* to obtain a solution. The two questions above now become "musts."

From our previous discussion of the uniform line charge, it is evident that only the radial component of **D** is present, or

$$\mathbf{D} = D_\rho \mathbf{a}_\rho$$

and this component is a function of ρ only.

$$D_\rho = f(\rho)$$

The choice of a closed surface is now simple, for a cylindrical surface is the only surface to which D_ρ is everywhere normal and it may be closed by plane surfaces normal to the z axis. A closed right circular cylindrical of radius ρ extending from $z = 0$ to $z = L$ is shown in Fig. 3.4.

We apply Gauss's law,

$$Q = \oint_{\text{cyl}} \mathbf{D}_S \cdot d\mathbf{S} = D_S \int_{\text{sides}} dS + 0 \int_{\text{top}} dS + 0 \int_{\text{bottom}} dS$$

$$= D_S \int_{z=0}^{L} \int_{\phi=0}^{2\pi} \rho \, d\phi \, dz = D_S 2\pi\rho L$$

and obtain

$$D_S = D_\rho = \frac{Q}{2\pi\rho L}$$

In terms of the charge density ρ_L, the total charge enclosed is

$$Q = \rho_L L$$

giving

$$D_\rho = \frac{\rho_L}{2\pi\rho}$$

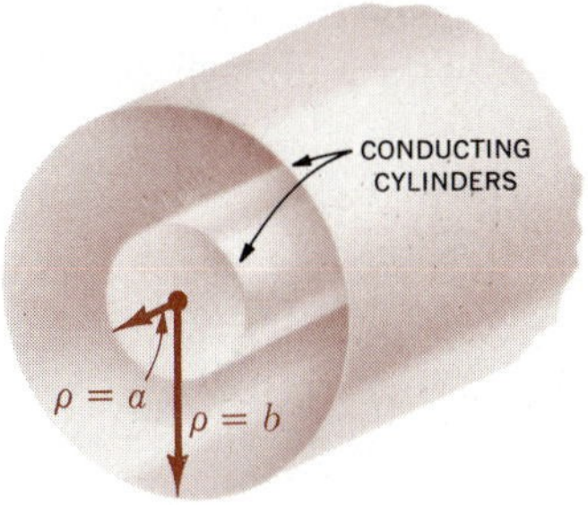

FIGURE 3.5
The two coaxial cylindrical conductors forming a coaxial cable provide an electric flux density within the cylinders, given by $D_\rho = a\rho_S/\rho$.

or

$$E_\rho = \frac{\rho_L}{2\pi\epsilon_0\rho}$$

Comparison with Sec. 2.4, Eq. (20), shows that the correct result has been obtained and with much less work. Once the appropriate surface has been chosen, the integration usually amounts only to writing down the area of the surface at which **D** is normal.

The problem of a coaxial cable is almost identical with that of the line charge and is an example which is extremely difficult to solve from the standpoint of Coulomb's law. Suppose that we have two coaxial cylindrical conductors, the inner of radius a and the outer of radius b, each infinite in extent (Fig. 3.5). We shall assume a charge distribution of ρ_S on the outer surface of the inner conductor.

Symmetry considerations show us that only the D_ρ component is present and that it can be a function only of ρ. A right circular cylinder of length L and radius ρ, where $a < \rho < b$, is necessarily chosen as the gaussian surface, and we quickly have

$$Q = D_S 2\pi\rho L$$

The total charge on a length L of the inner conductor is

$$Q = \int_{z=0}^{L}\int_{\phi=0}^{2\pi} \rho_S a \, d\phi \, dz = 2\pi a L \rho_S$$

from which we have

$$D_S = \frac{a\rho_S}{\rho} \qquad \mathbf{D} = \frac{a\rho_S}{\rho}\mathbf{a}_\rho \qquad (a < \rho < b)$$

This result might be expressed in terms of charge per unit length, because the inner conductor has $2\pi a\rho_S$ coulombs on a meter length, and hence, letting $\rho_L = 2\pi a\rho_S$,

$$\boxed{\mathbf{D} = \frac{\rho_L}{2\pi\rho}\mathbf{a}_\rho}$$

and the solution has a form identical with that of the infinite line charge.

Since every line of electric flux starting from the charge on the inner cylinder must terminate on a negative charge on the inner surface of the outer cylinder, the total charge on that surface must be

$$Q_{\text{outer cyl}} = -2\pi a L \rho_{S,\text{ inner cyl}}$$

and the surface charge on the outer cylinder is found as

$$2\pi b L \rho_{S,\text{ outer cyl}} = -2\pi a L \rho_{S,\text{ inner cyl}}$$

or

$$\rho_{S,\text{ outer cyl}} = -\frac{a}{b}\rho_{S,\text{ inner cyl}}$$

What would happen if we should use a cylinder of radius ρ, $\rho > b$, for the gaussian surface? The total charge enclosed would then be zero, for there are equal and opposite charges on each conducting cylinder. Hence

$$0 = D_S 2\pi\rho L \qquad (\rho > b)$$

$$D_S = 0 \qquad (\rho > b)$$

An identical result would be obtained for $\rho < a$. Thus the coaxial cable or capacitor has no external field (we have proved that the outer conductor is a "shield"), and there is no field within the center conductor.

Our result is also useful for a *finite* length of coaxial cable, open at both ends, provided the length L is many times greater than the radius b so that the unsymmetrical conditions at the two ends do not appreciably affect the solution. Such a device is also termed a *coaxial capacitor*. Both the coaxial cable and the coaxial capacitor will appear frequently in the work that follows.

As a numerical example illustrating these results, let us consider a 50-cm length of coaxial cable having an inner radius of 1 mm and an outer radius of 4 mm. The space between conductors is assumed to be filled with air. The total charge on the inner conductor is 30 nC. Thus,

$$\rho_{S,\text{ inner cyl}} = \frac{Q_{\text{inner cyl}}}{2\pi a L} = \frac{30 \times 10^{-9}}{2\pi(10^{-3})(0.5)} = 9.55 \qquad \mu\text{C/m}^2$$

The internal fields are

$$D_\rho = \frac{a\rho_S}{\rho} = \frac{10^{-3}(9.55 \times 10^{-6})}{\rho} = \frac{9.55}{\rho} \qquad \text{nC/m}^2$$

and

$$E_\rho = \frac{D_\rho}{\epsilon_0} = \frac{9.55 \times 10^{-9}}{8.854 \times 10^{-12}\rho} = \frac{1079}{\rho} \qquad \text{V/m}$$

Both of these expressions apply to the region where $1 < \rho < 4$ mm. For $\rho < 1$ mm or $\rho > 4$ mm, **E** and **D** are zero.

D3.5. Surface charge densities of 200, -50, and ρ_{Sx} $\mu C/m^2$ are located at $r = 3$, 5, and 7 cm, respectively. Find $\mathbf{D}$ at $r =$: (*a*) 2 cm; (*b*) 4 cm; (*c*) 6 cm. (*d*) Find ρ_{Sx} if $\mathbf{D} = 0$ at $r = 7.32$ cm.

Ans. 0; $112.5\mathbf{a}_r$ $\mu C/m^2$; $15.28\mathbf{a}_r$ $\mu C/m^2$; -11.22 $\mu C/m^2$

3.4 APPLICATION OF GAUSS'S LAW: DIFFERENTIAL VOLUME ELEMENT

We are now going to apply the methods of Gauss's law to a slightly different type of problem—one which does not possess any symmetry at all. At first glance it might seem that our case is hopeless, for without symmetry a simple gaussian surface cannot be chosen such that the normal component of $\mathbf{D}$ is constant or zero everywhere on the surface. Without such a surface, the integral cannot be evaluated. There is only one way to circumvent these difficulties, and that is to choose such a very small closed surface that $\mathbf{D}$ is *almost* constant over the surface, and the small change in $\mathbf{D}$ may be adequately represented by using the first two terms of the Taylor's-series expansion for $\mathbf{D}$. The result will become more nearly correct as the volume enclosed by the gaussian surface decreases, and we intend eventually to allow this volume to approach zero.

This example also differs from the preceding ones in that we shall not obtain the value of $\mathbf{D}$ as our answer, but instead receive some extremely valuable information about the way $\mathbf{D}$ varies in the region of our small surface. This leads directly to one of Maxwell's four equations, which are basic to all electromagnetic theory.

Let us consider any point P, shown in Fig. 3.6, located by a cartesian coordinate system. The value of $\mathbf{D}$ at the point P may be expressed in cartesian components, $\mathbf{D}_0 = D_{x0}\mathbf{a}_x + D_{y0}\mathbf{a}_y + D_{z0}\mathbf{a}_z$. We choose as our closed surface the small rectangular box, centered at P, having sides of lengths Δx, Δy, and Δz, and

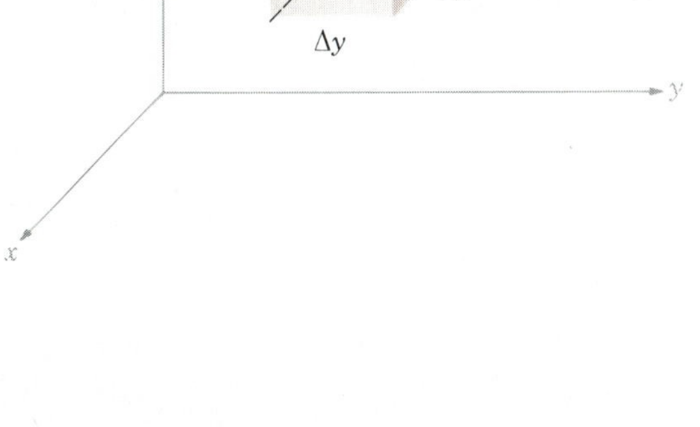

FIGURE 3.6
A differential-sized gaussian surface about the point P is used to investigate the space rate of change of $\mathbf{D}$ in the neighborhood of P.

apply Gauss's law,

$$\oint_S \mathbf{D} \cdot d\mathbf{S} = Q$$

In order to evaluate the integral over the closed surface, the integral must be broken up into six integrals, one over each face,

$$\oint_S \mathbf{D} \cdot d\mathbf{S} = \int_{\text{front}} + \int_{\text{back}} + \int_{\text{left}} + \int_{\text{right}} + \int_{\text{top}} + \int_{\text{bottom}}$$

Consider the first of these in detail. Since the surface element is very small, **D** is essentially constant (over *this* portion of the entire closed surface) and

$$\begin{aligned}\int_{\text{front}} &\doteq \mathbf{D}_{\text{front}} \cdot \Delta\mathbf{S}_{\text{front}} \\ &\doteq \mathbf{D}_{\text{front}} \cdot \Delta y\, \Delta z\, \mathbf{a}_x \\ &\doteq D_{x,\,\text{front}}\, \Delta y\, \Delta z\end{aligned}$$

where we have only to approximate the value of D_x at this front face. The front face is at a distance of $\Delta x/2$ from P, and hence

$$\begin{aligned}D_{x,\,\text{front}} &\doteq D_{x0} + \frac{\Delta x}{2} \times \text{rate of change of } D_x \text{ with } x \\ &\doteq D_{x0} + \frac{\Delta x}{2}\frac{\partial D_x}{\partial x}\end{aligned}$$

where D_{x0} is the value of D_x at P, and where a partial derivative must be used to express the rate of change of D_x with x, since D_x in general also varies with y and z. This expression could have been obtained more formally by using the constant term and the term involving the first derivative in the Taylor's-series expansion for D_x in the neighborhood of P.

We have now

$$\int_{\text{front}} \doteq \left(D_{x0} + \frac{\Delta x}{2}\frac{\partial D_x}{\partial x}\right) \Delta y\, \Delta z$$

Consider now the integral over the back surface,

$$\begin{aligned}\int_{\text{back}} &\doteq \mathbf{D}_{\text{back}} \cdot \Delta\mathbf{S}_{\text{back}} \\ &\doteq \mathbf{D}_{\text{back}} \cdot (-\Delta y\, \Delta z\, \mathbf{a}_x) \\ &\doteq -D_{x,\,\text{back}}\, \Delta y\, \Delta z\end{aligned}$$

and

$$D_{x,\,\text{back}} \doteq D_{x0} - \frac{\Delta x}{2}\frac{\partial D_x}{\partial x}$$

giving

$$\int_{\text{back}} \doteq \left(-D_{x0} + \frac{\Delta x}{2} \frac{\partial D_x}{\partial x} \right) \Delta y \, \Delta z$$

If we combine these two integrals, we have

$$\int_{\text{front}} + \int_{\text{back}} \doteq \frac{\partial D_x}{\partial x} \Delta x \, \Delta y \, \Delta z$$

By exactly the same process we find that

$$\int_{\text{right}} + \int_{\text{left}} \doteq \frac{\partial D_y}{\partial y} \Delta x \, \Delta y \, \Delta z$$

and

$$\int_{\text{top}} + \int_{\text{bottom}} \doteq \frac{\partial D_z}{\partial z} \Delta x \, \Delta y \, \Delta z$$

and these results may be collected to yield

$$\oint_S \mathbf{D} \cdot d\mathbf{S} \doteq \left(\frac{\partial D_x}{\partial x} + \frac{\partial D_y}{\partial y} + \frac{\partial D_z}{\partial z} \right) \Delta x \, \Delta y \, \Delta z$$

or

$$\oint_S \mathbf{D} \cdot d\mathbf{S} = Q \doteq \left(\frac{\partial D_x}{\partial x} + \frac{\partial D_y}{\partial y} + \frac{\partial D_z}{\partial z} \right) \Delta v \tag{7}$$

The expression is an approximation which becomes better as Δv becomes smaller, and in the following section we shall let the volume Δv approach zero. For the moment, we have applied Gauss's law to the closed surface surrounding the volume element Δv and have as a result the approximation (7) stating that

Charge enclosed in volume Δv

$$\doteq \left(\frac{\partial D_x}{\partial x} + \frac{\partial D_y}{\partial y} + \frac{\partial D_z}{\partial z} \right) \times \text{volume } \Delta v \tag{8}$$

For example, if $\mathbf{D} = e^{-x} \sin y \, \mathbf{a}_x - e^{-x} \cos y \, \mathbf{a}_y + 2z\mathbf{a}_z$ C/m^2, we see that

$$\frac{\partial D_x}{\partial x} = -e^{-x} \sin y$$

$$\frac{\partial D_y}{\partial y} = e^{-x} \sin y$$

$$\frac{\partial D_z}{\partial z} = 2$$

Adding these terms at the origin, for instance, we find that the charge enclosed in a small volume element there must be appproximately $2\,\Delta v$. If Δv is 10^{-9} m^3, then we have enclosed about 2 nC.

D3.6. Let $\mathbf{D} = y^2z^3\mathbf{a}_x + 2xyz^3\mathbf{a}_y + 3xy^2z^2\mathbf{a}_z$ pC/m^2 in free space. (*a*) Find the total electric flux passing through the surface $x = 3$, $0 \le y \le 2$, $0 \le z \le 1$ in a direction away from the origin. (*b*) Find $|\mathbf{E}|$ at $P(3,2,1)$. (*c*) Find the total charge contained in an incremental sphere having a radius of 2 μm centered at $P(3,2,1)$.

Ans. 0.667 pC; 4.31 V/m; 2.61×10^{-27} C

3.5 DIVERGENCE

We shall now obtain an exact relationship from (7), by allowing the volume element Δv to shrink to zero. We write this equation as

$$\frac{\partial D_x}{\partial x} + \frac{\partial D_y}{\partial y} + \frac{\partial D_z}{\partial z} \doteq \frac{\oint_S \mathbf{D} \cdot d\mathbf{S}}{\Delta v} = \frac{Q}{\Delta v}$$

or, as a limit

$$\frac{\partial D_x}{\partial x} + \frac{\partial D_y}{\partial y} + \frac{\partial D_z}{\partial z} = \lim_{\Delta v \to 0} \frac{\oint_S \mathbf{D} \cdot d\mathbf{S}}{\Delta v} = \lim_{\Delta v \to 0} \frac{Q}{\Delta v}$$

where the approximation has been replaced by an equality. It is evident that the last term is the volume charge density ρ_v, and hence that

$$\frac{\partial D_x}{\partial x} + \frac{\partial D_y}{\partial y} + \frac{\partial D_z}{\partial z} = \lim_{\Delta v \to 0} \frac{\oint_S \mathbf{D} \cdot d\mathbf{S}}{\Delta v} = \rho_v \tag{9}$$

This equation contains too much information to discuss all at once, and we shall write it as two separate equations,

$$\frac{\partial D_x}{\partial x} + \frac{\partial D_y}{\partial y} + \frac{\partial D_z}{\partial z} = \lim_{\Delta v \to 0} \frac{\oint_S \mathbf{D} \cdot d\mathbf{S}}{\Delta v} \tag{10}$$

and

$$\frac{\partial D_x}{\partial x} + \frac{\partial D_y}{\partial y} + \frac{\partial D_z}{\partial z} = \rho_v \tag{11}$$

where we shall save (11) for consideration in the next section.

Equation (10) does not involve charge density, and the methods of the previous section could have been used on any vector $\mathbf{A}$ to find $\oint_S \mathbf{A} \cdot d\mathbf{S}$ for a small closed surface, leading to

$$\frac{\partial A_x}{\partial x} + \frac{\partial A_y}{\partial y} + \frac{\partial A_z}{\partial z} = \lim_{\Delta v \to 0} \frac{\oint_S \mathbf{A} \cdot d\mathbf{S}}{\Delta v} \tag{12}$$

where **A** could represent velocity, temperature gradient, force, or any other vector field.

This operation appeared so many times in physical investigations in the last century that it received a descriptive name, *divergence*. The divergence of **A** is defined as

$$\text{Divergence of } \mathbf{A} = \text{div } \mathbf{A} = \lim_{\Delta v \to 0} \frac{\oint_S \mathbf{A} \cdot d\mathbf{S}}{\Delta v} \tag{13}$$

and is usually abbreviated div **A**. The physical interpretation of the divergence of a vector is obtained by describing carefully the operations implied by the right-hand side of (13), where we shall consider **A** as a member of the flux-density family of vectors in order to aid the physical interpretation.

The divergence of the vector flux density **A** *is the outflow of flux from a small closed surface per unit volume as the volume shrinks to zero.*

The physical interpretation of divergence afforded by this statement is often useful in obtaining qualitative information about the divergence of a vector field without resorting to a mathematical investigation. For instance, let us consider the divergence of the velocity of water in a bathtub after the drain has been opened. The net outflow of water through *any* closed surface lying entirely within the water must be zero, for water is essentially incompressible and the water entering and leaving different regions of the closed surface must be equal. Hence the divergence of this velocity is zero.

If, however, we consider the velocity of the air in an inner tube which has just been punctured by a nail, we realize that the air is expanding as the pressure drops, and that consequently there is a net outflow from any closed surface lying within the inner tube. The divergence of this velocity is therefore greater than zero.

A positive divergence for any vector quantity indicates a *source* of that vector quantity at that point. Similarly, a negative divergence indicates a *sink*. Since the divergence of the water velocity above is zero, no source or sink exists.[3] The expanding air, however, produces a positive divergence of the velocity, and each interior point may be considered a source.

Writing (10) with our new term, we have

$$\text{div } \mathbf{D} = \frac{\partial D_x}{\partial x} + \frac{\partial D_y}{\partial y} + \frac{\partial D_z}{\partial z} \tag{14}$$

[3] Having chosen a differential element of volume within the water, the gradual decrease in water level with time will eventually cause the volume element to lie above the surface of the water. At the instant the surface of the water intersects the volume element, the divergence is positive and the small volume is a source. This complication is avoided above by specifying an internal point.

This expression is again of a form which does not involve the charge density. It is the result of applying the definition of divergence (13) to a differential volume element in *cartesian coordinates.*

If a differential volume unit $\rho\, d\rho\, d\phi\, dz$ in cylindrical coordinates, or $r^2 \sin\theta\, dr\, d\theta\, d\phi$ in spherical coordinates, had been chosen, expressions for divergence involving the components of the vector in the particular coordinate system and involving partial derivatives with respect to the variables of that system would have been obtained. These expressions are obtained in Appendix A and are given here for convenience:

$$\text{div}\,\mathbf{D} = \frac{\partial D_x}{\partial x} + \frac{\partial D_y}{\partial y} + \frac{\partial D_z}{\partial z} \qquad \text{(cartesian)} \tag{15}$$

$$\text{div}\,\mathbf{D} = \frac{1}{\rho}\frac{\partial}{\partial \rho}(\rho D_\rho) + \frac{1}{\rho}\frac{\partial D_\phi}{\partial \phi} + \frac{\partial D_z}{\partial z} \qquad \text{(cylindrical)} \tag{16}$$

$$\text{div}\,\mathbf{D} = \frac{1}{r^2}\frac{\partial}{\partial r}(r^2 D_r) + \frac{1}{r\sin\theta}\frac{\partial}{\partial \theta}(\sin\theta\, D_\theta) + \frac{1}{r\sin\theta}\frac{\partial D_\phi}{\partial \phi} \qquad \text{(spherical)} \tag{17}$$

These relationships are also shown inside the back cover for easy reference.

It should be noted that the divergence is an operation which is performed on a vector, but that the result is a scalar. We should recall that, in a somewhat similar way, the dot, or scalar, product was a multiplication of two vectors which yielded a scalar product.

For some reason it is a common mistake on meeting divergence for the first time to impart a vector quality to the operation by scattering unit vectors around in the partial derivatives. Divergence merely tells us *how much* flux is leaving a small volume on a per-unit-volume basis; no direction is associated with it.

We can illustrate the concept of divergence by continuing with the example at the end of the previous section. If $\mathbf{D} = e^{-x}\sin y\,\mathbf{a}_x - e^{-x}\cos y\,\mathbf{a}_y + 2z\mathbf{a}_z$, then

$$\text{div}\,\mathbf{D} = \frac{\partial D_x}{\partial x} + \frac{\partial D_y}{\partial y} + \frac{\partial D_z}{\partial z}$$

$$= -e^{-x}\sin y + e^{-x}\sin y + 2 = 2$$

If the units of $\mathbf{D}$ are C/m^2, then the units of div $\mathbf{D}$ are C/m^3. This is a volume charge density, a concept on which we shall expound in the next section.

D3.7. Find a numerical value for the divergence of $\mathbf{D}$ at the point indicated if (*a*) $\mathbf{D} = 20xy^2(z+1)\mathbf{a}_x + 20x^2y(z+1)\mathbf{a}_y + 10x^2y^2\mathbf{a}_z\ \text{C/m}^2$ at $P_A(0.3, 0.4, 0.5)$; (*b*) $\mathbf{D} = 4\rho z\sin\phi\,\mathbf{a}_\rho + 2\rho z\cos\phi\,\mathbf{a}_\phi + 2\rho^2\sin\phi\,\mathbf{a}_z\ \text{C/m}^2$ at $P_B(1, \pi/2, 2)$; (*c*) $\mathbf{D} = \sin\theta\cos\phi\,\mathbf{a}_r + \cos\theta\cos\phi\,\mathbf{a}_\theta - \sin\phi\,\mathbf{a}_\phi\ \text{C/m}^2$ at $P_C(2, \theta = \pi/3, \phi = \pi/6)$.

Ans. 7.5 C/m^3; 12 C/m^3; 0

3.6 MAXWELL'S FIRST EQUATION (ELECTROSTATICS)

We now wish to consolidate the gains of the last two sections and to provide an interpretation of the divergence operation as it relates to electric flux density. The expressions developed there may be written as

$$\text{div } \mathbf{D} = \lim_{\Delta v \to 0} \frac{\oint_S \mathbf{D} \cdot d\mathbf{S}}{\Delta v} \tag{18}$$

$$\text{div } \mathbf{D} = \frac{\partial D_x}{\partial x} + \frac{\partial D_y}{\partial y} + \frac{\partial D_z}{\partial z} \tag{19}$$

and

$$\text{div } \mathbf{D} = \rho_v \tag{20}$$

The first equation is the definition of divergence, the second is the result of applying the definition to a differential volume element in cartesian coordinates, giving us an equation by which the divergence of a vector expressed in cartesian coordinates may be evaluated, and the third is merely (11) written using the new term div **D**. Equation (20) is almost an obvious result if we have achieved any familiarity at all with the concept of divergence as defined by (18), for given Gauss's law,

$$\oint_S \mathbf{D} \cdot d\mathbf{S} = Q$$

per unit volume

$$\frac{\oint_S \mathbf{D} \cdot d\mathbf{S}}{\Delta v} = \frac{Q}{\Delta v}$$

As the volume shrinks to zero,

$$\lim_{\Delta v \to 0} \frac{\oint_S \mathbf{D} \cdot d\mathbf{S}}{\Delta v} = \lim_{\Delta v \to 0} \frac{Q}{\Delta v}$$

we should see div·**D** on the left and volume charge density on the right,

$$\boxed{\text{div } \mathbf{D} = \rho_v} \tag{20}$$

This is the first of Maxwell's four equations as they apply to electrostatics and steady magnetic fields, and it states that the electric flux per unit volume leaving a vanishingly small volume unit is exactly equal to the volume charge density there. This equation is aptly called the *point form of Gauss's law*. Gauss's law relates the flux leaving any closed surface to the charge enclosed, and Maxwell's first equation makes an identical statement on a per-unit-volume basis for a vanishingly small volume, or at a point. Remembering that the divergence may be expressed as the sum of three partial derivatives, Maxwell's first equation is also described as the differential-equation form of Gauss's law, and conversely, Gauss's law is recognized as the integral form of Maxwell's first equation.

As a specific illustration, let us consider the divergence of **D** in the region about a point charge Q located at the origin. We have the field

$$\mathbf{D} = \frac{Q}{4\pi r^2}\mathbf{a}_r$$

and make use of (17), the expression for divergence in spherical coordinates given in the previous section:

$$\text{div } \mathbf{D} = \frac{1}{r^2}\frac{\partial}{\partial r}(r^2 D_r) + \frac{1}{r \sin\theta}\frac{\partial}{\partial\theta}(D_\theta \sin\theta) + \frac{1}{r\sin\theta}\frac{\partial D_\phi}{\partial\phi}$$

Since D_θ and D_ϕ are zero, we have

$$\text{div } \mathbf{D} = \frac{1}{r^2}\frac{d}{dr}\left(r^2 \frac{Q}{4\pi r^2}\right) = 0 \qquad (\text{if } r \neq 0)$$

Thus, $\rho_v = 0$ everywhere except at the origin where it is infinite.

The divergence operation is not limited to electric flux density; it can be applied to any vector field. We shall apply it to several other electromagnetic fields in the coming chapters.

D3.8. Find the volume charge density that is associated with each of the following fields: (*a*) $\mathbf{D} = xy^2\mathbf{a}_x + yx^2\mathbf{a}_y + z\mathbf{a}_z$ C/m^2; (*b*) $\mathbf{D} = \rho z^2 \sin^2\phi\, \mathbf{a}_\rho + \rho z^2 \sin\phi \cos\phi\, \mathbf{a}_\phi + \rho^2 z \sin^2\phi\, \mathbf{a}_z$ C/m^2; (*c*) $\mathbf{a}_r$ C/m^2.

Ans. $x^2 + y^2 + 1$ C/m^3; $z^2 + \rho^2 \sin^2\phi$ C/m^3; $2/r$ C/m^3

3.7 THE VECTOR OPERATOR ∇ AND THE DIVERGENCE THEOREM

If we remind ourselves again that divergence is an operation on a vector yielding a scalar result, just as the dot product of two vectors gives a scalar result, it seems possible that we can find something which may be dotted formally with **D** to yield the scalar

$$\frac{\partial D_x}{\partial x} + \frac{\partial D_y}{\partial y} + \frac{\partial D_z}{\partial z}$$

Obviously, this cannot be accomplished by using a dot *product;* the process must be a dot *operation.*

With this in mind, we define the *del operator* ∇ as a *vector operator,*

$$\nabla = \frac{\partial}{\partial x}\mathbf{a}_x + \frac{\partial}{\partial y}\mathbf{a}_y + \frac{\partial}{\partial z}\mathbf{a}_z \qquad (21)$$

Similar *scalar operators* appear in several methods of solving differential equations where we often let D replace d/dx, D^2 replace d^2/dx^2, and so forth.[4] We agree on defining ∇ (pronounced "del") that it shall be treated in every way as an ordinary vector with the one important exception that partial derivatives result instead of products of scalars.

Consider $\nabla \cdot \mathbf{D}$, signifying

$$\nabla \cdot \mathbf{D} = \left(\frac{\partial}{\partial x} \mathbf{a}_x + \frac{\partial}{\partial y} \mathbf{a}_y + \frac{\partial}{\partial z} \mathbf{a}_z \right) \cdot (D_x \mathbf{a}_x + D_y \mathbf{a}_y + D_z \mathbf{a}_z)$$

We first consider the dot products of the unit vectors, discarding the six zero terms and having left

$$\nabla \cdot \mathbf{D} = \frac{\partial}{\partial x}(D_x) + \frac{\partial}{\partial y}(D_y) + \frac{\partial}{\partial z}(D_z)$$

where the parentheses are now removed by operating or differentiating:

$$\nabla \cdot \mathbf{D} = \frac{\partial D_x}{\partial x} + \frac{\partial D_y}{\partial y} + \frac{\partial D_z}{\partial z}$$

This is recognized as the divergence of **D**, so that we have

$$\boxed{\operatorname{div} \mathbf{D} = \nabla \cdot \mathbf{D} = \frac{\partial D_x}{\partial x} + \frac{\partial D_y}{\partial y} + \frac{\partial D_z}{\partial z}}$$

The use of $\nabla \cdot \mathbf{D}$ is much more prevalent than that of div **D**, although both usages have their advantages. Writing $\nabla \cdot \mathbf{D}$ allows us to obtain simply and quickly the correct partial derivatives, but only in cartesian coordinates, as we shall see below. On the other hand, div **D** is an excellent reminder of the physical interpretation of divergence. We shall use the operator notation $\nabla \cdot \mathbf{D}$ from now on to indicate the divergence operation.

The vector operator ∇ not only is used with divergence, but will appear in several other very important operations later. One of these is ∇u, where u is any scalar field, and leads to

$$\nabla u = \left(\frac{\partial}{\partial x} \mathbf{a}_x + \frac{\partial}{\partial y} \mathbf{a}_y + \frac{\partial}{\partial z} \mathbf{a}_z \right) u = \frac{\partial u}{\partial x} \mathbf{a}_x + \frac{\partial u}{\partial y} \mathbf{a}_y + \frac{\partial u}{\partial z} \mathbf{a}_z$$

The ∇ operator does not have a specific form in other coordinate systems. If we are considering **D** in cylindrical coordinates, then $\nabla \cdot \mathbf{D}$ still indicates the divergence of **D**, or

$$\nabla \cdot \mathbf{D} = \frac{1}{\rho} \frac{\partial}{\partial \rho}(\rho D_\rho) + \frac{1}{\rho} \frac{\partial D_\phi}{\partial \phi} + \frac{\partial D_z}{\partial z}$$

[4] This scalar operator D, which will not appear again, is not to be confused with the electric flux density.

where this expression has been taken from Sec. 3.5. We have no form for ∇ itself to help us obtain this sum of partial derivatives. This means that ∇u, as yet unnamed but easily written above in cartesian coordinates, cannot be expressed by us at this time in cylindrical coordinates. Such an expression will be obtained when ∇u is defined in Chap. 4.

We shall close our discussion of divergence by presenting a theorem which will be needed several times in later chapters, the *divergence theorem*. This theorem applies to any vector field for which the appropriate partial derivatives exist, although it is easiest for us to develop it for the electric flux density. We have actually obtained it already and now have little more to do than point it out and name it, for starting from Gauss's law,

$$\oint_S \mathbf{D} \cdot d\mathbf{S} = Q$$

and letting

$$Q = \int_{\text{vol}} \rho_v \, dv$$

and then replacing ρ_v by its equal,

$$\nabla \cdot \mathbf{D} = \rho_v$$

we have

$$\oint_S \mathbf{D} \cdot d\mathbf{S} = Q = \int_{\text{vol}} \rho_v \, dv = \int_{\text{vol}} \nabla \cdot \mathbf{D} \, dv$$

The first and last expressions constitute the divergence theorem,

$$\boxed{\oint_S \mathbf{D} \cdot d\mathbf{S} = \int_{\text{vol}} \nabla \cdot \mathbf{D} \, dv} \tag{22}$$

which may be stated as follows:

> *The integral of the normal component of any vector field over a closed surface is equal to the integral of the divergence of this vector field throughout the volume enclosed by the closed surface.*

Again, we emphasize that the divergence theorem is true for any vector field, although we have obtained it specifically for the electric flux density **D**, and we shall have occasion later to apply it to several different fields. Its benefits derive from the fact that it relates a triple integration *throughout some volume* to a double integration *over the surface* of that volume. For example, it is much easier to look for leaks in a bottle full of some agitated liquid by an inspection of the surface than by calculating the velocity at every internal point.

The divergence theorem becomes obvious physically if we consider a volume v, shown in cross section in Fig. 3.7, which is surrounded by a closed surface S. Division of the volume into a number of small compartments of differential size

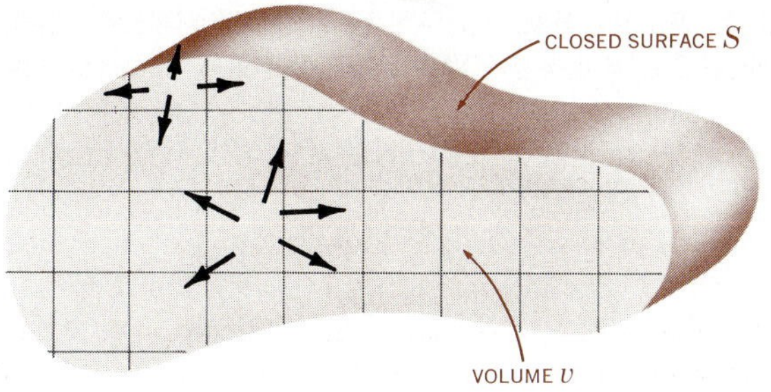

FIGURE 3.7
The divergence theorem states that the total flux crossing the closed surface is equal to the integral of the divergence of the flux density throughout the enclosed volume. The volume is shown here in cross section.

and consideration of one cell show that the flux diverging from such a cell *enters*, or *converges* on, the adjacent cells unless the cell contains a portion of the outer surface. In summary, the divergence of the flux density throughout a volume leads, then, to the same result as determining the net flux crossing the enclosing surface.

As a simple check on the validity of this theorem, consider the field $\mathbf{D} = 2xy\mathbf{a}_x + x^2\mathbf{a}_y$ C/m^2 and the rectangular parallelepiped formed by the planes $x = 0$ and 1, $y = 0$ and 2, and $z = 0$ and 3. Evaluating the surface integral first, we note that $\mathbf{D}$ is parallel to the surfaces at $z = 0$ and 3, so $\mathbf{D} \cdot d\mathbf{S} = 0$ there. For the remaining four surfaces we have

$$\begin{aligned}\oint_S \mathbf{D} \cdot d\mathbf{S} &= \int_0^3 \int_0^2 (\mathbf{D})_{x=0} \cdot (-dy\,dz\,\mathbf{a}_x) + \int_0^3 \int_0^2 (\mathbf{D})_{x=1} \cdot (dy\,dz\,\mathbf{a}_x) \\ &\quad + \int_0^3 \int_0^1 (\mathbf{D})_{y=0} \cdot (-dx\,dz\,\mathbf{a}_y) + \int_0^3 \int_0^1 (\mathbf{D})_{y=2} \cdot (dx\,dz\,\mathbf{a}_y) \\ &= -\int_0^3 \int_0^2 (D_x)_{x=0}\,dy\,dz + \int_0^3 \int_0^2 (D_x)_{x=1}\,dy\,dz \\ &\quad - \int_0^3 \int_0^1 (D_y)_{y=0}\,dx\,dz + \int_0^3 \int_0^1 (D_y)_{y=2}\,dx\,dz\end{aligned}$$

However, $(D_x)_{x=0} = 0$, and $(D_y)_{y=0} = (D_y)_{y=2}$, which leaves only

$$\begin{aligned}\oint_S \mathbf{D} \cdot d\mathbf{S} &= \int_0^3 \int_0^2 (D_x)_{x=1}\,dy\,dz = \int_0^3 \int_0^2 2y\,dy\,dz \\ &= \int_0^3 4\,dz = 12\end{aligned}$$

Since

$$\nabla \cdot \mathbf{D} = \frac{\partial}{\partial x}(2xy) + \frac{\partial}{\partial y}(x^2) = 2y$$

the volume integral becomes

$$\begin{aligned}\int_{\text{vol}} \nabla \cdot \mathbf{D}\,dv &= \int_0^3 \int_0^2 \int_0^1 2y\,dx\,dy\,dz = \int_0^3 \int_0^2 2y\,dy\,dz \\ &= \int_0^3 4\,dz = 12\end{aligned}$$

and the check is accomplished. Remembering Gauss's law, we see that we have also determined that a total charge of 12 C lies within this parallelepiped.

D3.9. Given the flux density $\mathbf{D} = (2\cos\theta/r^3)\mathbf{a}_r + (\sin\theta/r^3)\mathbf{a}_\theta$ C/m^2, evaluate both sides of the divergence theorem for the region defined by $1 < r < 2$, $0 < \theta < \pi/2$, $0 < \phi < \pi/2$.

Ans. 0; 0

SUGGESTED REFERENCES

1. Kraus, J. D.: "Electromagnetics," 3d ed., McGraw-Hill Book Company, New York, 1984. The static electric field in free space is introduced in chap. 2.
2. Plonsey, R., and R. E. Collin: "Principles and Applications of Electromagnetic Fields," McGraw-Hill Book Company, New York, 1961. The level of this text is somewhat higher than the one we are reading now, but it is an excellent text to read *next*. Gauss's law appears in the second chapter.
3. Plonus, M. A.: "Applied Electromagnetics," McGraw-Hill Book Company, New York, 1978. This book contains rather detailed descriptions of many practical devices that illustrate electromagnetic applications. For example, see the discussion of xerography on pp. 95–98 as an electrostatics application.
4. Skilling, H. H.: "Fundamentals of Electric Waves," 2d ed., John Wiley & Sons, Inc., New York, 1948. The operations of vector calculus are well illustrated. Divergence is discussed on pp. 22 and 38. Chapter 1 is interesting reading.
5. Thomas, G. B., Jr., and R. L. Finney: (see Suggested References for Chap. 1). The divergence theorem is developed and illustrated from several different points of view on pp. 976–980.

PROBLEMS

1. An empty metal paint can is placed on a marble table, the lid is removed, and both parts are discharged (honorably) by touching them to ground. An insulating nylon thread is glued to the center of the lid, and a penny, a nickel, and a dime are glued to the thread so that they are not touching each other. The penny is given a charge of +5 nC, and the nickel and dime are discharged. The assembly is lowered into the can so that the coins hang clear of all walls, and the lid is secured. The outside of the can is again touched momentarily to ground. The device is carefully disassembled with insulating gloves and tools. (*a*) What charges are found on each of the five metallic pieces? (*b*) If the penny had been given a charge of +5 nC, the dime a charge of −2 nC, and the nickel a charge of −1 nC, what would the final charge arrangement have been?

2. Point charges of 15 nC each are symmetrically located at (4,4,0), (4,−4,0), (−4,4,0), and (−4,−4,0), and a uniform line charge of 50 nC/m lies at $x = 0$, $y = 8$, all z. (*a*) Find $\mathbf{D}$ at the origin. (*b*) How much electric flux crosses the $y = 0$ plane? (*c*) How much electric flux leaves the surface of a sphere of radius 5 m centered at (0,6,0)?

3. A nonuniform surface charge density of $5\rho/(\rho^2 + 1)$ nC/m^2 lies in the plane $z = 2$ wherever $\rho < 5$; $\rho_S = 0$ for $\rho > 5$. (*a*) How much electric flux leaves the circular region $\rho < 5$, $z = 2$? (*b*) How much electric flux crosses the $z = 0$ plane in the $-\mathbf{a}_z$ direction? (*c*) How much electric flux leaves the cylinder $\rho = 3$ in the $\mathbf{a}_\rho$ direction?

4. The spherical surfaces $r = 1$, 2, and 3 carry surface charge densities of 20, -9, and 2 nC/m^2, respectively. (*a*) How much electric flux leaves the surface $r = 5$? (*b*) Find $\mathbf{D}$ at $P(1,-1,2)$.

5. A point charge of 6 μC is located at the origin, a uniform line charge density of 180 nC/m lies along the x axis, and a uniform sheet of charge equal to 25 nC/m^2 lies in the $z = 0$ plane. (*a*) Find $\mathbf{D}$ at $A(0,0,4)$. (*b*) Find $\mathbf{D}$ at $B(1,2,4)$. (*c*) Calculate the total electric flux leaving the surface of a sphere of 4-m radius centered at the origin.

6. If $\mathbf{D} = 3y^2\mathbf{a}_x + 3x^2y\mathbf{a}_y + 5\mathbf{a}_z$ C/m^2, find the total charge enclosed within the region $0 < x,y,z < 2$ by evaluating one or more surface integrals.

7. Uniform line charges of 20 nC/m each lie in the $z = 0$ plane at $x = 0, \pm 1, \pm 2, \ldots, \pm 5$. (*a*) Find the total electric flux leaving the spherical surface $r = 2.5$. (*b*) Find $\mathbf{D}$ at $P(0,2.5,4)$.

8. A point charge of 90 nC is located at the origin, and there are also three uniform surface charge distributions: -8 nC/m^2 at $r = 1$, 4.5 nC/m^2 at $r = 2$, and 1.2 nC/m^2 at $r = 3$. (*a*) Find $\mathbf{D}$ everywhere. (*b*) Plot $|\mathbf{D}|$ vs. r, $0 < r < 4$.

9. The spherical region, $0 < r < 10$ cm, contains a uniform volume charge density $\rho_v =$ 4 μC/m^3. (*a*) Find Q_{tot}, $0 < r < 10$ cm. (*b*) Find D_r, $0 < r < 10$ cm. (*c*) The nonuniform volume charge density, $\rho_v = -3/(r^3 + 0.001)$ nC/m^3, exists for 10 cm $< r < r_0$. Find r_0 so that the total charge, $0 < r < r_0$, is zero.

10. A uniform sheet of charge ρ_{S0} lies in that portion of the $z = 0$ plane for which $|x| < 5$. Find $\mathbf{D}(x,0,z)$ for $z > 0$.

11. Let $\rho_v = 0$ for $\rho < 1$ cm and also for $\rho > 3$ cm. In the region $1 < \rho < 3$ cm, $\rho_v =$ $10^{-8} \cos 50\pi\rho$ C/m^3. Find $\mathbf{D}$ everywhere.

12. A uniform volume charge density, $\rho_v = 60\rho$ C/m^3, lies within the cylinder $\rho = 1$ m, and $\rho_v = 0$ elsewhere. (*a*) Find D_ρ everywhere. (*b*) What surface charge density ρ_{S2} should be on the cylinder $\rho = 2$ so that $D_\rho|_{\rho=2^-} = 2D_\rho|_{\rho=2^+}$? (*c*) Plot D_ρ vs. ρ for $0 < \rho < 5$ with both charge distributions present.

13. Cylindrical surfaces at $\rho = 2$, 4, and 6 m carry uniform charge densities of 20 nC/m^2, -4 nC/m^2, and ρ_{S0}, respectively. (*a*) Find $\mathbf{D}$ at $\rho = 1$, 3, and 5 m. (*b*) Determine ρ_{S0} such that $\mathbf{D} = 0$ at $\rho = 7$ m.

14. Let $\rho_v = 8z(1 - z)$ C/m^3 for $0 < z < 1$ m, $8z(1 + z)$ C/m^3 for $-1 < z < 0$, and 0 for $|z| > 1$. (*a*) Find $\mathbf{D}$ everywhere. (*b*) Sketch D_z vs. z, $-2 < z < 2$.

15. Uniform surface charge densities of 100, -60, and 50 μC/m^2 lie in the $x = 0$, $y = 0$, and $z = 0$ planes, respectively. Find $|\mathbf{D}|$ at points 2.5 m from every coordinate plane in each of the eight octants.

16. Find $\mathbf{D}$ and plot D_r vs. r for each of the following charge distributions: (*a*) 1 C at the origin; (*b*) $1/(4\pi)$ C/m^2 at $r = 1$ m; (*c*) $3/(4\pi)$ C/m^3 for $0 < r < 1$, 0 elsewhere.

17. Let $\mathbf{D} = (8x + 4x^3)\mathbf{a}_x - 2y\mathbf{a}_y + 2z\mathbf{a}_z$ C/m^2. (*a*) Use Gauss's law to determine the charge enclosed in the cubical region, $-a < x,y,z < a$. (*b*) Use Eq. (8) to obtain an approximate value for the above charge. Evaluate the derivatives at the center of the volume. (*c*) Show that your answers agree in the limit as a becomes very small.

18. Let $\mathbf{D} = (10xyz^2 + 4x)\mathbf{a}_x + 5x^2z^2\mathbf{a}_y + 10x^2yz\mathbf{a}_z$ nC/m^2. (*a*) Find the total charge enclosed in a cube having a volume of 10^{-10} m^3 located at (2,3,4). (*b*) How much flux leaves this volume? (*c*) What is the average volume charge density within this volume?

19. Let $\mathbf{D} = 12xyz\mathbf{a}_x + (6x^2z + 6yz)\mathbf{a}_y + (6x^2y + 3y^2)\mathbf{a}_z$ C/m^2 and find the incremental amount of charge in a volume of 10^{-10} m^3 located at: (*a*) (0,0,0); (*b*) (4,2,−3); (*c*) (4,y,−3). (*d*) At what location in the cubical region $0 \le x,y,z \le 2$ should the small volume be located to contain a maximum charge?

20. A cubical region of space is defined by the surfaces $x = 1.8$, $y = 1.8$, $z = 1.8$, $x = 2$, $y = 2$, and $z = 2$. If $\mathbf{D} = 3y^2\mathbf{a}_x + 3x^2y\mathbf{a}_y$ C/m^2: (*a*) find the exact value of the total charge enclosed within the cube by surface integration; (*b*) find an approximate value for the enclosed charge by evaluating the derivatives of Eq. (8) at the center of the cube.

21. Calculate the divergence of $\mathbf{G}$ at $P(2,-3,4)$ if $\mathbf{G} =$: (*a*) $x\mathbf{a}_x + y\mathbf{a}_y + z\mathbf{a}_z$; (*b*) $\rho\mathbf{a}_\rho$; (*c*) $r\mathbf{a}_r$; (*d*) $6r^2 \sin\theta\, \mathbf{a}_r + 2r^2 \cos\theta\, \mathbf{a}_\theta$.

22. If $\mathbf{D} = 12x^2\mathbf{a}_x - 3z^3\mathbf{a}_y - 9yz^2\mathbf{a}_z$ C/m^2 in free space, specify the point within the cube, $1 \le x,y,z \le 2$ at which the following quantity is a maximum and give that maximum value: (*a*) $|\mathbf{D}|$; (*b*) $|\rho_v|$; (*c*) ρ_v.

23. The electric flux density $\mathbf{D} = 10r^3\mathbf{a}_r$ C/m^2 in the region $0 \le r < 1$ m. (*a*) Find ρ_v at $r = 0.4$ m. (*b*) Find D_r at $r = 0.4$ m. (*c*) What total charge is contained in the region $0 \le r \le 0.4$ m? (*d*) If $\rho_v = 0$ for $r > 1$ m, find D_r for $r > 1$ m.

24. Let $\mathbf{D} = (10r^2 + 5e^{-r})\mathbf{a}_r$ C/m^2. (*a*) Find ρ_v as a function of r. (*b*) Find the total charge lying within a sphere of radius a centered at the origin.

25. If $\mathbf{D} = 4x^3\mathbf{a}_x - 2z\mathbf{a}_y - 2y\mathbf{a}_z$ C/m^2, find: (*a*) $\nabla \cdot \mathbf{D}$; (*b*) ρ_v at $P(x,y,z)$; (*c*) the total charge lying within the region $-1 < x,y,z < 1$; (*d*) the total charge lying within that region without finding ρ_v first.

26. (*a*) A point charge Q lies at the origin. Show that $\nabla \cdot \mathbf{D} = 0$ everywhere except at the origin. (*b*) Replace the point charge with a uniform volume charge density ρ_{v0} for $0 \le r \le a$. Relate ρ_{v0} to Q and a so that the total charge is the same. Find $\nabla \cdot \mathbf{D}$ everywhere.

27. Within the cylindrical region $\rho \le 4$ m, the electric flux density is given as $5\rho^3\mathbf{a}_\rho$ C/m^2. (*a*) What is the volume charge density at $\rho = 3$ m? (*b*) What is the electric flux density at $\rho = 3$? (*c*) How much electric flux leaves the cylinder, $\rho = 3$, $|z| \le 2.5$? (*d*) How much charge is contained within the cylinder, $\rho = 3$, $|z| \le 2.5$?

28. Inside the sphere $r \le 4$ m, the electric flux density is given as $5r^3\mathbf{a}_r$ C/m^2. (*a*) What is the volume charge density at $r = 3$ m? (*b*) What is the electric flux density at $r = 3$? (*c*) How much electric flux leaves the sphere $r = 3$? (*d*) How much charge is contained within the sphere $r = 3$?

29. If $\mathbf{D} = 0.1r\mathbf{a}_r$ C/m^2 for $r \le 0.4$ m, and $\mathbf{D} = (0.0064/r^2)\mathbf{a}_r$ for $r \ge 0.4$: (*a*) find ρ_v at $r = 0.2$ and 0.5 m. (*b*) What point charge could be placed at the origin to cause $\mathbf{D}$ to be zero for $r \ge 0.4$?

30. Let $\mathbf{D} = 6xyz^2\mathbf{a}_x + 3x^2z^2\mathbf{a}_y + 6x^2yz\mathbf{a}_z$ C/m^2. Find the total charge lying within the region bounded by $x = 1$ and 3, $y = 0$ and 1, and $z = -1$ and 1 by separately evaluating each side of the divergence theorem.

31. Given the field $\mathbf{D} = (20/\rho^2)(-\sin^2\phi\, \mathbf{a}_\rho + \sin 2\phi\, \mathbf{a}_\phi)$ C/m^2, find the total charge lying within the volume, $1 < \rho < 2$, $0 < \phi < \pi/2$, $0 < z < 1$ by two different methods.

32. If $\mathbf{G} = 5r \sin^2\theta \cos^2\phi\, \mathbf{a}_r$, evaluate both sides of the divergence theorem for the region $r \le 2$.

33. Let $\mathbf{D} = x\mathbf{a}_x$ and find the value of $\oint_S \mathbf{D} \cdot d\mathbf{S}$ over the surface of the sphere $r = 1$.

CHAPTER 4

ENERGY AND POTENTIAL

In the previous two chapters we have become acquainted with Coulomb's law and its use in finding the electric field about several simple distributions of charge, and also with Gauss's law and its application in determining the field about some symmetrical charge arrangements. The use of Gauss's law was invariably easier for these highly symmetrical distributions, because the problem of integration always disappeared when the proper closed surface was chosen.

However, if we had attempted to find a slightly more complicated field, such as that of two unlike point charges separated by a small distance, we would have found it impossible to choose a suitable gaussian surface and obtain an answer. Coulomb's law, however, is more powerful and enables us to solve problems for which Gauss's law is not applicable. The application of Coulomb's law is laborious, detailed, and often quite complex, the reason for this being precisely the fact that the electric field intensity, a vector field, must be found directly from the charge distribution. Three different integrations are needed in general, one for each component, and the resolution of the vector into components usually adds to the complexity of the integrals.

Certainly it would be desirable if we could find some as yet undefined scalar function with a single integration and then determine the electric field from this scalar by some simple straightforward procedure, such as differentiation.

This scalar function does exist and is known as the potential or potential field. We shall find that it has a very real physical interpretation and is more familiar to most of us than is the electric field which it will be used to find.

We should expect, then, to be equipped soon with a third method of finding electric fields—a single scalar integration, although not always as simple as we might wish, followed by a pleasant differentiation.

The remaining difficult portion of the task, the integration, we intend to remove in Chap. 7.

4.1 ENERGY EXPENDED IN MOVING A POINT CHARGE IN AN ELECTRIC FIELD

The electric field intensity was defined as the force on a unit test charge at that point at which we wish to find the value of this vector field. If we attempt to move the test charge against the electric field, we have to exert a force equal and opposite to that exerted by the field, and this requires us to expend energy, or do work. If we wish to move the charge in the direction of the field, our energy expenditure turns out to be negative; we do not do the work, the field does.

Suppose we wish to move a charge Q a distance $d\mathbf{L}$ in an electric field $\mathbf{E}$. The force on Q due to the electric field is

$$\boxed{\mathbf{F}_E = Q\mathbf{E}} \tag{1}$$

where the subscript reminds us that this force is due to the field. The component of this force in the direction $d\mathbf{L}$ which we must overcome is

$$F_{EL} = \mathbf{F}_E \cdot \mathbf{a}_L = Q\mathbf{E} \cdot \mathbf{a}_L$$

where $\mathbf{a}_L$ = a unit vector in the direction of $d\mathbf{L}$.

The force which we must apply is equal and opposite to the force due to the field,

$$F_{\text{appl}} = -Q\mathbf{E} \cdot \mathbf{a}_L$$

and our expenditure of energy is the product of the force and distance,

$$\begin{aligned}\text{Differential work done by external source moving } Q &= -Q\mathbf{E} \cdot \mathbf{a}_L\, dL \\ &= -Q\mathbf{E} \cdot d\mathbf{L}\end{aligned}$$

or

$$\boxed{dW = -Q\mathbf{E} \cdot d\mathbf{L}} \tag{2}$$

where we have replaced $\mathbf{a}_L\, dL$ by the simpler expression $d\mathbf{L}$.

This differential amount of work required may be zero under several conditions determined easily from (2). There are the trivial conditions for which $\mathbf{E}$, Q, or $d\mathbf{L}$ is zero, and a much more important case in which $\mathbf{E}$ and $d\mathbf{L}$ are perpendicular. Here the charge is moved always in a direction at right angles to the electric field. We can draw on a good analogy between the electric field and the gravitational field, where, again, energy must be expended to move against the field. Sliding a mass around with constant velocity on a frictionless uneven surface is an effortless process if the mass is moved along a constant elevation contour; positive or negative work must be done in moving it to a higher or lower elevation, respectively.

Returning to the charge in the electric field, the work required to move the charge a finite distance must be determined by integrating,

$$W = -Q \int_{\text{init}}^{\text{final}} \mathbf{E} \cdot d\mathbf{L} \tag{3}$$

where the path must be specified before the integral can be evaluated. The charge is assumed to be at rest at both its initial and final positions.

This definite integral is basic to field theory, and we shall devote the following section to its interpretation and evaluation.

D4.1. An electric field is given as $\mathbf{E} = 6y^2z\mathbf{a}_x + 12xyz\mathbf{a}_y + 6xy^2\mathbf{a}_z$ V/m. An incremental path is represented by $\Delta\mathbf{L} = -3\mathbf{a}_x + 5\mathbf{a}_y - 2\mathbf{a}_z$ μm. Find the work done in moving a 2-μC charge along this path if the location of the path is at: (*a*) $P_A(0,2,5)$; (*b*) $P_B(1,1,1)$; (*c*) $P_C(-0.7,-2,-0.3)$.

Ans. 720 pJ; -60 pJ; -60 pJ

4.2 THE LINE INTEGRAL

The integral expression for the work done in moving a point charge Q from one position to another, Eq. (3), is an example of a line integral, which in vector-analysis notation always takes the form of the integral along some prescribed path of the dot product of a vector field and a differential vector path length $d\mathbf{L}$. Without using vector analysis we should have to write

$$W = -Q \int_{\text{init}}^{\text{final}} E_L \, dL$$

where E_L = component of $\mathbf{E}$ along $d\mathbf{L}$.

A line integral is like many other integrals which appear in advanced analysis, including the surface integral appearing in Gauss's law, in that it is essentially descriptive. We like to look at it much more than we like to work it out. It tells us to choose a path, break it up into a large number of very small segments, multiply the component of the field along each segment by the length of the segment, and then add the results for all the segments. This is a summation, of course, and the integral is obtained exactly only when the number of segments becomes infinite.

This procedure is indicated in Fig. 4.1, where a path has been chosen from an initial position B to a final position[1] A and a *uniform electric field* selected for simplicity. The path is divided into six segments, $\Delta\mathbf{L}_1, \Delta\mathbf{L}_2, \ldots, \Delta\mathbf{L}_6$, and the components of $\mathbf{E}$ along each segment denoted by $E_{L1}, E_{L2}, \ldots, E_{L6}$. The work involved

[1] The final position is given the designation A to correspond with the convention for potential difference, as discussed in the following section.

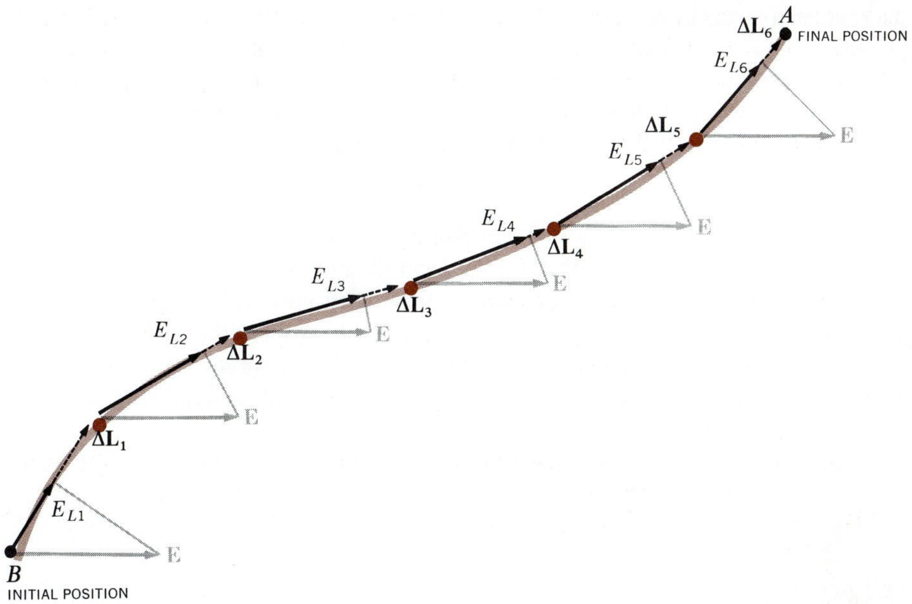

FIGURE 4.1
A graphical interpretation of a line integral in a uniform field. The line integral of **E** between points B and A is independent of the path selected, even in a nonuniform field; this result is not, in general, true for time-varying fields.

in moving a charge Q from B to A is then approximately

$$W = -Q(E_{L1}\,\Delta L_1 + E_{L2}\,\Delta L_2 + \cdots + E_{L6}\,\Delta L_6)$$

or, using vector notation,

$$W = -Q(\mathbf{E}_1 \cdot \Delta\mathbf{L}_1 + \mathbf{E}_2 \cdot \Delta\mathbf{L}_2 + \cdots + \mathbf{E}_6 \cdot \Delta\mathbf{L}_6)$$

and since we have assumed a uniform field,

$$\mathbf{E}_1 = \mathbf{E}_2 = \cdots = \mathbf{E}_6$$

$$W = -Q\mathbf{E} \cdot (\Delta\mathbf{L}_1 + \Delta\mathbf{L}_2 + \cdots + \Delta\mathbf{L}_6)$$

What is this sum of vector segments in the parentheses above? Vectors add by the parallelogram law, and the sum is just the vector directed from the initial point B to the final point A, $\mathbf{L}_{BA}$. Therefore

$$W = -Q\mathbf{E} \cdot \mathbf{L}_{BA} \qquad \text{(uniform } \mathbf{E}\text{)} \tag{4}$$

Remembering the summation interpretation of the line integral, this result for the uniform field can be obtained rapidly now from the integral expression

$$W = -Q \int_B^A \mathbf{E} \cdot d\mathbf{L} \tag{5}$$

as applied to a uniform field

$$W = -Q\mathbf{E} \cdot \int_B^A d\mathbf{L}$$

where the last integral becomes $\mathbf{L}_{BA}$ and

$$W = -Q\mathbf{E} \cdot \mathbf{L}_{BA} \qquad \text{(uniform } \mathbf{E}\text{)}$$

For this special case of a uniform electric field intensity, we should note that the work involved in moving the charge depends only on Q, $\mathbf{E}$, and $\mathbf{L}_{BA}$, a vector drawn from the initial to the final point of the path chosen. It does not depend on the particular path we have selected along which to carry the charge. We may proceed from B to A on a straight line or via the Old Chisholm Trail; the answer is the same. We shall show in Sec. 4.5 that an identical statement may be made for any nonuniform (static) $\mathbf{E}$ field.

In order to illustrate the mechanics of setting up the line integral (5) let us select the nonuniform field

$$\mathbf{E} = y\mathbf{a}_x + x\mathbf{a}_y + 2\mathbf{a}_z$$

and determine the work expended in carrying 2 C from $B(1,0,1)$ to $A(0.8,0.6,1)$ along the shorter arc of the circle

$$x^2 + y^2 = 1 \qquad z = 1$$

Working in cartesian coordinates, the differential path $d\mathbf{L}$ is $dx\,\mathbf{a}_x + dy\,\mathbf{a}_y + dz\,\mathbf{a}_z$, and the integral becomes

$$\begin{aligned} W &= -Q \int_B^A \mathbf{E} \cdot d\mathbf{L} \\ &= -2 \int_B^A (y\mathbf{a}_x + x\mathbf{a}_y + 2\mathbf{a}_z) \cdot (dx\,\mathbf{a}_x + dy\,\mathbf{a}_y + dz\,\mathbf{a}_z) \\ &= -2 \int_1^{0.8} y\,dx - 2 \int_0^{0.6} x\,dy - 4 \int_1^1 dz \end{aligned}$$

where the limits on the integrals have been chosen to agree with the initial and final values of the appropriate variable of integration. Using the equation of the circular path (and selecting the sign of the radical which is correct for the quadrant involved), we have

$$\begin{aligned} W &= -2 \int_1^{0.8} \sqrt{1 - x^2}\,dx - 2 \int_0^{0.6} \sqrt{1 - y^2}\,dy - 0 \\ &= -\left[x\sqrt{1 - x^2} + \sin^{-1} x\right]_1^{0.8} - \left[y\sqrt{1 - y^2} + \sin^{-1} y\right]_0^{0.6} \\ &= -(0.48 + 0.927 - 0 - 1.571) - (0.48 + 0.644 - 0 - 0) \\ &= -0.96 \text{ J} \end{aligned}$$

If we now select the straight-line path from B to A, then we must determine the equations of the straight line. Any two of the following three equations for planes passing through the line are sufficient to define the line:

$$y - y_B = \frac{y_A - y_B}{x_A - x_B}(x - x_B)$$

$$z - z_B = \frac{z_A - z_B}{y_A - y_B}(y - y_B)$$

$$x - x_B = \frac{x_A - x_B}{z_A - z_B}(z - z_B)$$

From the first equation above we have

$$y = -3(x - 1)$$

and from the second we obtain

$$z = 1$$

Thus,

$$\begin{aligned} W &= -2\int_1^{0.8} y\,dx - 2\int_0^{0.6} x\,dy - 4\int_1^1 dz \\ &= 6\int_1^{0.8}(x-1)\,dx - 2\int_0^{0.6}\left(1 - \frac{y}{3}\right)dy \\ &= -0.96 \text{ J} \end{aligned}$$

This is the same answer we found using the circular path between the same two points, and it again demonstrates the statement (unproved) that the work done is independent of the path taken in any electrostatic field.

It should be noted that the equations of the straight line show that $dy = -3\,dx$ and $dx = -\frac{1}{3}dy$. These substitutions may be made in the first two integrals above, along with a change in limits, and the answer may be obtained by evaluating the new integrals. This method is often simpler if the integrand is a function of only one variable.

Note that the expressions for $d\mathbf{L}$ in our three coordinate systems utilize the differential lengths obtained in the first chapter:

$$d\mathbf{L} = dx\,\mathbf{a}_x + dy\,\mathbf{a}_y + dz\,\mathbf{a}_z \qquad \text{(cartesian)} \tag{6}$$

$$d\mathbf{L} = d\rho\,\mathbf{a}_\rho + \rho\,d\phi\,\mathbf{a}_\phi + dz\,\mathbf{a}_z \qquad \text{(cylindrical)} \tag{7}$$

$$d\mathbf{L} = dr\,\mathbf{a}_r + r\,d\theta\,\mathbf{a}_\theta + r\sin\theta\,d\phi\,\mathbf{a}_\phi \qquad \text{(spherical)} \tag{8}$$

The interrelationships among the several variables in each expression are determined from the specific equations for the path.

As a final example illustrating the evaluation of the line integral, let us investigate several paths which we might take near an infinite line charge. The field has

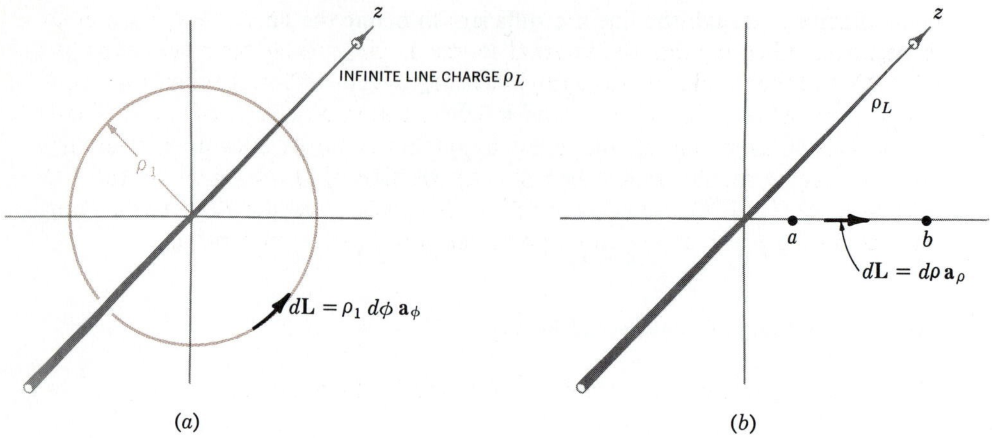

FIGURE 4.2
(*a*) A circular path and (*b*) a radial path along which a charge of Q is carried in the field of an infinite line charge. No work is expended in the former case.

been obtained several times and is entirely in the radial direction,

$$\mathbf{E} = E_\rho \mathbf{a}_\rho = \frac{\rho_L}{2\pi\epsilon_0 \rho}\mathbf{a}_\rho$$

Let us first find the work done in carrying the positive charge Q about a circular path of radius ρ_1, centered at the line charge, as illustrated in Fig. 4.2*a*. Without lifting a pencil, we see that the work must be nil, for the path is always perpendicular to the electric field intensity, or the force on the charge is always exerted at right angles to the direction in which we are moving it. For practice, however, let us set up the integral and obtain the answer.

The differential element $d\mathbf{L}$ is chosen in cylindrical coordinates, and the circular path selected demands that $d\rho$ and dz be zero, so $d\mathbf{L} = \rho_1\, d\phi\, \mathbf{a}_\phi$. The work is then

$$W = -Q\int_{\text{init}}^{\text{final}} \frac{\rho_L}{2\pi\epsilon_0 \rho_1}\mathbf{a}_\rho \cdot \rho_1\, d\phi\, \mathbf{a}_\phi$$

$$= -Q\int_0^{2\pi} \frac{\rho_L}{2\pi\epsilon_0}\, d\phi\, \mathbf{a}_\rho \cdot \mathbf{a}_\phi = 0$$

Let us now carry the charge from $\rho = a$ to $\rho = b$ along a radial path (Fig. 4.2*b*). Here $d\mathbf{L} = d\rho\, \mathbf{a}_\rho$ and

$$W = -Q\int_{\text{init}}^{\text{final}} \frac{\rho_L}{2\pi\epsilon_0 \rho}\mathbf{a}_\rho \cdot d\rho\, \mathbf{a}_\rho = -Q\int_a^b \frac{\rho_L}{2\pi\epsilon_0}\frac{d\rho}{\rho}$$

or

$$W = -\frac{Q\rho_L}{2\pi\epsilon_0}\ln\frac{b}{a}$$

Since b is larger than a, ln (b/a) is positive, and we see that the work done is negative, indicating that the external source moving the charge receives energy.

One of the pitfalls in evaluating line integrals is a tendency to use too many minus signs when a charge is moved in the direction of a *decreasing* coordinate value. This is taken care of completely by the limits on the integral, and no misguided attempt should be made to change the sign of $d\mathbf{L}$. Suppose we carry Q from b to a (Fig. 4.2b). We still have $d\mathbf{L} = d\rho\, \mathbf{a}_\rho$ and show the different direction by recognizing $\rho = b$ as the initial point and $\rho = a$ as the final point,

$$W = -Q \int_b^a \frac{\rho_L}{2\pi\epsilon_0} \frac{d\rho}{\rho} = \frac{Q\rho_L}{2\pi\epsilon_0} \ln \frac{b}{a}$$

This is the negative of the previous answer and is obviously correct.

D4.2. Find the work done in moving a 5-μC charge from the origin to $P(2,-1,4)$ through the field $\mathbf{E} = 2xyz\mathbf{a}_x + x^2z\mathbf{a}_y + x^2y\mathbf{a}_z$ V/m via the path: (a) straight line segments: (0,0,0) to (2,0,0) to (2,−1,0) to (2,−1,4); (b) straight line: $x = -2y$, $z = 2x$; (c) curve: $x = -2y^3$, $z = 4y^2$.

Ans. 80 μJ; 80 μJ; 80 μJ

D4.3. We shall see later that a time-varying **E** field need not be conservative. Let $\mathbf{E} = x\mathbf{a}_y$ V/m at a certain instant of time, and calculate the work required to move a 3-C point charge from (1,3,5) to (2,0,3) along the straight-line segments joining: (a) (1,3,5) to (2,3,5) to (2,0,5) to (2,0,3); (b) (1,3,5) to (1,3,3) to (1,0,3) to (2,0,3).

Ans. 18 J; 9 J

4.3 DEFINITION OF POTENTIAL DIFFERENCE AND POTENTIAL

We are now ready to define a new concept from the expression for the work done by an external source in moving a charge Q from one point to another in an electric field **E**,

$$W = -Q \int_{\text{init}}^{\text{final}} \mathbf{E} \cdot d\mathbf{L}$$

In much the same way as we defined the electric field intensity as the force on a *unit* test charge, we now define *potential difference* V as the work done (by an external source) in moving a *unit* positive charge from one point to another in an electric field,

$$\text{Potential difference} = V = -\int_{\text{init}}^{\text{final}} \mathbf{E} \cdot d\mathbf{L} \tag{9}$$

We shall have to agree on the direction of movement, as implied by our language, and we do this by stating that V_{AB} signifies the potential difference between points A and B and is the work done in moving the unit charge from B (last named) to A (first named). Thus, in determining V_{AB}, B is the initial point and A is the final point. The reason for this somewhat peculiar definition will become clearer shortly, when it is seen that the initial point B is often taken at infinity, whereas the final point A represents the fixed position of the charge; point A is thus inherently more significant.

Potential difference is measured in joules per coulomb, for which the *volt* is defined as a more common unit, abbreviated as V. Hence the potential difference between points A and B is

$$\boxed{V_{AB} = -\int_B^A \mathbf{E} \cdot d\mathbf{L} \qquad \text{V}} \tag{10}$$

and V_{AB} is positive if work is done in carrying the positive charge from B to A.

From the line-charge example of the last section we found that the work done in taking a charge Q from $\rho = b$ to $\rho = a$ was

$$W = \frac{Q\rho_L}{2\pi\epsilon_0} \ln \frac{b}{a}$$

Thus, the potential difference between points at $\rho = a$ and $\rho = b$ is

$$V_{ab} = \frac{W}{Q} = \frac{\rho_L}{2\pi\epsilon_0} \ln \frac{b}{a} \tag{11}$$

We can try out this definition by finding the potential difference between points A and B at radial distances r_A and r_B from a point charge Q. Choosing an origin at Q,

$$\mathbf{E} = E_r \mathbf{a}_r = \frac{Q}{4\pi\epsilon_0 r^2} \mathbf{a}_r$$

and

$$d\mathbf{L} = dr\, \mathbf{a}_r$$

we have

$$V_{AB} = -\int_B^A \mathbf{E} \cdot d\mathbf{L} = -\int_{r_B}^{r_A} \frac{Q}{4\pi\epsilon_0 r^2}\, dr = \frac{Q}{4\pi\epsilon_0}\left(\frac{1}{r_A} - \frac{1}{r_B}\right) \tag{12}$$

If $r_B > r_A$, the potential difference V_{AB} is positive, indicating that energy is expended by the external source in bringing the positive charge from r_B to r_A. This agrees with the physical picture showing the two like charges repelling each other.

It is often convenient to speak of the *potential*, or *absolute potential*, of a point, rather than the potential difference between two points, but this means only that we agree to measure every potential difference with respect to a specified reference point which we consider to have zero potential. Common agreement must be reached on the zero reference before a statement of the potential has any significance. A person having one hand on the deflection plates of a cathode-ray tube which are "at a potential of 50 V" and the other hand on the cathode terminal would probably be too shaken up to understand that the cathode is not the zero reference, but that all potentials in that circuit are customarily measured with respect to the metallic shield about the tube. The cathode may be several thousands of volts negative with respect to the shield.

Perhaps the most universal zero reference point in experimental or physical potential measurements is "ground," by which we mean the potential of the surface region of the earth itself. Theoretically, we usually represent this surface by an infinite plane at zero potential, although some large-scale problems, such as those involving propagation across the Atlantic Ocean, require a spherical surface at zero potential.

Another widely used reference "point" is infinity. This usually appears in theoretical problems approximating a physical situation in which the earth is relatively far removed from the region in which we are interested, such as the static field near the wing tip of an airplane which has acquired a charge in flying through a thunderhead, or the field inside an atom. Working with the *gravitational* potential field on earth, the zero reference is normally taken at sea level; for an interplanetary mission, however, the zero reference is more conveniently selected at infinity.

A cylindrical surface of some definite radius may occasionally be used as a zero reference when cylindrical symmetry is present and infinity proves inconvenient. In a coaxial cable the outer conductor is selected as the zero reference for potential. And, of course, there are numerous special problems, such as those for which a two-sheeted hyperboloid or an oblate spheroid must be selected as the zero-potential reference, but these need not concern us immediately.

If the potential at point A is V_A and that at B is V_B, then

$$\boxed{V_{AB} = V_A - V_B} \tag{13}$$

where we necessarily agree that V_A and V_B shall have the same zero reference point.

D4.4. Let $\mathbf{E} = (-6y/x^2)\mathbf{a}_x + (6/x)\mathbf{a}_y + 5\mathbf{a}_z$ V/m and calculate: (*a*) V_{PQ} given $P(-7,2,1)$ and $Q(4,1,2)$; (*b*) V_P if $V = 0$ at Q; (*c*) V_P if $V = 0$ at $(2,0,-1)$.

Ans. 8.21 V; 8.21 V; −8.29 V

4.4 THE POTENTIAL FIELD OF A POINT CHARGE

In the previous section we found an expression (12) for the potential difference between two points located at $r = r_A$ and $r = r_B$ in the field of a point charge Q placed at the origin,

$$V_{AB} = \frac{Q}{4\pi\epsilon_0}\left(\frac{1}{r_A} - \frac{1}{r_B}\right) = V_A - V_B \tag{14}$$

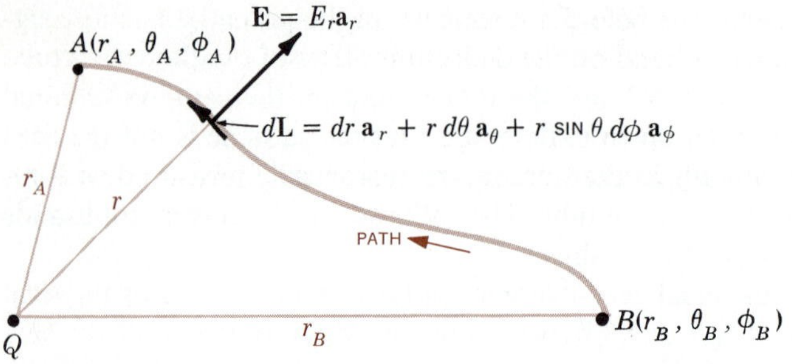

FIGURE 4.3
A general path between general points B and A in the field of a point charge Q at the origin. The potential difference V_{AB} is independent of the path selected.

It was assumed there that the two points lay on the same radial line or had the same θ and ϕ coordinate values, allowing us to set up a simple path on this radial line along which to carry our positive charge. We now should ask whether different θ and ϕ coordinate values for the initial and final position will affect our answer and whether we could choose more complicated paths between the two points without changing the results. Let us answer both questions at once by choosing two general points A and B (Fig. 4.3) at radial distances of r_A and r_B, and any values for the other coordinates.

The differential path length $d\mathbf{L}$ has r, θ, and ϕ components, and the electric field has only a radial component. Taking the dot product then leaves us only

$$V_{AB} = -\int_{r_B}^{r_A} E_r\,dr = -\int_{r_B}^{r_A} \frac{Q}{4\pi\epsilon_0 r^2}\,dr = \frac{Q}{4\pi\epsilon_0}\left(\frac{1}{r_A} - \frac{1}{r_B}\right)$$

We obtain the same answer and see, therefore, that the potential difference between two points in the field of a point charge depends only on the distance of each point from the charge and does not depend on the particular path used to carry our unit charge from one point to the other.

How might we conveniently define a zero reference for potential? The simplest possibility is to let $V = 0$ at infinity. If we let the point at $r = r_B$ recede to infinity the potential at r_A becomes

$$V_A = \frac{Q}{4\pi\epsilon_0 r_A}$$

or, since there is no reason to identify this point with the A subscript,

$$\boxed{V = \frac{Q}{4\pi\epsilon_0 r}} \qquad (15)$$

This expression defines the potential at any point distant r from a point charge Q at the origin, the potential at infinite radius being taken as the zero reference. Returning to a physical interpretation, we may say that $Q/4\pi\epsilon_0 r$ joules of work must be done in carrying a 1-C charge from infinity to any point r meters from the charge Q.

A convenient method to express the potential without selecting a specific zero reference entails identifying r_A as r once again and letting $Q/4\pi\epsilon_0 r_B$ be a constant. Then

$$V = \frac{Q}{4\pi\epsilon_0 r} + C_1 \tag{16}$$

and C_1 may be selected so that $V = 0$ at any desired value of r. We could also select the zero reference indirectly by electing to let V be V_0 at $r = r_0$.

It should be noted that the *potential difference* between two points is not a function of C_1.

Equation (15) or (16) represents the potential field of a point charge. The potential is a scalar field and does not involve any unit vectors.

Let us now define an *equipotential surface* as a surface composed of all those points having the same value of potential. No work is involved in moving a unit charge around on an equipotential surface, for, by definition, there is no potential difference between any two points on this surface.

The equipotential surfaces in the potential field of a point charge are spheres centered at the point charge.

An inspection of the form of the potential field of a point charge shows that it is an inverse-distance field, whereas the electric field intensity was found to be an inverse-square-law relationship. A similar result occurs for the gravitational force field of a point mass (inverse-square law) and the gravitational potential field (inverse distance). The gravitational force exerted by the earth on an object one million miles from it is four times that exerted on the same object two million miles away. The kinetic energy given to a freely falling object starting from the end of the universe with zero velocity, however, is only twice as much at one million miles as it is at two million miles.

D4.5. A point charge of 6 nC is located at the origin in free space. Find V_P if point P is located at $P(0.2, -0.4, 0.4)$ and: (*a*) $V = 0$ at infinity; (*b*) $V = 0$ at $(1,0,0)$; (*c*) $V = 20$ V at $(-0.5, 1, -1)$.

Ans. 89.9 V; 36.0 V; 73.9 V

4.5 THE POTENTIAL FIELD OF A SYSTEM OF CHARGES: CONSERVATIVE PROPERTY

The potential at a point has been defined as the work done in bringing a unit positive charge from the zero reference to the point, and we have suspected that this work, and hence the potential, is independent of the path taken. If it were not, potential would not be a very useful concept.

Let us now prove our assertion. We shall do so by beginning with the potential field of the single point charge for which we showed, in the last section, the independence with regard to the path, noting that the field is linear with respect to charge so that superposition is applicable. It will then follow that the potential of a system of charges has a value at any point which is independent of the path taken in carrying the test charge to that point.

Thus the potential field of a single point charge, which we shall identify as Q_1 and locate at $\mathbf{r}_1$, involves only the distance $|\mathbf{r} - \mathbf{r}_1|$ from Q_1 to the point at $\mathbf{r}$ where we are establishing the value of the potential. For a zero reference at infinity, we have

$$V(\mathbf{r}) = \frac{Q_1}{4\pi\epsilon_0|\mathbf{r} - \mathbf{r}_1|}$$

The potential due to two charges, Q_1 at $\mathbf{r}_1$ and Q_2 at $\mathbf{r}_2$, is a function only of $|\mathbf{r} - \mathbf{r}_1|$ and $|\mathbf{r} - \mathbf{r}_2|$, the distances from Q_1 and Q_2 to the field point, respectively,

$$V(\mathbf{r}) = \frac{Q_1}{4\pi\epsilon_0|\mathbf{r} - \mathbf{r}_1|} + \frac{Q_2}{4\pi\epsilon_0|\mathbf{r} - \mathbf{r}_2|}$$

Continuing to add charges, we find that the potential due to n point charges is

$$V(\mathbf{r}) = \frac{Q_1}{4\pi\epsilon_0|\mathbf{r} - \mathbf{r}_1|} + \frac{Q_2}{4\pi\epsilon_0|\mathbf{r} - \mathbf{r}_2|} + \cdots + \frac{Q_n}{4\pi\epsilon_0|\mathbf{r} - \mathbf{r}_n|}$$

or

$$V(\mathbf{r}) = \sum_{m=1}^{n} \frac{Q_m}{4\pi\epsilon_0|\mathbf{r} - \mathbf{r}_m|} \tag{17}$$

If each point charge is now represented as a small element of a continuous volume charge distribution $\rho_v \, \Delta v$, then

$$V(\mathbf{r}) = \frac{\rho_v(\mathbf{r}_1)\,\Delta v_1}{4\pi\epsilon_0|\mathbf{r} - \mathbf{r}_1|} + \frac{\rho_v(\mathbf{r}_2)\,\Delta v_2}{4\pi\epsilon_0|\mathbf{r} - \mathbf{r}_2|} + \cdots + \frac{\rho_v(\mathbf{r}_n)\,\Delta v_n}{4\pi\epsilon_0|\mathbf{r} - \mathbf{r}_n|}$$

As we allow the number of elements to become infinite, we obtain the integral expression

$$V(\mathbf{r}) = \int_{\text{vol}} \frac{\rho_v(\mathbf{r}')\,dv'}{4\pi\epsilon_0|\mathbf{r} - \mathbf{r}'|} \tag{18}$$

We have come quite a distance from the potential field of the single point charge, and it might be helpful to examine (18) and refresh ourselves as to the meaning of each term. The potential $V(\mathbf{r})$ is determined with respect to a zero reference potential at infinity and is an exact measure of the work done in bringing a unit charge from infinity to the field point at $\mathbf{r}$ where we are finding the potential. The volume charge density $\rho_v(\mathbf{r}')$ and differential volume element dv' combine to represent a differential amount of charge $\rho_v(\mathbf{r}')\,dv'$ located at $\mathbf{r}'$. The distance

$|\mathbf{r}-\mathbf{r}'|$ is that distance from the source point to the field point. The integral is a multiple (volume) integral.

If the charge distribution takes the form of a line charge or a surface charge, the integration is along the line or over the surface:

$$V(\mathbf{r}) = \int \frac{\rho_L(\mathbf{r}')\,dL'}{4\pi\epsilon_0|\mathbf{r}-\mathbf{r}'|} \tag{19}$$

$$V(\mathbf{r}) = \int_S \frac{\rho_S(\mathbf{r}')\,dS'}{4\pi\epsilon_0|\mathbf{r}-\mathbf{r}'|} \tag{20}$$

The most general expression for potential is obtained by combining (17), (18), (19), and (20).

These integral expressions for potential in terms of the charge distribution should be compared with similar expressions for the electric field intensity, such as (18) in Sec. 2.3:

$$\mathbf{E}(\mathbf{r}) = \int_{\text{vol}} \frac{\rho_v(\mathbf{r}')\,dv'}{4\pi\epsilon_0|\mathbf{r}-\mathbf{r}'|^2}\,\frac{\mathbf{r}-\mathbf{r}'}{|\mathbf{r}-\mathbf{r}'|}$$

The potential again is inverse distance, and the electric field intensity, inverse-square law. The latter, of course, is also a vector field.

To illustrate the use of one of these potential integrals, let us find V on the z axis for a uniform line charge ρ_L in the form of a ring, $\rho = a$, in the $z = 0$ plane, as shown in Fig. 4.4. Working with (19), we have $dL' = a\,d\phi'$, $\mathbf{r} = z\mathbf{a}_z$, $\mathbf{r}' = a\mathbf{a}_\rho$, $|\mathbf{r}-\mathbf{r}'| = \sqrt{a^2+z^2}$, and

$$V = \int_0^{2\pi} \frac{\rho_L a\,d\phi'}{4\pi\epsilon_0\sqrt{a^2+z^2}} = \frac{\rho_L a}{2\epsilon_0\sqrt{a^2+z^2}}$$

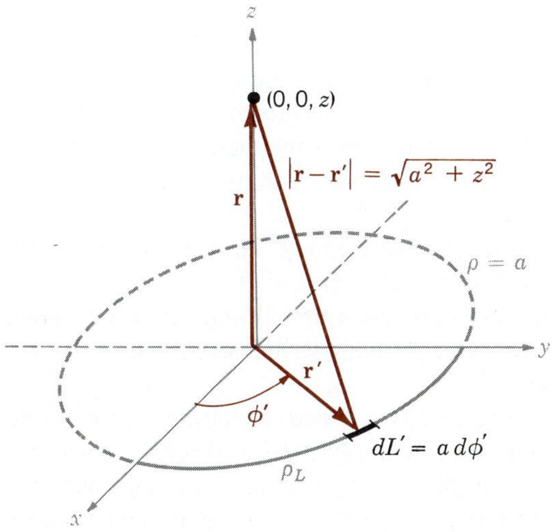

FIGURE 4.4
The potential field of a ring of uniform line charge density is easily obtained from $V = \int \rho_L(\mathbf{r}')\,dL'/(4\pi\epsilon_0|\mathbf{r}-\mathbf{r}'|)$.

For a zero reference at infinity, then:

1. The potential due to a single point charge is the work done in carrying a unit positive charge from infinity to the point at which we desire the potential, and the work is independent of the path chosen between those two points.
2. The potential field in the presence of a number of point charges is the sum of the individual potential fields arising from each charge.
3. The potential due to a number of point charges or any continuous charge distribution may therefore be found by carrying a unit charge from infinity to the point in question along any path we choose.

In other words, the expression for potential (zero reference at infinity),

$$V_A = -\int_{\infty}^{A} \mathbf{E} \cdot d\mathbf{L}$$

or potential difference,

$$V_{AB} = V_A - V_B = -\int_{B}^{A} \mathbf{E} \cdot d\mathbf{L}$$

is not dependent on the path chosen for the line integral, regardless of the source of the **E** field.

This result is often stated concisely by recognizing that no work is done in carrying the unit charge around any *closed path*, or

$$\boxed{\oint \mathbf{E} \cdot d\mathbf{L} = 0} \tag{21}$$

A small circle is placed on the integral sign to indicate the closed nature of the path. This symbol also appeared in the formulation of Gauss's law, where a closed *surface* integral was used.

Equation (21) is true for *static* fields, but we shall see in Chap. 10 that Faraday demonstrated it was incomplete when time-varying magnetic fields were present. One of Maxwell's greatest contributions to electromagnetic theory was in showing that a time-varying electric field produces a magnetic field, and therefore we should expect to find later that (21) is not correct when either **E** or **H** varies with time.

Restricting our attention to the static case where **E** does not change with time, consider the dc circuit shown in Fig. 4.5. Two points, A and B, are marked, and (21) states that no work is involved in carrying a unit charge from A through R_2 and R_3 to B and back to A through R_1, or that the sum of the potential differences around any closed path is zero.

Equation (21) is therefore just a more general form of Kirchhoff's circuital law for voltages, more general in that we can apply it to any region where an electric field exists and we are not restricted to a conventional circuit composed of wires, resistances and batteries. Equation (21) will have to be amended before

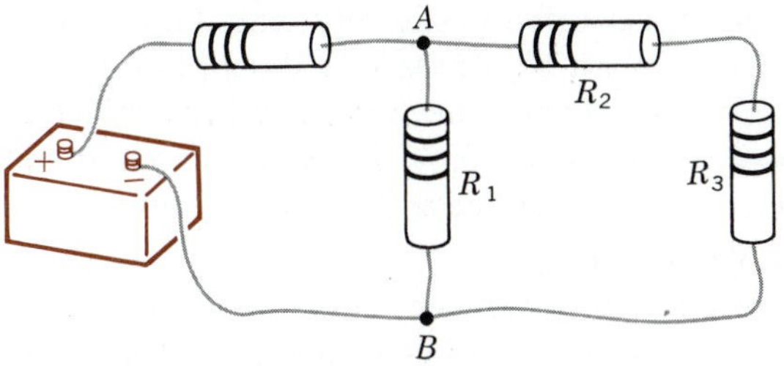

FIGURE 4.5
A simple dc-circuit problem which must be solved by applying $\oint \mathbf{E} \cdot d\mathbf{L} = 0$ in the form of Kirchhoff's voltage law.

we can apply it to time-varying fields. We shall take care of this in Chap. 10, and in Chap. 13 we will then be able to establish the general form of Kirchhoff's voltage law for circuits in which currents and voltages vary with time.

Any field that satisfies an equation of the form of (21), i.e., where the closed line integral of the field is zero, is said to be a *conservative field*. The name arises from the fact that no work is done (or that energy is *conserved*) around a closed path. The gravitational field is also conservative, for any energy expended in moving (raising) an object against the field is recovered exactly when the object is returned (lowered) to its original position. A nonconservative gravitational field could solve our energy problems forever.

Given a *nonconservative* field, it is of course possible that the line integral may be zero for certain closed paths. For example, consider the force field, $\mathbf{F} = \sin \pi\rho \, \mathbf{a}_\phi$. Around a circular path of radius $\rho = \rho_1$, we have $d\mathbf{L} = \rho_1 \, d\phi \, \mathbf{a}_\phi$, and

$$\oint \mathbf{F} \cdot d\mathbf{L} = \int_0^{2\pi} \sin \pi\rho_1 \, \mathbf{a}_\phi \cdot \rho_1 \, d\phi \, \mathbf{a}_\phi = \int_0^{2\pi} \rho_1 \sin \pi\rho_1 \, d\phi$$
$$= 2\pi\rho_1 \sin \pi\rho_1$$

The integral is zero if $\rho_1 = 1, 2, 3, \ldots$, etc., but it is not zero for other values of ρ_1, or for most other closed paths, and the given field is not conservative. A conservative field must yield a zero value for the line integral around every possible closed path.

D4.6. Assume a zero reference at infinity, and find the potential at $P(0,0,10)$ that is caused by this charge configuration in free space: (*a*) 20 nC at the origin; (*b*) 10 nC/m along the line $x = 0$, $z = 0$, $-1 < y < 1$; (*c*) 10 nC/m along the line $x = 0$, $y = 0$, $-1 < z < 1$.

Ans. 17.98 V; 17.95 V; 18.04 V.

4.6 POTENTIAL GRADIENT

We now have two methods of determining potential, one directly from the electric field intensity by means of a line integral, and another from the basic charge distribution itself by a volume integral. Neither method is very helpful in determining the fields in most practical problems, however, for as we shall see later, neither the electric field intensity nor the charge distribution is very often known. Preliminary information is much more more apt to consist of a description of two equipotential surfaces, such as the statement that we have two parallel conductors of circular cross section at potentials of 100 and -100 V. Perhaps we wish to find the capacitance between the conductors, or the charge and current distribution on the conductors from which losses may be calculated.

These quantities may be easily obtained from the potential field, and our immediate goal will be a simple method of finding the electric field intensity from the potential.

We already have the general line-integral relationship between these quantities,

$$\boxed{V = -\int \mathbf{E} \cdot d\mathbf{L}} \tag{22}$$

but this is much easier to use in the reverse direction: given $\mathbf{E}$, find V.

However, (22) may be applied to a very short element of length $\Delta\mathbf{L}$ along which $\mathbf{E}$ is essentially constant, leading to an incremental potential difference ΔV,

$$\Delta V \doteq -\mathbf{E} \cdot \Delta\mathbf{L} \tag{23}$$

Let us see first if we can determine any new information about the relation of V to $\mathbf{E}$ from this equation. Consider a general region of space, as shown in Fig. 4.6, in which $\mathbf{E}$ and V both change as we move from point to point. Equation (23) tells us to choose an incremental vector element of length $\Delta\mathbf{L} = \Delta L\, \mathbf{a}_L$ and multiply its magnitude by the component of $\mathbf{E}$ in the direction of $\mathbf{a}_L$ (one interpretation of the dot product) to obtain the small potential difference between the final and initial points of $\Delta\mathbf{L}$.

If we designate the angle between $\Delta\mathbf{L}$ and $\mathbf{E}$ as θ, then

$$\Delta V \doteq -E\,\Delta L \cos\theta$$

We now wish to pass to the limit and consider the derivative dV/dL. In order to do this, we need to show that V may be interpreted as a *function* $V(x,y,z)$. So far, V is merely the result of the line integral (22). If we assume a specified starting

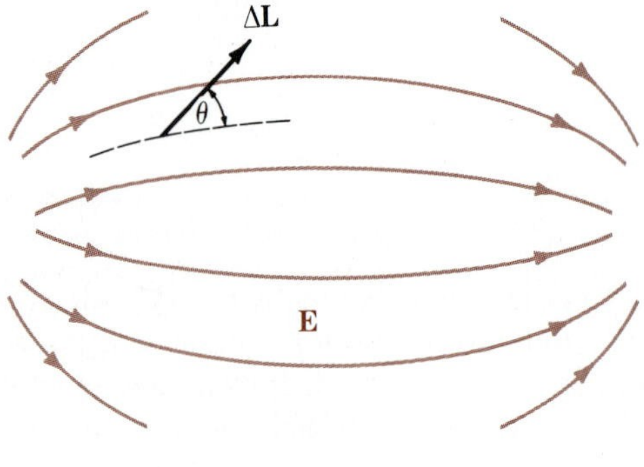

FIGURE 4.6
A vector incremental element of length $\Delta\mathbf{L}$ is shown making an angle of θ with an $\mathbf{E}$ field, indicated by its steamlines. The sources of the field are not shown.

point or zero reference and then let our end point be (x,y,z), we know that the result of the integration is a unique function of the end point (x,y,z) because **E** is a conservative field. Therefore V is a single-valued function $V(x,y,z)$. We may then pass to the limit and obtain

$$\frac{dV}{dL} = -E\cos\theta$$

In which direction should $\Delta\mathbf{L}$ be placed to obtain a maximum value of ΔV? Remember that **E** is a definite value at the point at which we are working and is independent of the direction of $\Delta\mathbf{L}$. The magnitude ΔL is also constant, and our variable is $\mathbf{a}_L$, the unit vector showing the direction of $\Delta\mathbf{L}$. It is obvious that the maximum positive increment of potential, ΔV_{max}, will occur when $\cos\theta$ is -1, or $\Delta\mathbf{L}$ points in the direction *opposite* to **E**. For this condition,

$$\left.\frac{dV}{dL}\right|_{max} = E$$

This little exercise shows us two characteristics of the relationship between **E** and V at any point:

1. The magnitude of the electric field intensity is given by the maximum value of the rate of change of potential with distance.
2. This maximum value is obtained when the direction of the distance increment is opposite to **E** or, in other words, the direction of **E** is *opposite* to the direction in which the potential is *increasing* the most rapidly.

Let us now illustrate these relationships in terms of potential. Figure 4.7 is intended to show the information we have been given about some potential field. It does this by showing the equipotential surfaces (shown as lines in the two-dimensional sketch). We desire information about the electric field intensity at point P. Starting at P, we lay off a small incremental distance $\Delta\mathbf{L}$ in various directions, hunting for that direction in which the potential is changing (increasing) the most

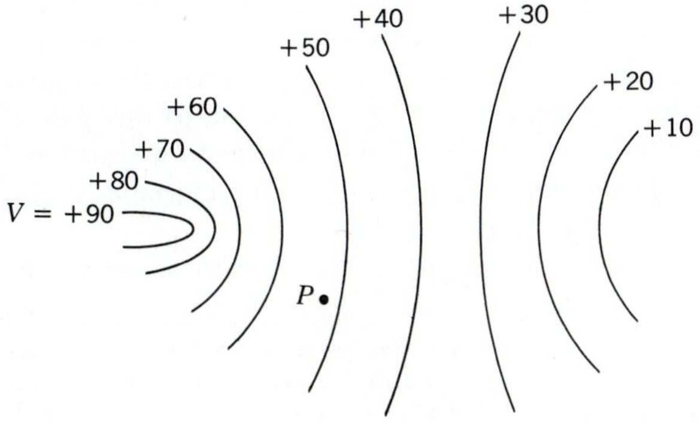

FIGURE 4.7
A potential field is shown by its equipotential surfaces. At any point the **E** field is normal to the equipotential surface passing through that point and is directed toward the more negative surfaces.

rapidly. From the sketch, this direction appears to be to the left and slightly upward. From our second characteristic above, the electric field intensity is therefore oppositely directed, or to the right and slightly downward at P. Its magnitude is given by dividing the small increase in potential by the small element of length.

It seems likely that the direction in which the potential is increasing the most rapidly is perpendicular to the equipotentials (in the direction of *increasing* potential), and this is correct, for if $\Delta\mathbf{L}$ is directed along an equipotential, $\Delta V = 0$ by our definition of an equipotential surface. But then

$$\Delta V = -\mathbf{E} \cdot \Delta\mathbf{L} = 0$$

and since neither $\mathbf{E}$ nor $\Delta\mathbf{L}$ is zero, $\mathbf{E}$ must be perpendicular to this $\Delta\mathbf{L}$ or perpendicular to the equipotentials.

Since the potential field information is more likely to be determined first, let us describe the direction of $\Delta\mathbf{L}$ which leads to a maximum increase in potential mathematically in terms of the potential field rather than the electric field intensity. We do this by letting $\mathbf{a}_N$ be a unit vector normal to the equipotential surface and directed toward the higher potentials. The electric field intensity is then expressed in terms of the potential,

$$\mathbf{E} = -\left.\frac{dV}{dL}\right|_{\max} \mathbf{a}_N \tag{24}$$

which shows that the magnitude of $\mathbf{E}$ is given by the maximum space rate of change of V and the direction of $\mathbf{E}$ is *normal* to the equipotential surface (in the direction of *decreasing* potential).

Since $dV/dL|_{\max}$ occurs when $\Delta\mathbf{L}$ is in the direction of $\mathbf{a}_N$, we may remind ourselves of this fact by letting

$$\left.\frac{dV}{dL}\right|_{\max} = \frac{dV}{dN}$$

and

$$\mathbf{E} = -\frac{dV}{dN}\mathbf{a}_N \tag{25}$$

Equation (24) or (25) serves to provide a physical interpretation of the process of finding the electric field intensity from the potential. Both are descriptive of a general procedure, and we do not intend to use them directly to obtain quantitative information. This procedure leading from V to $\mathbf{E}$ is not unique to this pair of quantities, however, but has appeared as the relationship between a scalar and a vector field in hydraulics, thermodynamics, and magnetics, and indeed in almost every field to which vector analysis has been applied.

The operation on V by which $-\mathbf{E}$ is obtained is known as the *gradient*, and the gradient of a scalar field T is defined as

$$\boxed{\text{Gradient of } T = \text{grad } T = \frac{dT}{dN}\mathbf{a}_N} \tag{26}$$

where $\mathbf{a}_N$ is a unit vector normal to the equipotential surfaces, and that normal is chosen which points in the direction of increasing values of T.

Using this new term, we now may write the relationship between V and $\mathbf{E}$ as

$$\boxed{\mathbf{E} = -\text{grad } V} \tag{27}$$

Since we have shown that V is a unique function of x, y, and z, we may take its total differential

$$dV = \frac{\partial V}{\partial x} dx + \frac{\partial V}{\partial y} dy + \frac{\partial V}{\partial z} dz$$

But we also have

$$dV = -\mathbf{E} \cdot d\mathbf{L} = -E_x dx - E_y dy - E_z dz$$

Since both expressions are true for any dx, dy, and dz, then

$$E_x = -\frac{\partial V}{\partial x}$$

$$E_y = -\frac{\partial V}{\partial y}$$

$$E_z = -\frac{\partial V}{\partial z}$$

These results may be combined vectorially to yield

$$\boxed{\mathbf{E} = -\left(\frac{\partial V}{\partial x}\mathbf{a}_x + \frac{\partial V}{\partial y}\mathbf{a}_y + \frac{\partial V}{\partial z}\mathbf{a}_z\right)} \tag{28}$$

and comparison of (27) and (28) provides us with an expression which may be used to evaluate the gradient in cartesian coordinates,

$$\boxed{\text{grad } V = \frac{\partial V}{\partial x}\mathbf{a}_x + \frac{\partial V}{\partial y}\mathbf{a}_y + \frac{\partial V}{\partial z}\mathbf{a}_z} \tag{29}$$

The gradient of a scalar is a vector, and old tests show that the unit vectors which are often incorrectly added to the divergence expression appear to be those which were incorrectly removed from the gradient. Once the physical interpretation of the gradient, expressed by (26), is grasped as showing the maximum space rate of change of a scalar quantity and *the direction in which this maximum occurs*, the vector nature of the gradient should be self-evident.

The vector operator

$$\nabla = \frac{\partial}{\partial x}\mathbf{a}_x + \frac{\partial}{\partial y}\mathbf{a}_y + \frac{\partial}{\partial z}\mathbf{a}_z$$

may be used formally as an operator on a scalar, T, ∇T, producing

$$\nabla T = \frac{\partial T}{\partial x}\mathbf{a}_x + \frac{\partial T}{\partial y}\mathbf{a}_y + \frac{\partial T}{\partial z}\mathbf{a}_z$$

from which we see that

$$\boxed{\nabla T = \text{grad } T}$$

This allows us to use a very compact expression to relate $\mathbf{E}$ and V,

$$\boxed{\mathbf{E} = -\nabla V} \tag{30}$$

The gradient may be expressed in terms of partial derivatives in other coordinate systems through application of its definition (26). These expressions are derived in Appendix A and repeated below for convenience when dealing with problems having cylindrical or spherical symmetry. They also appear inside the back cover.

$$\boxed{\nabla V = \frac{\partial V}{\partial x}\mathbf{a}_x + \frac{\partial V}{\partial y}\mathbf{a}_y + \frac{\partial V}{\partial z}\mathbf{a}_z \qquad \text{(cartesian)}} \tag{31}$$

$$\boxed{\nabla V = \frac{\partial V}{\partial \rho}\mathbf{a}_\rho + \frac{1}{\rho}\frac{\partial V}{\partial \phi}\mathbf{a}_\phi + \frac{\partial V}{\partial z}\mathbf{a}_z \qquad \text{(cylindrical)}} \tag{32}$$

$$\boxed{\nabla V = \frac{\partial V}{\partial r}\mathbf{a}_r + \frac{1}{r}\frac{\partial V}{\partial \theta}\mathbf{a}_\theta + \frac{1}{r\sin\theta}\frac{\partial V}{\partial \phi}\mathbf{a}_\phi \qquad \text{(spherical)}} \tag{33}$$

Note that the denominator of each term has the form of one of the components of $d\mathbf{L}$ in that coordinate system, except that partial differentials replace ordinary differentials; for example, $r\sin\theta\, d\phi$ becomes $r\sin\theta\, \partial\phi$.

Let us illustrate the gradient concept by inspecting the potential field, $V = 2x^2y - 5z$. We may evaluate the potential at a specific point $P(-4,3,6)$:

$$V_P = 2(-4)^2(3) - 5(6) = 66 \qquad \text{V}$$

Next, we may obtain the electric field intensity,

$$\mathbf{E} = -\nabla V = -4xy\mathbf{a}_x - 2x^2\mathbf{a}_y + 5\mathbf{a}_z \qquad \text{V/m}$$

The value of $\mathbf{E}$ at point P is

$$\mathbf{E}_P = 48\mathbf{a}_x - 32\mathbf{a}_y + 5\mathbf{a}_z \qquad \text{V/m}$$

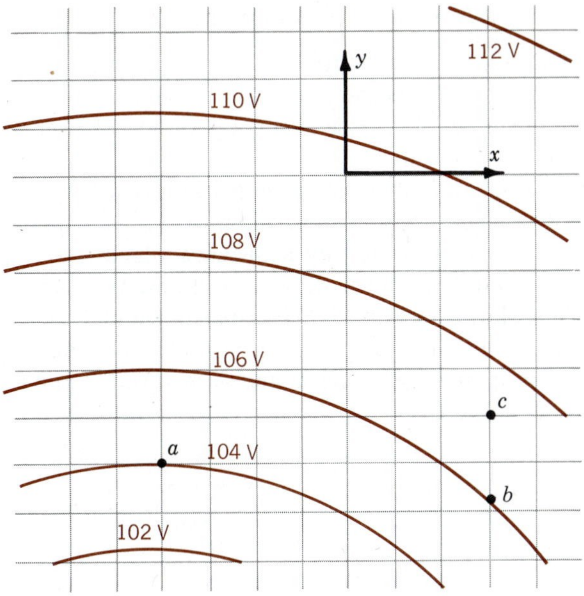

FIGURE 4.8
See Prob. D4.7.

and

$$|\mathbf{E}_P| = \sqrt{48^2 + (-32)^2 + 5^2} = 57.9 \qquad \text{V/m}$$

The direction of **E** at P is given by the unit vector

$$\mathbf{a}_{E,P} = (48\mathbf{a}_x - 32\mathbf{a}_y + 5\mathbf{a}_z)/57.9$$
$$= 0.829\mathbf{a}_x - 0.553\mathbf{a}_y + 0.086\mathbf{a}_z$$

If we assume these fields exist in free space, then

$$\mathbf{D} = \epsilon_0 \mathbf{E} = -35.4xy\mathbf{a}_x - 17.71x^2\mathbf{a}_y + 44.3\mathbf{a}_z \qquad \text{pC/m}^2$$

Finally, we may use the divergence relationship to find the volume charge density that establishes the given potential field,

$$\rho_v = \nabla \cdot \mathbf{D} = -35.4y \qquad \text{pC/m}^3$$

D4.7. A portion of a two-dimensional ($E_z = 0$) potential field is shown in Fig. 4.8. The grid lines are 1 mm apart in the actual field. Determine approximate values for **E** in cartesian coordinates at: (*a*) *a*; (*b*) *b*; (*c*) *c*.

Ans. $-1075\mathbf{a}_y$ V/m; $-600\mathbf{a}_x - 700\mathbf{a}_y$ V/m; $-500\mathbf{a}_x - 650\mathbf{a}_y$ V/m

D4.8. If $V = (60 \sin \theta)/r^2$ V in free space and point P is located at $r = 3$ m, $\theta = 60°$, $\phi = 25°$, find: (*a*) V_P; (*b*) $\mathbf{E}_P$; (*c*) dV/dN at P; (*d*) $\mathbf{a}_N$ at P; (*e*) ρ_v at P.

Ans. 5.77 V; $3.85\mathbf{a}_r - 1.111\mathbf{a}_\theta$ V/m; 4.01 V/m; $-0.961\mathbf{a}_r + 0.277\mathbf{a}_\theta$; -7.57 pC/m^3

4.7 THE DIPOLE

The dipole fields which we shall develop in this section are quite important in that they form the basis for the behavior of dielectric materials in electric fields, as discussed in part of the following chapter, as well as justifying the use of images, as described in Sec. 5.5 of the next chapter. Moreover, this development will serve to illustrate the importance of the potential concept presented in this chapter.

An *electric dipole*, or simply a *dipole*, is the name given to two point charges of equal magnitude and opposite sign, separated by a distance which is small compared to the distance to the point P at which we want to know the electric and potential fields. The dipole is shown in Fig. 4.9*a*. The distant point P is described by the spherical coordinates r, θ, and $\phi = 90°$, in view of the azimuthal symmetry. The positive and negative point charges have separation d and cartesian coordinates $(0,0,\frac{1}{2}d)$ and $(0,0,-\frac{1}{2}d)$, respectively.

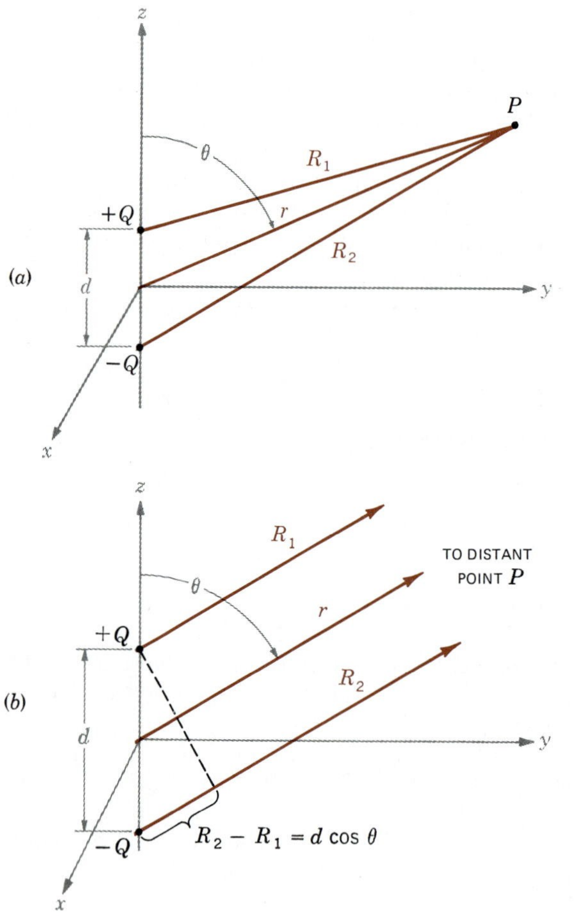

FIGURE 4.9
(*a*) The geometry of the problem of an electric dipole. The dipole moment $\mathbf{p} = Q\mathbf{d}$ is in the $\mathbf{a}_z$ direction. (*b*) For a distant point P, R_1 is essentially parallel to R_2, and we find that $R_2 - R_1 = d\cos\theta$.

So much for the geometry. What should we do next? Should we find the total electric field intensity by adding the known fields of each point charge? Would it be easier to find the total potential field first? In either case, having found one, we shall find the other from it before calling the problem solved.

If we choose to find **E** first, we shall have two components to keep track of in spherical coordinates (symmetry shows E_ϕ is zero), and then the only way to find V from **E** is by use of the line integral. This last step includes establishing a suitable zero reference for potential, since the line integral gives us only the potential difference between the two points at the ends of the integral path.

On the other hand, the determination of V first, which is a single quantity, not a vector, and has a slightly simpler expression in the case of one point charge, followed by the gradient operation to find **E**, seems like a much easier problem.

Choosing this simpler method, we let the distances from Q and $-Q$ to P be R_1 and R_2, respectively, and write the total potential as

$$V = \frac{Q}{4\pi\epsilon_0}\left(\frac{1}{R_1} - \frac{1}{R_2}\right) = \frac{Q}{4\pi\epsilon_0}\frac{R_2 - R_1}{R_1 R_2}$$

Note that the plane $z = 0$, midway between the two point charges, is the locus of points for which $R_1 = R_2$, and is therefore at zero potential, as are all points at infinity.

For a distant point, $R_1 \doteq R_2$, and the $R_1 R_2$ product in the denominator may be replaced by r^2. The approximation may not be made in the numerator, however, without obtaining the trivial answer that the potential field approaches zero as we go very far away from the dipole. Coming back a little closer to the dipole, we see from Fig. 4.9*b* that $R_2 - R_1$ may be approximated very easily if R_1 and R_2 are assumed to be parallel,

$$R_2 - R_1 \doteq d \cos \theta$$

The final result is then

$$\boxed{V = \frac{Qd \cos \theta}{4\pi\epsilon_0 r^2}} \tag{34}$$

Again we note that the plane $z = 0$ $(\theta = 90°)$ is at zero potential.

Using the gradient relationship in spherical coordinates,

$$\mathbf{E} = -\nabla V = -\left(\frac{\partial V}{\partial r}\mathbf{a}_r + \frac{1}{r}\frac{\partial V}{\partial \theta}\mathbf{a}_\theta + \frac{1}{r \sin \theta}\frac{\partial V}{\partial \phi}\mathbf{a}_\phi\right)$$

we obtain

$$\mathbf{E} = -\left(-\frac{Qd \cos \theta}{2\pi\epsilon_0 r^3}\mathbf{a}_r - \frac{Qd \sin \theta}{4\pi\epsilon_0 r^3}\mathbf{a}_\theta\right) \tag{35}$$

or

$$\mathbf{E} = \frac{Qd}{4\pi\epsilon_0 r^3}(2\cos\theta\,\mathbf{a}_r + \sin\theta\,\mathbf{a}_\theta) \tag{36}$$

These are the desired distant fields of the dipole, obtained with a very small amount of work. Any student who has several hours to spend may try to work the problem in the reverse direction—the author considers the process too long and detailed to include even for effect.

To obtain a plot of the potential field, we may choose a dipole such that $Qd/(4\pi\epsilon_0) = 1$, and then $\cos\theta = Vr^2$. The colored lines in Fig. 4.10 indicate equipotentials for which $V = 0$, ± 0.2, ± 0.4, ± 0.6, ± 0.8, and ± 1, as indicated. The dipole axis is vertical with the positive charge on the top. The streamlines for the electric field are obtained by applying the methods of Sec. 2.6 in spherical

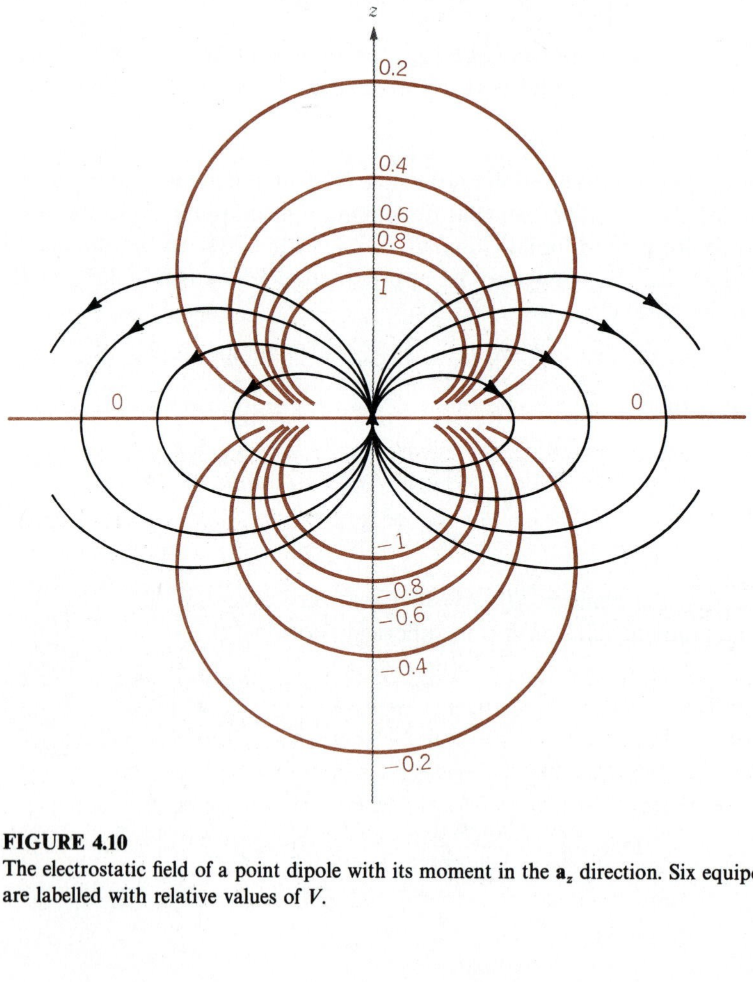

FIGURE 4.10
The electrostatic field of a point dipole with its moment in the $\mathbf{a}_z$ direction. Six equipotential surfaces are labelled with relative values of V.

coordinates,

$$\frac{E_\theta}{E_r} = \frac{r\,d\theta}{dr} = \frac{\sin\theta}{2\cos\theta}$$

or

$$\frac{dr}{r} = 2\cot\theta\,d\theta$$

from which we obtain

$$r = C_1 \sin^2\theta$$

The black streamlines shown in Fig. 4.10 are for $C_1 = 1$, 1.5, 2, and 2.5.

The potential field of the dipole, Eq. (34), may be simplified by making use of the dipole moment. Let us first identify the vector length directed from $-Q$ to $+Q$ as $\mathbf{d}$ and then define the *dipole moment* as $Q\mathbf{d}$ and assign it the symbol $\mathbf{p}$. Thus

$$\boxed{\mathbf{p} = Q\mathbf{d}} \tag{37}$$

and since $\mathbf{d} \cdot \mathbf{a}_r = d\cos\theta$, we then have

$$V = \frac{\mathbf{p} \cdot \mathbf{a}_r}{4\pi\epsilon_0 r^2} \tag{38}$$

This result may be generalized as

$$\boxed{V = \frac{1}{4\pi\epsilon_0|\mathbf{r} - \mathbf{r}'|^2}\,\mathbf{p} \cdot \frac{\mathbf{r} - \mathbf{r}'}{|\mathbf{r} - \mathbf{r}'|}} \tag{39}$$

where $\mathbf{r}$ locates the field point P, and $\mathbf{r}'$ determines the dipole center. Equation (39) is independent of any coordinate system.

The dipole moment $\mathbf{p}$ will appear again when we discuss dielectric materials. Since it is equal to the product of the charge and the separation, neither the dipole moment nor the potential will change as Q increases and $\mathbf{d}$ decreases, provided the product remains constant. The limiting case of a *point dipole* is achieved when we let $\mathbf{d}$ approach zero and Q approach infinity such that the product $\mathbf{p}$ is finite.

Turning our attention to the resultant fields, it is interesting to note that the potential field now decreases as the inverse *square* of the distance, and the electric field intensity decreases as the inverse *cube* of the distance from the dipole. Each field falls off faster than the corresponding field for the point charge, but this is no more than we should expect for the opposite charges appear to be closer together at greater distances and to act more like a single point charge of 0 C.

Symmetrical arrangements of larger numbers of point charges produce fields decreasing as the inverse of higher and higher powers of r. These charge distributions are called *multipoles*, and they are used in infinite series to approximate more unwieldy charge configurations.

D4.9. A dipole of moment $\mathbf{p} = -4\mathbf{a}_x + 5\mathbf{a}_y + 3\mathbf{a}_z$ nC·m is located at $D(1,2,-1)$ in free space. Find V at: (*a*) $P_A(0,0,0)$; (*b*) $P_B(1,2,0)$; (*c*) $P_C(1,2,-2)$; (*d*) $P_D(2,6,1)$.

Ans. -1.835 V; 27.0 V; -27.0 V; 2.05 V

D4.10. Point charges of $+3\ \mu$C and $-3\ \mu$C are located at (0,0,1 mm) and (0,0,-1 mm), respectively, in free space. (*a*) Find **p**. (*b*) Find **E** in spherical components at $P(r = 2, \theta = 40°, \phi = 50°)$. (*c*) Find **E** in spherical components at (1,2,1.5).

Ans. $6\mathbf{a}_z$ nC·m; $10.33\mathbf{a}_r + 4.33\mathbf{a}_\theta$ V/m; $3.08\mathbf{a}_r + 2.29\mathbf{a}_\theta$ V/m

4.8 ENERGY DENSITY IN THE ELECTROSTATIC FIELD

We have introduced the potential concept by considering the work done, or energy expended, in moving a point charge around in an electric field, and now we must tie up the loose ends of that discussion by tracing the energy flow one step further.

Bringing a positive charge from infinity into the field of another positive charge requires work, the work being done by the external source moving the charge. Let us imagine that the external source carries the charge up to a point near the fixed charge and then holds it there. Energy must be conserved, and the energy expended in bringing this charge into position now represents potential energy, for if the external source released its hold on the charge, it would accelerate away from the fixed charge, acquiring kinetic energy of its own and the capability of doing work.

In order to find the potential energy present in a system of charges, we must find the work done by an external source in positioning the charges.

We may start by visualizing an empty universe. Bringing a charge Q_1 from infinity to any position requires no work, for there is no field present.[2] The positioning of Q_2 at a point in the field of Q_1 requires an amount of work given by the product of the charge Q_2 and the potential at that point due to Q_1. If we represent this potential as $V_{2,1}$, where the first subscript indicates the location and the second subscript the source, that is, $V_{2,1}$ is the potential at the location of Q_2 due to Q_1, then

$$\text{Work to position } Q_2 = Q_2 V_{2,1}$$

[2] However, somebody in the workshop at infinity had to do an infinite amount of work to create the point charge in the first place! How much energy is required to bring two half-charges into coincidence to make a unit charge?

Similarly, we may express the work required to position each additional charge in the field of all those already present:

$$\text{Work to position } Q_3 = Q_3V_{3,1} + Q_3V_{3,2}$$

$$\text{Work to position } Q_4 = Q_4V_{4,1} + Q_4V_{4,2} + Q_4V_{4,3}$$

and so forth. The total work is obtained by adding each contribution:

$$\begin{aligned}\text{Total positioning work} &= \text{potential energy of field} = W_E \\ &= Q_2V_{2,1} + Q_3V_{3,1} + Q_3V_{3,2} + Q_4V_{4,1} + Q_4V_{4,2} \\ &\quad + Q_4V_{4,3} + \cdots \end{aligned} \tag{40}$$

Noting the form of a representative term in the above equation,

$$Q_3V_{3,1} = Q_3 \frac{Q_1}{4\pi\epsilon_0 R_{13}} = Q_1 \frac{Q_3}{4\pi\epsilon_0 R_{31}}$$

where R_{13} and R_{31} each represent the scalar distance between Q_1 and Q_3, we see that it might equally well have been written as $Q_1V_{1,3}$. If each term of the total energy expression is replaced by its equal, we have

$$W_E = Q_1V_{1,2} + Q_1V_{1,3} + Q_2V_{2,3} + Q_1V_{1,4} + Q_2V_{2,4} + Q_3V_{3,4} + \cdots \tag{41}$$

Adding the two energy expressions (40) and (41) gives us a chance to simplify the result a little:

$$\begin{aligned}2W_E = &\ Q_1(V_{1,2} + V_{1,3} + V_{1,4} + \cdots) \\ &+ Q_2(V_{2,1} + V_{2,3} + V_{2,4} + \cdots) \\ &+ Q_3(V_{3,1} + V_{3,2} + V_{3,4} + \cdots) \\ &+ \cdots\end{aligned}$$

because each sum of potentials in parentheses is the combined potential due to all the charges except for the charge at the point where this combined potential is being found. In other words,

$$V_{1,2} + V_{1,3} + V_{1,4} + \cdots = V_1$$

the potential at the location of Q_1 due to the presence of $Q_2, Q_3, \ldots$. We therefore have

$$W_E = \frac{1}{2}(Q_1V_1 + Q_2V_2 + Q_3V_3 + \cdots) = \frac{1}{2}\sum_{m=1}^{m=N} Q_mV_m \tag{42}$$

In order to obtain an expression for the energy stored in a region of continuous charge distribution, each charge is replaced by $\rho_v\,dv$, and the summation becomes an integral,

$$W_E = \frac{1}{2}\int_{\text{vol}} \rho_v V\,dv \tag{43}$$

Equations (42) and (43) allow us to find the total potential energy present in a system of point charges or distributed volume charge density. Similar expressions may be easily written in terms of line or surface charge density. Usually we prefer to use (43) and let it represent all the various types of charge which may have to be considered. This may always be done by considering point charges, line charge density, or surface charge density as continuous distributions of volume charge density over very small regions. We shall illustrate such a procedure with an example shortly.

Before we undertake any interpretation of this result, we should consider a few lines of more difficult vector analysis and obtain an expression equivalent to (43) but written in terms of **E** and **D**.

We begin by making the expression a little bit longer. Using Maxwell's first equation, replace ρ_v by its equal $\nabla \cdot \mathbf{D}$ and make use of a vector identity which is true for any scalar function V and any vector function **D**,

$$\nabla \cdot (V\mathbf{D}) \equiv V(\nabla \cdot \mathbf{D}) + \mathbf{D} \cdot (\nabla V) \tag{44}$$

which may be proved readily by expansion in cartesian coordinates. We then have, successively,

$$W_E = \frac{1}{2}\int_{\text{vol}} \rho_v V\,dv = \frac{1}{2}\int_{\text{vol}} (\nabla \cdot \mathbf{D})V\,dv$$

$$= \frac{1}{2}\int_{\text{vol}} [\nabla \cdot (V\mathbf{D}) - \mathbf{D} \cdot (\nabla V)]\,dv$$

Using the divergence theorem from the last chapter, the first volume integral of the last equation is changed into a closed surface integral, where the closed surface surrounds the volume considered. This volume, first appearing in (43), must contain *every* charge, and there can then be no charges outside of the volume. We may therefore consider the volume as *infinite* in extent if we wish. We have

$$W_E = \frac{1}{2}\oint_S (V\mathbf{D}) \cdot d\mathbf{S} - \frac{1}{2}\int_{\text{vol}} \mathbf{D} \cdot (\nabla V)\,dv$$

The surface integral is equal to zero, for over this closed surface surrounding the universe we see that V is approaching zero at least as rapidly as $1/r$ (the charges look like a point charge from there), **D** is approaching zero at least as rapidly as $1/r^2$, while the differential element of surface, looking more and more like a portion of a sphere, is increasing only as r^2. The integrand therefore approaches zero at least as rapidly as $1/r$. In the limit the integrand and the integral are zero. Substituting $\mathbf{E} = -\nabla V$ in the remaining volume integral, we have our answer,

$$\boxed{W_E = \frac{1}{2}\int_{\text{vol}} \mathbf{D} \cdot \mathbf{E}\,dv = \frac{1}{2}\int_{\text{vol}} \epsilon_0 E^2\,dv} \tag{45}$$

Let us now use this last expression to calculate the energy stored in the electrostatic field of a section of a coaxial cable or capacitor of length L. From the previous chapter we have

$$\mathbf{E} = \frac{a\rho_S}{\epsilon_0 \rho}\mathbf{a}_\rho$$

where ρ_S is the surface charge density on the inner conductor, whose radius is a. Thus,

$$W_E = \frac{1}{2}\int_0^L \int_0^{2\pi} \int_a^b \epsilon_0 \frac{a^2\rho_S^2}{\epsilon_0^2\rho^2}\rho\,d\rho\,d\phi\,dz = \frac{\pi L a^2 \rho_S^2}{\epsilon_0}\ln\frac{b}{a}$$

This same result may be obtained from (43). We choose the outer conductor as our zero-potential reference, and the potential of the inner cylinder is then

$$V_a = -\int_b^a E_\rho\,d\rho = -\int_b^a \frac{a\rho_S}{\epsilon_0\rho}\,d\rho = \frac{a\rho_S}{\epsilon_0}\ln\frac{b}{a}$$

The surface charge density ρ_S at $\rho = a$ can be interpreted as a volume charge density $\rho_v = \rho_S/t$, extending from $\rho = a - \frac{1}{2}t$ to $\rho = a + \frac{1}{2}t$, where $t \ll a$. The integrand in (43) is therefore zero everywhere between the cylinders (where the volume charge density is zero), as well as at the outer cylinder (where the potential is zero). The integration is therefore carried out only within the thin cylindrical shell at $\rho = a$,

$$W_E = \frac{1}{2}\int_{\text{vol}} \rho_v V\,dV = \frac{1}{2}\int_0^L \int_0^{2\pi} \int_{a-t/2}^{a+t/2} \frac{\rho_S}{t}\frac{a\rho_S}{\epsilon_0}\ln\frac{b}{a}\,\rho\,d\rho\,d\phi\,dz$$

from which

$$W_E = \frac{a^2\rho_S^2\ln(b/a)}{\epsilon_0}\pi L$$

once again.

This expression takes on a more familiar form if we recognize the total charge on the inner conductor as $Q = 2\pi a L \rho_S$. Combining this with the potential difference between the cylinders, V_a, we see that

$$W_E = \tfrac{1}{2}QV_a$$

which should be familiar as the energy stored in a capacitor.

The question of where the energy is stored in an electric field has not yet been answered. Potential energy can never be pinned down precisely in terms of physical location. Someone lifts a pencil, and the pencil acquires potential energy. Is the energy stored in the molecules of the pencil, in the gravitational field between the pencil and the earth, or in some obscure place? Is the energy in a capacitor stored in the charges themselves, in the field, or where? No one can offer any proof for his or her own private opinion, and the matter of deciding may be left to the philosophers.

Electromagnetic field theory makes it easy to believe that the energy of an electric field or a charge distribution is stored in the field itself, for if we take (45), an exact and rigorously correct expression,

$$W_E = \frac{1}{2}\int_{\text{vol}} \mathbf{D}\cdot\mathbf{E}\,dv$$

and write it on a differential basis,

$$dW_E = \tfrac{1}{2}\mathbf{D}\cdot\mathbf{E}\,dv$$

or

$$\boxed{\frac{dW_E}{dv} = \frac{1}{2}\mathbf{D}\cdot\mathbf{E}} \qquad (46)$$

we obtain a quantity $\frac{1}{2}\mathbf{D}\cdot\mathbf{E}$, which has the dimensions of an energy density, or joules per cubic meter. We know that if we integrate this energy density over the entire field-containing volume, the result is truly the total energy present, but we have no more justification for saying that the energy stored in each differential volume element dv is $\frac{1}{2}\mathbf{D}\cdot\mathbf{E}\,dv$ than we have for looking at (43) and saying that the stored energy is $\frac{1}{2}\rho_v V\,dv$. The interpretation afforded by (46), however, is a convenient one, and we shall use it until proved wrong.

D4.11. Find the energy stored in free space for the region, $0 < \rho < a$, $0 < \phi < \pi$, $0 < z < 2$, given the potential field $V =$: (*a*) $V_0\rho/a$; (*b*) $V_0(\rho/a)\cos^2\phi$.

Ans. $1.571\epsilon_0 V_0^2$; $1.374\epsilon_0 V_0^2$

SUGGESTED REFERENCES

1. Attwood, S. S.: "Electric and Magnetic Fields," 3d ed., John Wiley & Sons, Inc., New York, 1949. There are a large number of well-drawn field maps of various charge distributions, including the dipole field. Vector analysis is not used.
2. Skilling, H. H.: (see Suggested References for Chap. 3). Gradient is described on pp. 19–21.
3. Thomas, G. B., Jr., and R. L. Finney: (see Suggested References for Chap. 1). The directional derivative and the gradient are presented on pp. 823–830.

PROBLEMS

1. A field is given in spherical coordinates at $P(r = 5, \theta = 30°, \phi = 60°)$ as $\mathbf{E} = 20\mathbf{a}_r - 30\mathbf{a}_\theta + 60\mathbf{a}_\phi$ V/m. Find the incremental work done in moving a 10-μC charge a distance of 0.8 μm in the direction: (*a*) $\mathbf{a}_r$; (*b*) $\mathbf{a}_\theta$; (*c*) $\mathbf{a}_\phi$; (*d*) of $\mathbf{E}$; (*e*) of $\mathbf{G} = 2\mathbf{a}_x + 4\mathbf{a}_y - 3\mathbf{a}_z$.

2. The vector electric field intensity in the neighborhood of points P and Q is $\mathbf{E} = -5\mathbf{a}_x + 3\mathbf{a}_y - 2\mathbf{a}_z$ kV/m. Find an approximate value for the incremental work done in moving a 5-mC charge from $P(2,-1,3)$ to: (*a*) $Q(2.01,-1,3)$; (*b*) $Q(2.01,-0.98,2.97)$.

3. If $\mathbf{E} = -50y\mathbf{a}_x - 50x\mathbf{a}_y + 30\mathbf{a}_z$ V/m, find the differential amount of work done in moving a 2-μC charge a distance of 5 μm from: (*a*) $P(1,2,3)$ toward $Q(2,4,1)$; (*b*) $Q(2,4,1)$ toward $P(1,2,3)$.

4. Find the value of $\int_{P_1}^{P_2} \mathbf{G} \cdot d\mathbf{L}$ if $\mathbf{G} = x^2\mathbf{a}_x - xyz\mathbf{a}_z$, given points $P_1(1,2,3)$ and $P_2(5,0,4)$ and a path consisting of: (*a*) the straight line segments connecting P_1 to (5,2,3) to (5,0,3) to (5,0,4); (*b*) the straight line which is the intersection of the planes $x = 5 - 2y$ and $y = 8 - 2z$.

5. Compute the line integral of **E** from $B(0,0,0)$ to $A(2,1,3)$ if $\mathbf{E} = -4xy\mathbf{a}_x - (4y + 2x^2)\mathbf{a}_y + 8z\mathbf{a}_z$. Use the path: (*a*) straight line segments: (0,0,0) to (2,0,0) to (2,1,0) to (2,1,3); (*b*) direct straight line.

6. If $\mathbf{G} = 3x^2y\mathbf{a}_x + (2x - z)\mathbf{a}_y$, evaluate $\int \mathbf{G} \cdot d\mathbf{L}$ from an initial point $P(2,1,1)$ to a final point $Q(4,3,1)$ using the path: (*a*) a straight line: $y = x - 1$, $z = 1$; (*b*) a parabola: $6y = x^2 + 2$, $z = 1$.

7. If $\mathbf{E} = -8xy\mathbf{a}_x - 4x^2\mathbf{a}_y + \mathbf{a}_z$ V/m, find the work done in carrying a 6-C charge from $M(1,8,5)$ to $N(2,18,6)$ along the path: $y = 3x^2 + z$, $z = x + 4$.

8. A point charge Q_1 is located at the origin in free space. Find the work done in carrying a charge Q_2 from: (*a*) $B(r_B,\theta_B,\phi_B)$ to $C(r_A,\theta_B,\phi_B)$ with θ and ϕ held constant; (*b*) $C(r_A,\theta_B,\phi_B)$ to $D(r_A,\theta_A,\phi_B)$ with r and ϕ held constant; (*c*) $D(r_A,\theta_A,\phi_B)$ to $A(r_A,\theta_A,\phi_A)$ with r and θ held constant.

9. Each coordinate axis carries a uniform charge density of 20 nC/m in free space. Given the points $A(1,2,3)$ and $B(6,8,10)$, find V_{AB} due: (*a*) solely to ρ_L on the x axis; (*b*) solely to ρ_L on the y axis; (*c*) solely to ρ_L on the z axis; (*d*) to all three line charges.

10. Given a point charge of $200\pi\epsilon_0$ C at $(3,-1,2)$, a line charge of $40\pi\epsilon_0$ C/m on the x axis, and a surface charge of $8\epsilon_0$ C/m^2 on the plane $x = -3$, all in free space, find the potential at $P(5,6,7)$ if $V = 0$ at $Q(0,0,1)$.

11. In free space, a line charge $\rho_L = 80$ nC/m lies along the entire z axis, while point charges of 100 nC each are located at (1,0,0) and (0,1,0). Find the potential difference V_{PQ} given $P(2,1,0)$ and $Q(3,2,5)$.

12. A uniform sheet of charge, $\rho_{S1} = 50\epsilon_0$ C/m^2, is located at $z = 1.5$ m, while $\rho_{S2} = -50\epsilon_0$ C/m^2 is at $z = -0.5$ m. (*a*) Find **E** everywhere. (*b*) Find and sketch $V(z)$ as a function of z for $-0.5 < z < 1.5$ if $V = 0$ at $z = -0.5$ m. (*c*) Repeat if $V = 0$ at $z = 0$.

13. Four identical point charges of 4 pC each are located at the corners of a square 0.5 mm on a side in free space. How much work must be done to move one charge to a point equidistant from the other three and in the same plane?

14. Point charges of 6 nC each are located at $A(1,0,0)$, $B(0,1,0)$, $C(-1,0,0)$, and $D(0,-1,0)$ in free space. Assuming a zero reference at infinity: (*a*) find V at $P(0,0,z)$, and (*b*) give its maximum value for any z. (*c*) Use the result of part (*a*) to calculate $|dV/dz|$ on the z axis, and (*d*) give its maximum value.

15. A point charge of 16 nC is located at $Q(2,3,5)$ in free space, and a uniform line charge of 5 nC/m is at the intersection of the planes $x = 2$ and $y = 4$. If the potential at the origin is 100 V, find V at $P(4,1,3)$.

16. Given these uniform surface-charge distributions: 100 pC/m^2 at $r = 2$ m, 40 pC/m^2 at $r = 3$ m, and -47.5 pC/m^2 at $r = 4$ m, (*a*) find V at $r = 5$ m if $V = 0$ at infinity. (*b*) Plot V vs. r, $1 < r < 6$ m, if $V = 0$ at infinity.

17. A nonuniform linear charge density, $\rho_L = 10/z^2$ nC/m, lies along the z axis for $|z| > 1$; $\rho_L = 0$ for $|z| < 1$. Find the potential at $P(2,3,0)$ in free space if $V(\infty) = 0$.

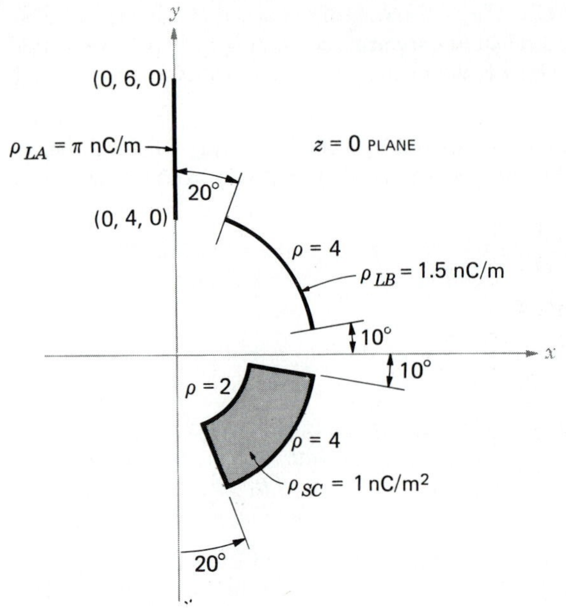

FIGURE 4.11
See Prob. 20.

18. The charge density $\rho_L = 10/|z|$ nC/m exists along the z axis for $|z| > 1$. Find V at $P(\rho,0,0)$ if $V = 0$ at $\rho = \infty$.

19. A circular disk, $\rho \leq 1$ cm, $z = 0$, carries a surface charge density $\rho_S = 1/\rho$ nC/m^2. Find V at $P(0,0,z)$ if $V = 0$ at infinity.

20. Figure 4.11 shows three separate charge distributions in the $z = 0$ plane in free space. (*a*) Find the total charge for each distribution. (*b*) Find the potential at $P(0,0,5)$ caused by each of the three charge distributions acting alone. (*c*) Find V_P.

21. Let $V = 2(x+1)^2(y+2)^2(z+3)^2$ V in free space. At $P(2,-1,4)$ find: (*a*) V; (*b*) $\mathbf{E}$; (*c*) $|\mathbf{E}|$; (*d*) $\mathbf{D}$; (*e*) $|\mathbf{D}|$; (*f*) ρ_v; (*g*) $\mathbf{a}_N$.

22. Given the potential field $V = 100\sqrt{r}$ V in free space, find: (*a*) $\mathbf{E}$; (*b*) $\mathbf{D}$. (*c*) How much charge lies within the surface $r = 0.5$?

23. For the point $P(3,60°,2)$ in cylindrical coordinates and the potential field $V = 10(\rho+1)z^2 \cos\phi$ V in free space, find at P: (*a*) V; (*b*) $\mathbf{E}$; (*c*) $\mathbf{D}$; (*d*) ρ_v; (*e*) dV/dN; (*f*) $\mathbf{a}_N$.

24. A potential field is expressed as $V = (50\rho^2 \cos\phi)/(z+1)$ V. Given a point $P(4,30°,2)$ in free space, find at P: (*a*) V; (*b*) $\mathbf{E}$; (*c*) ρ_v; (*d*) dV/dN; (*e*) $\mathbf{a}_N$.

25. Within the sphere $r = 1$, the potential is given as $V = 100 + 50r + 150r \sin\theta \sin\phi$ V. (*a*) Find V, $\mathbf{E}$, $\mathbf{D}$, and ρ_v at $r = 1$, $\theta = \pi/2$, $\phi = 0$. (*b*) How much charge lies within the sphere $r = 1$?

26. A dipole having $Qd/(4\pi\epsilon_0) = 100$ V$\cdot$m^2 is located at the origin in free space and aligned so that its moment is in the $\mathbf{a}_z$ direction. (*a*) Sketch $|V(r = 1, \theta, \phi = 0)|$ vs. θ on polar graph paper (homemade if you wish). (*b*) Sketch $|\mathbf{E}(r = 1, \theta, \phi = 0)|$ vs. θ on polar paper.

27. Point charges of 1 μC and -1 μC are located at (0,0,0.5) and (0,0,$-$0.5), respectively. Treating these two charges as a dipole at the origin, calculate: (*a*) V at $P(3,0,4)$; (*b*) $|\mathbf{E}|$ at P. Now find the exact values for: (*c*) V at P; (*d*) $|\mathbf{E}|$ at P.

28. In Fig. 4.9 let $Q = 50$ nC, $d = 0.5$ m, and the field point be $P(2,0,3)$ in cartesian coordinates. Find V_P by: (*a*) the approximate dipole relationship, Eq. (34); (*b*) exact methods.

29. A dipole having a moment $\mathbf{p} = 10\mathbf{a}_x - 3\mathbf{a}_y + 5\mathbf{a}_z$ nC·m is located at $Q(2,4,1)$ in free space. (*a*) Find V at $P(x,y,z)$. (*b*) Find V at (5,1,0).

30. A dipole at the origin in free space has $\mathbf{p} = 80\pi\epsilon_0\mathbf{a}_z$ C·m. (*a*) Give V at $P(x,y,z)$ in cartesian coordinates. (*b*) Give $\mathbf{E}$ at $P(x,y,z)$ in cartesian coordinates.

31. (*a*) Given the potential field $V = 50xyz$ V in free space, find the total energy stored within the cube $0 < x,y,z < 2$. (*b*) What value would be obtained by assuming a uniform energy density equal to the value at the center of the cube?

32. In the region of free space where $2 < \rho < 3$, $0 < \phi < \pi/2$, and $5 < z < 7$, let $\mathbf{E} = (k/\rho)\mathbf{a}_\phi$. (*a*) Find a positive value for k so that the total energy stored is 1 μJ. (*b*) Show that the surfaces $\phi = 0$ and $\phi = \pi/2$ are equipotential surfaces. (*c*) What is the potential difference between the surfaces at $\phi = 0$ and $\phi = \pi/2$?

33. A copper sphere of radius 4 cm carries a uniformly distributed total charge of 5 μC on its surface in free space. (*a*) Use Gauss's law to find $\mathbf{D}$ external to the sphere. (*b*) Calculate the total energy stored in the electrostatic field. (*c*) Use $W_E = Q^2/2C$ to calculate the capacitance of the isolated sphere.

34. A spherical surface in free space, $r = 4$ cm, contains a uniform surface charge density of 20 μC/m^2. Determine r_A if the region $0.06 < r < r_A$ contains precisely 1 mJ of energy.

35. Two concentric conducting spheres of radii $a = 6$ cm and $b = 16$ cm have equal and opposite charges, 10^{-8} C on the inner and -10^{-8} C on the outer. Assuming $\epsilon = \epsilon_0$ for the region between the spheres, find: (*a*) the maximum value of $|\mathbf{E}|$ between the spheres; (*b*) the potential difference between them; (*c*) the total energy stored.

CHAPTER 5

CONDUCTORS, DIELECTRICS, AND CAPACITANCE

In this chapter we intend to apply the laws and methods of the previous chapters to some of the materials with which an engineer must work. After defining current and current density and developing the fundamental continuity equation, we shall consider the conductor and present Ohm's law in both its microscopic and macroscopic forms. With these results we may calculate resistance values for a few of the simpler geometrical forms that resistors may assume. Conditions which must be met at conductor boundaries are obtained, and this knowledge enables us to introduce the use of images.

After a brief consideration of a general semiconductor, we shall investigate the polarization of dielectric materials and define relative permittivity, or the dielectric constant, an important engineering parameter. Having both conductors and dielectrics, we may then put them together to form capacitors. Most of the work of the previous chapters will be required to determine the capacitance of the several capacitors which we shall construct.

The fundamental electromagnetic principles on which resistors and capacitors depend are really the subject of this chapter; the inductor will not be introduced until Chap. 9.

5.1 CURRENT AND CURRENT DENSITY

Electric charges in motion constitute a *current*. The unit of current is the ampere (A), defined as a rate of movement of charge passing a given reference point (or crossing a given reference plane) of one coulomb per second. Current is symbolized by I, and therefore

$$I = \frac{dQ}{dt} \tag{1}$$

Current is thus defined as the motion of positive charges, although conduction in metals takes place through the motion of electrons, as we shall see shortly.

In field theory we are usually interested in events occurring at a point rather than within some large region, and we shall find the concept of *current density*, measured in amperes per square meter (A/m^2), more useful. Current density is a vector[1] represented by **J**.

The increment of current ΔI crossing an incremental surface ΔS normal to the current density is

$$\Delta I = J_N \, \Delta S$$

and in the case where the current density is not perpendicular to the surface,

$$\Delta I = \mathbf{J} \cdot \Delta \mathbf{S}$$

Total current is obtained by integrating,

$$\boxed{I = \int_S \mathbf{J} \cdot d\mathbf{S}} \tag{2}$$

Current density may be related to the velocity of volume charge density at a point. Consider the element of charge $\Delta Q = \rho_v \, \Delta v = \rho_v \, \Delta S \, \Delta L$, as shown in Fig. 5.1*a*. To simplify the explanation, let us assume that the charge element is oriented with its edges parallel to the coordinate axes, and that it possesses only an x component of velocity. In the time interval Δt, the element of charge has moved a distance Δx, as indicated in Fig. 5.1*b*. We have therefore moved a charge $\Delta Q = \rho_v \, \Delta S \, \Delta x$ through a reference plane perpendicular to the direction of motion in a time increment Δt, and the resultant current is

$$\Delta I = \frac{\Delta Q}{\Delta t} = \rho_v \, \Delta S \frac{\Delta x}{\Delta t}$$

---

[1] Current is not a vector, for it is easy to visualize a problem in which a total current I in a conductor of nonuniform cross section (such as a sphere) may have a different direction at each point of a given cross section. Current in an exceedingly fine wire, or a *filamentary current*, is occasionally defined as a vector, but we usually prefer to be consistent and give the direction to the filament, or path, and not to the current.

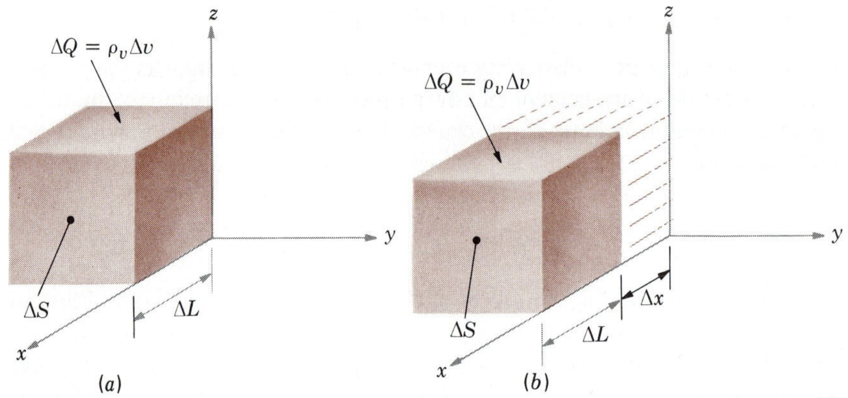

FIGURE 5.1
An increment of charge, $\Delta Q = \rho_v \Delta S \, \Delta L$, which moves a distance Δx in a time Δt, produces a component of current density in the limit of $J_x = \rho_v v_x$.

As we take the limit with respect to time, we have

$$\Delta I = \rho_v \Delta S \, v_x$$

where v_x represents the x component of the velocity $\mathbf{v}$.[2] In terms of current density, we find

$$J_x = \rho_v v_x$$

and in general

$$\boxed{\mathbf{J} = \rho_v \mathbf{v}} \tag{3}$$

This last result shows very clearly that charge in motion constitutes a current. We call this type of current a *convection current*, and $\rho_v \mathbf{v}$ is the *convection current density*. Note that the convection current density is related linearly to charge density as well as to velocity. The mass rate of flow of cars (cars per square foot per second) in the Holland Tunnel could be increased either by going to higher speeds or by raising the density of cars per cubic foot, if drivers were only capable of doing so.

D5.1. The vector current density is given as $\mathbf{J} = (4/r^2) \cos\theta \, \mathbf{a}_r + 20e^{-2r} \sin\theta \, \mathbf{a}_\theta - r \sin\theta \cos\phi \, \mathbf{a}_\phi$ A/m^2. (*a*) Find $\mathbf{J}$ at $r = 3$, $\theta = 0$, $\phi = \pi$. (*b*) Find the total current passing through the spherical cap $r = 3$, $0 < \theta < 20°$, $0 < \phi < 2\pi$, in the $\mathbf{a}_r$ direction.

Ans. $0.444\mathbf{a}_r$ A/m^2; 1.470 A

---

[2] The lowercase v is used both for volume and velocity. Note, however, that velocity always appears as a vector $\mathbf{v}$, a component v_x, or a magnitude $|\mathbf{v}|$, while volume appears only in differential form as dv or Δv.

5.2 CONTINUITY OF CURRENT

Although we are supposed to be studying static fields at this time, the introduction of the concept of current is logically followed by a discussion of the conservation of charge and the continuity equation. The principle of conservation of charge states simply that charges can be neither created nor destroyed, although equal amounts of positive and negative charge may be *simultaneously* created, obtained by separation, destroyed, or lost by recombination.

The continuity equation follows from this principle when we consider any region bounded by a closed surface. The current through the closed surface is

$$I = \oint_S \mathbf{J} \cdot d\mathbf{S}$$

and this *outward* flow of positive charge must be balanced by a decrease of positive charge (or perhaps an increase of negative charge) within the closed surface. If the charge inside the closed surface is denoted by Q_i, then the rate of decrease is $-dQ_i/dt$ and the principle of conservation of charge requires

$$I = \oint_S \mathbf{J} \cdot d\mathbf{S} = -\frac{dQ_i}{dt} \tag{4}$$

It might be well to answer here an often-asked question, "Isn't there a sign error? I thought $I = dQ/dt$." The presence or absence of a negative sign depends on what current and charge we consider. In circuit theory we usually associate the current flow *into* one terminal of a capacitor with the time rate of increase of charge on that plate. The current of (4), however, is an *outward-flowing* current.

Equation (4) is the integral form of the continuity equation, and the differential, or point, form is obtained by changing the surface integral to a volume integral by the divergence theorem,

$$\oint_S \mathbf{J} \cdot d\mathbf{S} = \int_{\text{vol}} (\nabla \cdot \mathbf{J})\, dv$$

and representing the enclosed charge Q_i by the volume integral of the charge density,

$$\int_{\text{vol}} (\nabla \cdot \mathbf{J})\, dv = -\frac{d}{dt} \int_{\text{vol}} \rho_v\, dv$$

If we agree to keep the surface constant, the derivative becomes a partial derivative and may appear within the integral,

$$\int_{\text{vol}} (\nabla \cdot \mathbf{J})\, dv = \int_{\text{vol}} -\frac{\partial \rho_v}{\partial t}\, dv$$

Since the expression is true for any volume, however small, it is true for an incremental volume,

$$(\nabla \cdot \mathbf{J})\, \Delta v = -\frac{\partial \rho_v}{\partial t}\, \Delta v$$

from which we have our point form of the continuity equation,

$$\boxed{\nabla \cdot \mathbf{J} = -\frac{\partial \rho_v}{\partial t}} \tag{5}$$

Remembering the physical interpretation of divergence, this equation indicates that the current, or charge per second, diverging from a small volume per unit volume is equal to the time rate of decrease of charge per unit volume at every point.

As a numerical example illustrating some of the concepts we have met in the last two sections, let us consider a current density that is directed radially outward,

$$\mathbf{J} = \frac{1}{r} e^{-t} \mathbf{a}_r \qquad \text{A/m}^2$$

Selecting an instant of time $t = 1$ s, we may calculate the total outward current at $r = 5$ m:

$$I = J_r S = (\tfrac{1}{5} e^{-1})(4\pi 5^2) = 23.1 \qquad \text{A}$$

At the same instant, but for a slightly larger radius, $r = 6$ m, we have

$$I = (\tfrac{1}{6} e^{-1})(4\pi 6^2) = 27.7 \qquad \text{A}$$

Thus, the total current is larger at $r = 6$ than it is at $r = 5$.

To see why this happens, we need to look at the volume charge density and the velocity. We use the continuity equation first:

$$-\frac{\partial \rho_v}{\partial t} = \nabla \cdot \mathbf{J} = \nabla \cdot \left(\frac{1}{r} e^{-t} \mathbf{a}_r \right) = \frac{1}{r^2} \frac{\partial}{\partial r} \left(r^2 \frac{1}{r} e^{-t} \right) = \frac{1}{r^2} e^{-t}$$

We next seek the volume charge density by integrating with respect to t. Since ρ_v is given by a partial derivative with respect to time, the "constant" of integration may be a function of r:

$$\rho_v = -\int \frac{1}{r^2} e^{-t} \, dt + K(r) = \frac{1}{r^2} e^{-t} + K(r)$$

If we assume that $\rho_v \to 0$ as $t \to \infty$, then $K(r) = 0$, and

$$\rho_v = \frac{1}{r^2} e^{-t} \qquad \text{C/m}^3$$

We may now use $\mathbf{J} = \rho_v \mathbf{v}$ to find the velocity,

$$v_r = \frac{J_r}{\rho_v} = \frac{\dfrac{1}{r} e^{-t}}{\dfrac{1}{r^2} e^{-t}} = r \qquad \text{m/s}$$

The velocity is greater at $r = 6$ than it is at $r = 5$, and we see that some (unspecified) force is accelerating the charge density in an outward direction.

In summary, we have a current density that is inversely proportional to r, a charge density that is inversely proportional to r^2, and a velocity and total current that are proportional to r. All quantities vary as e^{-t}.

D5.2. Assume that an electron beam carries a total current of $-500\ \mu\text{A}$ in the $\mathbf{a}_z$ direction, and has a current density J_z that is not a function of ρ or ϕ in the region $0 \le \rho < 10^{-4}$ m and is zero for $\rho > 10^{-4}$ m. If the electron velocities are given by $v_z = 8 \times 10^7 z$ m/s, calculate ρ_v at $\rho = 0$ and $z =$: (*a*) 1 mm; (*b*) 2 cm; (*c*) 1 m.

Ans. $-0.1989\ \text{C/m}^3$; $-9.95\ \text{mC/m}^3$; $-198.9\ \mu\text{C/m}^3$

5.3 METALLIC CONDUCTORS

Physicists today describe the behavior of the electrons surrounding the positive atomic nucleus in terms of the total energy of the electron with respect to a zero reference level for an electron at an infinite distance from the nucleus. The total energy is the sum of the kinetic and potential energies, and since energy must be given to an electron to pull it away from the nucleus, the energy of every electron in the atom is a negative quantity. Even though the picture has some limitations, it is convenient to associate these energy values with orbits surrounding the nucleus, the more negative energies corresponding to orbits of smaller radius. According to the quantum theory, only certain discrete energy levels, or energy states, are permissible in a given atom, and an electron must therefore absorb or emit discrete amounts of energy, or quanta, in passing from one level to another. A normal atom at absolute zero temperature has an electron occupying every one of the lower energy shells, starting outward from the nucleus and continuing until the supply of electrons is exhausted.

In a crystalline solid, such as a metal or a diamond, atoms are packed closely together, many more electrons are present, and many more permissible energy levels are available because of the interaction forces between adjacent atoms. We find that the energies which may be possessed by electrons are grouped into broad ranges, or "bands," each band consisting of very numerous, closely spaced, discrete levels. At a temperature of absolute zero, the normal solid also has every level occupied, starting with the lowest and proceeding in order until all the electrons are located. The electrons with the highest (least negative) energy levels, the valence electrons, are located in the *valence band*. If there are permissible higher-energy levels in the valence band, or if the valence band merges smoothly into a *conduction band*, then additional kinetic energy may be given to the valence electrons by an external field, resulting in an electron flow. The solid is called a *metallic conductor*. The filled valence band and the unfilled conduction band are shown for the conductor at 0 K in Fig. 5.2*a*.

If, however, the electron with the greatest energy occupies the top level in the valence band and a gap exists between the valence band and the conduction

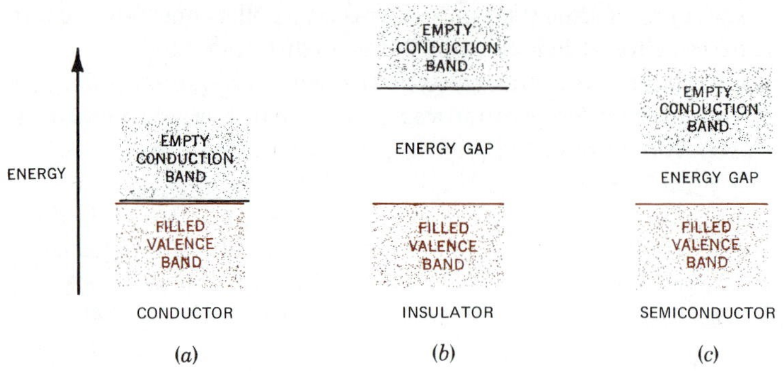

FIGURE 5.2
The energy-band structure in three different types of materials at 0 K. (*a*) The conductor exhibits no energy gap between the valence and conduction bands. (*b*) The insulator shows a large energy gap. (*c*) The semiconductor has only a small energy gap.

band, then the electron cannot accept additional energy in small amounts, and the material is an insulator. This band structure is indicated in Fig 5.2*b*. Note that if a relatively large amount of energy can be transferred to the electron, it may be sufficiently excited to jump the gap into the next band where conduction can occur easily. Here the insulator breaks down.

An intermediate condition occurs when only a small "forbidden region" separates the two bands, as indicated in Fig. 5.2*c*. Small amounts of energy in the form of heat, light, or an electric field may raise the energy of the electrons at the top of the filled band and provide the basis for conduction. These materials are insulators which display many of the properties of conductors and are called *semiconductors*.

Let us first consider the conductor. Here the valence electrons, or *conduction*, or *free*, electrons, move under the influence of an electric field. With a field **E**, an electron having a charge $Q = -e$ will experience a force

$$\mathbf{F} = -e\mathbf{E}$$

In free space the electron would accelerate and continually increase its velocity (and energy); in the crystalline material the progress of the electron is impeded by continual collisions with the thermally excited crystalline lattice structure, and a constant average velocity is soon attained. This velocity $\mathbf{v}_d$ is termed the *drift velocity*, and it is linearly related to the electric field intensity by the *mobility* of the electron in the given material. We designate mobility by the symbol μ (mu), and

$$\boxed{\mathbf{v}_d = -\mu_e \mathbf{E}} \qquad (6)$$

where μ_e is the mobility of an electron and is positive by definition. Note that the electron velocity is in a direction opposite to the direction of **E**. Equation (6) also shows that mobility is measured in the units of square meters per volt-second; typical values[3] are 0.0012 for aluminum, 0.0032 for copper, and 0.0056 for silver.

For these good conductors a drift velocity of a few inches per second is sufficient to produce a noticeable temperature rise and can cause the wire to melt if the heat cannot be quickly removed by thermal conduction or radiation.

Substituting (6) into Eq. (3) of Sec. 5.1, we obtain

$$\boxed{\mathbf{J} = -\rho_e \mu_e \mathbf{E}} \tag{7}$$

where ρ_e is the free-electron charge density, a negative value. The total charge density ρ_v is zero, since equal positive and negative charge is present in the neutral material.

The relationship between **J** and **E** for a metallic conductor, however, is also specified by the conductivity σ (sigma),

$$\boxed{\mathbf{J} = \sigma \mathbf{E}} \tag{8}$$

where σ is measured in mhos per meter (℧/m). A mho is one ampere per volt, and it (and the ohm) honors Georg Simon Ohm, a German physicist who first described the current-voltage relationship implied by (8). We call this equation the *point form of Ohm's law*; we shall look at the more common form of Ohm's law shortly.

First, however, it is informative to note the conductivity of several metallic conductors; typical values (in mhos per meter) are 3.82×10^7 for aluminum, 5.80×10^7 for copper, and 6.17×10^7 for silver. Data for other conductors may be found in Appendix C. On seeing data such as these, it is only natural to assume that we are being presented with *constant* values; this is essentially true. Metallic conductors obey Ohm's law quite faithfully, and it is a *linear* relationship; the conductivity is constant over wide ranges of current density and electric field intensity. Ohm's law and the metallic conductors are also described as *isotropic*, or having the same properties in every direction. A material which is not isotropic is called *anisotropic*, and we shall mention such a material a few pages from now.

The conductivity is a function of temperature, however. The resistivity, which is the reciprocal of the conductivity, varies almost linearly with temperature in the region of room temperature, and for aluminum, copper, and silver it increases about 0.4 percent for a 1 K rise in temperature.[4] For several metals the resistivity drops abruptly to zero at a temperature of a few kelvin; this property is termed *superconductivity*. Copper and silver are not superconductors, although aluminum is (for temperatures below 1.14 K).

---

[3] Wert and Thomson, p. 238, listed in the Suggested References at the end of this chapter.

[4] Copious temperature data for conducting materials are available in the "Standard Handbook for Electrical Engineers," listed among the Suggested References at the end of this chapter.

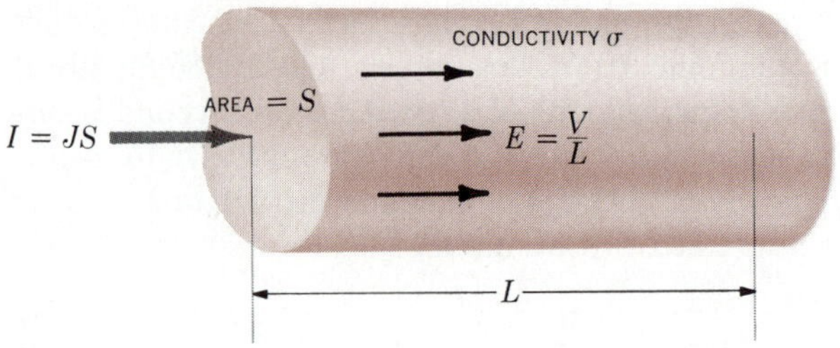

FIGURE 5.3
Uniform current density **J** and electric field intensity **E** in a cylindrical region of length L and cross-sectional area S. Here $V = IR$, where $R = L/\sigma S$.

If we now combine (7) and (8), the conductivity may be expressed in terms of the charge density and the electron mobility,

$$\boxed{\sigma = -\rho_e \mu_e} \tag{9}$$

From the definition of mobility (6), it is now satisfying to note that a higher temperature infers a greater crystalline lattice vibration, more impeded electron progress for a given electric field strength, lower drift velocity, lower mobility, lower conductivity from (9), and higher resistivity as stated.

The application of Ohm's law in point form to a macroscopic (visible to the naked eye) region leads to a more familiar form. Initially, let us assume that **J** and **E** are *uniform*, as they are in the cylindrical region shown in Fig. 5.3. Since they are uniform,

$$I = \int_S \mathbf{J} \cdot d\mathbf{S} = JS \tag{10}$$

and

$$\begin{aligned} V_{ab} &= -\int_b^a \mathbf{E} \cdot d\mathbf{L} = -\mathbf{E} \cdot \int_b^a d\mathbf{L} = -\mathbf{E} \cdot \mathbf{L}_{ba} \\ &= \mathbf{E} \cdot \mathbf{L}_{ab} \end{aligned} \tag{11}$$

or

$$V = EL$$

Thus

$$J = \frac{I}{S} = \sigma E = \sigma \frac{V}{L}$$

or

$$V = \frac{L}{\sigma S} I$$

The ratio of the potential difference between the two ends of the cylinder to the current entering the more positive end, however, is recognized from elementary circuit theory as the *resistance* of the cylinder, and therefore

$$\boxed{V = IR} \tag{12}$$

where

$$\boxed{R = \frac{L}{\sigma S}} \tag{13}$$

Equation (12) is, of course, known as *Ohm's law*, and (13) enables us to compute the resistance R, measured in ohms (abbreviated as Ω), of conducting objects which possess uniform fields. If the fields are not uniform, the resistance may still be defined as the ratio of V to I, where V is the potential difference between two specified equipotential surfaces in the material and I is the total current crossing the more positive surface into the material. From the general integral relationships in (10) and (11), and from Ohm's law (8), we may write this general expression for resistance when the fields are nonuniform,

$$R = \frac{V_{ab}}{I} = \frac{-\int_b^a \mathbf{E} \cdot d\mathbf{L}}{\int_S \sigma \mathbf{E} \cdot d\mathbf{S}} \tag{14}$$

The line integral is taken between two equipotential surfaces in the conductor, and the surface integral is evaluated over the more positive of these two equipotentials. We cannot solve these nonuniform problems at this time, but we should be able to solve several of them after perusing Chaps. 6 and 7.

As an example of the determination of the resistance of a cylinder, consider a #16 copper wire, which has a diameter of 0.0508 in, or 1.291×10^{-3} m. The area of the cross section is $\pi(1.291 \times 10^{-3}/2)^2 = 1.308 \times 10^{-6}\ \text{m}^2$, and the resistance of a wire 1 mi (1609 m) long is

$$R = \frac{1609}{(5.80 \times 10^7)(1.308 \times 10^{-6})} = 21.2 \qquad \Omega$$

This wire can safely carry about 10 A dc, corresponding to a current density of $10/(1.308 \times 10^{-6}) = 7.65 \times 10^6\ \text{A/m}^2$, or $7.65\ \text{A/mm}^2$. With this current the potential difference between the two ends of the wire is 212 V, the electric field intensity is 0.312 V/m, the drift velocity is 0.000 422 m/s, or a little more than one furlong a week, and the free-electron charge density is $-1.81 \times 10^{10}\ \text{C/m}^3$, or about one electron in a cube two angstroms on a side.

D5.3. Find the magnitude of the electric field intensity in a sample of silver having $\sigma = 6.17 \times 10^7\ \mho/\text{m}$ and $\mu_e = 0.0056\ \text{m}^2/\text{V}\cdot\text{s}$ if: (*a*) the drift velocity is 1 mm/s; (*b*) the current density is 10^7 A/m^2; (*c*) the sample is a cube, 3 mm on a side, carrying a total current of 80 A; (*d*) the sample is a cube, 3 mm on a side, having a potential difference of 0.5 mV between opposite faces.

Ans. 0.1786 V/m; 0.1621 V/m; 0.1441 V/m; 0.1667 V/m

D5.4. An aluminum conductor is 1000 ft long and has a circular cross section with a diameter of 0.8 in. If there is a dc voltage of 1.2 V between the ends, find: (*a*) the current density; (*b*) the current; (*c*) the power dissipated, using your vast knowledge of circuit theory.

Ans. 1.504×10^5 A/m^2; 48.8 A; 58.5 W

5.4 CONDUCTOR PROPERTIES AND BOUNDARY CONDITIONS

Once again we must temporarily depart from our assumed static conditions and let time vary for a few microseconds to see what happens when the charge distribution is suddenly unbalanced within a conducting material. Let us suppose, for the sake of the argument, that there suddenly appear a number of electrons in the interior of a conductor. The electric fields set up by these electrons are not counteracted by any positive charges, and the electrons therefore begin to accelerate away from each other. This continues until the electrons reach the surface of the conductor or until a number of electrons equal to the number injected have reached the surface.

Here the outward progress of the electrons is stopped, for the material surrounding the conductor is an insulator not possessing a convenient conduction band. No charge may remain within the conductor. If it did, the resulting electric field would force the charges to the surface.

Hence the final result within a conductor is zero charge density, and a surface charge density resides on the exterior surface. This is one of the two characteristics of a good conductor.

The other characteristic, stated for static conditions in which no current may flow, follows directly from Ohm's law: the electric field intensity within the conductor is zero. Physically, we see that if an electric field were present, the conduction electrons would move and produce a current, thus leading to a nonstatic condition.

Summarizing for electrostatics, no charge and no electric field may exist at any point *within* a conducting material. Charge may, however, appear on the surface as a surface charge density, and our next investigation concerns the fields *external* to the conductor.

We wish to relate these external fields to the charge on the surface of the conductor. The problem is a simple one, and we may first talk our way to the solution with little mathematics.

If the external electric field intensity is decomposed into two components, one tangential and one normal to the conductor surface, the tangential component is seen to be zero. If it were not zero, a tangential force would be applied to the elements of the surface charge, resulting in their motion and nonstatic conditions.

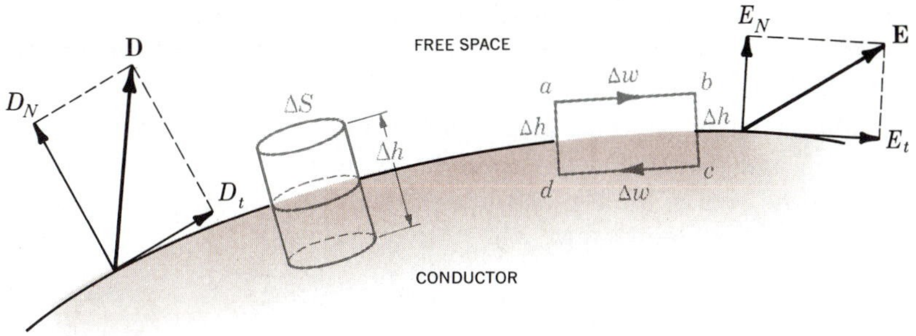

FIGURE 5.4
An appropriate closed path and gaussian surface are used to determine boundary conditions at a conductor–free space boundary; $E_t = 0$ and $D_N = \rho_S$.

Since static conditions are assumed, the tangential electric field intensity and electric flux density are zero.

Gauss's law answers our questions concerning the normal component. The electric flux leaving a small increment of surface must be equal to the charge residing on that incremental surface. The flux cannot leave the charge in the tangential direction, for this component is zero, and it cannot penetrate into the conductor, for the total field there is zero. It must then leave the surface normally. Quantitatively, we may say that the electric flux density in coulombs per square meter leaving the surface normally is equal to the surface charge density in coulombs per square meter, or $D_N = \rho_S$.

If we use some of our previously derived results in making a more careful analysis (and incidentally introducing a general method which we must use later), we should set up a conductor–free space boundary (Fig. 5.4) showing tangential and normal components of **D** and **E** on the free-space side of the boundary. Both fields are zero in the conductor. The tangential field may be determined by applying Sec. 4.5, Eq. (21),

$$\oint \mathbf{E} \cdot d\mathbf{L} = 0$$

around the small closed path *abcda*. The integral must be broken up into four parts

$$\int_a^b + \int_b^c + \int_c^d + \int_d^a = 0$$

Remembering that $\mathbf{E} = 0$ within the conductor, we let the length from a to b or c to d be Δw and from b to c or d to a be Δh, and obtain

$$E_t \Delta w - E_{N,\,\text{at}\,b} \tfrac{1}{2} \Delta h + E_{N,\,\text{at}\,a} \tfrac{1}{2} \Delta h = 0$$

As we allow Δh to approach zero, keeping Δw small but finite, it makes no difference whether or not the normal fields are equal at a and b, for Δh causes these products to become negligibly small. Hence

$$E_t \Delta w = 0$$

and therefore

$$E_t = 0$$

The condition on the normal field is found most readily by considering D_N rather than E_N and choosing a small cylinder as the gaussian surface. Let the height be Δh and the area of the top and bottom faces be ΔS. Again we shall let Δh approach zero. Using Gauss's law,

$$\oint_S \mathbf{D} \cdot d\mathbf{S} = Q$$

we integrate over the three distinct surfaces

$$\int_{\text{top}} + \int_{\text{bottom}} + \int_{\text{sides}} = Q$$

and find that the last two are zero (for different reasons). Then

$$D_N \Delta S = Q = \rho_S \Delta S$$

or

$$D_N = \rho_S$$

These are the desired *boundary conditions* for the conductor–free space boundary in electrostatics,

$$\boxed{D_t = E_t = 0} \tag{15}$$

$$\boxed{D_N = \epsilon_0 E_N = \rho_S} \tag{16}$$

The electric flux leaves the conductor in a direction normal to the surface, and the value of the electric flux density is numerically equal to the surface charge density.

An immediate and important consequence of a zero tangential electric field intensity is the fact that a conductor surface is an equipotential surface. The evaluation of the potential difference between any two points on the surface by the line integral leads to a zero result, because the path may be chosen on the surface itself where $\mathbf{E} \cdot d\mathbf{L} = 0$.

To summarize the principles which apply to conductors in electrostatic fields, we may state that

1. The static electric field intensity inside a conductor is zero.
2. The static electric field intensity at the surface of a conductor is everywhere directed normal to that surface.
3. The conductor surface is an equipotential surface.

Using these three principles, there are a number of quantities that may be calculated at a conductor boundary, given a knowledge of the potential field. Let us select the field,

$$V = 100(x^2 - y^2)$$

and stipulate that the point $P(2,-1,3)$ lies on a conductor–free space boundary.

The potential at point P is

$$V_P = 100[2^2 - (-1)^2] = 300 \qquad \text{V}$$

Since the conductor is an equipotential surface, the potential at the entire surface must be 300 V. Moreover, if the conductor is a solid object, then the potential everywhere in and on the conductor is 300 V, for $\mathbf{E} = 0$ within the conductor.

The equation representing the locus of all points having a potential of 300 V is

$$300 = 100(x^2 - y^2)$$

or

$$x^2 - y^2 = 3$$

This is therefore the equation of the conductor surface; it happens to be a hyperbolic cylinder, as shown in Fig. 5.5. Let us assume arbitrarily that the solid conductor lies above and to the right of the equipotential surface at point P, while free space is down and to the left.

Next, we find $\mathbf{E}$ by the gradient operation,

$$\mathbf{E} = -100\nabla(x^2 - y^2) = -200x\mathbf{a}_x + 200y\mathbf{a}_y$$

At point P,

$$\mathbf{E}_P = -400\mathbf{a}_x - 200\mathbf{a}_y \qquad \text{V/m}$$

Since $\mathbf{D} = \epsilon_0\mathbf{E}$, we have

$$\mathbf{D}_P = 8.854 \times 10^{-12}\mathbf{E}_P = -3.54\mathbf{a}_x - 1.771\mathbf{a}_y \qquad \text{nC/m}^2$$

The field is directed downward and to the left at P; it is normal to the equipotential surface. Therefore,

$$D_N = |\mathbf{D}_P| = 3.96 \qquad \text{nC/m}^2$$

Thus, the surface charge density at P is

$$\rho_{S,P} = D_N = 3.96 \qquad \text{nC/m}^2$$

Note that if we had taken the region to the left of the equipotential surface as the conductor, the $\mathbf{E}$ field would *terminate* on the surface charge and we would let $\rho_S = -3.96 \text{ nC/m}^2$.

As one last bit of information, let us determine the equation of the streamline passing through P. We see that

$$\frac{E_y}{E_x} = \frac{200y}{-200x} = -\frac{y}{x} = \frac{dy}{dx}$$

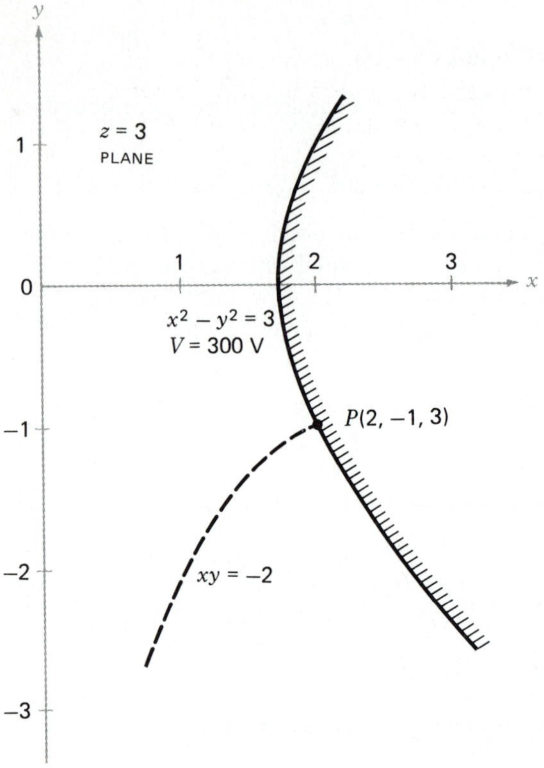

FIGURE 5.5
Given point $P(2,-1,3)$ and the potential field, $V = 100(x^2 - y^2)$, we find the equipotential surface through P is $x^2 - y^2 = 3$, and the streamline through P is $xy = -2$.

Thus,

$$\frac{dy}{y} + \frac{dx}{x} = 0$$

and

$$\ln y + \ln x = C_1$$

Therefore,

$$xy = C_2$$

The line (or surface) through P is obtained when $C_2 = (2)(-1) = -2$. Thus, the streamline is the trace of another hyperbolic cylinder,

$$xy = -2$$

This is also shown on Fig. 5.5.

D5.5. A potential field is given as $V = 100e^{-5x} \sin 3y \cos 4z$ V. Let point $P(0.1, \pi/12, \pi/24)$ be located at a conductor–free space boundary. At point P, find the magnitude of: (*a*) V; (*b*) $\mathbf{E}$; (*c*) E_N; (*d*) E_t; (*e*) ρ_S.

Ans. 37.1 V; 233 V/m; 233 V/m; 0; 2.06 nC/m^2

5.5 THE METHOD OF IMAGES

One important characteristic of the dipole field that we developed in the last chapter is the infinite plane at zero potential that exists midway between the two charges. Such a plane may be represented by a vanishingly thin conducting plane that is infinite in extent. The conductor is an equipotential surface at a potential $V = 0$, and the electric field intensity is therefore normal to the surface. Thus, if we replace the dipole configuration shown in Fig. 5.6*a* with the single charge and conducting plane shown in Fig. 5.6*b*, the fields in the upper half of each figure are the same. Below the conducting plane, all fields are zero since we have not provided any charges in that region. Of course, we might also substitute a single negative charge below a conducting plane for the dipole arrangement and obtain equivalence for the fields in the lower half of each region.

If we approach this equivalence from the opposite point of view, we begin with a single charge above a perfectly conducting plane and then see that we may maintain the same fields above the plane by removing the plane and locating a negative charge at a symmetrical location below the plane. This charge is called the *image* of the original charge, and it is the negative of that value.

If we can do this once, linearity allows us to do it again and again, and thus *any* charge configuration above an infinite ground plane may be replaced by an arrangement composed of the given charge configuration, its image, and no conducting plane. This is suggested by the two illustrations of Fig. 5.7. In many cases, the potential field of the new system is much easier to find since it does not contain the conducting plane with its unknown surface charge distribution.

As an example of the use of images, let us find the surface charge density at $P(2,5,0)$ on the conducting plane $z = 0$ if there is a line charge of 30 nC/m located at $x = 0$, $z = 3$, as shown in Fig. 5.8*a*. We remove the plane and install an image line charge of -30 nC/m at $x = 0$, $z = -3$, as illustrated in Fig. 5.8*b*. The field at P may now be obtained by superposition of the known fields of the line charges.

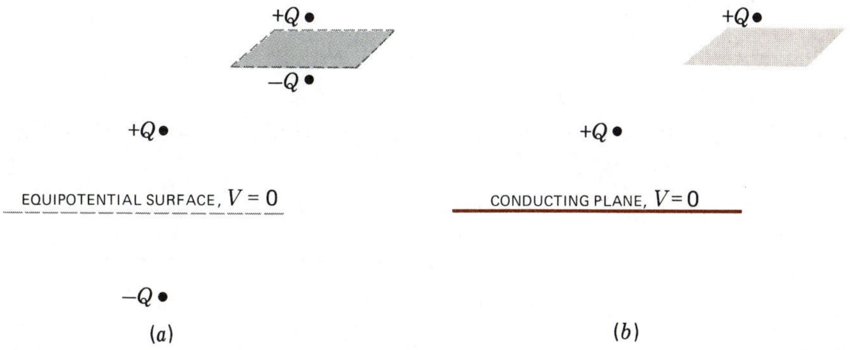

FIGURE 5.6
(*a*) Two equal but opposite charges may be replaced by (*b*) a single charge and a conducting plane without affecting the fields above the $V = 0$ surface.

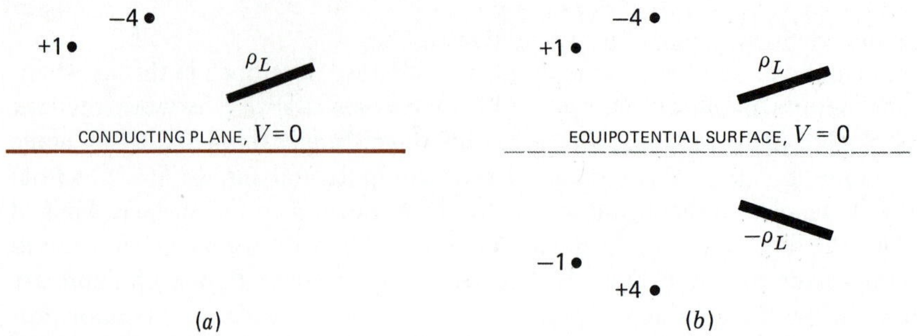

FIGURE 5.7
(*a*) A given charge configuration above an infinite conducting plane may be replaced by (*b*) the given charge configuration plus the image configuration, without the conducting plane.

The radial vector from the positive line charge to P is $\mathbf{R}_+ = 2\mathbf{a}_x - 3\mathbf{a}_z$, while $\mathbf{R}_- = 2\mathbf{a}_x + 3\mathbf{a}_z$. Thus, the individual fields are

$$\mathbf{E}_+ = \frac{\rho_L}{2\pi\epsilon_0 R_+}\mathbf{a}_{R+} = \frac{30 \times 10^{-9}}{2\pi\epsilon_0\sqrt{13}}\frac{2\mathbf{a}_x - 3\mathbf{a}_z}{\sqrt{13}}$$

and

$$\mathbf{E}_- = \frac{-30 \times 10^{-9}}{2\pi\epsilon_0\sqrt{13}}\frac{2\mathbf{a}_x + 3\mathbf{a}_z}{\sqrt{13}}$$

Adding these results, we have

$$\mathbf{E} = \frac{-180 \times 10^{-9}\mathbf{a}_z}{2\pi\epsilon_0(13)} = -249\mathbf{a}_z \qquad \text{V/m}$$

This then is the field at (or just above) P in both the configurations of Fig. 5.8, and it is certainly satisfying to note that the field is normal to the conducting

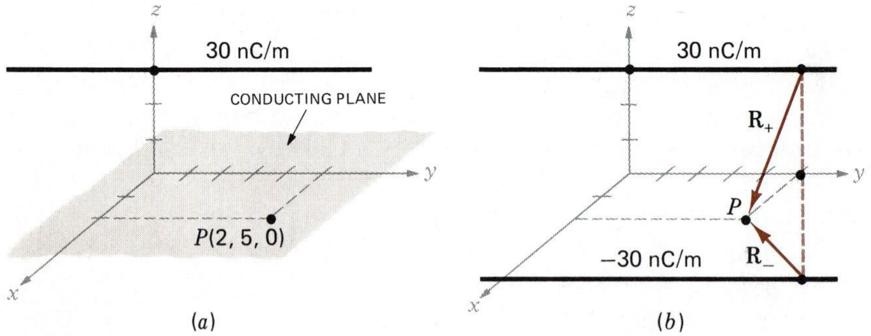

FIGURE 5.8
(*a*) A line charge above a conducting plane. (*b*) The conductor is removed, and the image of the line charge is added.

plane, as it must be. Thus, $\mathbf{D} = \epsilon_0 \mathbf{E} = -2.20\mathbf{a}_z$ nC/m^2, and since this is directed *toward* the conducting plane, ρ_S is negative and has a value of -2.20 nC/m^2 at P.

D5.6. A point charge of 25 nC is located in free space at $P(2,-3,5)$, and a perfectly conducting plane is at $z = 2$. Find: (*a*) V at (3,2,4); (*b*) $\mathbf{E}$ at (3,2,4); (*c*) ρ_S at (3,2,2).

Ans. 11.78 V; $0.985\mathbf{a}_x + 4.92\mathbf{a}_y - 4.69\mathbf{a}_z$ V/m; -57.6 pC/m^2

5.6 SEMICONDUCTORS

If we now turn our attention to an intrinsic semiconductor material, such as pure germanium or silicon, two types of current carriers are present, electrons and *holes*. The electrons are those from the top of the filled valence band which have received sufficient energy (usually thermal) to cross the relatively small forbidden band into the conduction band. The forbidden-band energy gap in typical semiconductors is of the order of one electronvolt. The vacancies left by these electrons represent unfilled energy states in the valence band which may also move from atom to atom in the crystal. The vacancy is called a *hole*, and many semiconductor properties may be described by treating the hole as if it had a positive charge of e, a mobility μ_h, and an effective mass comparable to that of the electron. Both carriers move in an electric field, and they move in opposite directions; hence each contributes a component of the total current which is in the same direction as that provided by the other. The conductivity is therefore a function of both hole and electron concentrations and mobilities,

$$\boxed{\sigma = -\rho_e \mu_e + \rho_h \mu_h} \qquad (17)$$

For pure, or *intrinsic*, silicon the electron and hole mobilities are 0.12 and 0.025, respectively, while for germanium, the mobilities are, respectively, 0.36 and 0.17. These values are given in square meters per volt-second and range from 10 to 100 times as large as those for aluminum, copper, silver, and other metallic conductors.[5] The mobilities listed above are given for a temperature of 300 K.

The electron and hole concentrations depend strongly on temperature. At 300 K the electron and hole volume charge densities are both 0.0024 C/m^3 in magnitude in intrinsic silicon and 3.0 C/m^3 in intrinsic germanium. These values lead to conductivities of 0.000 35 ℧/m in silicon and 1.6 ℧/m in germanium. As temperature increases, the mobilities decrease, but the charge densities increase very rapidly. As a result, the conductivity of silicon increases by a factor of 10 as the temperature increases from 300 to about 330 K and decreases by a factor of 10 as the temperature drops from 300 to about 275 K. Note that the conductivity of the intrinsic semiconductor increases with temperature, while that of a metallic conductor decreases with temperature; this is one of the characteristic differences between the metallic conductors and the intrinsic semiconductors.

[5] Mobility values for semiconductors are given in Refs. 2, 4, and 7 listed at the end of this chapter.

Intrinsic semiconductors also satisfy the point form of Ohm's law; that is, the conductivity is reasonably constant with current density and with the direction of the current density.

The number of charge carriers and the conductivity may both be increased dramatically by adding very small amounts of impurities. *Donor* materials provide additional electrons and form *n-type* semiconductors, while *acceptors* furnish extra holes and form *p-type* materials. The process is known as *doping*, and a donor concentration in silicon as low as one part in 10^7 causes an increase in conductivity by a factor of 10^5.

The range of value of the conductivity is extreme as we go from the best insulating materials to semiconductors and the finest conductors. In mhos per meter, σ ranges from 10^{-17} for fused quartz, 10^{-7} for poor plastic insulators, and roughly unity for semiconductors to almost 10^8 for metallic conductors at room temperature. These values cover the remarkably large range of some twenty-five orders of magnitude.

D5.7. Using the values given in this section for the electron and hole mobilities in germanium at 300 K, and assuming hole and electron concentrations of 2.7×10^{19} m^{-3}, find: (*a*) the component of the conductivity due to holes; (*b*) the component of the conductivity due to electrons; (*c*) the conductivity.

Ans. 0.735 ℧/m; 1.557 ℧/m; 2.29 ℧/m

5.7 THE NATURE OF DIELECTRIC MATERIALS

Although we have mentioned insulators and dielectric materials, we do not as yet have any quantitative relationships in which they are involved. We shall soon see, however, that a dielectric in an electric field can be viewed as a free-space arrangement of microscopic electric dipoles which are composed of positive and negative charges whose centers do not quite coincide.

These are not free charges, and they cannot contribute to the conduction process. Rather, they are bound in place by atomic and molecular forces and can only shift positions slightly in response to external fields. They are called *bound* charges, in contrast to the free charges that determine conductivity. The bound charges can be treated as any other sources of the electrostatic field. If we did not wish to, therefore, we would not need to introduce the dielectric constant as a new parameter or to deal with permittivities different from the permittivity of free space; however, the alternative would be to consider *every charge within a piece of dielectric material*. This is too great a price to pay for using all our previous equations in an unmodified form, and we shall therefore spend some time theorizing about dielectrics in a qualitative way; introducing polarization $\mathbf{P}$, permittivity ϵ, and relative permittivity ϵ_R; and developing some quantitative relationships involving these new quantities.

The characteristic which all dielectric materials have in common, whether they are solid, liquid, or gas, and whether or not they are crystalline in nature, is their ability to store electric energy. This storage takes place by means of a shift in the relative positions of the internal, bound positive and negative charges against the normal molecular and atomic forces.

This displacement against a restraining force is analogous to lifting a weight or stretching a spring and represents potential energy. The source of the energy is the external field, the motion of the shifting charges resulting perhaps in a transient current through a battery which is producing the field.

The actual mechanism of the charge displacement differs in the various dielectric materials. Some molecules, termed *polar* molecules, have a permanent displacement existing between the centers of "gravity" of the positive and negative charges, and each pair of charges acts as a dipole. Normally the dipoles are oriented in a random way throughout the interior of the material, and the action of the external field is to align these molecules, to some extent, in the same direction. A sufficiently strong field may even produce an additional displacement between the positive and negative charges.

A *nonpolar* molecule does not have this dipole arrangement until after a field is applied. The negative and positive charges shift in opposite directions against their mutual attraction and produce a dipole which is aligned with the electric field.

Either type of dipole may be described by its dipole moment **p**, as developed in Sec. 4.7, Eq. (37),

$$\mathbf{p} = Q\mathbf{d} \tag{18}$$

where Q is the positive one of the two bound charges composing the dipole, and **d** is the vector from the negative to the positive charge.

If there are n dipoles per unit volume and we deal with a volume Δv, then there are $n\,\Delta v$ dipoles, and the total dipole moment is obtained by the vector sum,

$$\mathbf{p}_{\text{total}} = \sum_{i=1}^{n\,\Delta v} \mathbf{p}_i$$

If the dipoles are aligned in the same general direction, $\mathbf{p}_{\text{total}}$ may have a significant value. However, a random orientation may cause $\mathbf{p}_{\text{total}}$ to be essentially zero.

We now define the polarization **P** as the *dipole moment per unit volume*,

$$\mathbf{P} = \lim_{\Delta v \to 0} \frac{1}{\Delta v} \sum_{i=1}^{n\,\Delta v} \mathbf{p}_i \tag{19}$$

with units of coulombs per square meter. We shall treat **P** as a typical continuous field, even though it is obvious that it is essentially undefined at points within an atom or molecule. Instead, we should think of its value at any point as an average value taken over a sample volume Δv—large enough to contain many molecules ($n\,\Delta v$ in number), but yet sufficiently small to be considered incremental in concept.

Our immediate goal is to show that the bound volume charge density acts like the free volume charge density in producing an external field; we shall obtain a result similar to Gauss's law.

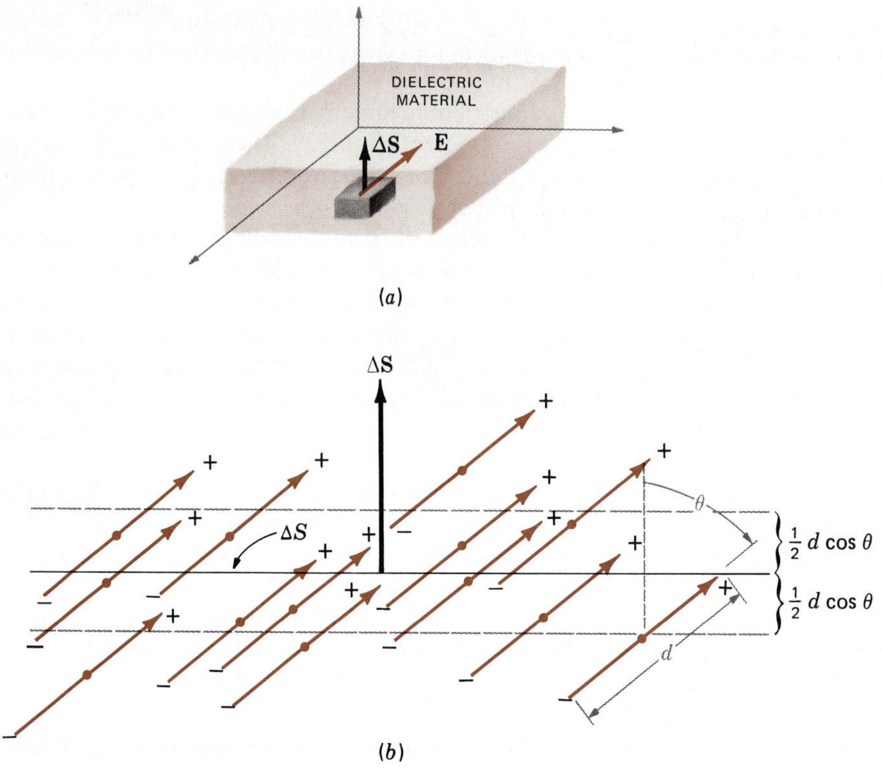

FIGURE 5.9
(*a*) An incremental surface element **ΔS** is shown in the interior of a dielectric in which an electric field **E** is present. (*b*) The nonpolar molecules form dipole moments **p** and a polarization **P**. There is a net transfer of bound charge across **ΔS**.

To be specific, let us assume that we have a dielectric containing nonpolar molecules. No molecule has a dipole moment, and $\mathbf{P} = 0$ throughout the material. Somewhere in the interior of the dielectric we select an incremental surface element $\Delta\mathbf{S}$, as shown in Fig. 5.9*a*, and apply an electric field $\mathbf{E}$. The electric field produces a moment $\mathbf{p} = Q\mathbf{d}$ in each molecule, such that $\mathbf{p}$ and $\mathbf{d}$ make an angle θ with $\Delta\mathbf{S}$, as indicated in Fig. 5.9*b*.

Now let us inspect the movement of bound charges across $\Delta\mathbf{S}$. Each of the charges associated with the creation of a dipole must have moved a distance $\frac{1}{2}d\cos\theta$ in the direction perpendicular to $\Delta\mathbf{S}$. Thus, any positive charges initially lying below the surface $\Delta\mathbf{S}$ and within the distance $\frac{1}{2}d\cos\theta$ of the surface must have crossed $\Delta\mathbf{S}$ going upward. Also, any negative charges initially lying above the surface and within that distance ($\frac{1}{2}d\cos\theta$) from $\Delta\mathbf{S}$ must have crossed $\Delta\mathbf{S}$ going downward. Therefore, since there are n molecules/m^3, the *net* total charge which crosses the elemental surface in an upward direction is $nQd\cos\theta\,\Delta S$, or

$$\Delta Q_b = nQ\mathbf{d} \cdot \Delta\mathbf{S}$$

where the subscript on Q_b reminds us that we are dealing with a bound charge and not a free charge. In terms of the polarization, we have

$$\Delta Q_b = \mathbf{P} \cdot \Delta \mathbf{S}$$

If we interpret $\Delta\mathbf{S}$ as an element of a *closed* surface inside the dielectric material, then the direction of $\Delta\mathbf{S}$ is outward, and the net increase in the bound charge *within* the closed surface is obtained through the integral

$$Q_b = -\oint_S \mathbf{P} \cdot d\mathbf{S} \tag{20}$$

This last relationship has some resemblance to Gauss's law, and we may now generalize our definition of electric flux density so that it applies to media other than free space. We first write Gauss's law in terms of $\epsilon_0 \mathbf{E}$ and Q_T, the *total* enclosed charge, bound plus free:

$$Q_T = \oint_S \epsilon_0 \mathbf{E} \cdot d\mathbf{S} \tag{21}$$

where

$$Q_T = Q_b + Q$$

and Q is the total *free* charge enclosed by the surface S. Note that the free charge appears without subscript since it is the most important type of charge and will appear in Maxwell's equations.

Combining these last three equations, we obtain an expression for the free charge enclosed,

$$Q = Q_T - Q_b = \oint_S (\epsilon_0 \mathbf{E} + \mathbf{P}) \cdot d\mathbf{S} \tag{22}$$

We may now define $\mathbf{D}$ in more general terms than we did in Chap. 3,

$$\boxed{\mathbf{D} = \epsilon_0 \mathbf{E} + \mathbf{P}} \tag{23}$$

There is thus an added term to $\mathbf{D}$ which appears when polarizable material is present. Thus,

$$Q = \oint_S \mathbf{D} \cdot d\mathbf{S} \tag{24}$$

where Q is the free charge enclosed.

Utilizing the several volume charge densities, we have

$$Q_b = \int_v \rho_b \, dv$$

$$Q = \int_v \rho_v \, dv$$

$$Q_T = \int_v \rho_T \, dv$$

With the help of the divergence theorem, we may therefore transform (20), (21), and (24) into the equivalent divergence relationships,

$$\nabla \cdot \mathbf{P} = -\rho_b$$

$$\nabla \cdot \epsilon_0 \mathbf{E} = \rho_T$$

$$\boxed{\nabla \cdot \mathbf{D} = \rho_v} \tag{25}$$

We shall emphasize only (24) and (25), the two expressions involving the free charge, in the work that follows.

In order to make any real use of these new concepts, it is necessary to know the relationship between the electric field intensity **E** and the polarization **P** which results. This relationship will, of course, be a function of the type of material, and we shall essentially limit our discussion to those isotropic materials for which **E** and **P** are linearly related. In an isotropic material the vectors **E** and **P** are always parallel, regardless of the orientation of the field. Although most engineering dielectrics are linear for moderate-to-large field strengths and are also isotropic, single crystals may be anisotropic. The periodic nature of crystalline materials causes dipole moments to be formed most easily along the crystal axes, and not necessarily in the direction of the applied field.

In *ferroelectric* materials the relationship between **P** and **E** is not only nonlinear, but also shows hysteresis effects; that is, the polarization produced by a given electric field intensity depends on the past history of the sample. Important examples of this type of dielectric are barium titanate and Rochelle salt.

The linear relationship between **P** and **E** is

$$\boxed{\mathbf{P} = \chi_e \epsilon_0 \mathbf{E}} \tag{26}$$

where χ_e (chi) is a dimensionless quantity called the *electric susceptibility* of the material.

Using this relationship in (23), we have

$$\mathbf{D} = \epsilon_0 \mathbf{E} + \chi_e \epsilon_0 \mathbf{E} = (\chi_e + 1)\epsilon_0 \mathbf{E}$$

The expression within the parentheses is now defined as

$$\epsilon_R = \chi_e + 1 \tag{27}$$

This is another dimensionless quantity and it is known as the *relative permittivity*, or *dielectric constant* of the material. Thus,

$$\mathbf{D} = \epsilon_0 \epsilon_R \mathbf{E} = \epsilon \mathbf{E} \tag{28}$$

where

$$\boxed{\epsilon = \epsilon_R \epsilon_0} \tag{29}$$

and ϵ is the *permittivity*. The dielectric constants are given for some representative materials in Appendix C.

Anisotropic dielectric materials cannot be described in terms of a simple susceptibility or permittivity parameter. Instead, we find that each component of **D** may be a function of every component of **E**, and $\mathbf{D} = \epsilon\mathbf{E}$ becomes a matrix equation where **D** and **E** are each 3 × 1 column matrices and ϵ is a 3 × 3 square matrix. Expanding the matrix equation gives

$$D_x = \epsilon_{xx}E_x + \epsilon_{xy}E_y + \epsilon_{xz}E_z$$

$$D_y = \epsilon_{yx}E_x + \epsilon_{yy}E_y + \epsilon_{yz}E_z$$

$$D_z = \epsilon_{zx}E_x + \epsilon_{zy}E_y + \epsilon_{zz}E_z$$

Note that the elements of the matrix depend on the selection of the coordinate axes in the anisotropic material. Certain choices of axis directions lead to simpler matrices.[6]

D and **E** (and **P**) are no longer parallel, and although $\mathbf{D} = \epsilon_0\mathbf{E} + \mathbf{P}$ remains a valid equation for anisotropic materials, we may continue to use $\mathbf{D} = \epsilon\mathbf{E}$ only by interpreting it as a matrix equation. We shall concentrate our attention on linear isotropic materials and reserve the general case for a more advanced text.

In summary, then, we now have a relationship between **D** and **E** which depends on the dielectric material present,

$$\boxed{\mathbf{D} = \epsilon\mathbf{E}} \qquad (28)$$

where

$$\boxed{\epsilon = \epsilon_R\epsilon_0} \qquad (29)$$

This electric flux density is still related to the free charge by either the point or integral form of Gauss's law:

$$\boxed{\nabla \cdot \mathbf{D} = \rho_v} \qquad (25)$$

$$\boxed{\oint_S \mathbf{D} \cdot d\mathbf{S} = Q} \qquad (24)$$

The use of the relative permittivity, as indicated by (29) above, makes consideration of the polarization, dipole moments, and bound charge unnecessary. However, when anisotropic or nonlinear materials must be considered, the relative permittivity, in the simple scalar form that we have discussed, is no longer applicable.

---

[6] A more complete discussion of this matrix may be found in the Ramo, Whinnery, and Van Duzer reference listed at the end of this chapter.

Let us now illustrate these new concepts by discussing an example in which there is a slab of Teflon in the region $0 \le x \le a$, and free space where $x < 0$ and $x > a$. Outside the Teflon there is a uniform field $\mathbf{E}_{\text{out}} = E_0 \mathbf{a}_x$ V/m. The dielectric constant of Teflon is 2.1, and thus the electric susceptibility is 1.1.

Outside the slab, we have immediately $\mathbf{D}_{\text{out}} = \epsilon_0 E_0 \mathbf{a}_x$. Also, since there is no dielectric material there, $\mathbf{P}_{\text{out}} = 0$. Now, any of the last four or five equations will enable us to relate the several fields inside the material to each other. Thus

$$\mathbf{D}_{\text{in}} = 2.1\epsilon_0 \mathbf{E}_{\text{in}} \qquad (0 \le x \le a)$$

$$\mathbf{P}_{\text{in}} = 1.1\epsilon_0 \mathbf{E}_{\text{in}} \qquad (0 \le x \le a)$$

As soon as we establish a value for any of these three fields within the dielectric, the other two can be found immediately. The difficulty lies in crossing over the boundary from the known fields external to the dielectric to the unknown ones within it. To do this we need a boundary condition, and this is the subject of the next exciting section. We shall complete this example then.

In the remainder of this text we shall describe polarizable materials in terms of $\mathbf{D}$ and ϵ rather than $\mathbf{P}$ and χ_e. We shall limit our discussion to isotropic materials.

D5.8. A certain homogeneous slab of lossless dielectric material is characterized by an electric susceptibility of 0.12 and carries a uniform electric flux density within it of 1.6 nC/m^2. Find: (*a*) *E*; (*b*) *P*; (*c*) the average dipole moment if there are 2×10^{19} dipoles per cubic meter; (*d*) the voltage between two equipotentials 1 in apart.

Ans. 161.3 V/m; 171.4 pC/m^2; 8.57×10^{-30} C·m; 4.10 V

5.8 BOUNDARY CONDITIONS FOR PERFECT DIELECTRIC MATERIALS

How do we attack a problem in which there are two different dielectrics, or a dielectric and a conductor? This is another example of a *boundary condition*, such as the condition at the surface of a conductor whereby the tangential fields are zero and the normal electric flux density is equal to the surface charge density on the conductor. Now we take the first step in solving a two-dielectric problem, or a dielectric-conductor problem, by determining the behavior of the fields at the dielectric interface.

Let us first consider the interface between two dielectrics having permittivities ϵ_1 and ϵ_2 and occupying regions 1 and 2, as shown in Fig. 5.10. We first examine the tangential components by using

$$\oint \mathbf{E} \cdot d\mathbf{L} = 0$$

around the small closed path on the left, obtaining

$$E_{\tan 1} \Delta w - E_{\tan 2} \Delta w = 0$$

The small contribution to the line integral by the normal component of $\mathbf{E}$ along the sections of length Δh becomes negligible as Δh decreases and the closed

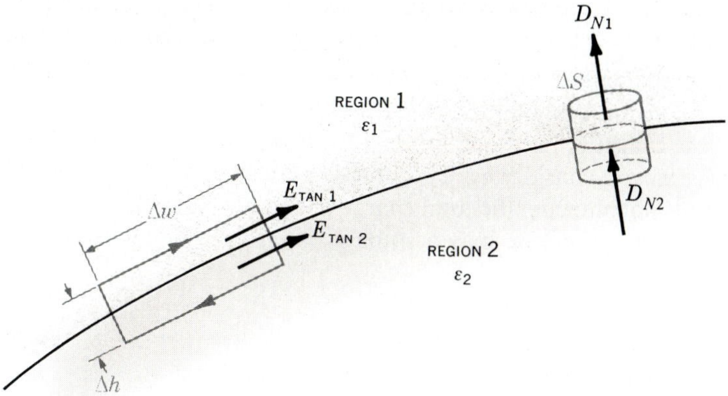

FIGURE 5.10
The boundary between perfect dielectrics of permittivities ϵ_1 and ϵ_2. The continuity of D_N is shown by the gaussian surface on the right, and the continuity of E_{tan} by the line integral about the closed path at the left.

path crowds the surface. Immediately, then,

$$\boxed{E_{\tan 1} = E_{\tan 2}} \tag{30}$$

and we might feel that Kirchhoff's voltage law is still applicable to this case. Certainly we have shown that the potential difference between any two points on the boundary that are separated by a distance Δw is the same immediately above or below the boundary.

If the tangential electric field intensity is continuous across the boundary, then tangential **D** is discontinuous, for

$$\frac{D_{\tan 1}}{\epsilon_1} = E_{\tan 1} = E_{\tan 2} = \frac{D_{\tan 2}}{\epsilon_2}$$

or

$$\frac{D_{\tan 1}}{D_{\tan 2}} = \frac{\epsilon_1}{\epsilon_2} \tag{31}$$

The boundary conditions on the normal components are found by applying Gauss's law to the small "pillbox" shown at the right in Fig. 5.10. The sides are again very short, and the flux leaving the top and bottom surfaces is the difference

$$D_{N1}\,\Delta S - D_{N2}\,\Delta S = \Delta Q = \rho_S\,\Delta S$$

from which

$$\boxed{D_{N1} - D_{N2} = \rho_S} \tag{32}$$

What is this surface charge density? It cannot be a *bound* surface charge density, because we are taking the polarization of the dielectric into effect by using a dielectric constant different from unity; that is, instead of considering bound charges in free space, we are using an increased permittivity. Also, it is extremely unlikely that any *free* charge is on the interface, for no free charge is available in the perfect dielectrics we are considering. This charge must then have been placed there deliberately, thus unbalancing the total charge in and on this dielectric body. Except for this special case, then, we may assume ρ_S is zero on the interface and

$$\boxed{D_{N1} = D_{N2}} \tag{33}$$

or the normal component of **D** is continuous. It follows that

$$\epsilon_1 E_{N1} = \epsilon_2 E_{N2} \tag{34}$$

and normal **E** is discontinuous.

These conditions may be combined to show the change in the vectors **D** and **E** at the surface. Let $\mathbf{D}_1$ (and $\mathbf{E}_1$) make an angle θ_1 with a normal to the surface (Fig. 5.11). Since the normal components of **D** are continuous,

$$D_{N1} = D_1 \cos\theta_1 = D_2 \cos\theta_2 = D_{N2} \tag{35}$$

The ratio of the tangential components is given by (31) as

$$\frac{D_{\tan 1}}{D_{\tan 2}} = \frac{D_1 \sin\theta_1}{D_2 \sin\theta_2} = \frac{\epsilon_1}{\epsilon_2}$$

or

$$\epsilon_2 D_1 \sin\theta_1 = \epsilon_1 D_2 \sin\theta_2 \tag{36}$$

and the division of this equation by (35) gives

$$\frac{\tan\theta_1}{\tan\theta_2} = \frac{\epsilon_1}{\epsilon_2} \tag{37}$$

In Fig. 5.11 we have assumed that $\epsilon_1 > \epsilon_2$, and therefore $\theta_1 > \theta_2$.

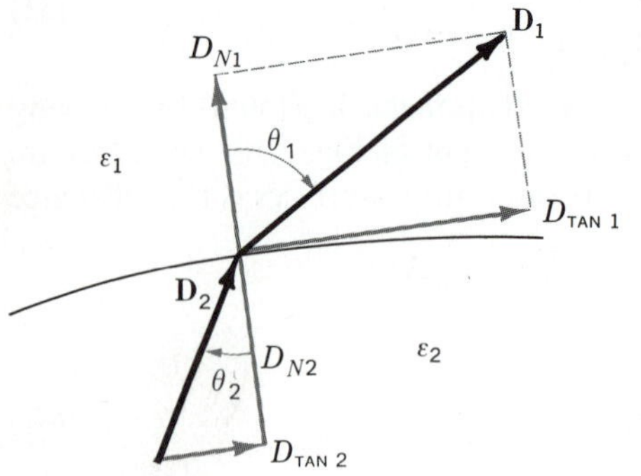

FIGURE 5.11
The refraction of **D** at a dielectric interface. For the case shown, $\epsilon_1 > \epsilon_2$; $\mathbf{E}_1$ and $\mathbf{E}_2$ are directed along $\mathbf{D}_1$ and $\mathbf{D}_2$, with $D_1 > D_2$ and $E_1 < E_2$.

The direction of **E** on each side of the boundary is identical with the direction of **D**, because $\mathbf{D} = \epsilon\mathbf{E}$.

The magnitude of **D** in region 2 may be found from (35) and (36),

$$D_2 = D_1 \sqrt{\cos^2\theta_1 + \left(\frac{\epsilon_2}{\epsilon_1}\right)^2 \sin^2\theta_1} \tag{38}$$

and the magnitude of $\mathbf{E}_2$ is

$$E_2 = E_1 \sqrt{\sin^2\theta_1 + \left(\frac{\epsilon_1}{\epsilon_2}\right)^2 \cos^2\theta_1} \tag{39}$$

An inspection of these equations shows that D is larger in the region of larger permittivity (unless $\theta_1 = \theta_2 = 0°$, where the magnitude is unchanged) and that E is larger in the region of smaller permittivity (unless $\theta_1 = \theta_2 = 90°$, where its magnitude is unchanged).

These boundary conditions, (30), (31), (33), and (34), or the magnitude and direction relations derived from them, (37) to (39), allow us to find quickly the field on one side of a boundary *if we know the field on the other side.* In the example we began at the end of the previous section, this was the case. Now let's finish up that problem. Recall that we had a slab of Teflon extending from $x = 0$ to $x = a$, as shown in Fig. 5.12, with free space on both sides of it and an external field $\mathbf{E}_{out} = E_0\mathbf{a}_x$. We also have $\mathbf{D}_{out} = \epsilon_0 E_0 \mathbf{a}_x$ and $\mathbf{P}_{out} = 0$.

Inside, the continuity of D_N at the boundary allows us to find that $\mathbf{D}_{in} = \mathbf{D}_{out} = \epsilon_0 E_0 \mathbf{a}_x$. This gives us $\mathbf{E}_{in} = \mathbf{D}_{in}/\epsilon = \epsilon_0 E_0 \mathbf{a}_x/(\epsilon_R \epsilon_0) = 0.476 E_0 \mathbf{a}_x$. To get the polarization field in the dielectric, we use $\mathbf{D} = \epsilon_0\mathbf{E} + \mathbf{P}$ and obtain

$$\mathbf{P}_{in} = \mathbf{D}_{in} - \epsilon_0\mathbf{E}_{in} = \epsilon_0 E_0 \mathbf{a}_x - 0.476\epsilon_0 E_0 \mathbf{a}_x = 0.524\epsilon_0 E_0 \mathbf{a}_x$$

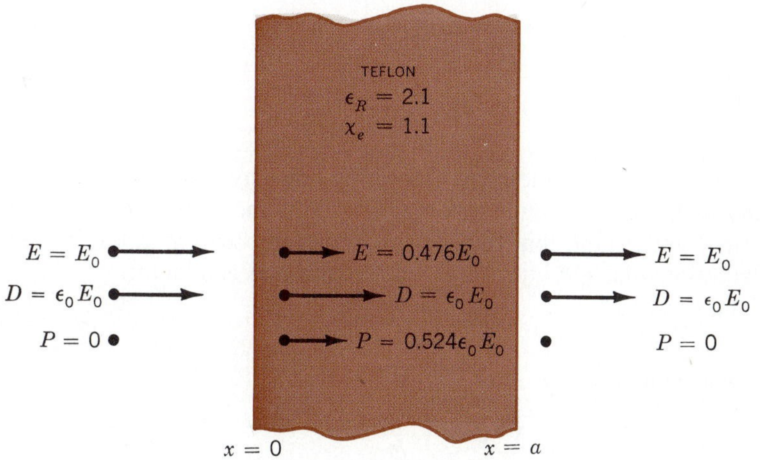

FIGURE 5.12
A knowledge of the electric field external to the dielectric enables us to find the remaining external fields first, and then to use the continuity of normal **D** to begin finding the internal fields.

Summarizing then gives

$$\mathbf{D}_{\text{in}} = \epsilon_0 E_0 \mathbf{a}_x \qquad (0 \le x \le a)$$

$$\mathbf{E}_{\text{in}} = 0.476 E_0 \mathbf{a}_x \qquad (0 \le x \le a)$$

$$\mathbf{P}_{\text{in}} = 0.524 \epsilon_0 E_0 \mathbf{a}_x \qquad (0 \le x \le a)$$

A practical problem most often does not provide us with a direct knowledge of the field on either side of the boundary. The boundary conditions must be used to help us determine the fields on both sides of the boundary from the other information which is given. A simple problem of this type will be considered in Sec. 5.10.

The boundary conditions existing at the interface between a conductor and a dielectric are much simpler than those above. First, we know that **D** and **E** are both zero inside the conductor. Second, the tangential **E** and **D** field components must both be zero to satisfy

$$\oint \mathbf{E} \cdot d\mathbf{L} = 0$$

and

$$\mathbf{D} = \epsilon \mathbf{E}$$

Finally, the application of Gauss's law,

$$\oint_S \mathbf{D} \cdot d\mathbf{S} = Q$$

shows once more that both **D** and **E** are normal to the conductor surface and that $D_N = \rho_S$ and $E_N = \rho_S/\epsilon$. We see, then, that the boundary conditions we developed previously for the conductor–free space boundary are valid for the conductor-dielectric boundary if we replace ϵ_0 by ϵ. Thus

$$\boxed{D_t = E_t = 0} \tag{40}$$

$$\boxed{D_N = \epsilon E_N = \rho_S} \tag{41}$$

It is interesting to spend a moment discovering how any charge that is introduced *internally* within a conducting material arrives at the surface as a surface charge. We should understand that this is not a common occurrence, but it does give us some additional insight into the characteristics of a conductor.

Given Ohm's law,

$$\mathbf{J} = \sigma \mathbf{E}$$

and the continuity equation,

$$\nabla \cdot \mathbf{J} = -\frac{\partial \rho_v}{\partial t}$$

in which **J** and ρ_v both involve only free charges, we have

$$\nabla \cdot \sigma \mathbf{E} = -\frac{\partial \rho_v}{\partial t}$$

or

$$\nabla \cdot \frac{\sigma}{\epsilon} \mathbf{D} = -\frac{\partial \rho_v}{\partial t}$$

If we assume that the medium is homogeneous, so that σ and ϵ are not functions of position,

$$\nabla \cdot \mathbf{D} = -\frac{\epsilon}{\sigma} \frac{\partial \rho_v}{\partial t}$$

Now we may use Maxwell's first equation to obtain

$$\rho_v = -\frac{\epsilon}{\sigma} \frac{\partial \rho_v}{\partial t}$$

Let us now make the simplifying assumption that σ is not a function of ρ_v. This is probably not a very good assumption, for we found in Sec. 5.3, Eq. (9), that σ depended on both ρ_v and the mobility, but it leads to an easy solution that at least permits us to compare different conductors. We simply rearrange and integrate directly, obtaining

$$\rho_v = \rho_0 e^{-(\sigma/\epsilon)t}$$

where ρ_0 = charge density at $t = 0$. This shows an exponential decay of charge density at every point with a time constant of ϵ/σ. This time constant, often called the *relaxation time*, may be calculated for a relatively poor conductor, such as distilled water, from the data in Appendix C, giving

$$\frac{\epsilon}{\sigma} = \frac{80 \times 8.854 \times 10^{-12}}{2 \times 10^{-4}} = 3.54 \qquad \mu\text{s}$$

In 3.54 μs any charge we place in the interior of a body of distilled water has dropped to about 37 percent of its initial value. This rapid decay is characteristic of good conductors and we see that, except for an extremely short transient period, we may safely consider the charge density to be zero within a good conductor.

With the physical materials with which we must work, no dielectric material is without some few free electrons; all have conductivities different from zero, and charge introduced internally in any of them will eventually reach the surface.

With the knowledge we now have of conducting materials, dielectric materials, and the necessary boundary conditions, we are ready to define and discuss capacitance.

D5.9. The region $y < 0$ contains a dielectric material for which $\epsilon_{R1} = 2.5$, while the region $y > 0$ is characterized by $\epsilon_{R2} = 4$. Let $\mathbf{E}_1 = -30\mathbf{a}_x + 50\mathbf{a}_y + 70\mathbf{a}_z$ V/m, and find: (*a*) E_{N1}; (*b*) $\mathbf{E}_{t1}$; (*c*) E_{t1}; (*d*) E_1; (*e*) θ_1.

Ans. 50.0 V/m; $-30\mathbf{a}_x + 70\mathbf{a}_z$ V/m; 76.2 V/m; 91.1 V/m; 56.7°

D5.10. Continue Prob. D5.9 by finding: (*a*) D_{N2}; (*b*) D_{t2}; (*c*) $\mathbf{D}_2$; (*d*) $\mathbf{P}_2$; (*e*) θ_2.

Ans. 1.107 nC/m^2; 2.70 nC/m^2; $-1.062\mathbf{a}_x + 1.107\mathbf{a}_y + 2.48\mathbf{a}_z$ nC/m^2; $-0.797\mathbf{a}_x + 0.830\mathbf{a}_y + 1.859\mathbf{a}_z$ nC/m^2; 67.7°

5.9 CAPACITANCE

Now let us consider two conductors embedded in a homogeneous dielectric (Fig. 5.13). Conductor M_2 carries a total positive charge Q, and M_1 carries an equal negative charge. There are no other charges present, and the *total* charge of the system is zero.

We now know that the charge is carried on the surface as a surface charge density and also that the electric field is normal to the conductor surface. Each conductor is, moreover, an equipotential surface. Since M_2 carries the positive charge, the electric flux is directed from M_2 to M_1, and M_2 is at the more positive potential. In other words, work must be done to carry a positive charge from M_1 to M_2.

Let us designate the potential difference between M_2 and M_1 as V_0. We may now define the *capacitance* of this two-conductor system as the ratio of the magnitudes of the total charge on either conductor to the potential difference between conductors.

$$\boxed{C = \frac{Q}{V_0}} \tag{42}$$

In general terms, we determine Q by a surface integral over the positive conductor, and we find V_0 by carrying a unit positive charge from the negative to the positive surface,

$$C = \frac{\oint_S \epsilon \mathbf{E} \cdot d\mathbf{S}}{-\int_-^+ \mathbf{E} \cdot d\mathbf{L}} \tag{43}$$

The capacitance is independent of the potential and total charge, for their ratio is constant. If the charge density is increased by a factor of N, Gauss's law

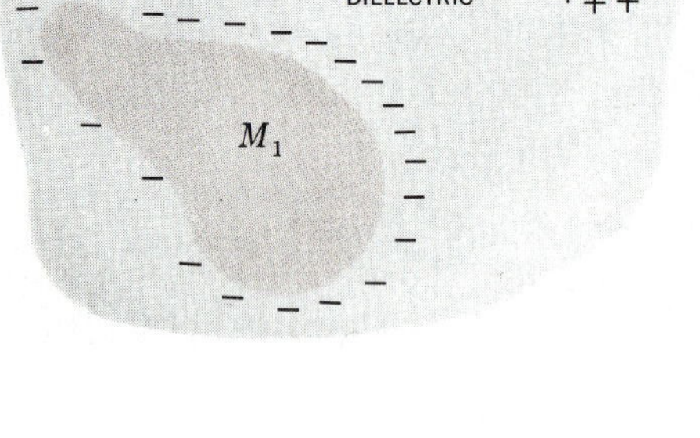

FIGURE 5.13
Two oppositely charged conductors M_1 and M_2 surrounded by a uniform dielectric. The ratio of the magnitude of the charge on either conductor to the magnitude of the potential difference between them is the capacitance C.

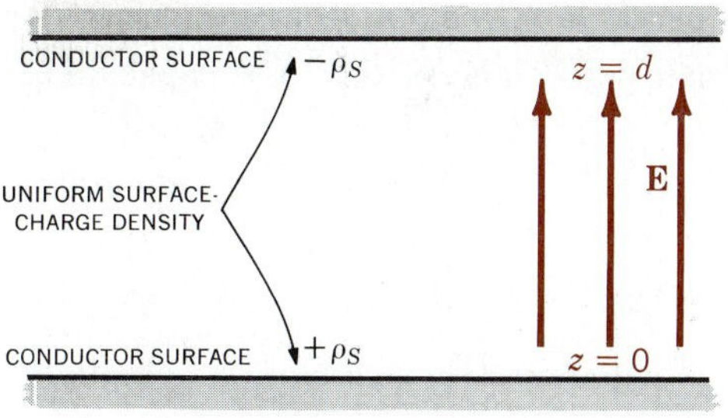

FIGURE 5.14
The problem of the parallel-plate capacitor. The capacitance per square meter of surface area is ϵ/d.

indicates that the electric flux density or electric field intensity also increases by N, as does the potential difference. The capacitance is a function only of the physical dimensions of the system of conductors and of the permittivity of the homogeneous dielectric.

Capacitance is measured in *farads* (F), where a farad is defined as one coulomb per volt. Common values of capacitance are apt to be very small fractions of a farad, and consequently more practical units are the microfarad (μF) and picofarad (pF).

We can apply the definition of capacitance to a simple two-conductor system in which the conductors are identical, infinite parallel planes with separation d (Fig. 5.14). Choosing the lower conducting plane at $z = 0$ and the upper one at $z = d$, a uniform sheet of surface charge $\pm\rho_S$ on each conductor leads to the uniform field [Sec. 2.5, Eq. (22)]

$$\mathbf{E} = \frac{\rho_S}{\epsilon}\mathbf{a}_z$$

where the permittivity of the homogeneous dielectric is ϵ, and

$$\mathbf{D} = \rho_S\mathbf{a}_z$$

The charge on the lower plane must then be positive, since $\mathbf{D}$ is directed upward, and the normal value of $\mathbf{D}$,

$$D_N = D_z = \rho_S$$

is equal to the surface charge density there. On the upper plane,

$$D_N = -D_z$$

and the surface charge there is the negative of that on the lower plane.

The potential difference between lower and upper planes is

$$V_0 = -\int_{\text{upper}}^{\text{lower}} \mathbf{E}\cdot d\mathbf{L} = -\int_d^0 \frac{\rho_S}{\epsilon}\,dz = \frac{\rho_S}{\epsilon}d$$

Since the total charge on either plane is infinite, the capacitance is infinite. A more practical answer is obtained by considering planes, each of area S, whose linear dimensions are much greater than their separation d. The electric field and charge distribution are then almost uniform at all points not adjacent to the edges, and this latter region contributes only a small percentage of the total capacitance, allowing us to write the familiar result

$$Q = \rho_S S$$

$$V_0 = \frac{\rho_S}{\epsilon} d$$

$$\boxed{C = \frac{Q}{V_0} = \frac{\epsilon S}{d}} \tag{44}$$

More rigorously, we might consider (44) as the capacitance of a portion of the infinite-plane arrangement having a surface area S. Methods of calculating the effect of the unknown and nonuniform distribution near the edges must wait until we are able to solve more complicated potential problems.

As an example, consider a capacitor having a mica dielectric, $\epsilon_R = 6$, a plate area of 10 in^2, and a separation of 0.01 in. Therefore,

$$S = 10 \times 0.0254^2 = 6.45 \times 10^{-3} \text{ m}^2$$

$$d = 0.01 \times 0.0254 = 2.54 \times 10^{-4} \text{ m}$$

and

$$C = \frac{6 \times 8.854 \times 10^{-12} \times 6.45 \times 10^{-3}}{2.54 \times 10^{-4}} = 1.349 \qquad \text{nF}$$

The large plate area is obtained in capacitors of small physical dimensions by stacking smaller plates in 50- or 100-decker sandwiches, or by rolling up foil plates separated by a flexible dielectric.

Table C.1 in Appendix C also indicates that materials are available having dielectric constants greater than 1000.

If more than two conductors are involved, *partial capacitances* between each pair of conductors must be defined. This is interestingly discussed in Maxwell's works.[7]

Finally, the total energy stored in the capacitor is

$$W_E = \frac{1}{2}\int_{\text{vol}} \epsilon E^2\, dv = \frac{1}{2}\int_0^S \int_0^d \frac{\epsilon \rho_S^2}{\epsilon^2}\, dz\, dS = \frac{1}{2}\frac{\rho_S^2}{\epsilon} Sd = \frac{1}{2}\frac{\epsilon S}{d}\frac{\rho_S^2 d^2}{\epsilon^2}$$

or

$$\boxed{W_E = \frac{1}{2} C V_0^2 = \frac{1}{2} Q V_0 = \frac{1}{2}\frac{Q^2}{C}} \tag{45}$$

[7] See the Suggested References at the end of the chapter.

which are all familiar expressions. Equation (45) also indicates that the energy stored in a capacitor with a fixed potential difference across it increases as the dielectric constant of the medium increases.

D5.11. Find the relative permittivity of the dielectric material used in a parallel-plate capacitor if: (*a*) $C = 40$ nF, $d = 0.1$ mm, and $S = 0.15\ \text{m}^2$; (*b*) $d = 0.2$ mm, $E = 500$ kV/m, and $\rho_S = 10\ \mu\text{C/m}^2$; (*c*) $D = 50\ \mu\text{C/m}^2$ and the energy density is $20\ \text{J/m}^3$.

Ans. 3.01; 2.26; 7.06

5.10 SEVERAL CAPACITANCE EXAMPLES

As a first brief example we choose a coaxial cable or coaxial capacitor of inner radius a, outer radius b, and length L. No great derivational struggle is required, because the potential difference is given as Eq. (11) in Sec. 4.3, and this is simply divided by the total charge $\rho_L L$ in the length L. Thus,

$$\boxed{C = \frac{2\pi\epsilon L}{\ln(b/a)}} \tag{46}$$

Next we consider a spherical capacitor formed of two concentric spherical conducting shells of radius a and b, $b > a$. The expression for the electric field was obtained previously by Gauss's law,

$$E_r = \frac{Q}{4\pi\epsilon r^2}$$

where the region between the spheres is a dielectric with permittivity ϵ. The expression for potential difference was found from this by the line integral. Thus,

$$V_{ab} = \frac{Q}{4\pi\epsilon}\left(\frac{1}{a} - \frac{1}{b}\right)$$

Here Q represents the total charge on the inner sphere, and the capacitance becomes

$$\boxed{C = \frac{Q}{V_{ab}} = \frac{4\pi\epsilon}{\dfrac{1}{a} - \dfrac{1}{b}}} \tag{47}$$

If we allow the outer sphere to become infinitely large, we obtain the capacitance of an isolated spherical conductor,

$$\boxed{C = 4\pi\epsilon a} \tag{48}$$

For a diameter of 1 cm, or a sphere about the size of a marble,

$$C = 0.556 \qquad \text{pF}$$

in free space.

Coating this sphere with a different dielectric layer, for which $\epsilon = \epsilon_1$, extending from $r = a$ to $r = r_1$,

$$D_r = \frac{Q}{4\pi r^2}$$

$$E_r = \frac{Q}{4\pi\epsilon_1 r^2} \qquad (a < r < r_1)$$

$$= \frac{Q}{4\pi\epsilon_0 r^2} \qquad (r_1 < r)$$

and the potential difference is

$$V_a - V_\infty = -\int_{r_1}^{a} \frac{Q\,dr}{4\pi\epsilon_1 r^2} - \int_{\infty}^{r_1} \frac{Q\,dr}{4\pi\epsilon_0 r^2}$$

$$= \frac{Q}{4\pi}\left[\frac{1}{\epsilon_1}\left(\frac{1}{a} - \frac{1}{r_1}\right) + \frac{1}{\epsilon_0 r_1}\right]$$

Therefore,

$$C = \frac{4\pi}{\dfrac{1}{\epsilon_1}\left(\dfrac{1}{a} - \dfrac{1}{r_1}\right) + \dfrac{1}{\epsilon_0 r_1}} \tag{49}$$

In order to look at the problem of multiple dielectrics a little more thoroughly, let us consider a parallel-plate capacitor of area S and spacing d, with the usual assumption that d is small compared to the linear dimensions of the plates. The capacitance is $\epsilon_1 S/d$, using a dielectric of permittivity ϵ_1. Now let us replace a part of this dielectric by another of permittivity ϵ_2, placing the boundary between the two dielectrics parallel to the plates (Fig. 5.15).

$$C = \frac{1}{\dfrac{d_1}{\epsilon_1 S} + \dfrac{d_2}{\epsilon_2 S}}$$

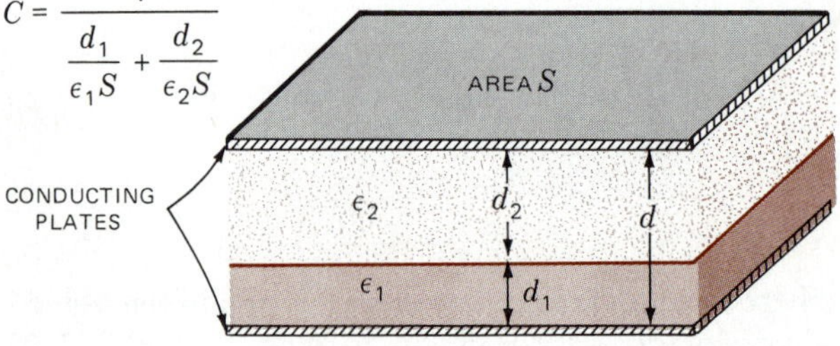

FIGURE 5.15
A parallel-plate capacitor containing two dielectrics with the dielectric interface parallel to the conducting plates.

Some of us may immediately suspect that this combination is effectively two capacitors in series, yielding a total capacitance of

$$C = \frac{1}{\dfrac{1}{C_1} + \dfrac{1}{C_2}}$$

where $C_1 = \epsilon_1 S/d_1$ and $C_2 = \epsilon_2 S/d_2$. This is the correct result, but we can obtain it using less intuition and a more basic approach.

Since our capacitance definition, $C = Q/V$, involves a charge and a voltage, we may assume either and then find the other in terms of it. The capacitance is not a function of either, but only of the dielectrics and the geometry. Suppose we assume a potential difference V_0 between the plates. The electric field intensities in the two regions, E_2 and E_1, are both uniform, and $V_0 = E_1 d_1 + E_2 d_2$. At the dielectric interface, E is normal and $D_{N1} = D_{N2}$, or $\epsilon_1 E_1 = \epsilon_2 E_2$. Eliminating E_2 in our V_0 relation, we have

$$E_1 = \frac{V_0}{d_1 + d_2(\epsilon_1/\epsilon_2)}$$

and the surface charge density therefore has the magnitude

$$\rho_{S1} = D_1 = \epsilon_1 E_1 = \frac{V_0}{\dfrac{d_1}{\epsilon_1} + \dfrac{d_2}{\epsilon_2}}$$

Since $D_1 = D_2$, the magnitude of the surface charge is the same on each plate. The capacitance is then

$$C = \frac{Q}{V_0} = \frac{\rho_S S}{V_0} = \frac{1}{\dfrac{d_1}{\epsilon_1 S} + \dfrac{d_2}{\epsilon_2 S}} = \frac{1}{\dfrac{1}{C_1} + \dfrac{1}{C_2}}$$

As an alternate (and slightly simpler) solution, we might assume a charge Q on one plate, leading to a charge density Q/S and a value of D that is also Q/S. This is true in both regions, as $D_{N1} = D_{N2}$ and D is normal. Then $E_1 = D/\epsilon_1 = Q/(\epsilon_1 S)$, $E_2 = D/\epsilon_2 = Q/(\epsilon_2 S)$, and the potential differences across the regions are $V_1 = E_1 d_1 = Qd_1/(\epsilon_1 S)$, and $V_2 = E_2 d_2 = Qd_2/(\epsilon_2 S)$. The capacitance is

$$C = \frac{Q}{V} = \frac{Q}{V_1 + V_2} = \frac{1}{\dfrac{d_1}{\epsilon_1 S} + \dfrac{d_2}{\epsilon_2 S}} \tag{50}$$

How would the method of solution or the answer change if there were a third conducting plane along the interface? We would now expect to find surface charge on each side of this conductor, and the magnitudes of these charges should be equal. In other words, we think of the electric lines not as passing directly from one outer plate to the other, but as terminating on one side of this interior plane and then continuing on the other side. The capacitance is unchanged, provided, of course, that the added conductor is of negligible thickness. The addition of a thick conducting plate will increase the capacitance if the separation of the outer plates is kept constant, and this is an example of a more general theorem which states that the replacement of any portion of the dielectric by a conducting body will cause an increase in the capacitance.

If the dielectric boundary were placed *normal* to the two conducting plates and the dielectrics occupied areas of S_1 and S_2, then an assumed potential difference V_0 would produce field strengths $E_1 = E_2 = V_0/d$. These are tangential fields at the interface, and they must be equal. Then we may find in succession D_1, D_2, ρ_{S1}, ρ_{S2}, and Q, obtaining a capacitance

$$C = \frac{\epsilon_1 S_1 + \epsilon_2 S_2}{d} = C_1 + C_2 \tag{51}$$

as we should expect.

At this time we can do very little with a capacitor in which two dielectrics are used in such a way that the interface is not everywhere normal or parallel to the fields. Certainly we know the boundary conditions at each conductor and at the dielectric interface; however, we do not know the fields to which to apply the boundary conditions. Such a problem must be put aside until our knowledge of field theory has increased and we are willing and able to use more advanced mathematical techniques.

D5.12. Find the capacitance of: (*a*) 20 cm of 58C/U coaxial cable having an inner conductor 0.0295 in in diameter, an outer conductor having an inside diameter of 0.116 in, and a polyethylene dielectric; (*b*) a conducting sphere 1 cm in diameter, covered with a layer of polyethylene 1 cm thick, in free space; (*c*) a conducting sphere 1 cm in diameter, covered with a layer of polyethylene 1 cm thick, and surrounded by a concentric conducting sphere 1.5 cm in radius.

Ans. 18.37 pF; 0.885 pF; 1.886 pF

D5.13. A parallel-plate capacitor similar to that shown in Fig. 5.15 contains three dielectric layers. Let $\epsilon_{R1} = 1$, $d_1 = 0.2$ mm, $\epsilon_{R2} = 2$, $d_2 = 0.3$ mm, $\epsilon_{R3} = 3$, $d_3 = 0.4$ mm, and $S = 20\ \text{cm}^2$. (*a*) Find C. (*b*) Find the percentage of the total stored energy located in each of the three regions.

Ans. 36.6 pF; 41.4%, 31.0%, 27.6%

5.11 CAPACITANCE OF A TWO-WIRE LINE

We conclude this chapter with the problem of the two-wire line. The final configuration will consist of two parallel conducting cylinders, each of circular cross section, and we shall be able to find complete information about the electric field intensity, the potential field, the surface-charge-density distribution, and the capacitance. This arrangement is an important type of transmission line, as is the coaxial cable we have discussed several times before.

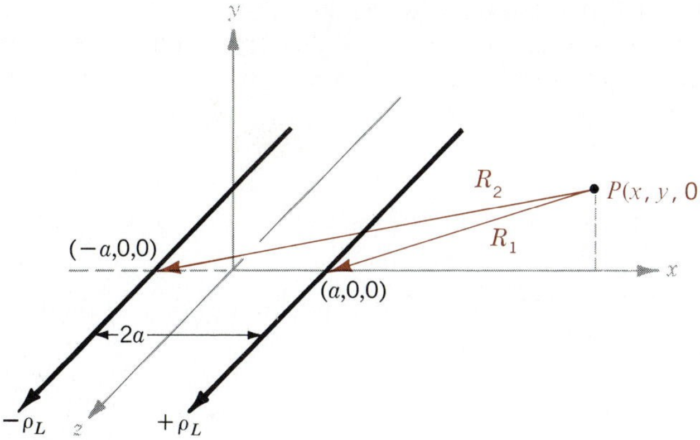

FIGURE 5.16
Two parallel infinite line charges carrying opposite charge. The positive line is at $x = a$, $y = 0$, and the negative line is at $x = -a$, $y = 0$. A general point $P(x,y,0)$ in the xy plane is radially distant R_1 and R_2 from the positive and negative lines, respectively. The equipotential surfaces are circular cylinders.

We begin by investigating the potential field of two infinite line charges. Figure 5.16 shows a positive line charge in the xz plane at $x = a$ and a negative line charge at $x = -a$. The potential of a single line charge with zero reference at a radius of R_0 is

$$V = \frac{\rho_L}{2\pi\epsilon} \ln \frac{R_0}{R}$$

We now write the expression for the combined potential field in terms of the radial distances from the positive and negative lines, R_1 and R_2, respectively,

$$V = \frac{\rho_L}{2\pi\epsilon}\left(\ln \frac{R_{10}}{R_1} - \ln \frac{R_{20}}{R_2}\right) = \frac{\rho_L}{2\pi\epsilon} \ln \frac{R_{10}R_2}{R_{20}R_1}$$

We choose $R_{10} = R_{20}$, thus placing the zero reference at equal distances from each line. This surface is the $x = 0$ plane. Expressing R_1 and R_2 in terms of x and y,

$$V = \frac{\rho_L}{2\pi\epsilon} \ln \sqrt{\frac{(x+a)^2 + y^2}{(x-a)^2 + y^2}} = \frac{\rho_L}{4\pi\epsilon} \ln \frac{(x+a)^2 + y^2}{(x-a)^2 + y^2} \tag{52}$$

In order to recognize the equipotential surfaces and adequately understand the problem we are going to solve, some algebraic manipulations are necessary. Choosing an equipotential surface $V = V_1$, we define K_1 as a dimensionless parameter that is a function of the potential V_1,

$$K_1 = e^{4\pi\epsilon V_1/\rho_L} \tag{53}$$

so that

$$K_1 = \frac{(x+a)^2 + y^2}{(x-a)^2 + y^2}$$

After multiplying and collecting like powers, we obtain

$$x^2 - 2ax\frac{K_1 + 1}{K_1 - 1} + y^2 + a^2 = 0$$

We next complete the square,

$$\left(x - a\frac{K_1 + 1}{K_1 - 1}\right)^2 + y^2 = \left(\frac{2a\sqrt{K_1}}{K_1 - 1}\right)^2$$

showing that the $V = V_1$ equipotential surface is independent of z (or is a cylinder) and intersects the xy plane in a circle of radius b,

$$b = \frac{2a\sqrt{K_1}}{K_1 - 1}$$

which is centered at $x = h$, $y = 0$, where

$$h = a\frac{K_1 + 1}{K_1 - 1}$$

Now let us attack a physical problem by considering a zero-potential conducting plane located at $x = 0$, and a conducting cylinder of radius b and potential V_0 with its axis located a distance h from the plane. We solve the last two equations for a and K_1 in terms of the dimensions b and h,

$$a = \sqrt{h^2 - b^2} \tag{54}$$

and

$$\sqrt{K_1} = \frac{h + \sqrt{h^2 - b^2}}{b} \tag{55}$$

But the potential of the cylinder is V_0, so (53) leads to

$$\sqrt{K_1} = e^{2\pi\epsilon V_0/\rho_L}$$

Therefore,

$$\rho_L = \frac{4\pi\epsilon V_0}{\ln K_1} \tag{56}$$

Thus, given h, b, and V_0, we may determine a, ρ_L, and the parameter K_1. The capacitance between the cylinder and plane is now available. For a length L in the z direction, we have

$$C = \frac{\rho_L L}{V_0} = \frac{4\pi\epsilon L}{\ln K_1} = \frac{2\pi\epsilon L}{\ln\sqrt{K_1}}$$

or

$$C = \frac{2\pi\epsilon L}{\ln\left[\left(h + \sqrt{h^2 - b^2}\right)/b\right]} = \frac{2\pi\epsilon L}{\cosh^{-1}(h/b)} \tag{57}$$

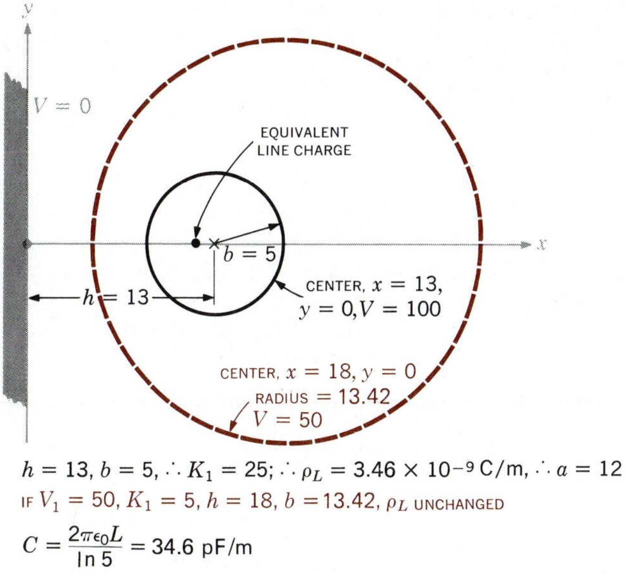

FIGURE 5.17
A numerical example of the capacitance, linear charge density, position of an equivalent line charge, and characteristics of the mid-equipotential surface for a cylindrical conductor of 5-m radius at a potential of 100 V, parallel to and 13 m from a conducting plane at zero potential.

The heavy black circle in Fig. 5.17 shows the cross section of a cylinder of 5-m radius at a potential of 100 V in free space, with its axis 13 m distant from a plane at zero potential. Thus, $b = 5$, $h = 13$, $V_0 = 100$, and we rapidly find the location of the equivalent line charge from (54),

$$a = \sqrt{h^2 - b^2} = \sqrt{13^2 - 5^2} = 12 \qquad \text{m}$$

the value of the potential parameter K_1 from (55),

$$\sqrt{K_1} = \frac{h + \sqrt{h^2 - b^2}}{b} = \frac{13 + 12}{5} = 5, \qquad K_1 = 25$$

the strength of the equivalent line charge from (56),

$$\rho_L = \frac{4\pi\epsilon V_0}{\ln K_1} = \frac{4\pi \times 8.854 \times 10^{-12} \times 100}{\ln 25} = 3.46 \qquad \text{nC/m}$$

and the capacitance between cylinder and plane from (57),

$$C = \frac{2\pi\epsilon}{\cosh^{-1}(h/b)} = \frac{2\pi \times 8.854 \times 10^{-12}}{\cosh^{-1}(13/5)} = 34.6 \qquad \text{pF/m}$$

We may also identify the cylinder representing the 50-V equipotential surface by finding new values for K_1, h, and b. We first use (53) to obtain

$$K_1 = e^{4\pi\epsilon V_1/\rho_L} = e^{4\pi \times 8.854 \times 10^{-12} \times 50/3.46 \times 10^{-9}} = 5.00$$

Then the new radius is

$$b = \frac{2a\sqrt{K_1}}{K_1 - 1} = \frac{2 \times 12\sqrt{5}}{5 - 1} = 13.42 \qquad \text{m}$$

and the corresponding value of h becomes

$$h = a\frac{K_1 + 1}{K_1 - 1} = 12\frac{5 + 1}{5 - 1} = 18 \qquad \text{m}$$

This cylinder is shown in color in Fig. 5.17.

The electric field intensity can be found by taking the gradient of the potential field, as given by (52),

$$\mathbf{E} = -\nabla\left[\frac{\rho_L}{4\pi\epsilon}\ln\frac{(x + a)^2 + y^2}{(x - a)^2 + y^2}\right]$$

Thus,

$$\mathbf{E} = -\frac{\rho_L}{4\pi\epsilon}\left[\frac{2(x + a)\mathbf{a}_x + 2y\mathbf{a}_y}{(x + a)^2 + y^2} - \frac{2(x - a)\mathbf{a}_x + 2y\mathbf{a}_y}{(x - a)^2 + y^2}\right]$$

and

$$\mathbf{D} = \epsilon\mathbf{E} = -\frac{\rho_L}{2\pi}\left[\frac{(x + a)\mathbf{a}_x + y\mathbf{a}_y}{(x + a)^2 + y^2} - \frac{(x - a)\mathbf{a}_x + y\mathbf{a}_y}{(x - a)^2 + y^2}\right]$$

If we evaluate D_x at $x = h - b$, $y = 0$, we may obtain $\rho_{S,\,\max}$:

$$\rho_{S,\,\max} = -D_{x,\,x=h-b,\,y=0} = \frac{\rho_L}{2\pi}\left[\frac{h - b + a}{(h - b + a)^2} - \frac{h - b - a}{(h - b - a)^2}\right]$$

For our example,

$$\rho_{S,\,\max} = \frac{3.46 \times 10^{-9}}{2\pi}\left[\frac{13 - 5 + 12}{(13 - 5 + 12)^2} - \frac{13 - 5 - 12}{(13 - 5 - 12)^2}\right] = 0.1650 \qquad \text{nC/m}^2$$

Similarly, $\rho_{S,\,\min} = D_{x,\,x=h+b,\,y=0}$, and

$$\rho_{S,\,\min} = -\frac{3.46 \times 10^{-9}}{2\pi}\left[\frac{13 + 5 + 12}{30^2} - \frac{13 + 5 - 12}{6^2}\right] = 0.0734 \qquad \text{nC/m}^2$$

Thus,

$$\rho_{S,\,\max} = 2.25\rho_{S,\,\min}$$

If we apply (57) to the case of a conductor for which $b \ll h$, then $[\ln(h + \sqrt{h^2 - b^2})/b] \doteq \ln[(h + h)/b] \doteq \ln(2h/b)$ and

$$C = \frac{2\pi\epsilon L}{\ln(2h/b)} \qquad (b \ll h) \tag{58}$$

The capacitance between two circular conductors separated by a distance $2h$ is one-half the capacitance given by (57) or (58). This last answer is of interest because it gives us an expression for the capacitance of a section of two-wire transmission line, one of the types of transmission lines studied later in Chap. 12.

D5.14. Find the capacitance between a circular conducting cylinder in air, 2 mm in radius, and: (*a*) a conducting plane, 1 cm distant from the cylinder axis, by Eq. (57), Sec. 5.11; (*b*) a conducting plane, 1 cm distant from the cylinder axis, by Eq. (58), Sec. 5.11; (*c*) a similar cylinder, axes separated by 1 cm.

Ans. 24.27 pF/m; 24.16 pF/m; 17.75 pF/m

SUGGESTED REFERENCES

1. Adler, R. B., A. C. Smith, and R. L. Longini: "Introduction to Semiconductor Physics," John Wiley & Sons, Inc., New York, 1964. Semiconductor theory is treated at an undergraduate level.
2. Dekker, A. J.: "Electrical Engineering Materials," Prentice-Hall, Inc., Englewood Cliffs, N.J., 1959. This admirable little book covers dielectrics, conductors, semiconductors, and magnetic materials.
3. Fano, R. M., L. J. Chu, and R. B. Adler: "Electromagnetic Fields, Energy, and Forces," John Wiley & Sons, Inc., New York, 1960. Polarization in dielectrics is discussed in the first part of chap. 5. This junior-level text presupposes a full-term physics course in electricity and magnetism, and it is therefore slightly more advanced in level. The introduction beginning on p. 1 should be read.
4. Fink, D. G., and H. W. Beaty: "Standard Handbook for Electrical Engineers," 12th ed., McGraw-Hill Book Company, New York, 1987.
5. Matsch, L. W.: "Capacitors, Magnetic Circuits, and Transformers," Prentice-Hall, Inc., Englewood Cliffs, N.J., 1964. Many of the practical aspects of capacitors are discussed in chap. 2.
6. Maxwell, J. C.: "A Treatise on Electricity and Magnetism," 3d ed., Oxford University Press, New York, 1904, or an inexpensive paperback edition, Dover Publications, Inc., New York, 1954.
7. Ramo, S., J. R. Whinnery, and T. Van Duzer: "Fields and Waves in Communications Electronics," 2d ed., John Wiley & Sons, Inc., New York, 1984. This book is essentially the fourth edition of the senior authors' popular texts of 1944 and 1953. Although it is directed primarily toward beginning graduate students, it may be profitably read by anyone who is familiar with basic electromagnetic concepts. Anisotropic dielectric materials are discussed on pp. 689–690.
8. Wert, C. A., and R. M. Thomson: "Physics of Solids," 2d ed., McGraw-Hill Book Company, New York, 1970. This is an advanced undergraduate-level text that covers metals, semiconductors, and dielectrics.

PROBLEMS

1. The current density in a certain region is expressed as $\mathbf{J} = -10^5 \nabla V$ A/m^2, where $V = 10e^{-x} \sin y$ V. (*a*) Find the current in the $\mathbf{a}_x$ direction crossing the surface $x = 1$ and bounded by $y = 0$, $y = 1$, $z = 0$, and $z = 1$. Find the total current leaving the cube $0 < x,y,z < 1$ by: (*b*) integrating $\mathbf{J} \cdot d\mathbf{S}$ over the surface of the cube; (*c*) using the divergence theorem on the closed surface integral.

2. Within the region $1 < \rho < 5$ cm, $0 < \phi < 0.3\pi$, $0 < z < 2$ cm, current density is given as $\mathbf{J} = (200 \cos \phi)\mathbf{a}_\phi/(\rho + 0.01)$ A/m^2. What total current in the $\mathbf{a}_\phi$ direction crosses the surface: (*a*) $\phi = 0$, $1 < \rho < 5$ cm, $0 < z < 2$ cm? (*b*) $\phi = 0.3\pi$, $1 < \rho < 5$ cm, and $0 < z < 2$ cm? (*c*) Calculate $\nabla \cdot \mathbf{J}$ at $\rho = 2$ cm, $\phi = 0.15\pi$, $z = 1$ cm.

3. A certain current density is given as $\mathbf{J} = [100 \cos \theta/(r^2 + 1)]\mathbf{a}_r$ A/m^2. (*a*) How much current flows through the spherical cap, $r = 3, 0 < \theta < \pi/6, 0 < \phi < 2\pi$? (*b*) What value would $\mathbf{J}_{av}$ have if the same total current were produced by a uniform current density $\mathbf{J}_{av}$ over the cap? (*c*) The same total current flows through the cap, $r = 10$, $0 < \theta < \alpha$, $0 < \phi < 2\pi$. Find α.

4. Electrons are emitted from a planar cathode at $z = 0$ with zero initial velocity. The region $z > 0$ is a vacuum in which $\mathbf{E} = -2.5 \times 10^6 \mathbf{a}_z$ V/m. Letting $e = 1.602 \times 10^{-19}$ C and $m = 9.11 \times 10^{-31}$ kg: (*a*) find $v(t)$ for an electron leaving the cathode at $t = 0$; (*b*) find $z(t)$; (*c*) find $v(z)$. (*d*) Assuming that electrons are emitted continuously as a beam of radius 0.2 mm and total current 100 μA, find $\mathbf{J}(z)$ and $\rho_v(z)$.

5. Let $V = 150x^{4/3}$ V for $x > 0$ with $\epsilon = \epsilon_0$. (*a*) Find **E**, **D**, and ρ_v as functions of x. (*b*) If the velocity of the charge density is $v_x = 6 \times 10^6 x^{2/3}$ m/s, find J_x at $x = 0$. (*b*) Repeat for $x = 1$ m.

6. Current density in a certain region is given as $\mathbf{J} = (5/\rho)\mathbf{a}_\rho + [10/(\rho^2 + 1)]\mathbf{a}_z$ A/m^2. (*a*) Find the total current crossing the surface $z = 3$, $\rho < 6$, in the $\mathbf{a}_z$ direction. (*b*) Find $\partial\rho_v/\partial t$. (*c*) Calculate the total current crossing the closed surface bounded by $z = 0$, $z = 3$, $\rho = 1$, and $\rho = 6$. (*d*) Is $\int_{vol} \nabla \cdot \mathbf{J}\, dv = \oint_S \mathbf{J} \cdot d\mathbf{S}$?

7. Assuming that there is no transformation of mass to energy or vice versa, it is possible to write a continuity equation for mass. (*a*) If we use the continuity equation for charge as our model, what quantities correspond to **J** and ρ_v? (*b*) Given a cube 1 cm on a side, experimental data show that the rates at which mass is leaving each of the six faces are 10.25, −9.85, 1.75, −2.00, −4.05, and 4.45 mg/s. If we assume that the cube is an incremental volume element, determine an approximate value for the time rate of change of density at its center.

8. After setting up a cartesian coordinate system inside a star, Captain Kirk and his intrepid crew make measurements over the faces of a cube centered at the origin with edges 40 km long and parallel to the coordinate axes. They find the mass rate of flow of material outward across the six faces to be −1112, 1183, 201, −196, 1989, and −1920 kg/km^2·s. (*a*) Estimate the divergence of the mass rate of flow at the origin. (*b*) Estimate the rate of change of the density at the origin.

9. (*a*) Using tabulated data from Appendix C, determine the length of 0.3-mm-diameter Nichrome wire required to dissipate an average power of 500 W when 120-V rms at 60 Hz is applied to it. (*b*) Find the maximum amplitude of the electric field intensity in the wire. (*c*) What is the rms current density inside the wire?

10. A conductor of circular cross section is constructed of steel ($\sigma = 6 \times 10^6$ ℧/m) in the region $0 < \rho < 1$ mm, and copper ($\sigma = 5.8 \times 10^7$ ℧/m), $1 < \rho < 1.5$ mm. The total current carried by the conductor is 100 A. Calculate: (*a*) J_{Cu}; (*b*) J_{st}; (*c*) E_{Cu}; (*d*) E_{st}; (*e*) the voltage drop in 1 km of conductor.

11. Two perfectly conducting spherical surfaces are located at $r = 2$ and $r = 10$ cm. The total current passing radially outward through the medium between the spheres is 2.5 A.

(*a*) Find the voltage and resistance between the spheres, and **E** in the region between the spheres, if a conducting material having $\sigma = 0.02$ ℧/m is present for $2 < r < 10$ cm. (*b*) Repeat if $\sigma = 0.001/r$ for $2 < r < 10$ cm. (*c*) Show that integrating the power dissipated per unit volume in part (*b*) over the volume gives the total dissipated power.

12. The cylindrical surfaces, $\rho = 2$ and 6 cm, are perfectly conducting, and the region between them is filled with a conducting material for which $\sigma = 80$ ℧/m. If the current density is $\mathbf{J} = (10/\pi\rho)\mathbf{a}_\rho$ A/m^2 for $2 < \rho < 6$ cm, find: (*a*) the current flowing from one perfect conductor to the other per meter length; (*b*) **E**; (*c*) the potential difference between perfect conductors; (*d*) the total power being dissipated in the conducting material per meter length.

13. The spherical surfaces, $r = 2$ and 6 cm, are perfectly conducting, and the region between them is filled with a conducting material for which $\sigma = 80$ ℧/m. If the current density is $\mathbf{J} = (10/\pi r^2)\mathbf{a}_r$ A/m^2 for $2 < r < 6$ cm, find: (*a*) the current flowing from one perfect conductor to the other; (*b*) **E**; (*c*) the potential difference between perfect conductors; (*d*) the total power being dissipated in the conducting material.

14. Find the magnitude of the current density in a conductor if: (*a*) the electric field intensity is 0.09 V/m, the volume charge density is -2×10^{10} C/m^3, and the electron mobility is 2.5×10^{-3} m^2/V·s; (*b*) there are 5×10^{28} conduction electrons/m^3 and the electron drift velocity is 0.05 mm/s; (*c*) $E = 0.05$ V/m and the resistivity is 2×10^{-8} Ω·m.

15. Given the potential field $V = (200 \sin\theta \cos\phi)/r^2$ V: (*a*) find the equation of the conductor surface on which $V = 100$ V; (*b*) find **E** at point $P(r,30°,30°)$ on the conductor surface; (*c*) find ρ_S at point P. Assume $\epsilon = \epsilon_0$ adjacent to the surface.

16. The potential field, $V = 2x^2 + 4y - 2z^2$ V, exists in the free space surrounding a perfectly conducting surface. Point $P(4,3,2)$ lies on the surface. (*a*) Give the equation of the surface. (*b*) Find a unit outward normal to the surface at P, assuming the origin is inside the surface. (*c*) Find ρ_S on the surface at P. (*d*) Find $|\mathbf{E}_P|$.

17. Given the potential field $V = 10y(x^3 + 5)$ V: (*a*) find **E** at the surface $y = 0$. (*b*) Show that the surface $y = 0$ is an equipotential surface. (*c*) If it is known that the surface $y = 0$ is a conductor, find the total charge in the region, $0 < x < 2$, $y = 0$, $0 < z < 1$. Assume that $\epsilon = \epsilon_0$ and that $V > 0$ in the region outside the conductor.

18. Given the field, $\mathbf{E} = 8xy(z+1)\mathbf{a}_x + 4x^2(z+1)\mathbf{a}_y + 4x^2y\mathbf{a}_z$ V/m in free space, and the information that point $P(2,3,5)$ lies on a conducting surface, find: (*a*) the magnitude of ρ_S at P; (*b*) ρ_v adjacent to the conductor surface at P; (*c*) V_{PQ} if Q is at the origin and **E** is a conservative field.

19. Given the potential field $V = 100e^{-50x} \sin 50y$ V, in free space: (*a*) show that $\nabla \cdot \mathbf{D} = 0$; (*b*) show that $y = 0$ is an equipotential surface; (*c*) show that **E** is perpendicular to the $y = 0$ plane; (*d*) find the total charge on the $y = 0$ plane, $0 < x < \infty$, $0 < z < 1$. Assume that $y < 0$ is the conductor interior. (*e*) How much energy is stored in the cube, $0 < x,y,z < 1$?

20. A potential field in free space is given as $V \doteq 10xyz - z + 5x$ V. (*a*) Find the equations of the equipotential surfaces on which $V = 0$ and 100 V. (*b*) These surfaces are conducting surfaces. Find the surface charge density on the $V = 100$ V surface at the point where $x = 2$, $y = 3$. Assume that $0 \le V \le 100$ V in the field-containing region. (*c*) Give a unit vector at the above point, normal to the conducting surface, and directed toward the $V = 0$ surface.

21. An infinite uniform line charge, $\rho_L = 25$ nC/m, lies along the line $x = 0$, $z = 1$ m in free space. The surface $z = 0$ is a perfect conductor. (*a*) Find **E** at $P(1,2,3)$. (*b*) Find V at P if $V = 0$ at the origin. (*c*) What is the maximum magnitude of ρ_S on the conducting plane?

22. The $z = 0$ plane is a perfectly conducting surface. A point charge of 5 nC is located at $A(2,-5,4)$, and a point charge of -8 nC is at $B(4,3,1)$, all in free space. (*a*) Find V at a point midway between the two point charges. (*b*) Find y if $V = 0$ at $C(5,y,1)$.

23. A dipole with $\mathbf{p} = 0.1\mathbf{a}_z\ \mu$C·m is located at point $A(0,0,1)$ in free space, and the $z = 0$ plane is perfectly conducting. (*a*) Find V at $P(0,1,2)$. (*b*) Find the equation of the 200-V equipotential surface in cartesian coordinates.

24. A circular ring, $\rho = 2$ m, $z = 1$ m, carries a uniform charge density of 8 nC/m, and the plane $z = 0$ is perfectly conducting. Assume the region $z > 0$ is free space. Find V at $P(0,0,z)$, $z > 0$.

25. Let $\mu_e = 0.45\ \text{m}^2/\text{V}\cdot\text{s}$ and $\mu_h = 0.25\ \text{m}^2/\text{V}\cdot\text{s}$ in intrinsic germanium at a certain temperature. Both the electron and hole concentrations are $2.5 \times 10^{19}\ \text{m}^{-3}$. (*a*) Determine σ and the resistivity for the germanium sample. (*b*) Calculate the resistance of a germanium block that is 1 cm long with a 4 mm by 1 mm cross section.

26. Electron and hole concentrations increase with temperature. For pure silicon suitable expressions are $\rho_h = -\rho_e = 6200T^{1.5}e^{-7000/T}\ \text{C/m}^3$. The functional dependence of the mobilities on temperature is given by $\mu_h = 2.3 \times 10^5 T^{-2.7}\ \text{m}^2/\text{V}\cdot\text{s}$ and $\mu_e = 2.1 \times 10^5 T^{-2.5}\ \text{m}^2/\text{V}\cdot\text{s}$. Find σ at: (*a*) 0°C; (*b*) 25°C; (*c*) 50°C.

27. A small amount of phosphorus is added to silicon so that there are 2.5×10^{17} conduction electrons per cubic meter and 9.0×10^{14} holes per cubic meter. Let $\mu_e = 0.13\ \text{m}^2/\text{V}\cdot\text{s}$ for this sample, while $\mu_h = 0.05\ \text{m}^2/\text{V}\cdot\text{s}$. Determine the conductivity and resistivity.

28. At a certain temperature and pressure, atomic hydrogen contains $5.67 \times 10^{25}\ \text{atoms/m}^3$. If the electric susceptibility is 4.1×10^{-4} with $E = 3$ kV/m, find: (*a*) P; (*b*) p; (*c*) the effective dipole length d; (*d*) ϵ_R.

29. A certain linear, homogeneous, isotropic, dielectric material has a relative permittivity of 1.8. If $V = -4000y$ V in the material, find: (*a*) **E**, **D**, and **P**; (*b*) ρ_v, ρ_b, and ρ_T.

30. A coax with $a = 1$ mm and $b = 5$ mm has a Teflon dielectric (see Appendix C). If $E_\rho = 80/\rho$ V/m, neglect any losses in the dielectric, assume $4 \times 10^{19}\ \text{molecules/m}^3$ in the Teflon, and find: (*a*) **E**, **D**, **P**, and **p** as functions of ρ; (*b*) V_{ab}, ϵ_R, and χ_e.

31. The polarization within a region having $\epsilon_R = 2.7$ has the uniform value $\mathbf{P} = -0.2\mathbf{a}_x + 0.7\mathbf{a}_y + 0.3\mathbf{a}_z\ \mu\text{C/m}^2$. Find: (*a*) **E**; (*b*) **D**; (*c*) the magnitude of the voltage gradient.

32. A unit vector directed from region 1 to region 2 at the planar boundary between two perfect dielectrics is given as $\mathbf{a}_{N12} = -\frac{2}{7}\mathbf{a}_x + \frac{3}{7}\mathbf{a}_y + \frac{6}{7}\mathbf{a}_z$. Assume $\epsilon_{R1} = 3$, $\epsilon_{R2} = 2$, and $\mathbf{E}_1 = 100\mathbf{a}_x + 80\mathbf{a}_y + 60\mathbf{a}_z$ V/m. Find $\mathbf{E}_2$.

33. The region $z < 0$ contains a perfect dielectric for which $\epsilon_{R1} = 2.5$, while the region $z > 0$ is characterized by $\epsilon_{R2} = 4$. Let $\mathbf{E}_1 = -30\mathbf{a}_x + 50\mathbf{a}_y + 70\mathbf{a}_z$ V/m and find: (*a*) E_{N1}; (*b*) $\mathbf{E}_{t1}$; (*c*) E_{t1}; (*d*) E_1; (*e*) the angle $\theta_1 \leq 90°$ between $\mathbf{E}_1$ and a normal to the surface; (*f*) D_{N2}; (*g*) D_{t2}; (*h*) $\mathbf{D}_2$; (*i*) $\mathbf{P}_2$; (*j*) the angle θ_2 between $\mathbf{E}_2$ and a normal to the surface.

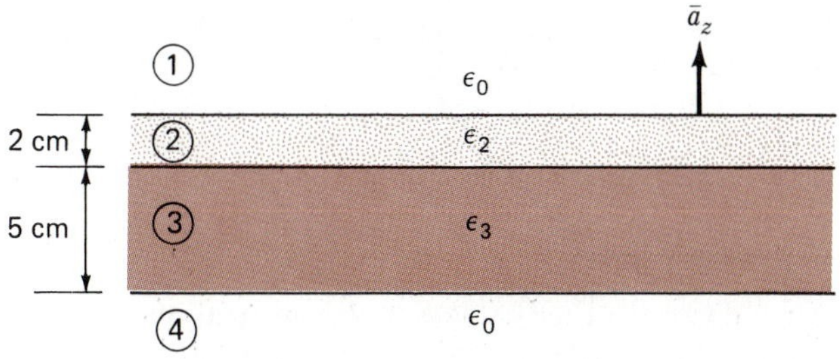

FIGURE 5.18
See Prob. 34.

34. Two infinite slabs of perfect dielectric are parallel and surrounded by free space, as shown in Fig. 5.18. In region 3, $\mathbf{E}_3 = 3\mathbf{a}_z$ V/m and $\mathbf{P}_3 = 15\mathbf{a}_z$ pC/m^2. If $\mathbf{P}_2 = 20\mathbf{a}_z$ pC/m^2, find: (*a*) ϵ_{R3}; (*b*) ϵ_{R2}; (*c*) $\mathbf{E}_1$; (*d*) the total voltage across regions 2 and 3.

35. In Fig. 5.19 let $\mathbf{D}_1 = 2\mathbf{a}_x + 5\mathbf{a}_y - 3\mathbf{a}_z$ nC/m^2 and find: (*a*) $\mathbf{D}_2$; (*b*) $\mathbf{D}_{N2}$; (*c*) $\mathbf{D}_{t2}$; (*d*) the energy density in each region; (*e*) the angle that $\mathbf{D}_2$ makes with $\mathbf{a}_z$; (*f*) D_2/D_1; (*g*) P_2/P_1.

36. Two concentric conducting spheres of radii $a = 5$ cm, $b = 20$ cm, have dielectrics $\epsilon_{R1} = 2, 0 < \theta < \pi/2$, and $\epsilon_{R2} = 5, \pi/2 < \theta < \pi$, separating them. Assume $E_{r1} = 200$ V/m at $r = 10$ cm, $\theta = 28°$, $\phi = 106°$, and find: (*a*) $\mathbf{E}_1$; (*b*) $\mathbf{E}_2$; (*c*) V_{ab}; (*d*) the total charge on the inner sphere.

37. A dielectric interface is defined by the plane, $4x - 2y + 3z = 4$. Region 1, which includes the point $P_1(0,0,100)$, has $\epsilon_{R1} = 3$, while region 2, which contains $P_2(0,0,-100)$, has $\epsilon_{R2} = 8$. (*a*) Find the unit vector that is perpendicular to the plane and directed into region 1. (*b*) Let point $A(1,3,z)$ lie on the plane. At A, $\mathbf{E}_1 = 10\mathbf{a}_x + 20\mathbf{a}_y + 40\mathbf{a}_z$ V/m. Find $\mathbf{E}_2$ at A.

38. A parallel-plate capacitor utilizes plates of area 0.06 m^2 and 0.4 mm separation. (*a*) If the region between the plates is filled with a dielectric for which $\epsilon_R = 200$ and $\sigma = 10^{-5}$ ℧/m, find C and the Q (quality factor) of the capacitor at $\omega = 10^7$ rad/s. (*b*) If the maximum voltage gradient allowable in the dielectric is 10^5 V/m, what is the breakdown voltage for the capacitor? (*c*) If 10 V dc is applied across the capacitor, find the energy stored and the power being dissipated.

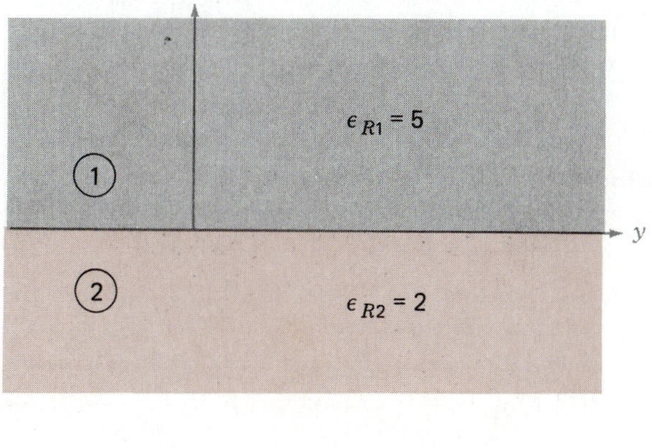

FIGURE 5.19
See Prob. 35.

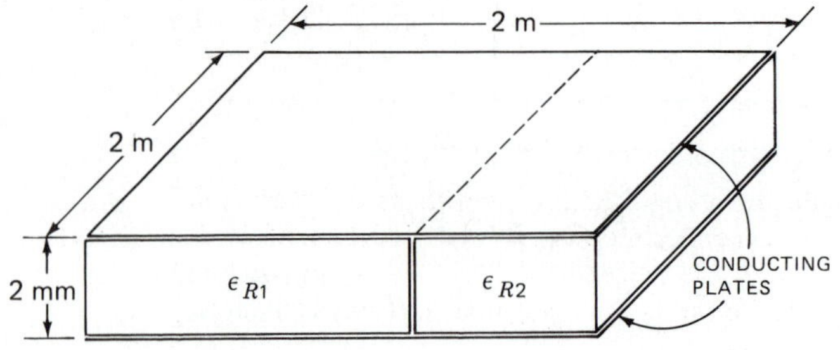

FIGURE 5.20
See Prob. 42.

39. Two parallel conducting plates are each 10 cm by 10 cm and separated 2 mm. The region between the plates is filled with a perfect dielectric for which $\epsilon_R = (1 + 500x)^2$, where x is the distance from one plate. Assume a uniform surface charge density of $10\ \text{nC/m}^2$ on the positive plate, and determine: (*a*) Q_{total}; (*b*) D_x; (*c*) E_x; (*d*) V_0; (*e*) C.

40. Consider a parallel-plate capacitor having $d = 5$ mm, $S = 80\ \text{cm}^2$, and $\epsilon_R = 10$. (*a*) Find the capacitance. (*b*) If a dc voltage source of 50 V is now placed across the capacitor, calculate E, D, Q, and the total stored energy. (*c*) The source is now disconnected and the dielectric is carefully removed from between the plates. Find Q, D, E, and the total stored energy. (*d*) Find V_0.

41. Capacitors tend to be more expensive as their capacitance C and maximum voltage V_{max} increase. The voltage V_{max} depends on the field strength at which the dielectric breaks down, E_{BD}. Which of these dielectrics will give the largest CV_{max} product for equal plate areas: (*a*) air: $\epsilon_R = 1$, $E_{BD} = 3 \times 10^6$ V/m; (*b*) barium titanate: $\epsilon_R = 1200$, $E_{BD} = 3 \times 10^6$ V/m; (*c*) silicon dioxide: $\epsilon_R = 3.78$, $E_{BD} = 1.6 \times 10^7$ V/m; (*d*) polyethylene: $\epsilon_R = 2.26$, $E_{BD} = 4.7 \times 10^6$?

42. (*a*) Find the dimensions of the two dielectric slabs shown in Fig. 5.20 if the total stored energy in region 1 is three times that stored in region 2, and $D_{1\,\text{max}} = 2D_{2\,\text{max}}$. (*b*) For these dimensions, find ϵ_{R1} and ϵ_{R2} if $C_{\text{total}} = 50$ nF.

43. Two conducting planes are located at $z = 0$ and 6 mm. In the region $0 < z < 2$ mm, there is a perfect dielectric for which $\epsilon_{R1} = 2$; for $2 < z < 5$ mm, let $\epsilon_{R2} = 5$. Find the capacitance per square meter of surface area if the region for which $5 < z < 6$ mm is: (*a*) air; (*b*) ϵ_{R1}; (*c*) ϵ_{R2}; (*d*) copper.

44. A conducting cylinder of 1-cm radius carries a uniformly distributed charge of 1 nC per meter length. The cylinder is coated with a material having a dielectric constant $\epsilon_{R1} = 2$ in a layer of thickness a_1. A second layer, ϵ_{R2} with thickness a_2, continues to the outer conductor of the coax, $b = 4$ cm in radius. Find ϵ_{R2}, a_1, and a_2 so that the voltage across each layer is the same and the capacitance of the coax is 75 pF per meter length.

45. Two conducting spherical shells have radii of $a = 2$ cm and $b = 5$ cm. The interior is a perfect dielectric for which $\epsilon_R = 10$. (*a*) Find C. (*b*) A portion of the dielectric is now removed so that $\epsilon_R = 1$, $0 < \theta < \pi/6$, and $\epsilon_R = 10$, $\pi/6 < \theta < \pi$. Find C.

46. A coaxial capacitor has dimensions $a = 3$ mm, $b = 12$ mm, and a length of 1 m. The region between the conducting cylinders contains three different dielectrics: $\epsilon_{R1} = 5$, $3 < \rho < 6$ mm; $\epsilon_{R2} = 3$, $6 < \rho < 9$ mm; and $\epsilon_{R3} = 1$, $9 < \rho < 12$ mm. (*a*) Calculate C. (*b*) Plot E_ρ vs. ρ if $E_{\rho(\text{max})} = 100$ V/m.

47. Change the conductor radius from 5 to 4 m in Fig. 5.17, maintaining $h = 13$ m and $V = 100$ V. Find new values for K_1, ρ_L, a, and C.

48. Given the potential function $V = -300 + 100 \ln \{[(8 + x)^2 + y^2]/[(8 - x)^2 + y^2]\}$ V, sketch the 0-V and 200-V equipotential surfaces in the region $x > 0$.

49. A 2-cm-diameter conductor is suspended in air with its axis 5 cm from a conducting plane. (*a*) Find the capacitance per unit length between cylinder and plane. (*b*) Let the potential of the cylinder be 2 V and that of the plane be 0 V, and find the shortest distance from the cylinder to the 1-V equipotential surface.

CHAPTER

6

EXPERIMENTAL MAPPING METHODS

We have seen in the last few chapters that the potential is the gateway to any information we desire about the electrostatic field at a point. The path is straight, and travel on it is easy in whichever direction we wish to go. The electric field intensity may be found from the potential by the gradient operation, which is a differentiation, and the electric field intensity may then be used to find the electric flux density by multiplying by the permittivity. The divergence of the flux density, again a differentiation, gives the volume charge density, and the surface charge density on any conductors in the field is quickly found by evaluating the flux density at the surface. Our boundary conditions show that it must be normal to such a surface.

Integration is still required if we need more information than the value of a field or charge density *at a point*. Finding the total charge on a conductor, the total energy stored in an electrostatic field, or a capacitance or resistance value are examples of such problems, each requiring an integration. These integrations cannot generally be avoided, no matter how extensive our knowledge of field theory becomes, and indeed, we should find that the greater this knowledge becomes, the more integrals we should wish to evaluate. Potential can do one important thing for us, and that is to furnish us quickly and easily with the quantity we must integrate.

Our goal, then, is to find the potential first. This cannot be done in terms of a charge configuration in a practical problem, because no one is kind enough to tell us exactly how the charges are distributed. Instead, we are usually given several conducting objects or conducting boundaries and the potential difference between them. Unless we happen to recognize the boundary surfaces as belonging to a simple problem we have already disposed of, we can do little now and must wait until Laplace's equation is discussed in the following chapter.

Although we thus postpone the mathematical solution to this important type of practical problem, we may acquaint ourselves with several experimental methods of finding the potential field. Some of these methods involve special equipment such as an electrolytic trough, a fluid-flow device, resistance paper and the associated bridge equipment, or rubber sheets; others use only pencil, paper, and a good supply of erasers. The *exact* potential can never be determined, but sufficient accuracy for engineering purposes can usually be attained. One other method, called the *iteration* method, does allow us to achieve any desired accuracy for the potential, but the number of calculations required increases very rapidly as the desired accuracy increases.

Several of the experimental methods to be described below are based on an analogy with the electrostatic field, rather than directly on measurements on this field itself.

Finally, we cannot introduce this subject of experimental methods of finding potential fields without emphasizing the fact that many practical problems possess such a complicated geometry that no exact method of finding that field is possible or feasible and experimental techniques are the only ones which can be used.

6.1 CURVILINEAR SQUARES

Our first method is a graphical one, requiring only pencil and paper. Besides being economical, it is also capable of yielding good accuracy if used skillfully and patiently. Fair accuracy (5 to 10 percent on a capacitance determination) may be obtained by a beginner who does no more than follow the few rules and hints of the art.

The method to be described is applicable only to fields in which no variation exists in the direction normal to the plane of the sketch. The procedure is based on several facts we have already demonstrated:

1. A conductor boundary is an equipotential surface.
2. The electric field intensity and electric flux density are both perpendicular to the equipotential surfaces.
3. **E** and **D** are therefore perpendicular to the conductor boundaries and possess zero tangential values.
4. The lines of electric flux, or streamlines, begin and terminate on charge and hence, in a charge-free, homogeneous dielectric, begin and terminate only on the conductor boundaries.

Let us consider the implications of these statements by drawing the streamlines on a sketch which already shows the equipotential surfaces. In Fig. 6.1*a* two conductor boundaries are shown, and equipotentials have been drawn with a constant potential difference between lines. We should remember that these lines are

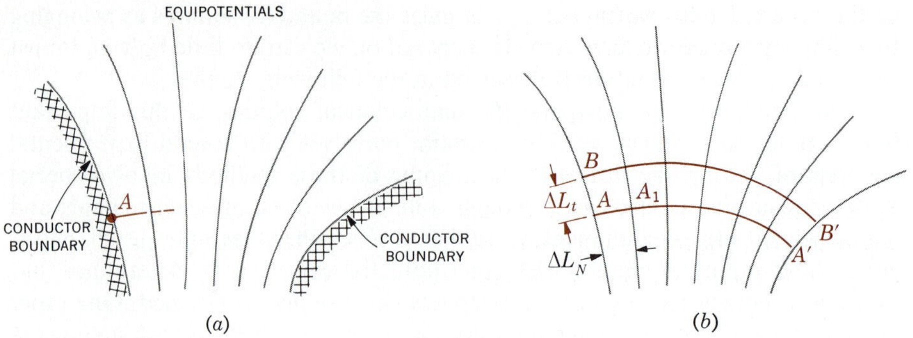

FIGURE 6.1
(*a*) Sketch of the equipotential surfaces between two conductors. The increment of potential between each of the two adjacent equipotentials is the same. (*b*) One flux line has been drawn from *A* to *A*′, and a second from *B* to *B*′.

only the cross sections of the equipotential surfaces, which are cylinders (although not circular), since no variation in the direction normal to the surface of the paper is permitted. We arbitrarily choose to begin a streamline, or flux line, at *A* on the surface of the more positive conductor. It leaves the surface normally and must cross at right angles the undrawn but very real equipotential surfaces between the conductor and the first surface shown. The line is continued to the other conductor, obeying the single rule that the intersection with each equipotential must be square. Turning the paper from side to side as the line progresses enables us to maintain perpendicularity more accurately. The line has been completed in Fig. 6.1*b*.

In a similar manner, we may start at *B* and sketch another streamline ending at *B*′. Before continuing, let us interpret the meaning of this pair of streamlines. The streamline, by definition, is everywhere tangent to the electric field intensity or to the electric flux density. Since the streamline is tangent to the electric flux density, the flux density is tangent to the streamline and no electric flux may cross any streamline. In other words, if there is a charge of 5 μC on the surface between *A* and *B* (and extending 1 m into the paper), then 5 μC of flux begins in this region and all must terminate between *A*′ and *B*′. Such a pair of lines is sometimes called a *flux tube*, because it physically seems to carry flux from one conductor to another without losing any.

We now wish to construct a third streamline, and both the mathematical and visual interpretations we may make from the sketch will be greatly simplified if we draw this line starting from some point *C* chosen so that the same amount of flux is carried in the tube *BC* as is contained in *AB*. How do we choose the position of *C*?

The electric field intensity at the midpoint of the line joining *A* to *B* may be found approximately by assuming a value for the flux in the tube *AB*, say $\Delta\Psi$, which allows us to express the electric flux density by $\Delta\Psi/\Delta L_t$, where the depth of the tube into the paper is 1 m and ΔL_t is the length of the line joining A to B. The magnitude of E is then

$$E = \frac{1}{\epsilon}\frac{\Delta\Psi}{\Delta L_t}$$

However, we may also find the magnitude of the electric field intensity by dividing the potential difference between points A and A_1, lying on two adjacent equipotential surfaces, by the distance from A to A_1. If this distance is designated ΔL_N and an increment of potential between equipotentials of ΔV is assumed, then

$$E = \frac{\Delta V}{\Delta L_N}$$

This value applies most accurately to the point at the middle of the line segment from A to A_1, while the previous value was most accurate at the midpoint of the line segment from A to B. If, however, the equipotentials are close together (ΔV small) and the two streamlines are close together ($\Delta\Psi$ small), the two values found for the electric field intensity must be approximately equal,

$$\frac{1}{\epsilon}\frac{\Delta\Psi}{\Delta L_t} = \frac{\Delta V}{\Delta L_N} \tag{1}$$

Throughout our sketch we have assumed a homogeneous medium (ϵ constant), a constant increment of potential between equipotentials (ΔV constant), and a constant amount of flux per tube ($\Delta\Psi$ constant). To satisfy all these conditions, (1) shows that

$$\boxed{\frac{\Delta L_t}{\Delta L_N} = \text{constant} = \frac{1}{\epsilon}\frac{\Delta\Psi}{\Delta V}} \tag{2}$$

A similar argument might be made at any point in our sketch, and we are therefore led to the conclusion that a constant ratio must be maintained between the distance between streamlines as measured along an equipotential, and the distance between equipotentials as measured along a streamline. It is this *ratio* which must have the same value at every point, not the individual lengths. Each length must decrease in regions of greater field strength, because ΔV is constant.

The simplest ratio we can use is unity, and the streamline from B to B' shown in Fig. 6.1*b* was started at a point for which $\Delta L_t = \Delta L_N$. Since the ratio of these distances is kept at unity, the streamlines and equipotentials divide the field-containing region into curvilinear squares, a term implying a planar geometric figure which differs from a true square in having slightly curved and slightly unequal sides but which approaches a square as its dimensions decrease. Those incremental surface elements in our three coordinate systems which are planar may also be drawn as curvilinear squares.

We may now rapidly sketch in the remainder of the streamlines by keeping each small box as square as possible. The complete sketch is shown in Fig. 6.2.

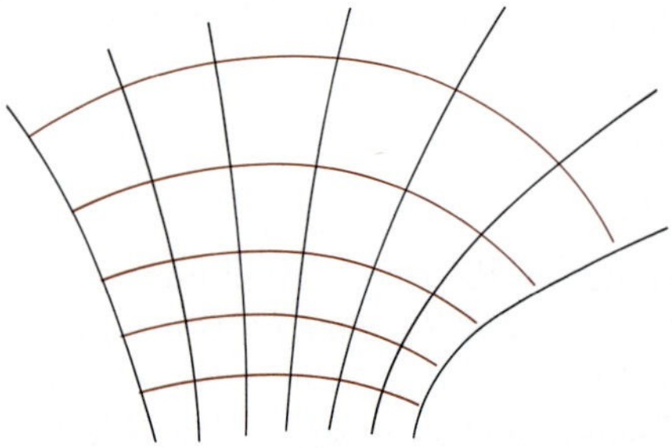

FIGURE 6.2
The remainder of the streamlines have been added to Fig. 6.1*b* by beginning each new line normally to the conductor and maintaining curvilinear squares throughout the sketch.

The only difference between this example and the production of a field map using the method of curvilinear squares is that the intermediate potential surfaces are not given. The streamlines and equipotentials must both be drawn on an original sketch which shows only the conductor boundaries. Only one solution is possible, as we shall prove later by the uniqueness theorem for Laplace's equation, and the rules we have outlined above are sufficient. One streamline is begun, an equipotential line is roughed in, another streamline is added, forming a curvilinear square, and the map is gradually extended throughout the desired region. Since none of us can ever expect to be perfect at this, we shall soon find that we can no longer make squares and also maintain right-angle corners. An error is accumulating in the drawing, and our present troubles should indicate the nature of the correction to make on some of the earlier work. It is usually best to start again on a fresh drawing, with the old one available as a guide.

The construction of a useful field map is an art; the science merely furnishes the rules. Proficiency in any art requires practice. A good problem for beginners is the coaxial cable or coaxial capacitor, since all the equipotentials are circles, while the flux lines are straight lines. The next sketch attempted should be two parallel circular conductors, where the equipotentials are again circles, but with a varying center. Each of these is given as a problem at the end of the chapter, and the accuracy of the sketch may be checked by a capacitance calculation as outlined below.

Figure 6.3 shows a completed map for a cable containing a square inner conductor surrounded by a circular conductor. The capacitance is found from $C = Q/V_0$ by replacing Q by $N_Q \Delta Q = N_Q \Delta\Psi$, where N_Q is the number of flux tubes joining the two conductors, and letting $V_0 = N_V \Delta V$, where N_V is the number of potential increments between conductors,

$$C = \frac{N_Q \Delta Q}{N_V \Delta V}$$

and then using (2),

$$\boxed{C = \frac{N_Q}{N_V} \epsilon \frac{\Delta L_t}{\Delta L_N} = \epsilon \frac{N_Q}{N_V}} \tag{3}$$

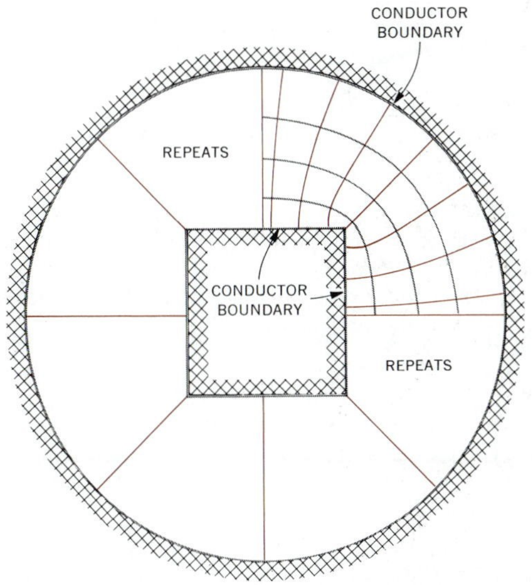

FIGURE 6.3
An example of a curvilinear-square field map. The side of the square is two-thirds the radius of the circle. $N_V = 4$ and $N_Q = 8 \times 3.25 = 26$, and therefore $C = \epsilon_0 N_Q/N_V = 57.6$ pF/m.

since $\Delta L_t/\Delta L_N = 1$. The determination of the capacitance from a flux plot merely consists of counting squares in two directions, between conductors and around either conductor. From Fig. 6.3 we obtain

$$C = \epsilon_0 \frac{8 \times 3.25}{4} = 57.6 \qquad \text{pF/m}$$

Ramo, Whinnery, and Van Duzer have an excellent discussion with examples of the construction of field maps by curvilinear squares. They offer the following suggestions:[1]

1. Plan on making a number of rough sketches, taking only a minute or so apiece, before starting any plot to be made with care. The use of transparent paper over the basic boundary will speed up this preliminary sketching.
2. Divide the known potential difference between electrodes into an equal number of divisions, say four or eight to begin with.
3. Begin the sketch of equipotentials in the region where the field is known best, as for example in some region where it approaches a uniform field. Extend the equipotentials according to your best guess throughout the plot. Note that they should tend to hug acute angles of the conducting boundary and be spread out in the vicinity of obtuse angles of the boundary.
4. Draw in the orthogonal set of field lines. As these are started, they should form curvilinear squares, but, as they are extended, the condition of orthogonality should be kept paramount, even though this will result in some rectangles with ratios other than unity.

[1] By permission from S. Ramo, J. R. Whinnery, and T. Van Duzer, pp. 54–55. See Suggested References at the end of Chap. 5. Curvilinear maps are discussed on pp. 53–58.

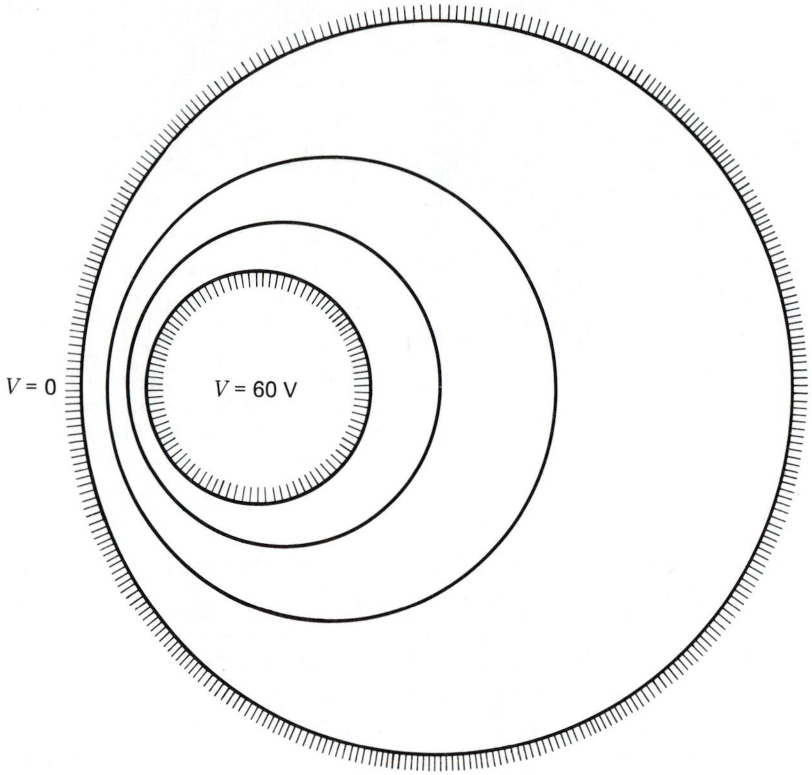

FIGURE 6.4
See Prob. D6.1.

5. Look at the regions with poor side ratios and try to see what was wrong with the first guess of equipotentials. Correct them and repeat the procedure until reasonable curvilinear squares exist throughout the plot.
6. In regions of low field intensity, there will be large figures, often of five or six sides. To judge the correctness of the plot in this region, these large units should be subdivided. The subdivisions should be started back away from the region needing subdivision, and each time a flux tube is divided in half, the potential divisions in this region must be divided by the same factor.

D6.1. Figure 6.4 shows the cross section of two circular cylinders at potentials of 0 and 60 V. The axes are parallel and the region between the cylinders is air-filled. Equipotentials at 20 V and 40 V are also shown. Prepare a curvilinear-square map on the figure and use it to establish suitable values for: (*a*) the capacitance per meter length; (*b*) E at the left side of the 60-V conductor if its true radius is 2 mm; (*c*) ρ_S at that point.

Ans. 69 pF/m; 60 kV/m; 550 nC/m^2

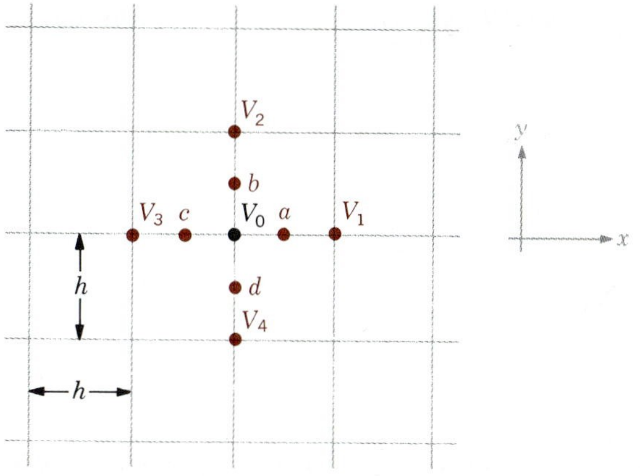

FIGURE 6.5
A portion of a region containing a two-dimensional potential field, divided into squares of side h. The potential V_0 is approximately equal to the average of the potentials at the four neighboring points.

6.2 THE ITERATION METHOD

In potential problems where the potential is completely specified on the boundaries of a given region, particularly problems in which the potential does not vary in one direction, i.e., two-dimensional potential distributions, there exists a pencil-and-paper repetitive method which is capable of yielding any desired accuracy. Digital computers should be used when the value of the potential is required with high accuracy; otherwise, the time required is prohibitive except in the simplest problems. The iterative method, to be described below, is well suited for calculation by any digital computer.

Let us assume a two-dimensional problem in which the potential does not vary with the z coordinate and divide the interior of a cross section of the region where the potential is desired into squares of length h on a side. A portion of this region is shown in Fig. 6.5. The unknown values of the potential at five adjacent points are indicated as V_0, V_1, V_2, V_3, and V_4. If the region is charge-free and contains a homogeneous dielectric, then $\nabla \cdot \mathbf{D} = 0$ and $\nabla \cdot \mathbf{E} = 0$, from which we have, in two dimensions,

$$\frac{\partial E_x}{\partial x} + \frac{\partial E_y}{\partial y} = 0$$

But the gradient operation gives $E_x = -\partial V/\partial x$ and $E_y = -\partial V/\partial y$, from which we obtain[2]

$$\frac{\partial^2 V}{\partial x^2} + \frac{\partial^2 V}{\partial y^2} = 0$$

[2] This is Laplace's equation in two dimensions. The three-dimensional form will be derived in the following chapter.

Approximate values for these partial derivatives may be obtained in terms of the assumed potentials, for

$$\left.\frac{\partial V}{\partial x}\right|_a \doteq \frac{V_1 - V_0}{h}$$

and

$$\left.\frac{\partial V}{\partial x}\right|_c \doteq \frac{V_0 - V_3}{h}$$

from which

$$\left.\frac{\partial^2 V}{\partial x^2}\right|_0 \doteq \frac{\left.\frac{\partial V}{\partial x}\right|_a - \left.\frac{\partial V}{\partial x}\right|_c}{h} \doteq \frac{V_1 - V_0 - V_0 + V_3}{h^2}$$

and similarly,

$$\left.\frac{\partial^2 V}{\partial y^2}\right|_0 \doteq \frac{V_2 - V_0 - V_0 + V_4}{h^2}$$

Combining, we have

$$\frac{\partial^2 V}{\partial x^2} + \frac{\partial^2 V}{\partial y^2} \doteq \frac{V_1 + V_2 + V_3 + V_4 - 4V_0}{h^2} = 0$$

or

$$\boxed{V_0 \doteq \tfrac{1}{4}(V_1 + V_2 + V_3 + V_4)} \tag{4}$$

The expression becomes exact as h approaches zero, and we shall write it without the approximation sign. It is intuitively correct, telling us that the potential is the average of the potential at the four neighboring points. The iterative method merely uses (4) to determine the potential at the corner of every square subdivision in turn, and then the process is repeated over the entire region as many times as is necessary until the values no longer change. The method is best shown in detail by an example.

For simplicity, consider a square region with conducting boundaries (Fig. 6.6). The potential of the top is 100 V and that of the sides and bottom is zero. The problem is two-dimensional, and the sketch is a cross section of the physical configuration. The region is divided first into 16 squares, and some estimate of the potential must now be made at every corner before applying the iterative method. The better the estimate, the shorter the solution, although the final result is independent of these initial estimates. When the computer is used for iteration, the initial potentials are usually set equal to zero to simplify the program. Reasonably accurate values could be obtained from a rough curvilinear-square map, or we could apply (4) to the large squares. At the center of the figure the potential estimate is then $\frac{1}{4}(100 + 0 + 0 + 0) = 25.0$.

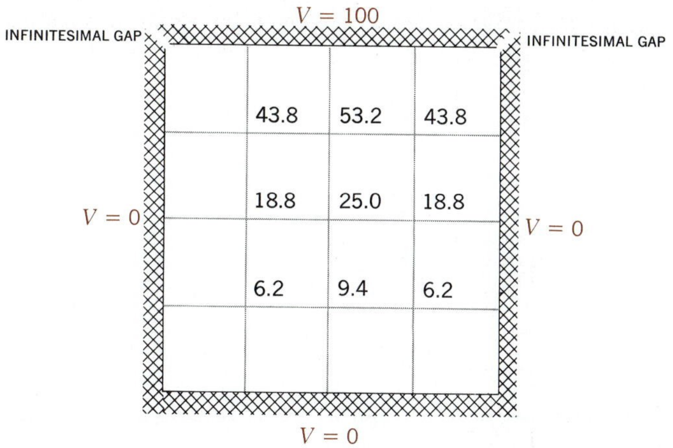

FIGURE 6.6
Cross section of a square trough with sides and bottom at zero potential and top at 100 V. The cross section has been divided into 16 squares, with the potential estimated at every corner. More accurate values may be determined by using the iteration method.

The potential may now be estimated at the centers of the four double-sized squares by taking the average of the potentials at the four corners or applying (4) along a diagonal set of axes. Use of this "diagonal average" is made only in preparing initial estimates. For the two upper double squares, we select a potential of 50 V for the gap (the average of 0 and 100), and then $V = \frac{1}{4}(50 + 100 + 25 + 0) = 43.8$ (to the nearest tenth of a volt[3]), and for the lower ones,

$$V = \tfrac{1}{4}(0 + 25 + 0 + 0) = 6.2$$

The potential at the remaining four points may now be obtained by applying (4) directly. The complete set of estimated values is shown in Fig. 6.6.

The initial traverse is now made to obtain a corrected set of potentials, beginning in the upper left corner (with the 43.8 value, not with the boundary where the potentials are known and fixed), working across the row to the right, and then dropping down to the second row and proceeding from left to right again. Thus the 43.8 value changes to $\frac{1}{4}(100 + 53.2 + 18.8 + 0) = 43.0$. The best or newest potentials are always used when applying (4), so both points marked 43.8 are changed to 43.0, because of the evident symmetry, and the 53.2 value becomes $\frac{1}{4}(100 + 43.0 + 25.0 + 43.0) = 52.8$.

Because of the symmetry, little would be gained by continuing across the top line. Each point of this line has now been improved once. Dropping down to the next line, the 18.8 value becomes

$$\tfrac{1}{4}(43.0 + 25.0 + 6.2 + 0) = 18.6$$

[3] When rounding off a decimal ending exactly with a five, the preceding digit should be made *even*; e.g., 42.75 becomes 42.8 and 6.25 becomes 6.2. This generally ensures a random process leading to better accuracy than would be obtained by always increasing the previous digit by 1.

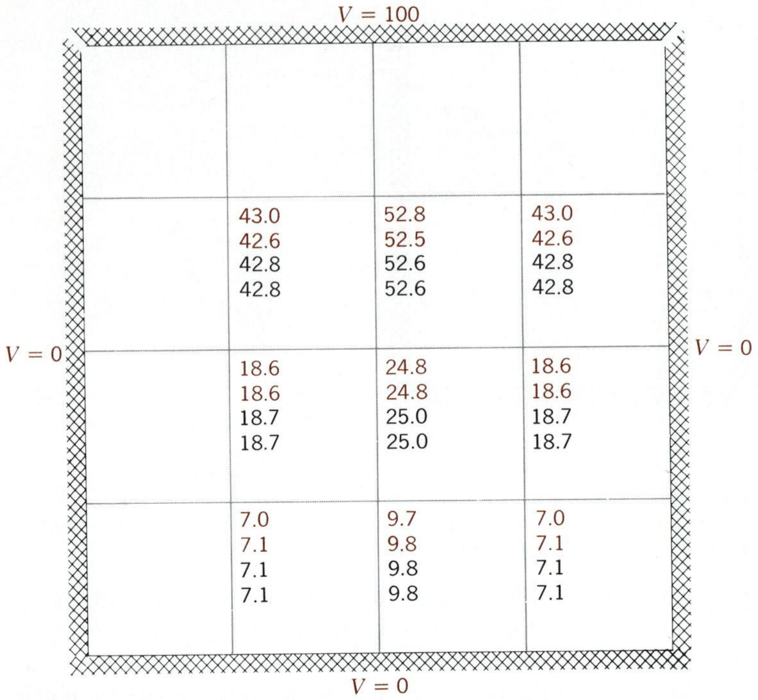

FIGURE 6.7
The results of each of the four necessary traverses of the problem of Fig. 6.6 are shown in order in the columns. The final values, unchanged in the last traverse, are at the bottom of each column.

and the traverse continues in this manner. The values at the end of this traverse are shown as the top numbers in each column of Fig. 6.7. Additional traverses must now be made until the value at each corner shows no change. The values for the successive traverses are usually entered below each other in column form, as shown in Fig. 6.7, and the final value is shown at the bottom of each column. Only four traverses are required in this example.

If each of the nine initial values were set equal to zero, it is interesting to note that ten traverses would be required. The cost of having a computer do these additional traverses is probably much less than the cost of the programming necessary to make decent initial estimates.

Since there is a large difference in potential from square to square, we should not expect our answers to be accurate to the tenth of a volt shown (and perhaps not to the nearest volt). Increased accuracy comes from dividing each square into four smaller squares, not from finding the potential to a larger number of significant figures at each corner.

In Fig. 6.8, which shows only one of the symmetrical halves plus an additional column, this subdivision is accomplished, and the potential at the newly created corners is estimated by applying (4) directly where possible and diagonally when

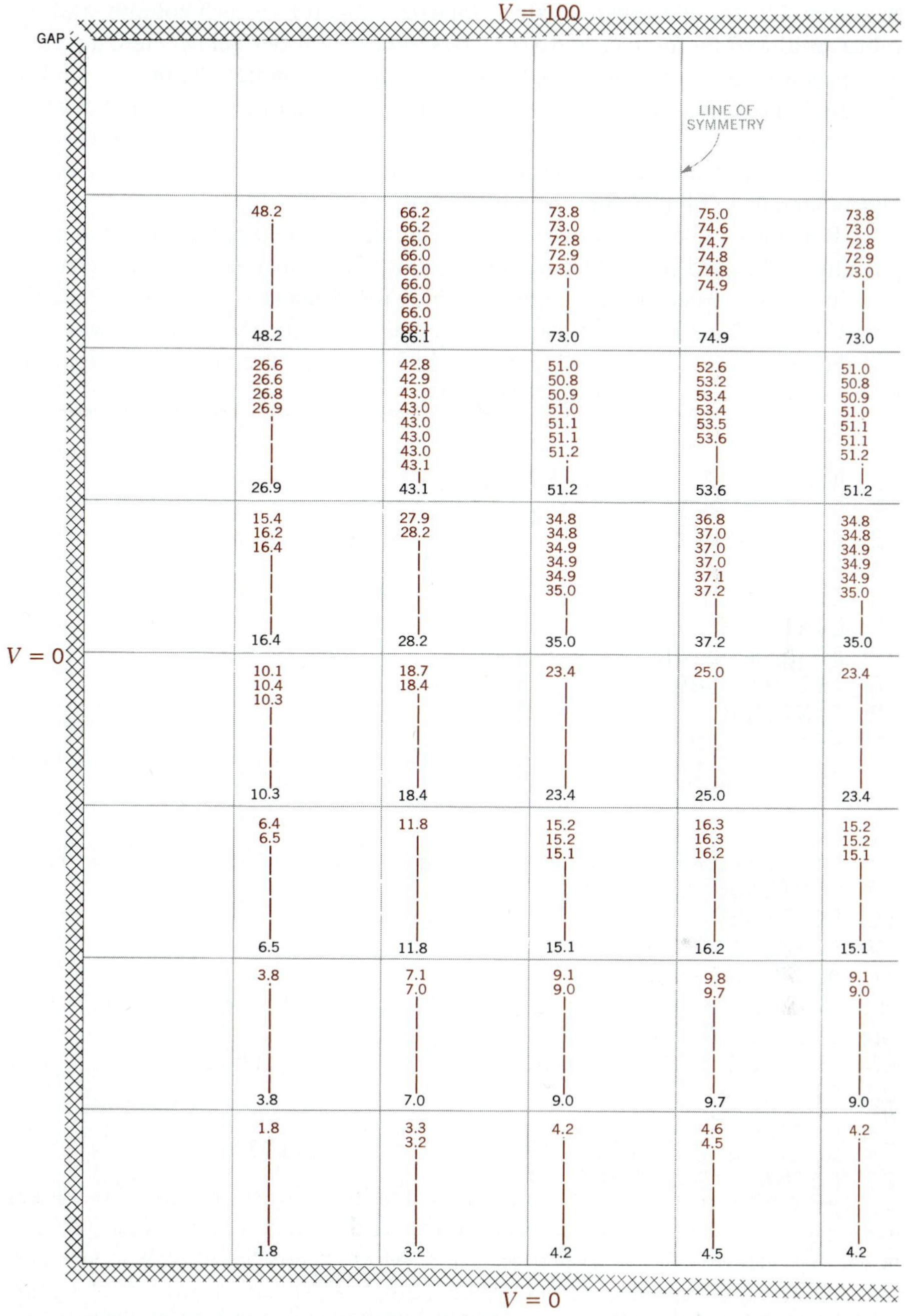

FIGURE 6.8
The problem of Figs. 6.6 and 6.7 is divided into smaller squares. Values obtained on the nine successive traverses are listed in order in the columns.

necessary. The set of estimated values appears at the top of each column, and the values produced by the successive traverses appear in order below. Here nine sets of values are required, and it might be noted that no values change on the last traverse (a necessary condition for the *last* traverse), and only one value changes on each of the preceding three traverses. No value in the bottom four rows changes after the second traverse; this results in a great saving in time, for if none of the four potentials in (4) changes, the answer is of course unchanged.

For this problem, it is possible to compare our final values with the exact potentials, obtained by evaluating some infinite series, as discussed at the end of the following chapter. At the point for which the original estimate was 53.2, the final value for the coarse grid was 52.6, the final value for the finer grid was 53.6, and the final value for a 16 × 16 grid is 53.93 V to two decimals, using data obtained with the following Fortran program:

```
 1 DIMENSION A(17,17),B(17,17)
 2 DO 6 I=2,17
 3 DO 5 J=1,17
 4 A(I,J)=0.
 5 CONTINUE
 6 CONTINUE
 7 DO 9 J=2,16
 8 A(1,J)=100.
 9 CONTINUE
10 A(1,1)=50.
11 A(1,17)=50.
12 DO 16 I=2,16
13 DO 15 J=2,16
14 A(I,J)=(A(I,J-1)+A(I-1,J)+A(I,J+1)+A(I+1,J))/4.
15 CONTINUE
16 CONTINUE
17 DO 23 I=2,16
18 DO 22 J=2,16
19 C=(A(I,J-1)+A(I-1,J)+A(I,J+1)+A(I+1,J))/4.
20 B(I,J)=A(I,J)-C
21 IF((ABS(B(I,J))-.00001).GT.0.) GO TO 12
22 CONTINUE
23 CONTINUE
24 WRITE(6,25)((A(I,J),J=1,17),I=1,17)
25 FORMAT (1H0,17F7.2)
26 STOP
27 END
```

Line 21 shows that the iteration is continued until the difference between successive traverses is less than 10^{-5}.

The exact potential obtained by a Fourier expansion is 54.05 V to two decimals. Two other points are also compared in tabular form, as shown in Table 6.1.

TABLE 6.1

Original estimate	53.2	25.0	9.4
4 × 4 grid	52.6	25.0	9.8
8 × 8 grid	53.6	25.0	9.7
16 × 16 grid	53.93	25.00	9.56
Exact	54.05	25.00	9.54

Computer flowcharts and programs for iteration solutions are given in chap. 24 of Boast[4] and in chap. 2 and the appendix of Silvester.[5]

Very few electrode configurations have a square or rectangular cross section that can be neatly subdivided into a square grid. Curved boundaries, acute- or obtuse-angled corners, reentrant shapes, and other irregularities require slight modifications of the basic method. An important one of these is described in Prob. 12 at the end of this chapter, and other irregular examples appear as Probs. 9, 11, and 13.

A refinement of the iteration method is known as the *relaxation method*. In general it requires less work but more care in carrying out the arithmetical steps.[6]

D6.2. In Fig. 6.9, a square grid is shown within an irregular potential trough. Using the iteration method to find the potential to the nearest volt, determine the final value at: (*a*) point *a*; (*b*) point *b*; (*c*) point *c*.

Ans. 26 V; 46 V; 78 V

6.3 CURRENT ANALOGIES

Several experimental methods depend upon an analogy between current density in conducting media and electric flux density in dielectric media. The analogy is easily demonstrated, for in a conducting medium Ohm's law and the gradient relationship are, for direct currents only,

$$\mathbf{J} = \sigma \mathbf{E}_\sigma$$

$$\mathbf{E}_\sigma = -\nabla V_\sigma$$

whereas in a homogeneous dielectric

$$\mathbf{D} = \epsilon \mathbf{E}_\epsilon$$

$$\mathbf{E}_\epsilon = -\nabla V_\epsilon$$

The subscripts serve to identify the analogous problems. It is evident that the potentials V_σ and V_ϵ, the electric field intensities $\mathbf{E}_\sigma$ and $\mathbf{E}_\epsilon$, the conductivity and

[4] See Suggested References at the end of Chap. 2.

[5] See Suggested References at the end of this chapter.

[6] A detailed description appears in Scarborough, and the basic procedure and one example are in Hayt. See Suggested References at the end of the chapter.

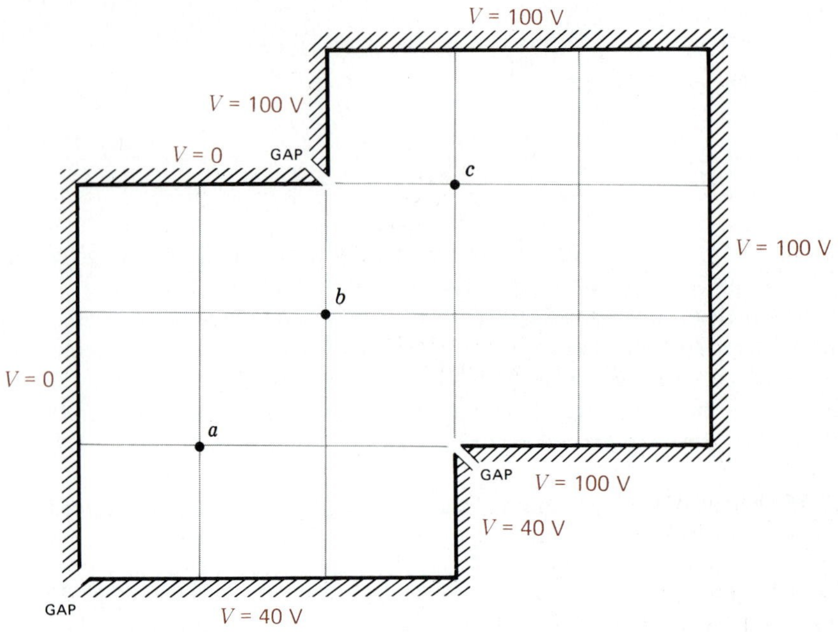

FIGURE 6.9
See Prob. D6.2.

permittivity σ and ϵ, and the current density and electric flux density **J** and **D** are analogous in pairs. Referring to a curvilinear-square map, we should interpret flux tubes as current tubes, each tube now carrying an incremental current which cannot leave the tube.

Finally, we must look at the boundaries. What is analogous to a conducting boundary which terminates electric flux normally and is an equipotential surface? The analogy furnishes the answer, and we see that the surface must terminate current density normally and again be an equipotential surface. This is the surface of a *perfect* conductor, although in practice it is necessary only to use one whose conductivity is many times that of the conducting medium.

Therefore, if we wished to find the field within a coaxial capacitor, which, as we have seen several times before, is a portion of the field of an infinite line charge, we might take two copper cylinders and fill the region between them with, for convenience, an electrolytic solution. Applying a potential difference between the cylinders, we may use a probe to establish the potential at any intermediate point, or to find all those points having the same potential. This is the essence of the electrolytic trough or tank. The greatest advantage of this method lies in the fact that it is not limited to two-dimensional problems. Practical suggestions for the construction and use of the tank are given in many places.[7]

---

[7] Weber is good. See Suggested References at the end of the chapter.

The determination of capacitance from electrolytic-trough measurements is particularly easy. The total current leaving the more positive conductor is

$$I = \oint_S \mathbf{J} \cdot d\mathbf{S} = \sigma \oint_S \mathbf{E}_\sigma \cdot d\mathbf{S}$$

where the closed surface integral is taken over the entire conductor surface. The potential difference is given by the negative line integral from the less to the more positive plate,

$$V_{\sigma 0} = -\int \mathbf{E}_\sigma \cdot d\mathbf{L}$$

and the total resistance is therefore

$$R = \frac{V_{\sigma 0}}{I} = \frac{-\int \mathbf{E}_\sigma \cdot d\mathbf{L}}{\sigma \oint_S \mathbf{E}_\sigma \cdot d\mathbf{S}}$$

The capacitance is given by the ratio of the total charge to the potential difference,

$$C = \frac{Q}{V_{\epsilon 0}} = \frac{\epsilon \oint_S \mathbf{E}_\epsilon \cdot d\mathbf{S}}{-\int \mathbf{E}_\epsilon \cdot d\mathbf{L}}$$

We now invoke the analogy by letting $V_{\epsilon 0} = V_{\sigma 0}$ and $\mathbf{E}_\epsilon = \mathbf{E}_\sigma$. The result is

$$\boxed{RC = \frac{\epsilon}{\sigma}} \tag{5}$$

Knowing the conductivity of the electrolyte and the permittivity of the dielectric, we may determine the capacitance by a simple resistance measurement.

A simpler technique is available for two-dimensional problems. Conducting paper is used as the base on which the conducting boundaries are drawn with silver paint. In the case of the coaxial capacitor, we should draw two circles of radii ρ_A and ρ_B, $\rho_B > \rho_A$, extending the paint a small distance outward from ρ_B and inward from ρ_A to provide sufficient area to make a good contact with wires to an external potential source. A probe is again used to establish potential values between the circles.

Conducting paper is described in terms of its *sheet resistance* R_s. The sheet resistance is the resistance between opposite edges of a square. Since the fields are uniform in such a square, we may apply Eq. (13) from Chap. 5,

$$R = \frac{L}{\sigma S}$$

to the case of a square of conducting paper having a width w and a thickness t, where $w = L$,

$$R_s = \frac{L}{\sigma t L} = \frac{1}{\sigma t} \quad \Omega \tag{6}$$

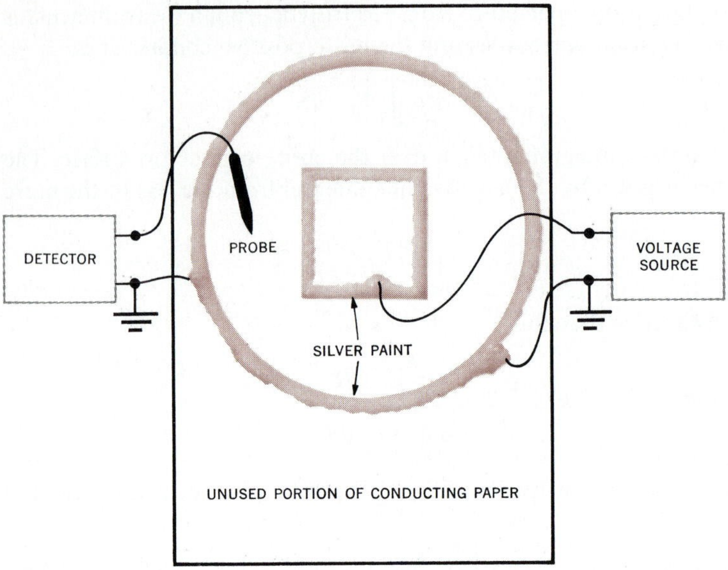

FIGURE 6.10
A two-dimensional, two-conductor problem similar to that of Fig. 6.3 is drawn on conducting paper. The probe may be used to trace out an equipotential surface.

Thus, if the conductive coating has a thickness of 0.2 mm and a conductivity of 2 ℧/m, its surface resistance is $1/(2 \times 0.2 \times 10^{-3}) = 2500\ \Omega$. The units of R_s are often given as "ohms per square" (but never as ohms per square meter).

Figure 6.10 shows the silver-paint boundaries that would be drawn on the conducting paper to determine the capacitance of a square-in-a-circle transmission line like that shown in Fig. 6.3. The generator and detector often operate at 1 kHz to permit use of a more sensitive tuned detector or bridge.

D6.3. Let us assume that the conducting paper shown in Fig. 6.10 has a sheet resistance of 2000 Ω per square. What resistance would be measured between the opposite edges of: (*a*) a square 10 cm on a side; (*b*) a square 8 cm on a side; (*c*) a rectangle 3 cm by 10 cm, across the longer dimension; (*d*) a rectangle 3 cm by 10 cm, across the shorter dimension? (*e*) What resistance would be measured between an inner circle of 1 cm radius and an outer circle of 4 cm radius drawn on this paper?

Ans. 2000 Ω; 2000 Ω; 6670 Ω; 600 Ω; 441 Ω

6.4 PHYSICAL MODELS

The analogy between the electric field and the gravitational field has been mentioned several times previously and may be used to construct physical models which are capable of yielding solutions to electrostatic problems of complicated geometry. The basis of the analogy is simply this: in the electrostatic field the potential difference between two points is the difference in the potential energy of unit positive charges at these points, and in a uniform gravitational field the difference in the potential energy of point masses at two points is proportional to their difference in height. In other words,

$$\Delta W_E = Q\,\Delta V \qquad \text{(electrostatic)}$$

$$\Delta W_G = Mg\,\Delta h \qquad \text{(gravitational)}$$

where M is the point mass and g is the acceleration due to gravity, essentially constant at the surface of the earth. For the same energy difference, then,

$$\Delta V = \frac{Mg}{Q}\,\Delta h = k\,\Delta h$$

where k is the constant of proportionality. This shows the direct analogy between difference in potential and difference in elevation.

This analogy allows us to construct a physical model of a known two-dimensional potential field by fabricating a surface, perhaps from wood, whose elevation h above any point (x,y) located in the zero-elevation zero-potential plane is proportional to the potential at that point. Note that three-dimensional fields cannot be handled.

The field of an infinite line charge,

$$V = \frac{\rho_L}{2\pi\epsilon} \ln \frac{\rho_B}{\rho}$$

is shown on such a model in Fig. 6.11, which provides an accurate picture of the variation of potential with radius between ρ_A and ρ_B. The potential and elevation at ρ_B are conveniently set equal to zero.

Such a model may be constructed for any two-dimensional potential field and enables us to visualize the field a little better. The construction of the models themselves is enormously simplified, both physically and theoretically, by the use

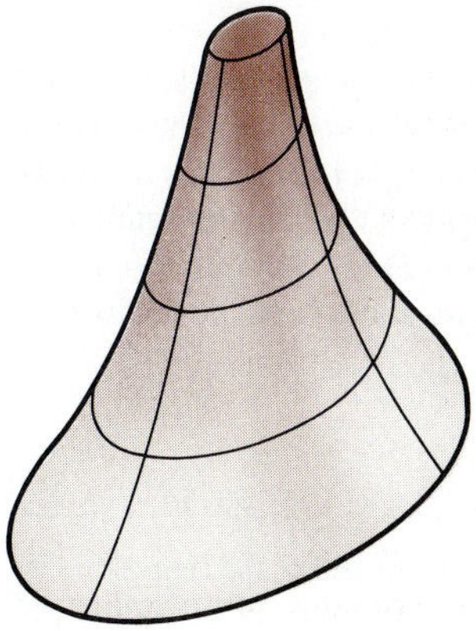

FIGURE 6.11
A model of the potential field of an infinite line charge. The difference in potential is proportional to the difference in elevation. Contour lines indicate equal potential increments.

of rubber sheets. The sheet is placed under moderate tension and approximates closely the *elastic membrane* of applied mechanics. It can be shown[8] that the vertical displacement h of the membrane satisfies the second-order partial differential equation

$$\frac{\partial^2 h}{\partial x^2} + \frac{\partial^2 h}{\partial y^2} = 0$$

if the surface slope is small. We shall see in the next chapter that every potential field in a charge-free region also satisfies this equation, Laplace's equation in two dimensions,

$$\frac{\partial^2 V}{\partial x^2} + \frac{\partial^2 V}{\partial y^2} = 0$$

We shall also prove a uniqueness theorem which assures us that if a potential solution in some specified region satisfies the above equation and also gives the correct potential on the boundaries of this region, then this solution is the only solution. Hence we need only force the elevation of the sheet to corresponding prescribed potential values on the boundaries, and the elevation at all other points is proportional to the potential.

For instance, the infinite-line-charge field may be displayed by recognizing the circular symmetry and fastening the rubber sheet at zero elevation around a circle by the use of a large clamping ring of radius ρ_B. Since the potential is constant at ρ_A, we raise that portion of the sheet to a greater elevation by pushing a cylinder of radius ρ_A up against the rubber sheet. The analogy breaks down for large surface slopes, and only a slight displacement at ρ_A is possible. The surface then represents the potential field, and marbles may be used to determine particle trajectories, in this case obviously radial lines as viewed from above.

There is also an analogy between electrostatics and hydraulics that is particularly useful in obtaining photographs of the streamlines or flux lines. This process is described completely by Moore in a number of publications[9] which include many excellent photographs.

D6.4. The potential field, $V = 100(xy + x - y - 1)$ V is illustrated by a plaster model with a scale of 1 vertical inch = 100 V, and 1 horizontal inch = 0.1 m. The model is prepared for the region $1 \leq x \leq 2$ m, $0 \leq y \leq 2$ m. In this region: (*a*) what is the maximum height of the model; (*b*) what is its minimum height? (*c*) What is the difference in height between points above $A(x = 15$ in, $y = 10$ in$)$ and $B(x = 14.99$ in, $y = 10$ in$)$? (*d*) What angle does the line connecting these two points make with the horizontal?

Ans. 3 in; 0; 0.002 in; 11.31°

---

[8] See, for instance, Spangenberg, pp. 75–76, in Suggested References at the end of the chapter.

[9] See Suggested References at the end of this chapter.

SUGGESTED REFERENCES

1. Hayt, W. H., Jr.: "Engineering Electromagnetics," 1st ed., McGraw-Hill Book Company, New York, 1958, pp. 150–152.
2. Moore, A. D.: Fields from Fluid Flow Mappers, *J. Appl. Phys.*, vol. 20, pp. 790–804, August 1949; Soap Film and Sandbed Mapper Techniques, *J. Appl. Mech.* (bound with *Trans. ASME*), vol. 17, pp. 291–298, September 1950; Four Electromagnetic Propositions, with Fluid Mapper Verifications, *Elec. Eng.*, vol. 69, pp. 607–610, July 1950; The Further Development of Fluid Mappers, *Trans. AIEE*, vol. 69, part II, pp. 1615–1624, 1950; Mapping Techniques Applied to Fluid Mapper Patterns, *Trans. AIEE*, vol. 71, part I, pp. 1–5, 1952.
3. Salvadori, M. G., and M. L. Baron: "Numerical Methods in Engineering," 2d ed., Prentice-Hall, Inc., Englewood Cliffs, N.J., 1961. Iteration and relaxation methods are discussed in chap. 1.
4. Scarborough, J. B.: "Numerical Mathematical Analysis," 6th ed., The Johns Hopkins Press, Baltimore, 1966. Describes iteration relaxation methods and gives several complete examples. Inherent errors are discussed.
5. Silvester, P.: "Modern Electromagnetic Fields," Prentice-Hall, Inc., Englewood Cliffs, N.J., 1968.
6. Soroka, W. W.: "Analog Methods in Computation and Simulation," McGraw-Hill Book Company, New York, 1954.
7. Spangenberg, K. R.: "Vacuum Tubes," McGraw-Hill Book Company, New York, 1948. Experimental mapping methods are discussed on pp. 75–82.
8. Weber, E.: "Electromagnetic Fields," vol. I, John Wiley & Sons, Inc., New York, 1950. Experimental mapping methods are discussed in chap. 5.

PROBLEMS

1. Construct a curvilinear-square map for a coaxial capacitor of 2.5-cm inner radius and 8-cm outer radius. These dimensions are suitable for the drawing. As a check on the accuracy, compute the capacitance per meter length both from your drawing and from the exact formula if $\epsilon_R = 1$.

2. Construct a curvilinear-square map of the potential field about two parallel circular cylinders, each of 2-cm radius, separated a center-to-center distance of 13 cm. These dimensions are suitable for the actual sketch if symmetry is considered. As a check, compute the capacitance per meter both from your sketch and from the exact formula. Assume $\epsilon_R = 1$.

3. Construct a curvilinear-square map of the potential field between two parallel circular cylinders, one of 40-mm radius inside one of 90-mm radius. The two axes are displaced by 25 mm. These dimensions are suitable for the drawing. As a check on the accuracy, compute the capacitance per meter from the sketch and from the exact expression,

$$C = 2\pi\epsilon \Bigg/ \left[\cosh^{-1}\left(\frac{a^2 + b^2 - D^2}{2ab}\right)\right] \qquad \text{F/m}$$

where a and b are the conductor radii and D is the axis separation. Let $\epsilon_R = 1$.

4. A solid conducting cylinder with a 0.08×0.12 m rectangular cross section is centered within a circular cylinder of 0.1-m radius. (*a*) Make a full-size sketch of one quadrant

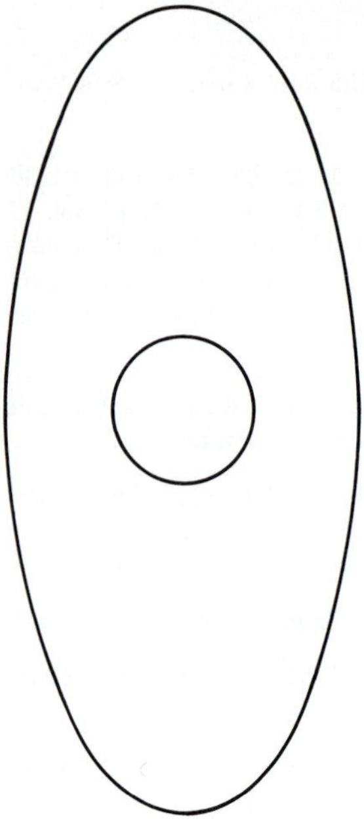

FIGURE 6.12
See Prob. 5.

of this configuration and construct a curvilinear-square map for its interior. (*b*) Assuming $\epsilon = \epsilon_0$, estimate C.

5. The "coaxial" transmission line shown in Fig. 6.12 was manufactured on a day when some of the machinery was slightly out of adjustment. Assuming an air dielectric: (*a*) prepare a curvilinear-square map of the interior and (*b*) estimate the capacitance of a 1-m length.

6. The inner conductor of the transmission line shown in Fig. 6.13 has a square cross section while the outer conductor is rectangular. (*a*) Construct a good-sized drawing of this transmission line, say with $a = 1$ in, and prepare a curvilinear-square plot of the electrostatic field between conductors. (*b*) Use your map to calculate the capacitance per meter length for $\epsilon = 1.6\epsilon_0$. (*c*) How would the answer to part (*b*) change if $a = 0.6$ cm?

7. A square conductor, 1 cm on a side, is parallel to and 1 cm from a perfectly conducting plane. (*a*) Prepare a curvilinear-square map, using symmetry, a scale of 2 to 1, and a manageable width. (*b*) Assuming free space around the conductor, calculate the capacitance per meter length of this transmission line.

8. Let the inner conductor of the transmission line shown in Fig. 6.13 be at a potential of 100 V, while the outer is at zero potential. Construct a grid, $0.5a$ on a side, and use iteration to find V at a point a units above the upper left corner of the inner conductor. Work to the nearest volt.

9. Use the iteration method to estimate the potential at points x and y in the triangular trough of Fig. 6.14. Work only to the nearest volt.

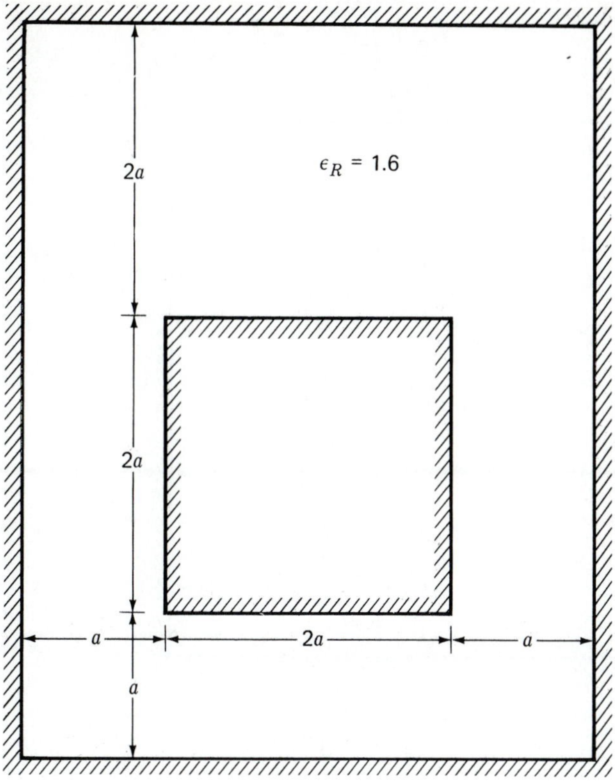

FIGURE 6.13
See Probs. 6, 8, and 16.

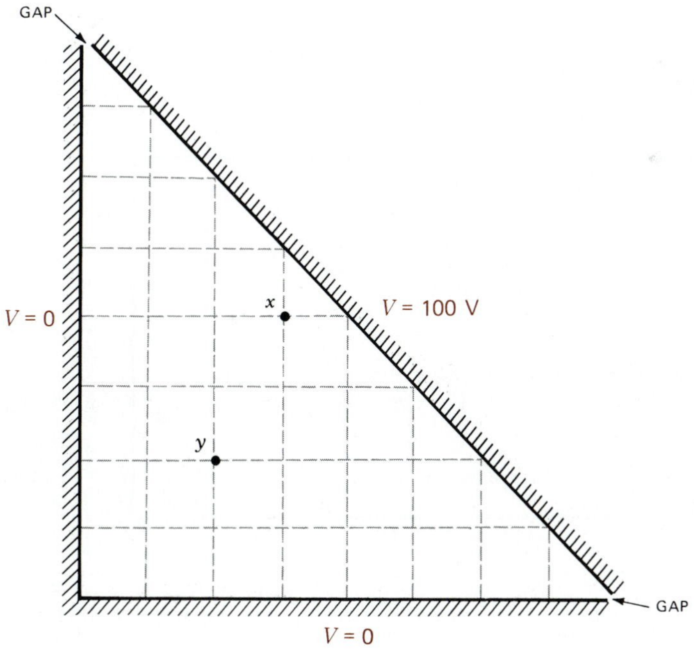

FIGURE 6.14
See Prob. 9.

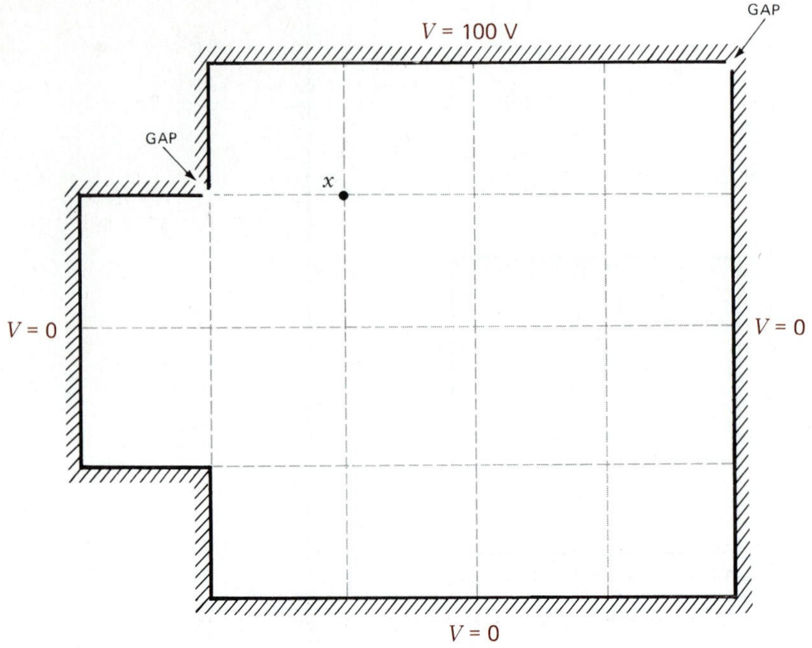

FIGURE 6.15
See Prob. 10.

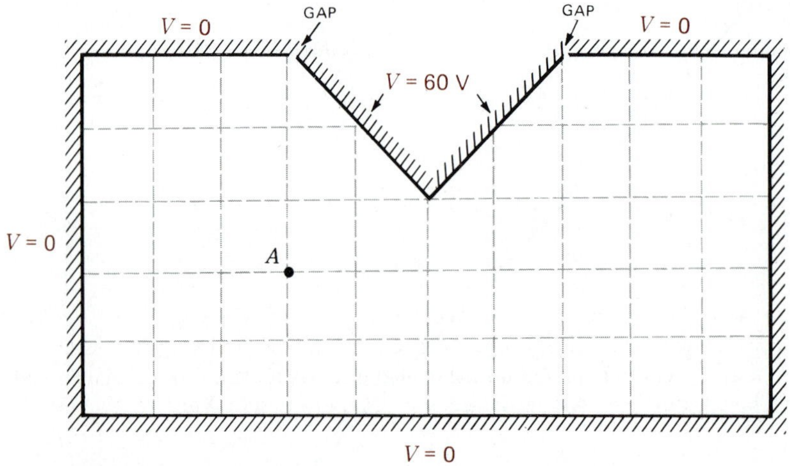

FIGURE 6.16
See Prob. 11.

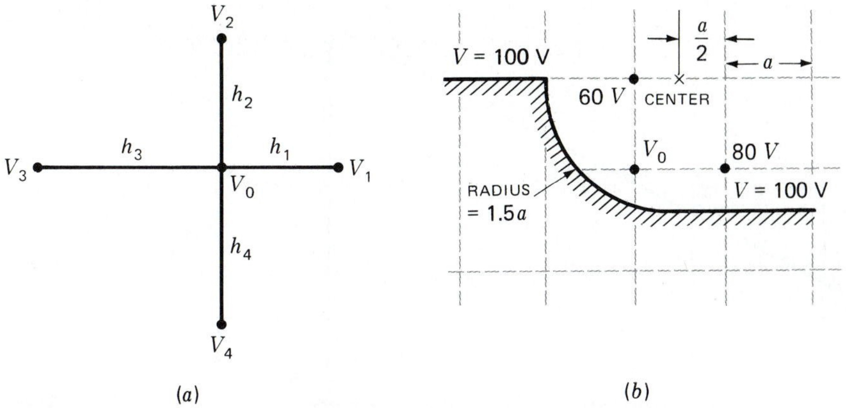

FIGURE 6.17
See Prob. 12.

10. Use iteration methods to estimate the potential at point x in the trough shown in Fig. 6.15. Working to the nearest volt is sufficient.

11. Using the grid indicated in Fig. 6.16, work to the nearest volt to estimate the potential at point A.

12. Conductors having boundaries that are curved or skewed usually do not permit every grid point to coincide with the actual boundary. Figure 6.17*a* illustrates the situation where the potential at V_0 is to be estimated in terms of V_1, V_2, V_3, V_4, and the unequal distances h_1, h_2, h_3, and h_4. (*a*) Show that

$$V_0 \doteq \frac{V_1}{\left(1+\dfrac{h_1}{h_3}\right)\left(1+\dfrac{h_1h_3}{h_4h_2}\right)} + \frac{V_2}{\left(1+\dfrac{h_2}{h_4}\right)\left(1+\dfrac{h_2h_4}{h_1h_3}\right)} + \frac{V_3}{\left(1+\dfrac{h_3}{h_1}\right)\left(1+\dfrac{h_3h_1}{h_2h_4}\right)}$$
$$+ \frac{V_4}{\left(1+\dfrac{h_4}{h_2}\right)\left(1+\dfrac{h_4h_2}{h_3h_1}\right)}$$

(*b*) Determine V_0 in Fig. 6.17*b*.

13. Use the grid indicated in Fig. 6.18 and the method described in Prob. 12 to estimate the potential at point x.

14. (*a*) After estimating potentials for the configuration of Fig. 6.19, use the iteration method with a square grid 1 cm on a side to find better estimates at the seven grid points. Work to the nearest volt. (*b*) Construct a 0.5-cm grid, establish new rough estimates, and then use the iteration method on the 0.5-cm grid. Again work to the nearest volt. (*c*) Use the computer to obtain values for a 0.25-cm grid. Work to the nearest 0.1 V.

15. Perfectly conducting concentric cylinders have radii of 2 and 6 cm; each is 50 cm in length. The region $2 < \rho < 3$ cm, $0 < z < 50$ cm is filled with a solid conducting material having $\sigma = 100$ ℧/m, while the portion for which $3 < \rho < 6$ cm, $0 < z < 50$ cm has

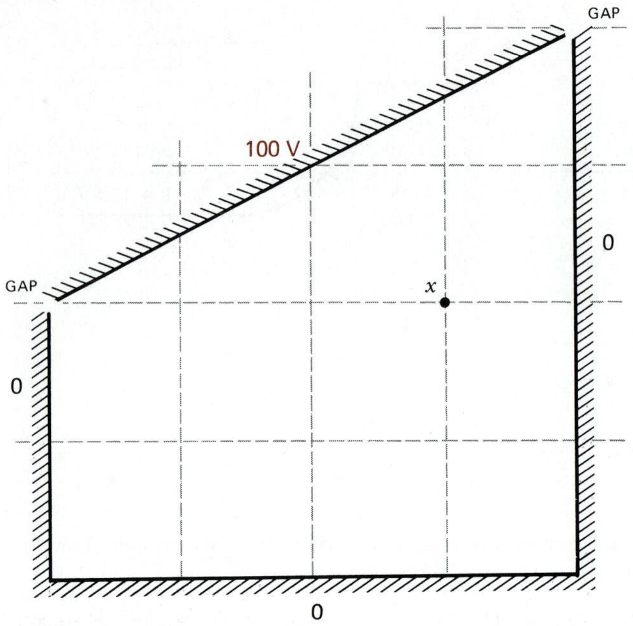

FIGURE 6.18
See Prob. 13.

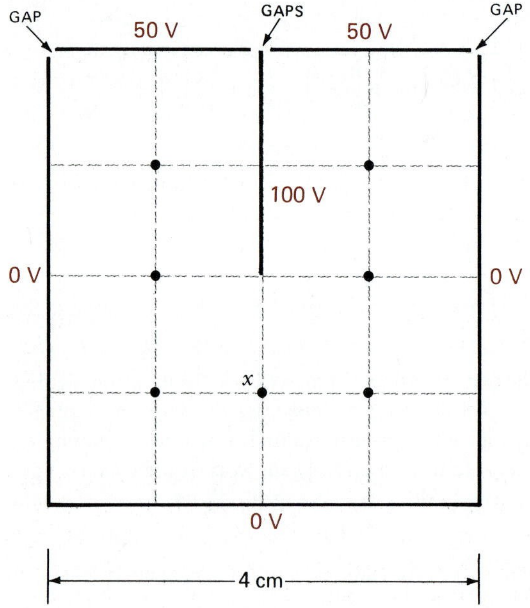

FIGURE 6.19
See Prob. 14.

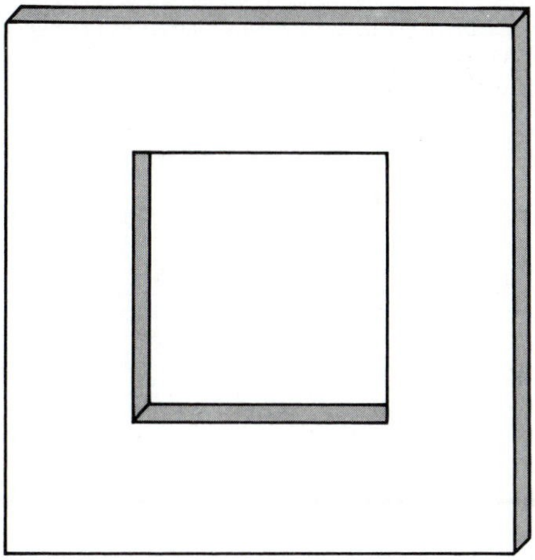

FIGURE 6.20
See Prob. 18.

$\sigma = 25$ ℧/m. The inner cylinder is held at 1 V while the outer is at $V = 0$. (*a*) Find **E** and **J** everywhere. (*b*) What resistance would be measured between the two cylinders? (*c*) What is V at $\rho = 3$ cm?

16. The cross section of the transmission line shown in Fig. 6.13 is drawn on a sheet of conducting paper with metallic paint. The sheet resistance is 2500 Ω per square and the dimension a is 1 in. (*a*) Assuming a result for Prob. 6*b* of 120 pF/m, what total resistance would be measured between the metallic conductors drawn on the conducting paper? (*b*) What would the total resistance be if $a = 0.6$ cm?

17. The cross section of a coaxial transmission line is drawn on a sheet of conducting paper using the dimensions $a = 0.5$ in and $b = 3$ in. Silver paint is used for the two conducting surfaces. If a resistance of 1000 Ω is measured between conductors: (*a*) what is the resistance per square of the paper? (*b*) If the conducting layer on the paper is 0.008 in thick, what is the conductivity of the resistive coating?

18. The square washer shown in Fig. 6.20 is 2 mm thick and has outer dimensions of 2 × 2 cm and inner dimensions of 1 × 1 cm. The inside and outside surfaces are perfectly conducting. If the material has a conductivity of 5 ℧/m, estimate the resistance offered between the inner and outer surfaces.

19. Two wires of circular cross section, each having a radius of 0.1 mm, are parallel with their axes separated by 1.2 mm. Assume the wires are perfectly conducting and embedded in a homogeneous medium of infinite extent for which $\epsilon_R = 2.5$ and $\sigma = 1$ m℧/m. A 100-V battery is connected from wire to wire. For each meter of length of the wires, determine: (*a*) the magnitude of the charge on either wire, and (*b*) the current flowing from wire to wire.

20. A rubber-sheet model is constructed for a coaxial transmission line. The horizontal dimensions are 50 times those of the actual line, for which $a = 0.8$ mm and $b = 4$ mm. The model is 6 cm high at the inner conductor and zero at the outer. (*a*) What angle does a normal to the rubber sheet make with the vertical at the inner and outer radii? (*b*) At what distance from the model axis is the height 3 cm?

CHAPTER 7

POISSON'S AND LAPLACE'S EQUATIONS

A study of the previous chapter shows that several of the analogies used to obtain experimental field maps involved demonstrating that the analogous quantity satisfies Laplace's equation. This is true for small deflections of an elastic membrane, and we might have proved the current analogy by showing that the direct-current density in a conducting medium also satisfies Laplace's equation. It appears that this is a fundamental equation in more than one field of science, and, perhaps without knowing it, we have spent the last chapter obtaining solutions for Laplace's equation by experimental, graphical, and numerical methods. Now we are ready to obtain this equation formally and discuss several methods by which it may be solved analytically.

It may seem that this material properly belongs before that of the previous chapter; as long as we are solving one equation by so many methods, would it not be fitting to see the equation first? The disadvantage of this more logical order lies in the fact that solving Laplace's equation is an exercise in mathematics, and unless we have the physical problem well in mind, we may easily miss the physical significance of what we are doing. A rough curvilinear map can tell us much about a field and then may be used later to check our mathematical solutions for gross errors or to indicate certain peculiar regions in the field which require special treatment.

With this explanation let us finally obtain the equations of Laplace and Poisson.

7.1 POISSON'S AND LAPLACE'S EQUATIONS

Obtaining Poisson's equation is exceedingly simple, for from the point form of Gauss's law,

$$\nabla \cdot \mathbf{D} = \rho_v \tag{1}$$

the definition of **D**,

$$\mathbf{D} = \epsilon \mathbf{E} \tag{2}$$

and the gradient relationship,

$$\mathbf{E} = -\nabla V \tag{3}$$

by substitution we have

$$\nabla \cdot \mathbf{D} = \nabla \cdot (\epsilon \mathbf{E}) = -\nabla \cdot (\epsilon \nabla V) = \rho_v$$

or

$$\nabla \cdot \nabla V = -\frac{\rho_v}{\epsilon} \tag{4}$$

for a homogeneous region in which ϵ is constant.

Equation (4) is *Poisson's equation*, but the "double ∇" operation must be interpreted and expanded, at least in cartesian coordinates, before the equation can be useful. In cartesian coordinates,

$$\nabla \cdot \mathbf{A} = \frac{\partial A_x}{\partial x} + \frac{\partial A_y}{\partial y} + \frac{\partial A_z}{\partial z}$$

$$\nabla V = \frac{\partial V}{\partial x}\mathbf{a}_x + \frac{\partial V}{\partial y}\mathbf{a}_y + \frac{\partial V}{\partial z}\mathbf{a}_z$$

and therefore

$$\nabla \cdot \nabla V = \frac{\partial}{\partial x}\left(\frac{\partial V}{\partial x}\right) + \frac{\partial}{\partial y}\left(\frac{\partial V}{\partial y}\right) + \frac{\partial}{\partial z}\left(\frac{\partial V}{\partial z}\right)$$

$$= \frac{\partial^2 V}{\partial x^2} + \frac{\partial^2 V}{\partial y^2} + \frac{\partial^2 V}{\partial z^2} \tag{5}$$

Usually the operation $\nabla \cdot \nabla$ is abbreviated ∇^2 (and pronounced "del squared"), a good reminder of the second-order partial derivatives appearing in (5), and we have

$$\boxed{\nabla^2 V = \frac{\partial^2 V}{\partial x^2} + \frac{\partial^2 V}{\partial y^2} + \frac{\partial^2 V}{\partial z^2} = -\frac{\rho_v}{\epsilon}} \tag{6}$$

in cartesian coordinates.

If $\rho_v = 0$, indicating zero *volume* charge density, but allowing point charges, line charge, and surface charge density to exist at singular locations as sources of the field, then

$$\boxed{\nabla^2 V = 0} \tag{7}$$

which is *Laplace's equation.* The ∇^2 operation is called the *Laplacian of V.*

In cartesian coordinates Laplace's equation is

$$\boxed{\nabla^2 V = \frac{\partial^2 V}{\partial x^2} + \frac{\partial^2 V}{\partial y^2} + \frac{\partial^2 V}{\partial z^2} = 0 \qquad \text{(cartesian)}} \tag{8}$$

and the form of $\nabla^2 V$ in cylindrical and spherical coordinates may be obtained by using the expressions for the divergence and gradient already obtained in those coordinate systems. For reference, the Laplacian in cylindrical coordinates is

$$\boxed{\nabla^2 V = \frac{1}{\rho}\frac{\partial}{\partial \rho}\left(\rho \frac{\partial V}{\partial \rho}\right) + \frac{1}{\rho^2}\left(\frac{\partial^2 V}{\partial \phi^2}\right) + \frac{\partial^2 V}{\partial z^2} \qquad \text{(cylindrical)}} \tag{9}$$

and in spherical coordinates is

$$\boxed{\nabla^2 V = \frac{1}{r^2}\frac{\partial}{\partial r}\left(r^2 \frac{\partial V}{\partial r}\right) + \frac{1}{r^2 \sin\theta}\frac{\partial}{\partial \theta}\left(\sin\theta \frac{\partial V}{\partial \theta}\right) + \frac{1}{r^2 \sin^2\theta}\frac{\partial^2 V}{\partial \phi^2} \qquad \text{(spherical)}} \tag{10}$$

These equations may be expanded by taking the indicated partial derivatives, but it is usually more helpful to have them in the forms given above; furthermore, it is much easier to expand them later if necessary than it is to put the broken pieces back together again.

Laplace's equation is all-embracing, for, applying as it does wherever volume charge density is zero, it states that every conceivable configuration of electrodes or conductors produces a field for which $\nabla^2 V = 0$. All these fields are different, with different potential values and different spatial rates of change, yet for each of them $\nabla^2 V = 0$. Since *every* field (if $\rho_v = 0$) satisfies Laplace's equation, how can we expect to reverse the procedure and use Laplace's equation to find one specific field in which we happen to have an interest? Obviously, more information is required, and we shall find that we must solve Laplace's equation subject to certain *boundary conditions.*

Every physical problem must contain at least one conducting boundary and usually contains two or more. The potentials on these boundaries are assigned values, perhaps $V_0, V_1, \ldots$, or perhaps numerical values. These definite equipotential surfaces will provide the boundary conditions for the type of problem to be solved in this chapter. In other types of problems, the boundary conditions take the form of specified values of E on an enclosing surface, or a mixture of known values of V and E.

Before using Laplace's equation or Poisson's equation in several examples, we must pause to show that if our answer satisfies Laplace's equation and also satisfies the boundary conditions, then it is the only possible answer. It would be very distressing to work a problem by solving Laplace's equation with two different approved methods and then to obtain two different answers. We shall show that the two answers must be identical.

D7.1. Find the potential and the volume charge density at $P(0.5,1.5,1)$ in free space given the potential field: (*a*) $V = 2x^2 - y^2 - z^2$ V; (*b*) $V = 6\rho\phi z$ V; (*c*) $V = 5(2r^2 - 7)\cos\theta\cos\phi$ V; (*d*) $V = 3x - y$ V.

Ans. -2.75 V, 0; 11.85 V, -42.0 pC/m^3; 0, -89.8 pC/m^3; 0, 0.

7.2 UNIQUENESS THEOREM

Let us assume that we have two solutions of Laplace's equation, V_1 and V_2, both general functions of the coordinates used. Therefore

$$\nabla^2 V_1 = 0$$

and

$$\nabla^2 V_2 = 0$$

from which

$$\nabla^2(V_1 - V_2) = 0$$

Each solution must also satisfy the boundary conditions, and if we represent the given potential values on the boundaries by V_b, then the value of V_1 on the boundary V_{1b} and the value of V_2 on the boundary V_{2b} must both be identical to V_b,

$$V_{1b} = V_{2b} = V_b$$

or

$$V_{1b} - V_{2b} = 0$$

In Sec. 4.8, Eq. (44), we made use of a vector identity,

$$\nabla \cdot (V\mathbf{D}) \equiv V(\nabla \cdot \mathbf{D}) + \mathbf{D} \cdot (\nabla V)$$

which holds for any scalar V and any vector $\mathbf{D}$. For the present application we shall select $V_1 - V_2$ as the scalar and $\nabla(V_1 - V_2)$ as the vector, giving

$$\nabla \cdot [(V_1 - V_2)\nabla(V_1 - V_2)] \equiv (V_1 - V_2)[\nabla \cdot \nabla(V_1 - V_2)] + \nabla(V_1 - V_2) \cdot \nabla(V_1 - V_2)$$

which we shall integrate throughout the volume *enclosed* by the boundary surfaces specified:

$$\int_{\text{vol}} \nabla \cdot [(V_1 - V_2)\nabla(V_1 - V_2)]\, dv$$
$$\equiv \int_{\text{vol}} (V_1 - V_2)[\nabla \cdot \nabla(V_1 - V_2)]\, dv + \int_{\text{vol}} [\nabla(V_1 - V_2)]^2\, dv \qquad (11)$$

The divergence theorem allows us to replace the volume integral on the left side of the equation by the closed surface integral over the surface surrounding the volume. This surface consists of the boundaries already specified on which $V_{1b} = V_{2b}$, and therefore

$$\int_{\text{vol}} \nabla \cdot [(V_1 - V_2)\nabla(V_1 - V_2)]\, dv = \oint_S [(V_{1b} - V_{2b})\nabla(V_{1b} - V_{2b})] \cdot d\mathbf{S} = 0$$

One of the factors of the first integral on the right side of (11) is $\nabla \cdot \nabla(V_1 - V_2)$, or $\nabla^2(V_1 - V_2)$, which is zero by hypothesis, and therefore that integral is zero. Hence the remaining volume integral must be zero:

$$\int_{\text{vol}} [\nabla(V_1 - V_2)]^2\, dv = 0$$

There are in general two reasons why an integral may be zero: either the integrand (the quantity under the integral sign) is everywhere zero, or the integrand is positive in some regions and negative in others, and the contributions cancel algebraically. In this case the first reason must hold because $[\nabla(V_1 - V_2)]^2$ cannot be negative. Therefore

$$[\nabla(V_1 - V_2)]^2 = 0$$

and

$$\nabla(V_1 - V_2) = 0$$

Finally, if the gradient of $V_1 - V_2$ is everywhere zero, then $V_1 - V_2$ cannot change with any coordinates and

$$V_1 - V_2 = \text{constant}$$

If we can show that this constant is zero, we shall have accomplished our proof. The constant is easily evaluated by considering a point on the boundary. Here $V_1 - V_2 = V_{1b} - V_{2b} = 0$, and we see that the constant is indeed zero, and therefore

$$V_1 = V_2$$

giving two identical solutions.

The uniqueness theorem also applies to Poisson's equation, for if $\nabla^2 V_1 = -\rho_v/\epsilon$ and $\nabla^2 V_2 = -\rho_v/\epsilon$, then $\nabla^2(V_1 - V_2) = 0$ as before. Boundary conditions still require that $V_{1b} - V_{2b} = 0$, and the proof is identical from this point.

This constitutes the proof of the uniqueness theorem. Viewed as the answer to a question, "How do two solutions of Laplace's or Poisson's equation compare if they both satisfy the same boundary conditions?" the uniqueness theorem should please us by its assurance that the answers are identical. Once we can find any method of solving Laplace's or Poisson's equation subject to given boundary conditions, we have solved our problem once and for all. No other method can ever give a different answer.

D7.2. Consider the two potential fields $V_1 = x$ and $V_2 = x + \sinh y \sin x$. (*a*) Is $\nabla^2 V_1 = 0$? (*b*) Is $\nabla^2 V_2 = 0$? (*c*) Is $V_1 = 0$ at $x = 0$? (*d*) Is $V_2 = 0$ at $x = 0$? (*e*) Is $V_1 = \pi$ at $x = \pi$? (*f*) Is $V_2 = \pi$ at $x = \pi$? (*g*) Are V_1 and V_2 identical? (*h*) Why does the uniqueness theorem not apply?

Ans. Yes; yes; yes; yes; yes; yes; no; boundary conditions not given for a *closed* surface

7.3 EXAMPLES OF THE SOLUTION OF LAPLACE'S EQUATION

Several methods have been developed for solving the second-order partial differential equation known as Laplace's equation. The first and simplest method is that of direct integration, and we shall use this technique to work several examples in various coordinate systems in this section. In Sec. 7.5 one other method will be used on a more difficult problem. Additional methods, requiring a more advanced mathematical knowledge, are described in the references given at the end of the chapter.

The method of direct integration is applicable only to problems which are "one-dimensional," or in which the potential field is a function of only one of the three coordinates. Since we are working with only three coordinate systems, it might seem, then, that there are nine problems to be solved, but a little reflection will show that a field which varies only with x is fundamentally the same as a field which varies only with y. Rotating the physical problem a quarter turn is no change. Actually, there are only five problems to be solved, one in cartesian coordinates, two in cylindrical, and two in spherical. We shall enjoy life to the fullest by solving them all.

Example 1

Let us assume that V is a function only of x and worry later about which physical problem we are solving when we have a need for boundary conditions. Laplace's equation reduces to

$$\frac{\partial^2 V}{\partial x^2} = 0$$

and the partial derivative may be replaced by an ordinary derivative, since V is not a function of y or z,

$$\frac{d^2 V}{dx^2} = 0$$

We integrate twice, obtaining

$$\frac{dV}{dx} = A$$

and

$$V = Ax + B \tag{12}$$

where A and B are constants of integration. Equation (12) contains two such constants, as we should expect for a second-order differential equation. These constants can be determined only from the boundary conditions.

What boundary conditions should we supply? They are our choice, since no physical problem has yet been specified, with the exception of the original hypothesis that the potential varied only with x. We should now attempt to visualize such a field. Most of us probably already have the answer, but it may be obtained by exact methods.

Since the field varies only with x and is not a function of y and z, then V is a constant if x is a constant or, in other words, the equipotential surfaces are described by setting x constant. These surfaces are parallel planes normal to the x axis. The field is thus that of a parallel-plate capacitor, and as soon as we specify the potential on any two planes, we may evaluate our constants of integration.

To be very general, let $V = V_1$ at $x = x_1$ and $V = V_2$ at $x = x_2$. These values are then substituted into (12), giving

$$V_1 = Ax_1 + B \qquad V_2 = Ax_2 + B$$

$$A = \frac{V_1 - V_2}{x_1 - x_2} \qquad B = \frac{V_2 x_1 - V_1 x_2}{x_1 - x_2}$$

and

$$V = \frac{V_1(x - x_2) - V_2(x - x_1)}{x_1 - x_2}$$

A simpler answer would have been obtained by choosing simpler boundary conditions. If we had fixed $V = 0$ at $x = 0$ and $V = V_0$ at $x = d$, then

$$A = \frac{V_0}{d} \qquad B = 0$$

and

$$\boxed{V = \frac{V_0 x}{d}} \tag{13}$$

Suppose our primary aim is to find the capacitance of a parallel-plate capacitor. We have solved Laplace's equation, obtaining (12) with the two constants A and B. Should they be evaluated or left alone? Presumably we are not interested in the potential field itself, but only in the capacitance, and we may continue successfully with A and B or we may simplify the algebra by a little foresight. Capacitance is given by the ratio of charge to potential difference, so we may choose now the potential *difference* as V_0, which is equivalent to one boundary condition, and then choose whatever second boundary condition seems to help the form of the equation the most. This is the essence of the second set of boundary conditions which produced (13). The potential difference was fixed as V_0 by choosing the potential of one plate zero and the other V_0; the location of these plates was made as simple as possible by letting $V = 0$ at $x = 0$.

Using (13), then, we still need the total charge on either plate before the capacitance can be found. We should remember that when we first solved this capacitor problem in Chap. 5, the sheet of charge provided our starting point. We did not have to work very hard to find the charge, for all the fields were expressed in terms of it. The work then was spent in finding potential difference. Now the problem is reversed (and simplified).

The necessary steps are these, after the choice of boundary conditions has been made:

1. Given V, use $\mathbf{E} = -\nabla V$ to find $\mathbf{E}$.
2. Use $\mathbf{D} = \epsilon \mathbf{E}$ to find $\mathbf{D}$.
3. Evaluate $\mathbf{D}$ at either capacitor plate, $\mathbf{D} = \mathbf{D}_S = D_N \mathbf{a}_N$.
4. Recognize that $\rho_S = D_N$.
5. Find Q by a surface integration over the capacitor plate, $Q = \int_S \rho_S \, dS$.

Here we have

$$V = V_0 \frac{x}{d}$$

$$\mathbf{E} = -\frac{V_0}{d} \mathbf{a}_x$$

$$\mathbf{D} = -\epsilon \frac{V_0}{d} \mathbf{a}_x$$

$$\mathbf{D}_S = \mathbf{D}\Big|_{x=0} = -\epsilon \frac{V_0}{d} \mathbf{a}_x$$

$$\mathbf{a}_N = \mathbf{a}_x$$

$$D_N = -\epsilon \frac{V_0}{d} = \rho_S$$

$$Q = \int_S \frac{-\epsilon V_0}{d} \, dS = -\epsilon \frac{V_0 S}{d}$$

and the capacitance is

$$\boxed{C = \frac{|Q|}{V_0} = \frac{\epsilon S}{d}} \tag{14}$$

We shall use this procedure several times in the examples to follow.

Example 2

Since no new problems are solved by choosing fields which vary only with y or with z in cartesian coordinates, we pass on to cylindrical coordinates for our next example. Variations with respect to z are again nothing new, and we next assume variation with respect to ρ only. Laplace's equation becomes

$$\frac{1}{\rho}\frac{\partial}{\partial \rho}\left(\rho\frac{\partial V}{\partial \rho}\right) = 0$$

or

$$\frac{1}{\rho}\frac{d}{d\rho}\left(\rho\frac{dV}{d\rho}\right) = 0$$

Noting the ρ in the denominator, we exclude $\rho = 0$ from our solution and then multiply by ρ and integrate,

$$\rho\frac{dV}{d\rho} = A$$

rearrange, and integrate again,

$$V = A \ln \rho + B \tag{15}$$

The equipotential surfaces are given by $\rho =$ constant and are cylinders, and the problem is that of the coaxial capacitor or coaxial transmission line. We choose a potential difference of V_0 by letting $V = V_0$ at $\rho = a$, $V = 0$ at $\rho = b$, $b > a$, and obtain

$$\boxed{V = V_0 \frac{\ln (b/\rho)}{\ln (b/a)}} \tag{16}$$

from which

$$\mathbf{E} = \frac{V_0}{\rho}\frac{1}{\ln (b/a)}\mathbf{a}_\rho$$

$$D_{N(\rho=a)} = \frac{\epsilon V_0}{a \ln (b/a)}$$

$$Q = \frac{\epsilon V_0 2\pi a L}{a \ln (b/a)}$$

$$\boxed{C = \frac{2\pi\epsilon L}{\ln (b/a)}} \tag{17}$$

which agrees with our results in Chap. 5.

Example 3

Now let us assume that V is a function only of ϕ in cylindrical coordinates. We might look at the physical problem first for a change and see that equipotential surfaces are given by ϕ = constant. These are radial planes. Boundary conditions might be $V = 0$ at $\phi = 0$ and $V = V_0$ at $\phi = \alpha$, leading to the physical problem detailed in Fig. 7.1.

Laplace's equation is now

$$\frac{1}{\rho^2}\frac{\partial^2 V}{\partial \phi^2} = 0$$

We exclude $\rho = 0$ and have

$$\frac{d^2 V}{d\phi^2} = 0$$

The solution is

$$V = A\phi + B$$

The boundary conditions determine A and B, and

$$\boxed{V = V_0 \frac{\phi}{\alpha}} \tag{18}$$

Taking the gradient of (18) produces the electric field intensity,

$$\boxed{\mathbf{E} = -\frac{V_0 \mathbf{a}_\phi}{\alpha\rho}} \tag{19}$$

and it is interesting to note that E is a function of ρ and not of ϕ. This does not contradict our original assumptions, which were restrictions only on the potential field. Note, however, that the *vector* field $\mathbf{E}$ is a function of ϕ.

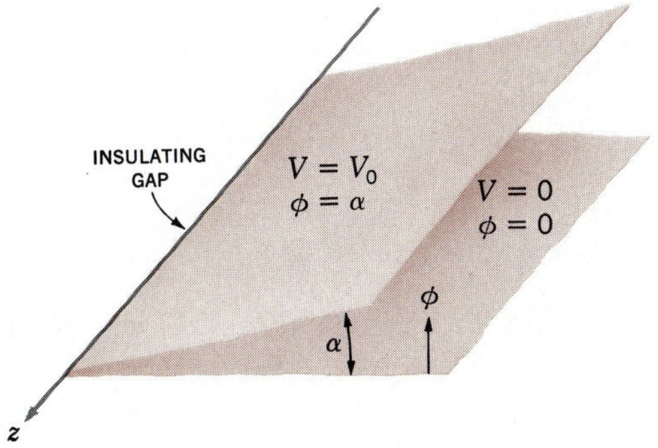

FIGURE 7.1
Two infinite radial planes with an interior angle α. An infinitesimal insulating gap exists at $\rho = 0$. The potential field may be found by applying Laplace's equation in cylindrical coordinates.

A problem involving the capacitance of these two radial planes is included at the end of the chapter.

Example 4

We now turn to spherical coordinates, dispose immediately of variations with respect to ϕ only as having just been solved, and treat first $V = V(r)$.

The details are left for a problem later, but the final potential field is given by

$$V = V_0 \frac{\dfrac{1}{r} - \dfrac{1}{b}}{\dfrac{1}{a} - \dfrac{1}{b}} \tag{20}$$

where the boundary conditions are evidently $V = 0$ at $r = b$ and $V = V_0$ at $r = a$, $b > a$. The problem is that of concentric spheres. The capacitance was found previously in Sec. 5.10 (by a somewhat different method) and is

$$C = \frac{4\pi\epsilon}{\dfrac{1}{a} - \dfrac{1}{b}} \tag{21}$$

Example 5

In spherical coordinates we now restrict the potential function to $V = V(\theta)$, obtaining

$$\frac{1}{r^2 \sin\theta} \frac{d}{d\theta}\left(\sin\theta \frac{dV}{d\theta}\right) = 0$$

We exclude $r = 0$ and $\theta = 0$ or π and have

$$\sin\theta \frac{dV}{d\theta} = A$$

The second integral is then

$$V = \int \frac{A\, d\theta}{\sin\theta} + B$$

which is not as obvious as the previous ones. From integral tables (or a good memory) we have

$$V = A \ln\left(\tan\frac{\theta}{2}\right) + B$$

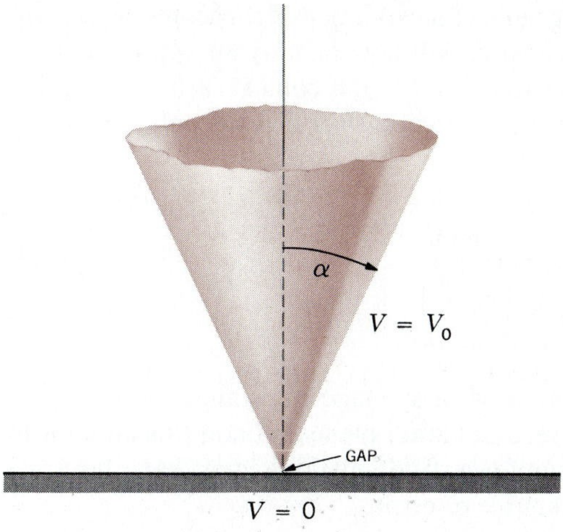

FIGURE 7.2
For the cone $\theta = \alpha$ at V_0 and the plane $\theta = \pi/2$ at $V = 0$, the potential field is given by $V = V_0[\ln(\tan\theta/2)]/[\ln(\tan\alpha/2)]$.

The equipotential surfaces are cones. Figure 7.2 illustrates the case where $V = 0$ at $\theta = \pi/2$ and $V = V_0$ at $\theta = \alpha$, $\alpha < \pi/2$. We obtain

$$V = V_0 \frac{\ln\left(\tan\dfrac{\theta}{2}\right)}{\ln\left(\tan\dfrac{\alpha}{2}\right)} \tag{22}$$

In order to find the capacitance between a conducting cone with its vertex separated from a conducting plane by an infinitesimal insulating gap and its axis normal to the plane, let us first find the field strength:

$$\mathbf{E} = -\nabla V = -\frac{1}{r}\frac{\partial V}{\partial \theta}\mathbf{a}_\theta = -\frac{V_0}{r\sin\theta\ln\left(\tan\dfrac{\alpha}{2}\right)}\mathbf{a}_\theta$$

The surface charge density on the cone is then

$$\rho_s = \frac{-\epsilon V_0}{r\sin\alpha\ln\left(\tan\dfrac{\alpha}{2}\right)}$$

producing a total charge Q,

$$Q = \frac{-\epsilon V_0}{\sin\alpha\ln\left(\tan\dfrac{\alpha}{2}\right)}\int_0^\infty\int_0^{2\pi}\frac{r\sin\alpha\,d\phi\,dr}{r}$$

$$= \frac{-2\pi\epsilon_0 V_0}{\ln\left(\tan\dfrac{\alpha}{2}\right)}\int_0^\infty dr$$

This leads to an infinite value of charge and capacitance, and it becomes necessary to consider a cone of finite size. Our answer will now be only an approximation, because the theoretical equipotential surface is $\theta = \alpha$, a conical surface extending from $r = 0$ to $r = \infty$, whereas our physical conical surface extends only from $r = 0$ to, say, $r = r_1$. The approximate capacitance is

$$\boxed{C \doteq \frac{2\pi\epsilon r_1}{\ln\left(\cot\frac{\alpha}{2}\right)}} \tag{23}$$

If we desire a more accurate answer, we may make an estimate of the capacitance of the base of the cone to the zero-potential plane and add this amount to our answer above. Fringing, or nonuniform, fields in this region have been neglected and introduce an additional source of error.

D7.3. Find V at $P(2,1,3)$ for the field of: (a) two coaxial conducting cylinders, $V = 50$ V at $\rho = 2$ m, and $V = 20$ V at $\rho = 3$ m; (b) two radial conducting planes, $V = 50$ V at $\phi = 10°$, and $V = 20$ at $\phi = 30°$; (c) two concentric conducting spheres, $V = 50$ V at $r = 3$ m, and $V = 20$ V at $r = 5$ m; (d) two coaxial conducting cones, $V = 50$ V at $\theta = 30°$, and $V = 20$ V at $\theta = 50°$.

Ans. 41.7 V; 25.2 V; 35.1 V; 38.4 V

7.4 EXAMPLE OF THE SOLUTION OF POISSON'S EQUATION

In order to select a reasonably simple problem which might illustrate the application of Poisson's equation, we shall have to assume that the volume charge density is specified. This is not usually the case, however; in fact, it is often the quantity about which we are seeking further information. The type of problem which we might encounter later would begin with a knowledge only of the boundary values of the potential, the electric field intensity, and the current density. From these we would have to apply Poisson's equation, the continuity equation, and some relationship expressing the forces on the charged particles, such as the Lorentz force equation or the diffusion equation, and solve the whole system of equations simultaneously. Such an ordeal is beyond the scope of this text, and we shall therefore assume a reasonably large amount of information.

As an example, let us select a *pn* junction between two halves of a semiconductor bar extending in the x direction. We shall assume that the region for $x < 0$ is doped p type and that the region for $x > 0$ is n type. The degree of doping is identical on each side of the junction. To review qualitatively some of the facts about the semiconductor junction, we note that initially there are excess holes to the left of the junction and excess electrons to the right. Each diffuses across the junction until an electric field is built up in such a direction that the diffusion current drops to zero. Thus, in order to prevent more holes from moving to the right, the electric field in the neighborhood of the junction must be directed to the left; E_x is negative there. This field must be produced by a net positive charge to the right of the junction and a net negative charge to the left. Note that the layer of positive charge consists of two parts—the holes which have crossed the junction and the positive donor ions from which the electrons have departed. The negative layer of charge is constituted in the opposite manner by electrons and negative acceptor ions.

The type of charge distribution which results is shown in Fig. 7.3*a*, and the negative field which it produces is shown in Fig. 7.3*b*. After looking at these two figures, one might read the previous paragraph again profitably.

A charge distribution of this form may be approximated by many different expressions. One of the simpler expressions is

$$\rho_v = 2\rho_{v0} \operatorname{sech} \frac{x}{a} \tanh \frac{x}{a} \tag{24}$$

which has a maximum charge density $\rho_{v,\max} = \rho_{v0}$ that occurs at $x = 0.881a$. The maximum charge density ρ_{v0} is related to the acceptor and donor concentrations N_a and N_d by noting that all the donor and acceptor ions in this region (the *depletion* layer) have been stripped of an electron or a hole, and thus

$$\rho_{v0} = eN_a = eN_d$$

Let us now solve Poisson's equation,

$$\nabla^2 V = -\frac{\rho_v}{\epsilon}$$

subject to the charge distribution assumed above,

$$\frac{d^2V}{dx^2} = -\frac{2\rho_{v0}}{\epsilon} \operatorname{sech} \frac{x}{a} \tanh \frac{x}{a}$$

in this one-dimensional problem in which variations with y and z are not present. We integrate once,

$$\frac{dV}{dx} = \frac{2\rho_{v0}a}{\epsilon} \operatorname{sech} \frac{x}{a} + C_1$$

and obtain the electric field intensity,

$$E_x = -\frac{2\rho_{v0}a}{\epsilon} \operatorname{sech} \frac{x}{a} - C_1$$

To evaluate the constant of integration C_1, we note that no net charge density and no fields can exist *far* from the junction. Thus, as $x \to \pm\infty$, E_x must approach zero. Therefore $C_1 = 0$, and

$$E_x = -\frac{2\rho_{v0}a}{\epsilon} \operatorname{sech} \frac{x}{a} \tag{25}$$

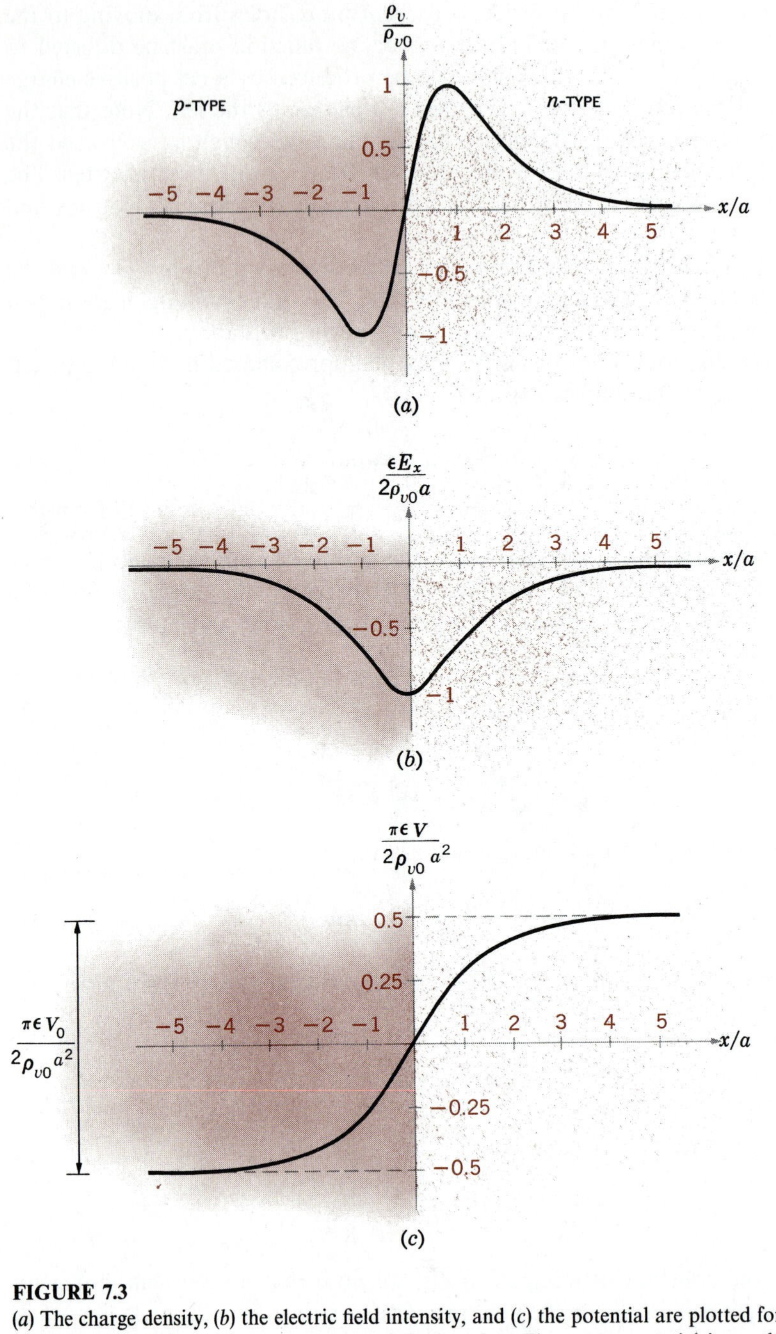

FIGURE 7.3
(*a*) The charge density, (*b*) the electric field intensity, and (*c*) the potential are plotted for a *pn* junction as functions of distance from the center of the junction. The *p*-type material is on the left, and the *n*-type is on the right.

Integrating again,

$$V = \frac{4\rho_{v0}a^2}{\epsilon} \tan^{-1} e^{x/a} + C_2$$

Let us arbitrarily select our zero reference of potential at the center of the junction, $x = 0$,

$$0 = \frac{4\rho_{v0}a^2}{\epsilon} \frac{\pi}{4} + C_2$$

and finally,

$$V = \frac{4\rho_{v0}a^2}{\epsilon}\left(\tan^{-1} e^{x/a} - \frac{\pi}{4}\right) \tag{26}$$

Figure 7.3 shows the charge distribution, electric field intensity, and the potential, as given by (24), (25), and (26), respectively.

The potential is constant once we are a distance of about $4a$ or $5a$ from the junction. The total potential difference V_0 across the junction is obtained from (26),

$$V_0 = V_{x\to\infty} - V_{x\to-\infty} = \frac{2\pi\rho_{v0}a^2}{\epsilon} \tag{27}$$

This expression suggests the possibility of determining the total charge on one side of the junction and then using (27) to find a junction capacitance. The total positive charge is

$$Q = S \int_0^\infty 2\rho_{v0} \operatorname{sech} \frac{x}{a} \tanh \frac{x}{a}\, dx = 2\rho_{v0}aS$$

where S is the area of the junction cross section. If we make use of (27) to eliminate the distance parameter a, the charge becomes

$$Q = S\sqrt{\frac{2\rho_{v0}\epsilon V_0}{\pi}} \tag{28}$$

Since the total charge is a function of the potential difference, we have to be careful in defining a capacitance. Thinking in "circuit" terms for a moment,

$$I = \frac{dQ}{dt} = C\frac{dV_0}{dt}$$

and thus

$$C = \frac{dQ}{dV_0}$$

By differentiating (28) we therefore have the capacitance,

$$C = \sqrt{\frac{\rho_{v0}\epsilon}{2\pi V_0}}\, S = \frac{\epsilon S}{2\pi a} \tag{29}$$

The first form of (29) shows that the capacitance varies inversely as the square root of the voltage. That is, a higher voltage causes a greater separation of the charge layers and a smaller capacitance. The second form is interesting in that it indicates that we may think of the junction as a parallel-plate capacitor with a "plate" separation of $2\pi a$. In view of the dimensions of the region in which the charge is concentrated, this is a logical result.

Poisson's equation enters into any problem involving volume charge density. Besides semiconductor diode and transistor models, we find that vacuum tubes, magnetohydrodynamic energy conversion, and ion propulsion require its use in constructing satisfactory theories.

D7.4. The volume charge density at a certain semiconductor junction is expressed as $\rho_v = 500 \operatorname{sech} 10^6 x \tanh 10^6 x$ C/m^3. If $\epsilon_R = 12$ for the semiconductor material and the area of the junction is 10^{-7} m^2, find: (*a*) V_0; (*b*) C; (*c*) E at the junction.

Ans. 14.78 V; 1.691 pF; 4.71 MV/m

D7.5. If $V = 2$ V at $x = 1$ mm, and $V = 0$ at $x = 0$, find E_x at $x = 1$ mm in free space for the volume charge density $\rho_v =$: (*a*) $-10^6\epsilon_0$ C/m^3; (*b*) $-3 \times 10^8\epsilon_0 x$ C/m^3.

Ans. -2500 V/m; -2100 V/m

7.5 PRODUCT SOLUTION OF LAPLACE'S EQUATION

In this section we are confronted with the class of potential fields which vary with more than one of the three coordinates. Although our examples are taken in the cartesian coordinate system, the general method is applicable to the other coordinate systems. We shall avoid those applications, however, because the potential fields are given in terms of more advanced mathematical functions, such as Bessel functions and spherical and cylindrical harmonics, and our interest now does not lie with new mathematical functions but with the techniques and methods of solving electrostatic field problems.

We may provide ourselves with a general class of problems by specifying merely that the potential is a function of x and y alone, so that

$$\frac{\partial^2 V}{\partial x^2} + \frac{\partial^2 V}{\partial y^2} = 0 \tag{30}$$

We now assume that the potential is expressible as the *product* of a function of x alone and a function of y alone. It might seem that this prohibits too many solutions, such as $V = x + y$, or any *sum* of a function of x and a function of y, but we should realize that Laplace's equation is linear and the sum of any two solutions is also a solution. We could treat $V = x + y$ as the sum of $V_1 = x$ and $V_2 = y$, where each of these latter potentials is now a (trivial) product solution.

Representing the function of x by X and the function of y by Y, we have

$$V = XY \tag{31}$$

which is substituted into (30),

$$Y \frac{\partial^2 X}{\partial x^2} + X \frac{\partial^2 Y}{\partial y^2} = 0$$

Since X does not involve y and Y does not involve x, ordinary derivatives may be used,

$$Y \frac{d^2 X}{dx^2} + X \frac{d^2 Y}{dy^2} = 0 \tag{32}$$

Equation (32) may be solved by separating the variables through division by XY, giving

$$\frac{1}{X}\frac{d^2 X}{dx^2} + \frac{1}{Y}\frac{d^2 Y}{dy^2} = 0$$

or

$$\frac{1}{X}\frac{d^2 X}{dx^2} = -\frac{1}{Y}\frac{d^2 Y}{dy^2}$$

Now we need one of the cleverest arguments of mathematics: since $(1/X)\,d^2X/dx^2$ involves no y and $-(1/Y)\,d^2Y/dy^2$ involves no x, and since the two quantities are equal, then $(1/X)\,d^2X/dx^2$ cannot be a function of x either, and similarly, $-(1/Y)\,d^2Y/dy^2$ cannot be a function of y! In other words, we have shown that each of these terms must be a constant. For convenience, let us call this constant α^2,

$$\frac{1}{X}\frac{d^2 X}{dx^2} = \alpha^2 \tag{33}$$

$$-\frac{1}{Y}\frac{d^2 Y}{dy^2} = \alpha^2 \tag{34}$$

The constant α^2 is called the *separation constant*, because its use results in separating one equation into two simpler equations.

Equation (33) may be written as

$$\frac{d^2 X}{dx^2} = \alpha^2 X \tag{35}$$

and must now be solved. There are several methods by which a solution may be obtained. The first method is experience, or recognition, which becomes more powerful with practice. We are just beginning and can barely recognize Laplace's equation itself. The second method might be that of direct integration, when applicable, of course. Applying it here, we should write

$$d\left(\frac{dX}{dx}\right) = \alpha^2 X\,dx$$

$$\frac{dX}{dx} = \alpha^2 \int X\,dx$$

and then pass on to the next method, for X is some unknown function of x, and the method of integration is not applicable here. The third method we might describe as intuition, common sense, or inspection. It involves taking a good look at the equation, perhaps putting the operation into words. This method will work on (35) for some of us if we ask ourselves, "What function has a second derivative which has the same form as the function itself, except for multiplication by a constant?" The answer is the exponential function, of course, and we could go on from here to construct the solution. Instead, let us work with those of us whose intuition is suffering from exposure and apply a very powerful but long method, the infinite-power-series substitution.

We assume hopefully that X may be represented by

$$X = \sum_{n=0}^{\infty} a_n x^n$$

and substitute into (35), giving

$$\frac{d^2X}{dx^2} = \sum_{0}^{\infty} n(n-1)a_n x^{n-2} = \alpha^2 \sum_{0}^{\infty} a_n x^n$$

If these two different infinite series are to be equal for all x, they must be identical, and the coefficients of like powers of x may be equated term by term. Thus

$$2 \times 1 \times a_2 = \alpha^2 a_0$$

$$3 \times 2 \times a_3 = \alpha^2 a_1$$

and in general we have the recurrence relationship

$$(n+2)(n+1)a_{n+2} = \alpha^2 a_n$$

The even coefficients may be expressed in terms of a_0 as

$$a_2 = \frac{\alpha^2}{1 \times 2} a_0$$

$$a_4 = \frac{\alpha^2}{3 \times 4} a_2 = \frac{\alpha^4}{4!} a_0$$

$$a_6 = \frac{\alpha^6}{6!} a_0$$

and, in general, for n even, as

$$a_n = \frac{\alpha^n}{n!} a_0 \qquad (n \text{ even})$$

For odd values of n, we have

$$a_3 = \frac{\alpha^2}{2 \times 3} a_1 = \frac{\alpha^3}{3!} \frac{a_1}{\alpha}$$

$$a_5 = \frac{\alpha^5}{5!} \frac{a_1}{\alpha}$$

and in general, for n odd,

$$a_n = \frac{\alpha^n}{n!}\frac{a_1}{\alpha} \qquad (n \text{ odd})$$

Substituting back into the original power series for X, we obtain

$$X = a_0 \sum_{0,\,\text{even}}^{\infty} \frac{\alpha^n}{n!} x^n + \frac{a_1}{\alpha} \sum_{1,\,\text{odd}}^{\infty} \frac{\alpha^n}{n!} x^n$$

or

$$X = a_0 \sum_{0,\,\text{even}}^{\infty} \frac{(\alpha x)^n}{n!} + \frac{a_1}{\alpha} \sum_{1,\,\text{odd}}^{\infty} \frac{(\alpha x)^n}{n!}$$

Although the sum of these two infinite series is the solution of the differential equation in x, the form of the solution may be improved immeasurably by recognizing the first series as the hyperbolic cosine,

$$\cosh \alpha x = \sum_{0,\,\text{even}}^{\infty} \frac{(\alpha x)^n}{n!} = 1 + \frac{(\alpha x)^2}{2!} + \frac{(\alpha x)^4}{4!} + \cdots$$

and the second series as the hyperbolic sine,

$$\sinh \alpha x = \sum_{1,\,\text{odd}}^{\infty} \frac{(\alpha x)^n}{n!} = \alpha x + \frac{(\alpha x)^3}{3!} + \frac{(\alpha x)^5}{5!} + \cdots$$

The solution may therefore be written as

$$X = a_0 \cosh \alpha x + \frac{a_1}{\alpha} \sinh \alpha x$$

or

$$X = A \cosh \alpha x + B \sinh \alpha x$$

where the slightly simpler terms A and B have replaced a_0 and a_1/α, respectively, and are the two constants which must be evaluated in terms of the boundary conditions. The separation constant is not an arbitrary constant as far as the solution of (35) is concerned, for it appears in that equation.

An alternate form of the solution is obtained by expressing the hyperbolic functions in terms of exponentials, collecting terms, and selecting new arbitrary constants, A' and B',

$$X = A'e^{\alpha x} + B'e^{-\alpha x}$$

Turning our attention now to (34), we see the solution proceeds along similar lines, leading to two power series representing the sine and cosine, and we have

$$Y = C \cos \alpha y + D \sin \alpha y$$

from which the potential is

$$V = XY = (A \cosh \alpha x + B \sinh \alpha x)(C \cos \alpha y + D \sin \alpha y) \tag{36}$$

Before describing a physical problem and forcing the constants appearing in (36) to fit the boundary conditions prescribed, let us consider the physical nature of the potential field given by a simple choice of these constants. Letting $A = 0$, $C = 0$, and $BD = V_1$, we have

$$V = V_1 \sinh \alpha x \sin \alpha y \tag{37}$$

The $\sinh \alpha x$ factor is zero at $x = 0$ and increases smoothly with x, soon becoming nearly exponential in form, since

$$\sinh \alpha x = \tfrac{1}{2}(e^{\alpha x} - e^{-\alpha x})$$

The $\sin \alpha y$ term causes the potential to be zero at $y = 0$, $y = \pi/\alpha$, $y = 2\pi/\alpha$, and so forth. We therefore may place zero-potential conducting planes at $x = 0$, $y = 0$, and $y = \pi/\alpha$. Finally, we can describe the V_1 equipotential surface by setting $V = V_1$ in (37), obtaining

$$\sinh \alpha x \sin \alpha y = 1$$

or

$$\alpha y = \sin^{-1} \frac{1}{\sinh \alpha x}$$

This is not a familiar equation, but a hand calculator or a set of tables can furnish enough numerical values to allow us to plot αy as a function of αx. Such a curve is shown in Fig. 7.4. Note that the curve is double-valued and symmetrical about the line $\alpha y = \pi/2$ when αy is restricted to the interval between 0 and π. The information of Fig. 7.4 is transferred directly to the $V = 0$ and $V = V_1$ equipotential conducting surfaces in Fig. 7.5. The surfaces are shown in cross section, since the potential is not a function of z.

It is very unlikely that we shall ever be asked to find the potential field of these peculiarly shaped electrodes, but we should bear in mind the possibility of combining a number of the fields having the form given by (36) or (37) and thus

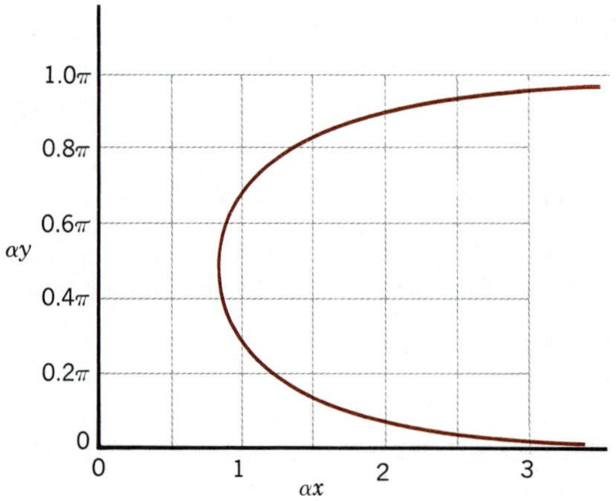

FIGURE 7.4
A graph of the double-valued function $\alpha y = \sin^{-1}(1/\sinh \alpha x)$, $0 < \alpha y < \pi$.

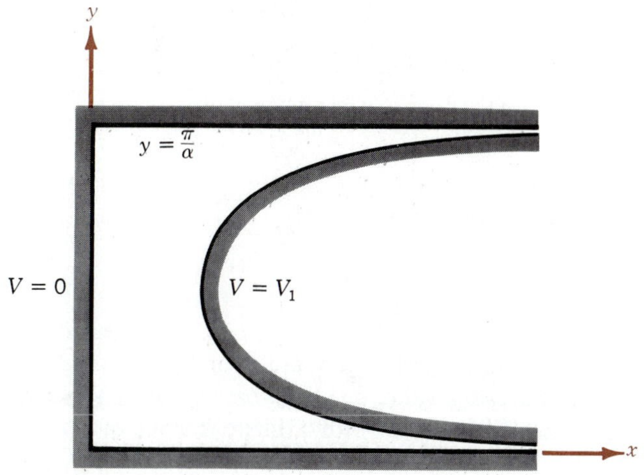

FIGURE 7.5
Cross section of the $V = 0$ and $V = V_1$ equipotential surfaces for the potential field $V = V_1 \sinh \alpha x \sin \alpha y$.

satisfying the boundary conditions of a more practical problem. We close this chapter with such an example.

The problem to be solved is that shown in Fig. 7.6. The boundary conditions shown are $V = 0$ at $x = 0$, $y = 0$, and $y = b$, and $V = V_0$ at $x = d$ for all y between 0 and b. It is immediately apparent that the potential field given by (37) and outlined in Fig. 7.5 satisfies two of the four boundary conditions. A third condition, $V = 0$ at $y = b$, may be satisfied by the choice of α, for the substitution of these values of (37) leads to the equation

$$0 = V_1 \sinh \alpha x \sin \alpha b$$

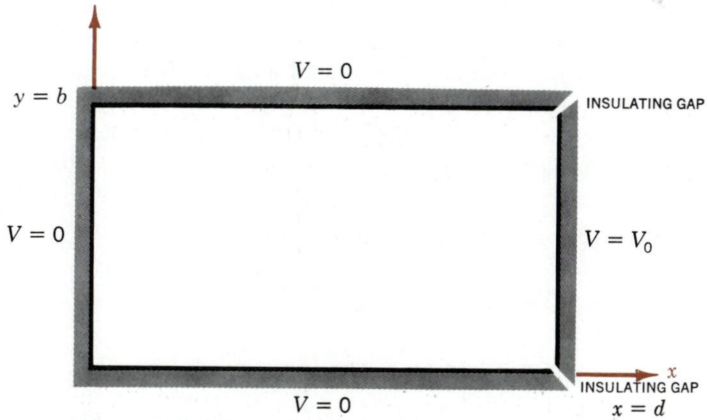

FIGURE 7.6
Potential problem requiring an infinite summation of fields of the form $V = V_1 \sinh \alpha x \sin \alpha y$. A similar configuration was analyzed by the iteration method in Chap. 6.

which may be satisfied by setting

$$\alpha b = m\pi \qquad (m = 1, 2, 3, \ldots)$$

or

$$\alpha = \frac{m\pi}{b}$$

The potential function

$$V = V_1 \sinh \frac{m\pi x}{b} \sin \frac{m\pi y}{b} \tag{38}$$

thus produces the correct potential at $x = 0$, $y = 0$, and $y = b$, regardless of the choice of m or the value of V_1. It is impossible to choose m or V_1 in such a way that $V = V_0$ at $x = d$ for each and every value of y between 0 and b. We must combine an infinite number of these fields, each with a different value of m and a corresponding value of V_1,

$$V = \sum_{m=0}^{\infty} V_{1m} \sinh \frac{m\pi x}{b} \sin \frac{m\pi y}{b}$$

The subscript on V_{1m} indicates that this amplitude factor will have a different value for each different value of m. Applying the last boundary condition now,

$$V_0 = \sum_{m=0}^{\infty} V_{1m} \sinh \frac{m\pi d}{b} \sin \frac{m\pi y}{b} \qquad (0 < y < b, m = 1, 2, \ldots)$$

Since $V_{1m} \sinh (m\pi d/b)$ is a function only of m, we may simplify the expression by replacing this factor by c_m,

$$V_0 = \sum_{m=0}^{\infty} c_m \sin \frac{m\pi y}{b} \qquad (0 < y < b, m = 1, 2, \ldots)$$

This is a Fourier sine series, and the c_m coefficients may be determined by the standard Fourier-series methods[1] if we can interpret V_0 as a periodic function of y. Since our physical problem is bounded by conducting planes at $y = 0$ and $y = b$, and our interest in the potential does not extend outside of this region, we may *define* the potential at $x = d$ for y *outside* of the range 0 to b in any manner we choose. Probably the simplest periodic expression is obtained by selecting the interval $0 < y < b$ as the half-period and choosing $V = -V_0$ in the adjacent half-period, or

$$V = V_0 \qquad (x = d, 0 < y < b)$$

$$V = -V_0 \qquad (x = d, b < y < 2b)$$

---

[1] Fourier series are discussed in almost every electrical engineering text on circuit theory. The author is partial to the Hayt and Kemmerly reference given in the Suggested References at the end of the chapter.

The c_m coefficients are then

$$c_m = \frac{1}{b}\left[\int_0^b V_0 \sin\frac{m\pi y}{b}\,dy + \int_b^{2b} (-V_0)\sin\frac{m\pi y}{b}\,dy\right]$$

leading to

$$c_m = \frac{4V_0}{m\pi} \qquad (m \text{ odd})$$

$$= 0 \qquad (m \text{ even})$$

However, $c_m = V_{1m} \sinh (m\pi d/b)$, and therefore

$$V_{1m} = \frac{4V_0}{m\pi \sinh (m\pi d/b)} \qquad (m \text{ odd only})$$

which may be substituted into (38) to give the desired potential function,

$$V = \frac{4V_0}{\pi} \sum_{1,\,\text{odd}}^{\infty} \frac{1}{m} \frac{\sinh (m\pi x/b)}{\sinh (m\pi d/b)} \sin\frac{m\pi y}{b} \tag{39}$$

The map of this field may be obtained by evaluating (39) at a number of points and drawing equipotentials by interpolation between these points. If we let $b = d$ and $V_0 = 100$, the problem is identical with that used as the example in the discussion of the iteration method. Checking one of the grid points in that problem, we let $x = d/4 = b/4$, $y = b/2 = d/2$, and $V_0 = 100$ and obtain

$$V = \frac{400}{\pi} \sum_{1,\,\text{odd}}^{\infty} \frac{1}{m} \frac{\sinh (m\pi/4)}{\sinh m\pi} \sin\frac{m\pi}{2}$$

$$= \frac{400}{\pi}\left(\frac{\sinh (\pi/4)}{\sinh \pi} - \frac{1}{3}\frac{\sinh (3\pi/4)}{\sinh 3\pi} + \frac{1}{5}\frac{\sinh (5\pi/4)}{\sinh 5\pi} - \cdots\right)$$

$$= \frac{400}{\pi}\left(\frac{0.8687}{11.549} - \frac{5.228}{3 \times 6195.8} + \cdots\right)$$

$$= 9.577 - 0.036 + \cdots$$

$$= 9.541 \text{ V}$$

The equipotentials are drawn for increments of 10 V in Fig. 7.7, and flux lines have been added graphically to produce a curvilinear map.

The material covered in this discussion of the product solution has been more difficult than much of the preceding work, and it has, moreover, presented three new ideas. The first new technique was the assumption that the potential might be expressed as the product of a function of x and a function of y, and the resultant separation of Laplace's equation into two simpler ordinary differential equations. The second new approach was employed when an infinite-power-series solution was assumed as the solution for one of the ordinary differential equations. Finally, we considered an example which required the combination of an infinite

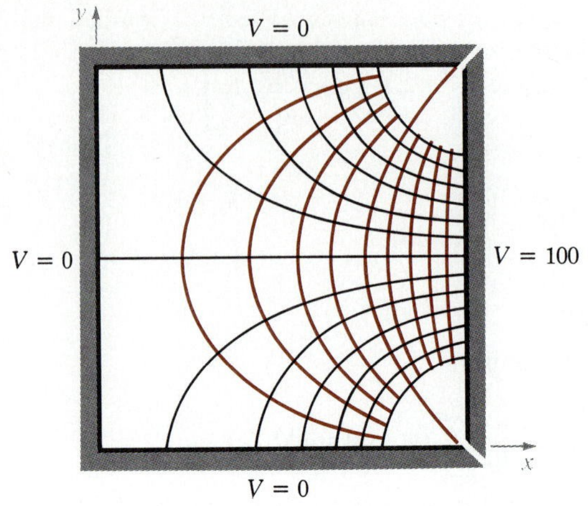

FIGURE 7.7
The field map corresponding to $V = \frac{4V_0}{\pi} \sum_{1,\text{odd}}^{\infty} \frac{1}{m} \frac{\sinh(m\pi x/b)}{\sinh(m\pi d/b)} \sin\frac{m\pi y}{b}$ with $b = d$ and $V_0 = 100$ V.

number of simpler product solutions, each having a different amplitude and a different variation in one of the coordinate directions. All these techniques are very powerful. They are useful in all coordinate systems, and they can be used in problems in which the potential varies with all three coordinates.

We have merely introduced the subject here, and more information can be obtained from the references listed at the end of the chapter, several of which devote hundreds of pages to the solution of Laplace's equation.

D7.6. The dimensions of a square trough are $b = d = 2$ cm. If $V_0 = 100$ V, find: (*a*) V at $x = y = 1.2$ cm; (*b*) **E** at $x = y = 1$ cm.

Ans. 33.2 V; $-4170\mathbf{a}_x$ V/m

D7.7. Given the potential function, $V = 6\cosh\alpha x\cos\alpha y$: (*a*) let $X = 3\cosh\alpha x$ and show that this potential function satisfies (32); (*b*) find α if $V = 0$ at $y = \pm 1$ and $V > 0$ for $-1 < y < 1$.

Ans. OK; 1.571

SUGGESTED REFERENCES

1. Dekker, A. J.: (see Suggested References for Chap. 5).
2. Hayt, W. H., Jr., and J. E. Kemmerly: "Engineering Circuit Analysis," 4th ed., McGraw-Hill Book Company, New York, 1986.
3. Pugh, E. M., and E. W. Pugh: "Principles of Electricity and Magnetism," 2d ed., Addison-Wesley Publishing Co., Reading, Mass., 1970. This text provides the physicist's view of electricity and magnetism, but electrical engineering students should find it easy to read. The solution to Laplace's equation by a number of methods is discussed in chap. 4.
4. Ramo, S., J. R. Whinnery, and T. Van Duzer: (see Suggested References for Chap. 6). A more complete and advanced discussion of methods of solving Laplace's equation is given in chap. 7.

5. Seeley, S., and A. D. Poularikas: "Electromagnetics: Classical and Modern Theory and Applications," Marcel Dekker, Inc., New York, 1979. Several examples of the solution of Laplace's equation by separation of variables appear in chap. 4.

6. Smythe, W. R.: "Static and Dynamic Electricity," 3d ed., McGraw-Hill Book Company, New York, 1968. An advanced treatment of potential theory is given in chap. 4.

7. Weber E.: (see Suggested References for Chap. 6). There are a tremendous number of potential solutions given with the original references.

PROBLEMS

1. In a certain region of a perfect dielectric, it is known that $\nabla^2 V = 0$ and $\rho_v = 0$. If the dielectric is not homogeneous, what constraints are placed on the variation of ϵ with x, y, and z?

2. In a charge-free region of free space, a potential field is given as $V(x,y) = 5x^3 + f(x) - 2y^2$. Find $f(x)$ if E_x and V are both zero at the origin.

3. If $\epsilon = \epsilon_0$ and $V = 8x^2yz$, find: (*a*) V at $P(2,-1,3)$; (*b*) ρ_v at P; (*c*) $\mathbf{E}$ at P; (*d*) the equation of the equipotential surface through P; (*e*) the equation of the streamline through P. (*f*) Does V satisfy Laplace's equation?

4. A potential field in free space is given as $V = 100[2\rho - \ln(\rho + 1)]$. For $0 < \rho < 5$, sketch curves of: (*a*) V vs. ρ; (*b*) E_ρ vs. ρ; (*c*) ρ_v vs. ρ.

5. Let $V = A \ln [B(1 - \cos\theta)/(1 + \cos\theta)]$ and: (*a*) show that V satisfies Laplace's equation in spherical coordinates; (*b*) select A and B so that $V = 100$ V and $|\mathbf{E}| = 500$ V/m at $r = 5$, $\theta = 90°$, $\phi = 60°$.

6. Given the potential field $V = 100[\rho - (1/\rho)] \cos\phi$ V: (*a*) show that $\nabla^2 V = 0$; (*b*) sketch the intersections of the 0-V, 100-V, and 200-V equipotential surfaces with the $z = 0$ plane for $1 \le \rho \le 5$ and $0 \le \phi \le 90°$.

7. Given the potential field $V = (50 \sin\theta)/r^2$ V in free space: (*a*) determine whether V satisfies Laplace's equation; (*b*) find the total charge stored inside the spherical shell $1 < r < 2$.

8. Let $V_1(r,\theta,\phi) = 6/r$ and $V_2(r,\theta,\phi) = 3$. (*a*) State whether V_1 and V_2 satisfy Laplace's equation. (*b*) Evaluate V_1 and V_2 at $r = 2$. (*c*) Conciliate your results with the statement of the uniqueness theorem.

9. The functions $V_1(\rho,z)$ and $V_2(\rho,z)$ both satisfy Laplace's equation in the region $a < \rho < b$, $0 < \phi < 2\pi$, $-L < z < L$; each is zero on the surfaces $\rho = b$ for $-L < z < L$; $z = -L$ for $a < \rho < b$; and $z = L$ for $a < \rho < b$; and each is 100 V on the surface $\rho = a$ for $-L < z < L$. (*a*) In the region specified above, is Laplace's equation satisfied by the functions $V_1 + V_2$, $V_1 - V_2$, $V_1 + 3$, and $V_1 V_2$? (*b*) On the boundary surfaces specified, are the potential values given above obtained from the functions $V_1 + V_2$, $V_1 - V_2$, $V_1 + 3$, and $V_1 V_2$? (*c*) Are the functions V_2, $V_1 + V_2$, $V_1 + 3$, and $V_1 V_2$ identical with V_1?

10. Conducting plates at $z = 1$ cm and $z = 5$ cm are held at potentials of -8 V and 6 V, respectively. If the region between the plates is a homogeneous dielectric with $\epsilon = 5\epsilon_0$, find: (*a*) the capacitance between the plates per unit area; (*b*) $V(z)$; (*c*) $\mathbf{D}(z)$.

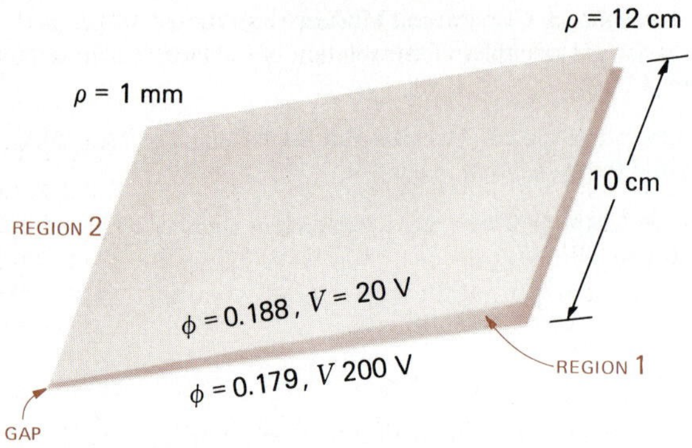

FIGURE 7.8
See Prob. 15.

11. The potential on the plane, $x - 2y + 5z = 2$, is 50 V. Point $P(2,3,-7)$ lies on a parallel conducting plane having a potential of -360 V. (*a*) Find V at $A(-1,4,6)$. (*b*) Find $\mathbf{E}(x,y,z)$.

12. Conducting cylinders at $\rho = 1.6$ cm and $\rho = 5$ cm in free space are held at potentials of 80 V and -40 V, respectively. (*a*) Find V and $\mathbf{E}$ at $\rho = 2$ cm. (*b*) Find the surface at which $V = 0$. (*c*) Find the surface charge density on each cylinder. (*d*) Determine the capacitance between the conducting cylinders per meter length.

13. Conducting cylinders at $\rho = 2$ cm and $\rho = 6$ cm are at potentials of 100 V and 0, respectively. The region between the cylinders is filled with an inhomogeneous perfect dielectric for which $\epsilon_R = 0.3/(\rho + 0.04)$. Find: (*a*) $\mathbf{D}(\rho)$; (*b*) $\mathbf{E}(\rho)$; (*c*) $V(\rho)$; (*d*) C per meter length.

14. Two semi-infinite planes are located at $\phi = \alpha$ and $\phi = \beta$. A narrow insulating strip separates them along the z axis. Points $P_\alpha(5,2,0)$ and $P_\beta(1,4,0)$ lie on the planes. Let $V_\alpha = 100$ V, while $E_\phi = 20$ V/m at P_β. (*a*) Find V_β. (*b*) Find E_ϕ at P_α.

15. The two conducting planes shown in Fig. 7.8 are defined by $0.001 < \rho < 0.120$ m, $0 < z < 0.1$ m, $\phi = 0.179$ and 0.188 rad. The medium surrounding the planes is air. For region 1, $0.179 < \phi < 0.188$, neglect fringing and find: (*a*) $V(\phi)$; (*b*) $\mathbf{E}(\rho)$; (*c*) $\mathbf{D}(\rho)$; (*d*) ρ_S on the upper surface of the lower plane; (*e*) Q on the upper surface of the lower plane. (*f*) Repeat (*a*) to (*c*) for region 2 by letting the location of the upper plane be $\phi = 0.188 - 2\pi$, and then find ρ_S and Q on the lower surface of the lower plane. (*g*) Find the total charge on the lower plane and the capacitance between the planes.

16. (*a*) Solve Laplace's equation for the potential field in the homogeneous region between two concentric conducting spheres with radii a and b, $b > a$, if $V = 0$ at $r = b$, and $V = V_0$ at $r = a$. (*b*) Find the capacitance between them.

17. Two concentric conducting spheres have radii of 3 and 5 cm. The region between them is filled with a homogeneous dielectric for which $\epsilon_R = 5$. If the potential of the inner sphere is 100 V while that of the outer is -100 V, find: (*a*) $V(r)$; (*b*) $E_r(r)$; (*c*) the potential midway between the conducting spheres; (*d*) the value of r at which $V = 0$; (*e*) Q on the inner sphere; (*f*) the capacitance between the two spheres.

18. Conducting spheres at $r = 2$ and 6 cm are at potentials of 100 V and 0, respectively. The region between the spheres is filled with an inhomogeneous dielectric for which $\epsilon_R = 0.3/(r + 0.04)$. Find (*a*) $\mathbf{D}(r)$; (*b*) $\mathbf{E}(r)$; (*c*) $V(r)$; (*d*) C.

19. Two coaxial conducting cones have their vertices at the origin and are defined by $\theta = 40^\circ$ and $\theta = 55^\circ$. The inner cone is at a potential of 2000 V, while the outer is at 500 V. A small bit of insulation keeps the cones from touching. (*a*) Find $V(\theta)$. (*b*) Find V at $P_1(r = 0.6, \theta = 45^\circ, \phi = 60^\circ)$. (*c*) Find $\mathbf{E}(r,\theta)$. (*d*) Find $\mathbf{E}$ at P_1. (*e*) Find the total charge on the surface $\theta = 40^\circ$, $0.1 < r < 0.5$, $100^\circ < \phi < 170^\circ$, assuming $\epsilon = \epsilon_0$ for $40^\circ < \theta < 55^\circ$.

20. In the potential field $V = A \ln[\tan(\theta/2)] + B$ it is specified that $V = 400$ V at $P(3, \theta = 60^\circ, \phi = 0^\circ)$, while $E_\theta = 2000$ V/m at $Q(4, 40^\circ, 90^\circ)$. (*a*) Find the surface at which $E_\theta = 1000$ V/m. (*b*) Find the surface at which $|E_z| = 500$ V/m.

21. Volume charge density in free space varies as $\rho_v = 100\epsilon_0/r^{2.5}$. (*a*) Use Poisson's equation to find $V(r)$ if it is assumed that $r^2 E_r \to 0$ when $r \to 0$, while $V \to 0$ as $r \to \infty$. (*b*) Now find $V(r)$ by using Gauss's law and a line integral.

22. The charge distribution near a semiconductor junction similar to that of Fig. 7.3*a* is represented as $\rho_v = \rho_{v0} a^3 x/(x^2 + a^2)^2$. (*a*) Determine $\rho_{v,\,\max}$ and $\rho_{v,\,\min}$ and their locations. (*b*) Find E_x and $V(x)$, assuming that $E_x \to 0$ as $x \to \infty$, and $V(0) = 0$. (*c*) Use a development similar to that in Sec. 7.4 to show that $C = dQ/dV_0 = \epsilon S/(2\pi a)$.

23. A rectangular trough is formed by four conducting planes located in air at $x = 0$ and 8 cm and $y = 0$ and 6 cm. The surface at $y = 6$ cm is at a potential of 100 V, the other three are at zero potential, and the necessary gaps are placed at two corners. Find the potential at $x = 3$ cm, $y = 4$ cm.

24. The four sides of a square trough are held at potentials of 0, -20, 40, and 50 V. Find the potential at the center of the trough.

25. The potential on the side of the rectangular trough shown in Fig. 7.6 varies sinusoidally as $V_{(x=d)} = 200 \sin(2\pi y/b)$ V. The other potentials and dimensions are the same as those of Fig. 7.6. (*a*) Find $V(x,y)$. (*b*) Find V at $x = 0.7d$, $y = 0.2b$, if $b = d$.

26. If X is a function only of x and $X'' + xX' - 2X = 0$, assume a solution in the form of an infinite power series and determine numerical values for a_2 to a_8 if $a_0 = 1$ and $a_1 = -1$.

27. It is known that $V = XY$ is a solution of Laplace's equation, where X is a function of x alone and Y is a function of y alone. Determine which of the following functions are also solutions of Laplace's equation: (*a*) $V_a = 100XY$; (*b*) $V_b = 100XY + 2x$; (*c*) $V_c = X + 3Y$; (*d*) $V_d = 2XY + y^2 - x^2$; (*e*) $V_e = (XY)^2$.

28. Assume a product solution of Laplace's equation in cylindrical coordinates, $V = PF$, where V is not a function of z, P is a function only of ρ, and F is a function only of ϕ. (*a*) Obtain the two separated equations if the separation constant is n^2. Select the sign of n^2 so that the solution of the ϕ equation leads to trigonometric functions. (*b*) Show that $P = A\rho^n + B\rho^{-n}$ satisfies the ρ equation. (*c*) Construct the solution $V(\rho,\phi)$. Functions of this form are called *circular harmonics*.

CHAPTER 8

THE STEADY MAGNETIC FIELD

At this point the concept of a field should be a familiar one. Since we first accepted the experimental law of forces existing between two point charges and defined electric field intensity as the force per unit charge on a test charge in the presence of a second charge, we have met numerous fields. These fields possess no real physical basis, for physical measurements must always be in terms of the forces on the charges in the detection equipment. Those charges which are the source cause measurable forces to be exerted on other charges, which we may think of as detector charges. The fact that we attribute a field to the source charges and then determine the effect of this field on the detector charges amounts merely to a division of the basic problem into two parts for our own convenience.

We shall begin our study of the magnetic field with a definition of the magnetic field itself and show how it arises from a current distribution. The effect of this field on other currents, or the second half of the physical problem, will be discussed in the following chapter. As we did with the electric field, we shall confine our initial discussion to free-space conditions, and the effect of material media will also be saved for discussion in the following chapter.

The relation of the steady magnetic field to its source is more complicated than is the relation of the electrostatic field to its source. We shall find it necessary to accept several laws temporarily on faith alone, relegating their proof to the (rather difficult) final section in this chapter. This section may well be omitted when meeting magnetic fields for the first time. It is included to make acceptance of the laws a little easier; the proof of the laws does exist and is available for the disbelievers or the more advanced student.

8.1 BIOT-SAVART LAW

The source of the steady magnetic field may be a permanent magnet, an electric field changing linearly with time, or a direct current. We shall largely ignore the permanent magnet and save the time-varying electric field for a later discussion. Our present relationships will concern the magnetic field produced by a differential dc element in free space.

We may think of this differential current element as a vanishingly small section of a current-carrying filamentary conductor, where a filamentary conductor is the limiting case of a cylindrical conductor of circular cross section as the radius approaches zero. We assume a current I flowing in a differential vector length of the filament $d\mathbf{L}$. The law of Biot-Savart[1] then states that at any point P the magnitude of the magnetic field intensity produced by the differential element is proportional to the product of the current, the magnitude of the differential length, and the sine of the angle lying between the filament and a line connecting the filament to the point P where the field is desired. The magnitude of the magnetic field intensity is inversely proportional to the square of the distance from the differential element to the point P. The direction of the magnetic field intensity is normal to the plane containing the differential filament and the line drawn from the filament to the point P. Of the two possible normals, that one is to be chosen which is in the direction of progress of a right-handed screw turned from $d\mathbf{L}$ through the smaller angle to the line from the filament to P. Using rationalized mks units, the constant of proportionality is $1/4\pi$.

The *Biot-Savart law*, described above in some 150 words, may be written concisely using vector notation as

$$d\mathbf{H} = \frac{I\,d\mathbf{L} \times \mathbf{a}_R}{4\pi R^2} = \frac{I\,d\mathbf{L} \times \mathbf{R}}{4\pi R^3} \tag{1}$$

The units of the *magnetic field intensity* $\mathbf{H}$ are evidently amperes per meter (A/m). The geometry is illustrated in Fig. 8.1. Subscripts may be used to indicate the point to which each of the quantities in (1) refers. If we locate the current element at point 1 and describe the point P at which the field is to be determined as point 2, then

$$\boxed{d\mathbf{H}_2 = \frac{I_1\,d\mathbf{L}_1 \times \mathbf{a}_{R12}}{4\pi R_{12}^2}} \tag{2}$$

[1] Biot and Savart were colleagues of Ampère, and all three were professors of physics at the Collège de France at one time or another. The Biot-Savart law was proposed in 1820.

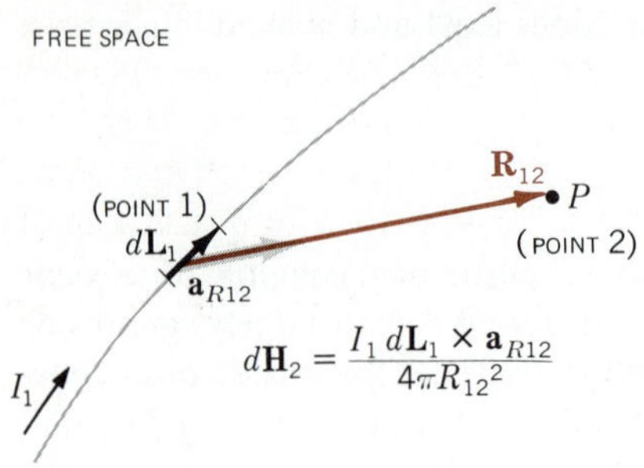

FIGURE 8.1
The law of Biot-Savart expresses the magnetic field intensity $d\mathbf{H}_2$ produced by a differential current element $I_1\, d\mathbf{L}_1$. The direction of $d\mathbf{H}_2$ is into the page.

The law of Biot-Savart is sometimes called *Ampère's law for the current element*, but we shall retain the former name because of possible confusion with Ampère's circuital law, to be met later.

In some aspects, the Biot-Savart law is reminiscent of Coulomb's law when that law is written for a differential element of charge,

$$d\mathbf{E}_2 = \frac{dQ_1\, \mathbf{a}_{R12}}{4\pi\epsilon_0 R_{12}^2}$$

Both show an inverse-square-law dependence on distance, and both show a linear relationship between source and field. The chief difference appears in the direction of the field.

It is impossible to check experimentally the law of Biot-Savart as expressed by (1) or (2) because the differential current element cannot be isolated. We have restricted our attention to direct currents only, so the charge density is not a function of time. The continuity equation in Sec. 5.2, Eq. (5),

$$\nabla \cdot \mathbf{J} = -\frac{\partial \rho_v}{\partial t}$$

therefore shows that

$$\nabla \cdot \mathbf{J} = 0$$

or upon applying the divergence theorem,

$$\oint_S \mathbf{J} \cdot d\mathbf{S} = 0$$

The total current crossing any closed surface is zero, and this condition may be satisfied only by assuming a current flow around a closed path. It is this current flowing in a closed circuit which must be our experimental source, not the differential element.

It follows that only the integral form of the Biot-Savart law can be verified experimentally,

$$\boxed{\mathbf{H} = \oint \frac{I\, d\mathbf{L} \times \mathbf{a}_R}{4\pi R^2}} \qquad (3)$$

Equation (1) or (2), of course, leads directly to the integral form (3), but other differential expressions also yield the same integral formulation. Any term may be added to (1) whose integral around a closed path is zero. That is, any conservative field could be added to (1). The gradient of any scalar field always yields a conservative field, and we could therefore add a term ∇G to (1), where G is a general scalar field, without changing (3) in the slightest. This qualification on (1) or (2) is mentioned to show that if we later ask some foolish questions, not subject to any experimental check, concerning the force exerted by one *differential* current element on another, we should expect foolish answers.

The Biot-Savart law may also be expressed in terms of distributed sources, such as current density $\mathbf{J}$ and *surface current density* $\mathbf{K}$. Surface current flows in a sheet of vanishingly small thickness, and the current density $\mathbf{J}$, measured in amperes per square meter, is therefore infinite. Surface current density, however, is measured in amperes per meter width and designated by $\mathbf{K}$. If the surface current density is uniform, the total current I in any width b is

$$I = Kb$$

where we have assumed that the width b is measured perpendicularly to the direction in which the current is flowing. The geometry is illustrated by Fig. 8.2. For a nonuniform surface current density, integration is necessary:

$$\boxed{I = \int K\, dN} \tag{4}$$

where dN is a differential element of the path *across* which the current is flowing. Thus the differential current element $I\,d\mathbf{L}$, where $d\mathbf{L}$ is in the direction of the current, may be expressed in terms of surface current density $\mathbf{K}$ or current density $\mathbf{J}$,

$$I\,d\mathbf{L} = \mathbf{K}\,dS = \mathbf{J}\,dv \tag{5}$$

and alternate forms of the Biot-Savart law obtained,

$$\boxed{\mathbf{H} = \int_S \frac{\mathbf{K} \times \mathbf{a}_R\, dS}{4\pi R^2}} \tag{6}$$

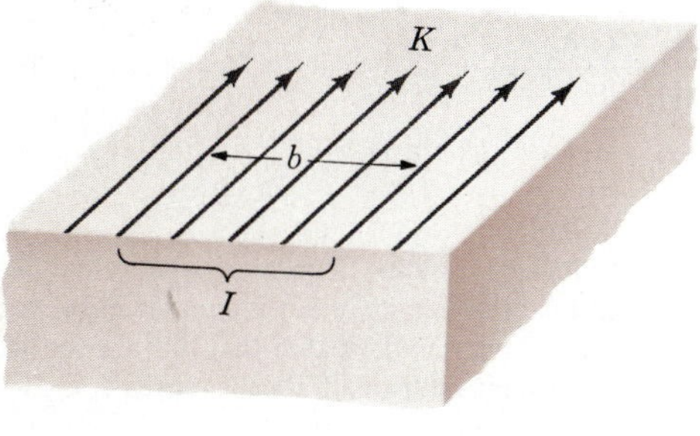

FIGURE 8.2
The total current I within a transverse width b in which there is a *uniform* surface current density K is Kb.

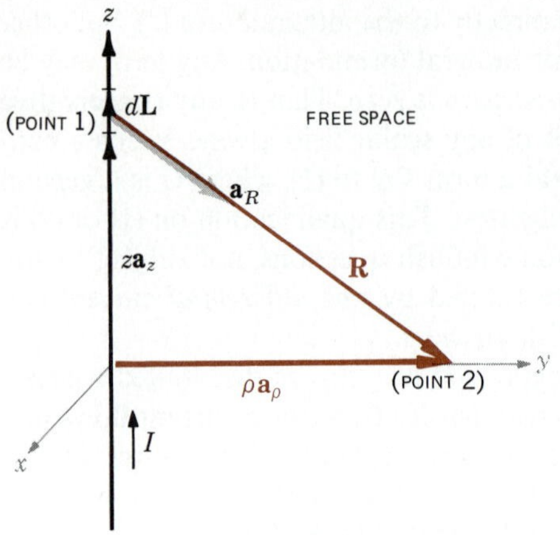

FIGURE 8.3
An infinitely long straight filament carrying a direct current I. The field at point 2 is $\mathbf{H} = (I/2\pi\rho)\mathbf{a}_\phi$.

and

$$\boxed{\mathbf{H} = \int_{\text{vol}} \frac{\mathbf{J} \times \mathbf{a}_R\, dv}{4\pi R^2}} \tag{7}$$

We may illustrate the application of the Biot-Savart law by considering an infinitely long straight filament. We shall apply (2) first and then integrate. This, of course, is the same as using the integral form (3) in the first place.[2]

Referring to Fig. 8.3, we should recognize the symmetry of this field. No variation with z or with ϕ can exist. Point 2, at which we shall determine the field, is therefore chosen in the $z = 0$ plane, and we have $\mathbf{r} = \rho\mathbf{a}_\rho$. The source is located at $\mathbf{r}' = z'\mathbf{a}_z$, and therefore

$$\mathbf{R}_{12} = \mathbf{r} - \mathbf{r}' = \rho\mathbf{a}_\rho - z'\mathbf{a}_z$$

so that

$$\mathbf{a}_{R12} = \frac{\rho\mathbf{a}_\rho - z'\mathbf{a}_z}{\sqrt{\rho^2 + z'^2}}$$

We take $d\mathbf{L} = dz'\,\mathbf{a}_z$ and (2) becomes

$$d\mathbf{H}_2 = \frac{I\,dz'\,\mathbf{a}_z \times (\rho\mathbf{a}_\rho - z'\mathbf{a}_z)}{4\pi(\rho^2 + z'^2)^{3/2}}$$

---

[2] The closed path for the current may be considered to include a return filament parallel to the first filament and infinitely far removed. An outer coaxial conductor of infinite radius is another theoretical possibility. Practically, the problem is an impossible one, but we should realize that our answer will be quite accurate near a very long straight wire having a distant return path for the current.

Since the current is directed toward increasing values of z', the limits are $-\infty$ and ∞ on the integral, and we have

$$\mathbf{H}_2 = \int_{-\infty}^{\infty} \frac{I\,dz'\,\mathbf{a}_z \times (\rho\mathbf{a}_\rho - z'\mathbf{a}_z)}{4\pi(\rho^2 + z'^2)^{3/2}}$$

$$= \frac{I}{4\pi} \int_{-\infty}^{\infty} \frac{\rho\,dz'\,\mathbf{a}_\phi}{(\rho^2 + z'^2)^{3/2}}$$

At this point the unit vector $\mathbf{a}_\phi$ under the integral sign should be investigated, for it is not always a constant, as are the unit vectors of the cartesian coordinate system. A vector is constant when its magnitude and direction are both constant. The unit vector certainly has constant magnitude, but its direction may change. Here $\mathbf{a}_\phi$ changes with the coordinate ϕ but not with ρ or z. Fortunately, the integration here is with respect to z', and $\mathbf{a}_\phi$ is a constant and may be removed from under the integral sign,

$$\mathbf{H}_2 = \frac{I\rho\mathbf{a}_\phi}{4\pi} \int_{-\infty}^{\infty} \frac{dz'}{(\rho^2 + z'^2)^{3/2}}$$

$$= \frac{I\rho\mathbf{a}_\phi}{4\pi} \left. \frac{z'}{\rho^2\sqrt{\rho^2 + z'^2}} \right|_{-\infty}^{\infty}$$

and

$$\boxed{\mathbf{H}_2 = \frac{I}{2\pi\rho}\mathbf{a}_\phi} \qquad (8)$$

The magnitude of the field is not a function of ϕ or z and it varies inversely as the distance from the filament. The direction of the magnetic-field-intensity vector is circumferential. The streamlines are therefore circles about the filament, and the field may be mapped in cross section as in Fig. 8.4.

The separation of the streamlines is proportional to the radius, or inversely proportional to the magnitude of **H**. To be specific, the streamlines have been drawn with curvilinear squares in mind. As yet we have no name for the family of lines[3] which are perpendicular to these circular streamlines, but the spacing of the streamlines has been adjusted so that the addition of this second set of lines will produce an array of curvilinear squares.

A comparison of Fig. 8.4 with the map of the *electric* field about an infinite line *charge* shows that the streamlines of the magnetic field correspond exactly to the equipotentials of the electric field, and the unnamed (and undrawn) perpendicular family of lines in the magnetic field corresponds to the streamlines of the electric field. This correspondence is not an accident, but there are several other

---

[3] If you can't wait, see Sec. 8.6.

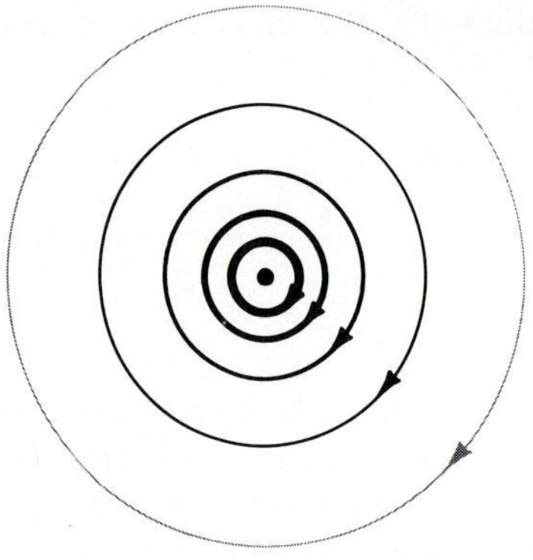

FIGURE 8.4
The streamlines of the magnetic field intensity about an infinitely long straight filament carrying a direct current I. The direction of I is into the page.

concepts which must be mastered before the analogy between electric and magnetic fields can be explored more thoroughly.

Using the Biot-Savart law to find **H** is in many respects similar to the use of Coulomb's law to find **E**. Each requires the determination of a moderately complicated integrand containing vector quantities, followed by an integration. When we were concerned with Coulomb's law we solved a number of examples, including the fields of the point charge, line charge, and sheet of charge. The law of Biot-Savart can be used to solve analogous problems in magnetic fields, and some of these problems now appear as exercises at the end of the chapter rather than as examples here.

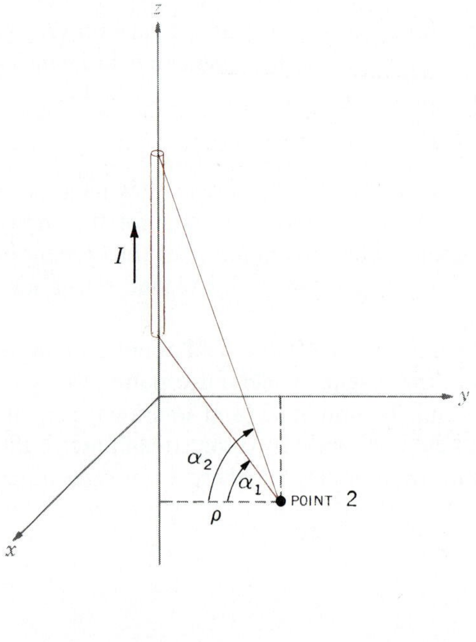

FIGURE 8.5
The magnetic field intensity caused by a finite-length current filament on the z axis is $(I/4\pi\rho)(\sin \alpha_2 - \sin \alpha_1)\mathbf{a}_\phi$.

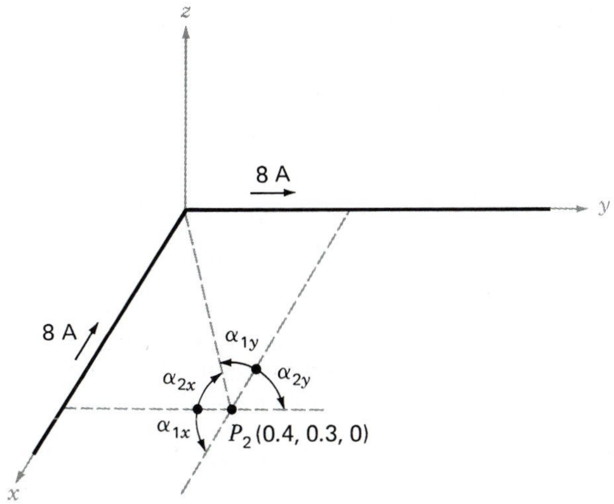

FIGURE 8.6
The individual fields of two semi-infinite current segments are found by (9) and added to obtain $\mathbf{H}_2$ at P_2.

One useful result is the field of the finite-length current element, shown in Fig. 8.5. It turns out (see Prob. 8 at the end of the chapter) that **H** is most easily expressed in terms of the angles α_1 and α_2, as identified in the figure. The result is

$$\mathbf{H} = \frac{I}{4\pi\rho}(\sin\alpha_2 - \sin\alpha_1)\mathbf{a}_\phi \tag{9}$$

If one or both ends are below point 2, α_1 or both α_1 and α_2 are negative.

Equation (9) may be used to find the magnetic field intensity caused by current filaments arranged as a sequence of straight line segments.

As a numerical example illustrating the use of (9), let us determine **H** at $P_2(0.4,0.3,0)$ in the field of an 8-A filamentary current directed inward from infinity to the origin on the positive x axis, and then outward to infinity along the y axis. This arrangement is shown in Fig. 8.6.

We first consider the semi-infinite current on the x axis, identifying the two angles, $\alpha_{1x} = -90^\circ$ and $\alpha_{2x} = \tan^{-1}(0.4/0.3) = 53.1^\circ$. The radial distance ρ is measured from the x axis, and we have $\rho_x = 0.3$. Thus, this contribution to $\mathbf{H}_2$ is

$$\mathbf{H}_{2(x)} = \frac{8}{4\pi(0.3)}(\sin 53.1^\circ + 1)\mathbf{a}_\phi = \frac{2}{0.3\pi}(1.8)\mathbf{a}_\phi = \frac{12}{\pi}\mathbf{a}_\phi$$

The unit vector $\mathbf{a}_\phi$ must also be referred to the x axis. We see that it becomes $-\mathbf{a}_z$. Therefore,

$$\mathbf{H}_{2(x)} = -\frac{12}{\pi}\mathbf{a}_z \qquad \text{A/m}$$

For the current on the y axis, we have $\alpha_{1y} = -\tan^{-1}(0.3/0.4) = -36.9°$, $\alpha_{2y} = 90°$, and $\rho_y = 0.4$. It follows that

$$\mathbf{H}_{2(y)} = \frac{8}{4\pi(0.4)}(1 + \sin 36.9°)(-\mathbf{a}_z) = -\frac{8}{\pi}\mathbf{a}_z \qquad \text{A/m}$$

Adding these results, we have

$$\mathbf{H}_2 = \mathbf{H}_{2(x)} + \mathbf{H}_{2(y)} = -\frac{20}{\pi}\mathbf{a}_z = -6.37\mathbf{a}_z \qquad \text{A/m}$$

D8.1. Find the incremental field $\Delta\mathbf{H}_2$ at P_2 caused by a source at P_1 of $I_1\,\Delta\mathbf{L}_1 =$: (*a*) $2\pi\mathbf{a}_z$ μA·m, given $P_1(4,0,0)$ and $P_2(0,3,0)$; (*b*) $2\pi\mathbf{a}_z$ μA·m, given $P_1(4,-2,3)$ and $P_2(0,3,0)$; (*c*) $2\pi(0.6\mathbf{a}_x - 0.8\mathbf{a}_y)$ μA·m, given $P_1(4,-2,3)$ and $P_2(1,3,2)$.

Ans. $-12\mathbf{a}_x - 16\mathbf{a}_y$ nA/m; $-7.07\mathbf{a}_x - 5.66\mathbf{a}_y$ nA/m; $1.932\mathbf{a}_x + 1.449\mathbf{a}_y + 1.449\mathbf{a}_z$ nA/m

D8.2. Use (9) to find the vector magnetic field intensity in cartesian coordinates at $P_2(1.5,2,3)$ caused by a current filament of 24 A in the $\mathbf{a}_z$ direction on the z axis and extending from: (*a*) $z = 0$ to $z = 6$; (*b*) $z = 6$ to $z = \infty$; (*c*) $z = -\infty$ to $z = \infty$.

Ans. $-0.939\mathbf{a}_x + 0.704\mathbf{a}_y$ A/m; $-0.1417\mathbf{a}_x + 0.1062\mathbf{a}_y$ A/m; $-1.222\mathbf{a}_x + 0.917\mathbf{a}_y$ A/m

8.2 AMPERE'S CIRCUITAL LAW

After solving a number of simple electrostatic problems with Coulomb's law, we found that the same problems could be solved much more easily by using Gauss's law whenever a high degree of symmetry was present. Again, an analogous procedure exists in magnetic fields. Here, the law that helps us solve problems more easily is known as *Ampère's circuital*[4] *law*, sometimes called Ampère's work law. This law may be derived from the Biot-Savart law, and the derivation is accomplished in Sec. 8.7. For the present we might agree to accept Ampère's circuital law temporarily as another law capable of experimental proof. Its use will also require careful consideration of the symmetry of the problem to determine which variables and components are present.

Ampère's circuital law states that the line integral of $\mathbf{H}$ about any *closed* path is exactly equal to the direct current enclosed by that path,

$$\oint \mathbf{H} \cdot d\mathbf{L} = I \qquad (10)$$

We define positive current as flowing in the direction of advance of a right-handed screw turned in the direction in which the closed path is traversed.

Referring to Fig. 8.7, which shows a circular wire carrying a direct current I, the line integral of $\mathbf{H}$ about the closed paths lettered a and b results in an answer

[4] The preferred pronunciation puts the accent on "circ-."

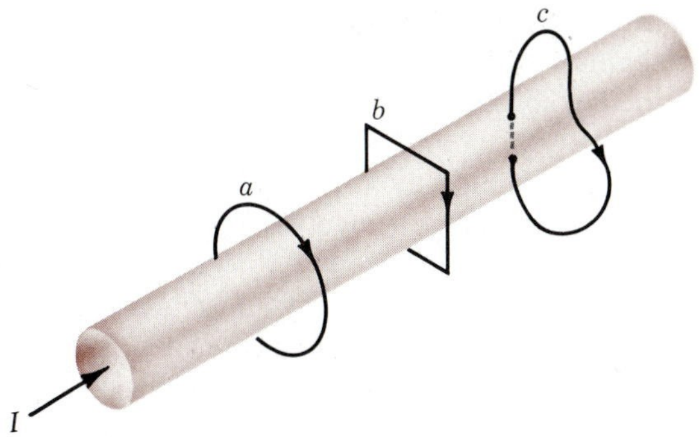

FIGURE 8.7
A conductor has a total current I. The line integral of $\mathbf{H}$ about the closed paths a and b is equal to I, and the integral around path c is less than I, since the entire current is not enclosed by the path.

of I; the integral about the closed path c which passes through the conductor gives an answer less than I and is exactly that portion of the total current which is enclosed by the path c. Although paths a and b give the same answer, the integrands are, of course, different. The line integral directs us to multiply the component of $\mathbf{H}$ in the direction of the path by a small increment of path length at one point of the path, move along the path to the next incremental length, and repeat the process, continuing until the path is completely traversed. Since $\mathbf{H}$ will in general vary from point to point, and since paths a and b are not alike, the contributions to the integral made by, say, each millimeter of path length are quite different. Only the final answers are the same.

We should also consider exactly what is meant by the expression "current enclosed by the path." Suppose we solder a circuit together after passing the conductor once through a rubber band, which we shall use to represent the closed path. Some strange and formidable paths can be constructed by twisting and knotting the rubber band, but if neither the rubber band nor the conducting circuit is broken, the current enclosed by the path is that carried by the conductor. Now let us replace the rubber band by a circular ring of spring steel across which is stretched a rubber sheet. The steel loop forms the closed path, and the current-carrying conductor must pierce the rubber sheet if the current is to be enclosed by the path. Again, we may twist the steel loop, and we may also deform the rubber sheet by pushing our fist into it or folding it in any way we wish. A single current-carrying conductor still pierces the sheet once, and this is the true measure of the current enclosed by the path. If we should thread the conductor once through the sheet from front to back and once from back to front, the total current enclosed by the path is the algebraic sum, which is zero.

In more general language, given a closed path, we recognize this path as the perimeter of an infinite number of surfaces (not closed surfaces). Any current-carrying conductor enclosed by the path must pass through every one of these surfaces once. Certainly some of the surfaces may be chosen in such a way that the conductor pierces them twice in one direction and once in the other direction, but the algebraic total current is still the same.

We shall find that the closed path is usually of an extremely simple nature and can be drawn on a plane. The simplest surface is, then, that portion of the plane enclosed by the path. We need merely find the total current passing through this region of the plane.

The application of Gauss's law involves finding the total charge enclosed by a closed surface; the application of Ampère's circuital law involves finding the total current enclosed by a closed path.

Let us again find the magnetic field intensity produced by an infinitely long filament carrying a current I. The filament lies on the z axis in free space (as in Fig. 8.3), and the current flows in the direction given by $\mathbf{a}_z$. Symmetry inspection comes first, showing that there is no variation with z or ϕ. Next we determine which components of $\mathbf{H}$ are present by using the Biot-Savart law. Without specifically using the cross product, we may say that the direction of $d\mathbf{H}$ is perpendicular to the plane containing $d\mathbf{L}$ and $\mathbf{R}$ and therefore is in the direction of $\mathbf{a}_\phi$. Hence the only component of $\mathbf{H}$ is H_ϕ, and it is a function only of ρ.

We therefore choose a path to any section of which $\mathbf{H}$ is either perpendicular or tangential and along which H is constant. The first requirement (perpendicularity or tangency) allows us to replace the dot product of Ampère's circuital law with the product of the scalar magnitudes, except along that portion of the path where $\mathbf{H}$ is normal to the path and the dot product is zero; the second requirement (constancy) then permits us to remove the magnetic field intensity from the integral sign. The integration required is usually trivial and consists of finding the length of that portion of the path to which $\mathbf{H}$ is parallel.

In our example the path must be a circle of radius ρ and Ampère's circuital law becomes

$$\oint \mathbf{H} \cdot d\mathbf{L} = \int_0^{2\pi} H_\phi \rho \, d\phi = H_\phi \rho \int_0^{2\pi} d\phi = H_\phi 2\pi\rho = I$$

or

$$H_\phi = \frac{I}{2\pi\rho}$$

as before.

As a second example of the application of Ampère's circuital law, consider an infinitely long coaxial transmission line carrying a uniformly distributed total current I in the center conductor and $-I$ in the outer conductor. The line is shown in Fig. 8.8*a*. Symmetry shows that H is not a function of ϕ or z. In order to determine the components present, we may use the results of the previous example by considering the solid conductors as being composed of a large number of filaments. No filament has a z component of $\mathbf{H}$. Furthermore, the H_ρ component at $\phi = 0°$, produced by one filament located at $\rho = \rho_1$, $\phi = \phi_1$, is canceled by the H_ρ component produced by a symmetrically located filament at $\rho = \rho_1$, $\phi = -\phi_1$. This symmetry is illustrated by Fig. 8.8*b*. Again we find only an H_ϕ component which varies with ρ.

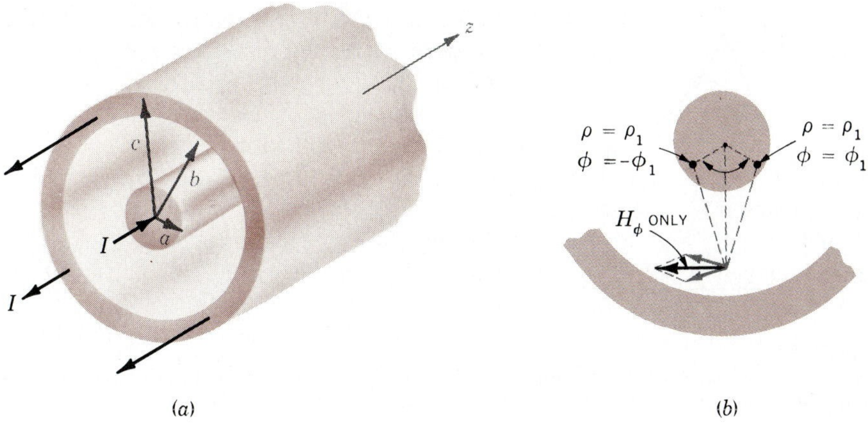

FIGURE 8.8
(*a*) Cross section of a coaxial cable carrying a uniformly distributed current I in the inner conductor and $-I$ in the outer conductor. The magnetic field at any point is most easily determined by applying Ampère's circuital law about a circular path. (*b*) Current filaments at $\rho = \rho_1$, $\phi = \pm\phi_1$, produces H_ρ components which cancel. For the total field, $\mathbf{H} = H_\phi \mathbf{a}_\phi$.

A circular path of radius ρ, where ρ is larger than the radius of the inner conductor but less than the inner radius of the outer conductor, then leads immediately to

$$H_\phi = \frac{I}{2\pi\rho} \qquad (a < \rho < b)$$

If we choose ρ smaller than the radius of the inner conductor, the current enclosed is

$$I_{\text{encl}} = I\frac{\rho^2}{a^2}$$

and

$$2\pi\rho H_\phi = I\frac{\rho^2}{a^2}$$

or

$$H_\phi = \frac{I\rho}{2\pi a^2} \qquad (\rho < a)$$

If the radius ρ is larger than the outer radius of the outer conductor, no current is enclosed and

$$H_\phi = 0 \qquad (\rho > c)$$

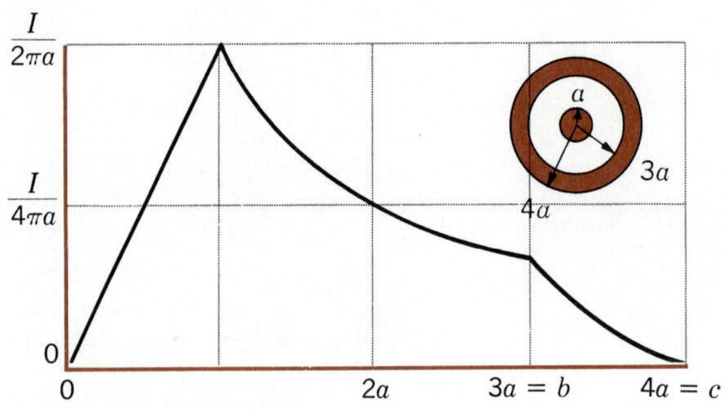

FIGURE 8.9
The magnetic field intensity as a function of radius in an infinitely long coaxial transmission line with the dimensions shown.

Finally, if the path lies within the outer conductor, we have

$$2\pi\rho H_\phi = I - I\left(\frac{\rho^2 - b^2}{c^2 - b^2}\right)$$

$$H_\phi = \frac{I}{2\pi\rho}\frac{c^2 - \rho^2}{c^2 - b^2} \qquad (b < \rho < c)$$

The magnetic-field-strength variation with radius is shown in Fig. 8.9 for a coaxial cable in which $b = 3a$, $c = 4a$. It should be noted that the magnetic field intensity **H** is continuous at all the conductor boundaries. In other words, a slight increase in the radius of the closed path does not result in the enclosure of a tremendously different current. The value of H_ϕ shows no sudden jumps.

The external field is zero. This, we see, results from equal positive and negative currents enclosed by the path. Each produces an external field of magnitude $I/2\pi\rho$, but complete cancellation occurs. This is another example of "shielding"; such a coaxial cable carrying large currents would not produce any noticeable effect in an adjacent circuit.

As a final example, let us consider a sheet of current flowing in the positive y direction and located in the $z = 0$ plane. We may think of the return current as equally divided between two distant sheets on either side of the sheet we are considering. A sheet of uniform surface current density $\mathbf{K} = K_y\mathbf{a}_y$ is shown in Fig. 8.10. **H** cannot vary with x or y. If the sheet is subdivided into a number of filaments, it is evident that no filament can produce an H_y component. Moreover, the Biot-Savart law shows that the contributions to H_z produced by a symmetrically located pair of filaments cancel. Thus, H_z is zero also; only an H_x component is present. We therefore choose the path 1-1′-2′-2-1 composed of straight-line segments which are either parallel or perpendicular to H_x. Ampère's circuital law gives

$$H_{x1}L + H_{x2}(-L) = K_yL$$

or

$$H_{x1} - H_{x2} = K_y$$

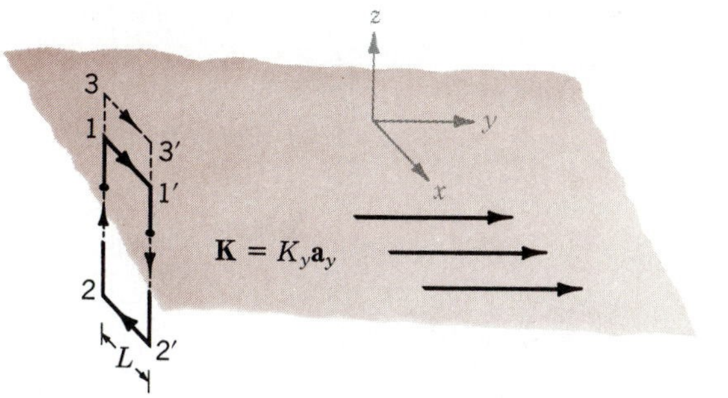

FIGURE 8.10
A uniform sheet of surface current $\mathbf{K} = K_y\mathbf{a}_y$ in the $z = 0$ plane. **H** may be found by applying Ampère's circuital law about the paths 1-1′-2′-2-1 and 3-3′-2′-2-3.

If the path 3-3′-2′-2-3 is now chosen, the same current is enclosed, and

$$H_{x3} - H_{x2} = K_y$$

and therefore

$$H_{x3} = H_{x1}$$

It follows that H_x is the same for all positive z. Similarly, H_x is the same for all negative z. Because of the symmetry, then, the magnetic field intensity on one side of the current sheet is the negative of that on the other. Above the sheet,

$$H_x = \tfrac{1}{2}K_y \qquad (z > 0)$$

while below it

$$H_x = -\tfrac{1}{2}K_y \qquad (z < 0)$$

Letting $\mathbf{a}_N$ be a unit vector normal (outward) to the current sheet, the result may be written in a form correct for all z as

$$\boxed{\mathbf{H} = \tfrac{1}{2}\mathbf{K} \times \mathbf{a}_N} \qquad (11)$$

If a second sheet of current flowing in the opposite direction, $\mathbf{K} = -K_y\mathbf{a}_y$, is placed at $z = h$, (11) shows that the field in the region between the current sheets is

$$\boxed{\mathbf{H} = \mathbf{K} \times \mathbf{a}_N \qquad (0 < z < h)} \qquad (12)$$

and is zero elsewhere,

$$\boxed{\mathbf{H} = 0 \qquad (z < 0, \quad z > h)} \qquad (13)$$

The most difficult part of the application of Ampère's circuital law is the determination of the components of the field which are present. The surest method is the logical application of the Biot-Savart law and a knowledge of the magnetic

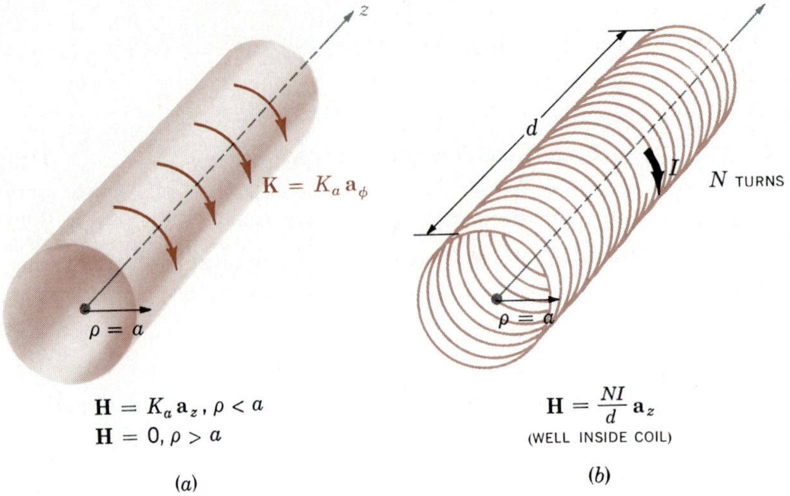

FIGURE 8.11
(*a*) An ideal solenoid of infinite length with a circular current sheet $\mathbf{K} = K_a\mathbf{a}_\phi$. (*b*) An N-turn solenoid of finite length d.

fields of simple form. Problem 14 at the end of this chapter outlines the steps involved in applying Ampère's circuital law to an infinitely long solenoid of radius a and uniform current density $K_a\mathbf{a}_\phi$, as shown in Fig. 8.11*a*. For reference, the result is

$$\mathbf{H} = K_a\mathbf{a}_z \qquad (\rho < a) \tag{14a}$$

$$\mathbf{H} = 0 \qquad (\rho > a) \tag{14b}$$

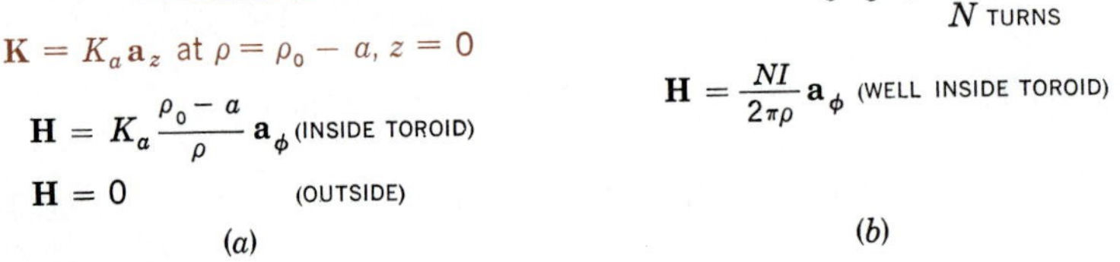

FIGURE 8.12
(*a*) An ideal toroid carrying a surface current **K** in the direction shown. (*b*) An N-turn toroid carrying a filamentary current I.

If the solenoid has a finite length d and consists of N closely wound turns of a filament that carries a current I, then the field at points well within the solenoid is given closely by

$$\mathbf{H} = \frac{NI}{d}\mathbf{a}_z \qquad \text{(well within the solenoid)} \tag{15}$$

The approximation is useful if it is not applied closer than two radii to the open ends, nor closer to the solenoid surface than twice the separation between turns.

For the toroids shown in Fig. 8.12, it can be shown that the magnetic field intensity for the ideal case, Fig. 8.12*a*, is

$$\mathbf{H} = K_a \frac{\rho_0 - a}{\rho}\mathbf{a}_\phi \qquad \text{(inside toroid)} \tag{16a}$$

$$\mathbf{H} = 0 \qquad \text{(outside)} \tag{16b}$$

For the N-turn toroid of Fig. 8.12*b*, we have the good approximations,

$$\mathbf{H} = \frac{NI}{2\pi\rho}\mathbf{a}_\phi \qquad \text{(inside toroid)} \tag{17a}$$

$$\mathbf{H} = 0 \qquad \text{(outside)} \tag{17b}$$

as long as we consider points removed from the toroidal surface by several times the separation between turns.

Toroids having square cross sections are also treated quite readily, as you can see for yourself by trying Prob. 15.

Accurate formulas for solenoids, toroids, and coils of other shapes are available in sec. 2 of the "Standard Handbook for Electrical Engineers" (see Suggested References for Chap. 5).

D8.3. Express the value of $\mathbf{H}$ in cartesian components at $P(0.01,0,0)$ in the field of: (*a*) two current filaments, 0.08 A in the $\mathbf{a}_z$ direction on the z axis, and 0.08 A in the $-\mathbf{a}_z$ direction at $x = 0.015$, $y = 0$; (*b*) coax, $a = 3$ mm, $b = 9$ mm, $c = 12$ mm, $I = 0.8$ A, centered on the z axis, $\mathbf{a}_z$ direction in center conductor; (*c*) three current sheets, $1.5\pi\mathbf{a}_y$ A/m at $x = 6$ mm, $-3\pi\mathbf{a}_y$ A/m at $x = 9$ mm, and $1.5\pi\mathbf{a}_y$ A/m at $x = 12$ mm; (*d*) infinite solenoid, $\mathbf{K} = 3\pi\mathbf{a}_\phi$ A/m at $\rho = 12$ mm; (*e*) toroid, centered at origin, axis on y axis, $\rho_0 = 12$ mm, $a = 3$ mm, $N = 250$ turns, $I = 2$ mA in $\mathbf{a}_y$ direction at outer radius.

Ans. $3.82\mathbf{a}_y$ A/m; $8.89\mathbf{a}_y$ A/m; $4.71\mathbf{a}_z$ A/m; $9.42\mathbf{a}_z$ A/m; $7.96\mathbf{a}_z$ A/m

8.3 CURL

We completed our study of Gauss's law by applying it to a differential volume element and were led to the concept of divergence. We now apply Ampère's circuital law to the perimeter of a differential surface element and meet the third and

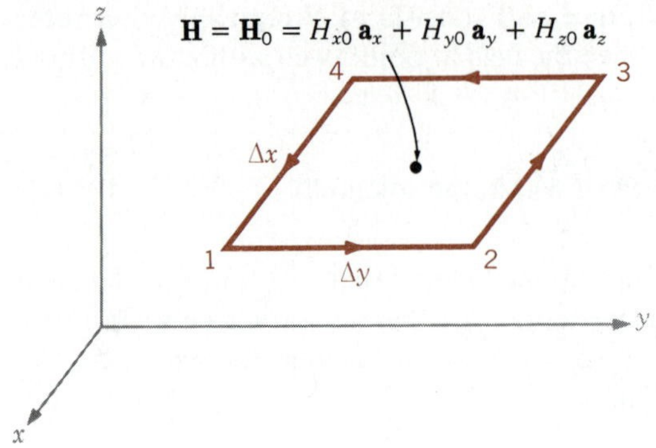

FIGURE 8.13
An incremental closed path in cartesian coordinates is selected for the application of Ampère's circuital law in order to determine the spatial rate of change of **H**.

last of the special derivatives of vector analysis, the curl. Our immediate objective is to obtain the point form of Ampère's circuital law.

Again we shall choose cartesian coordinates, and an incremental closed path of sides Δx and Δy is selected (Fig. 8.13). We assume that some current, as yet unspecified, produces a reference value for **H** at the *center* of this small rectangle,

$$\mathbf{H}_0 = H_{x0}\mathbf{a}_x + H_{y0}\mathbf{a}_y + H_{z0}\mathbf{a}_z$$

The closed line integral of **H** about this path is then approximately the sum of the four values of $\mathbf{H} \cdot \Delta\mathbf{L}$ on each side. We choose the direction of traverse as 1-2-3-4-1, which corresponds to a current in the $\mathbf{a}_z$ direction, and the first contribution is therefore

$$(\mathbf{H} \cdot \Delta\mathbf{L})_{1\text{-}2} = H_{y,1\text{-}2}\,\Delta y$$

The value of H_y *on this section* of the path may be given in terms of the reference value H_{y0} at the center of the rectangle, the rate of change of H_y with x, and the distance $\Delta x/2$ from the center to the midpoint of side 1-2:

$$H_{y,1\text{-}2} \doteq H_{y0} + \frac{\partial H_y}{\partial x}\left(\tfrac{1}{2}\Delta x\right)$$

Thus

$$(\mathbf{H} \cdot \Delta\mathbf{L})_{1\text{-}2} \doteq \left(H_{y0} + \frac{1}{2}\frac{\partial H_y}{\partial x}\Delta x\right)\Delta y$$

Along the next section of the path we have

$$(\mathbf{H} \cdot \Delta\mathbf{L})_{2\text{-}3} \doteq H_{x,2\text{-}3}(-\Delta x) \doteq -\left(H_{x0} + \frac{1}{2}\frac{\partial H_x}{\partial y}\Delta y\right)\Delta x$$

Continuing for the remaining two segments and adding the results,

$$\oint \mathbf{H} \cdot d\mathbf{L} \doteq \left(\frac{\partial H_y}{\partial x} - \frac{\partial H_x}{\partial y}\right)\Delta x\,\Delta y$$

By Ampère's circuital law, this result must be equal to the current enclosed by the path, or the current crossing any surface bounded by the path. If we assume a general current density **J**, the enclosed current is then $\Delta I \doteq J_z \Delta x \Delta y$, and

$$\oint \mathbf{H} \cdot d\mathbf{L} \doteq \left(\frac{\partial H_y}{\partial x} - \frac{\partial H_x}{\partial y} \right) \Delta x \Delta y \doteq J_z \Delta x \Delta y$$

or

$$\frac{\oint \mathbf{H} \cdot d\mathbf{L}}{\Delta x \Delta y} \doteq \frac{\partial H_y}{\partial x} - \frac{\partial H_x}{\partial y} \doteq J_z$$

As we cause the closed path to shrink, the above expression becomes more nearly exact, and in the limit we have the equality

$$\lim_{\Delta x, \Delta y \to 0} \frac{\oint \mathbf{H} \cdot d\mathbf{L}}{\Delta x \Delta y} = \frac{\partial H_y}{\partial x} - \frac{\partial H_x}{\partial y} = J_z \tag{18}$$

After beginning with Ampère's circuital law equating the closed line integral of **H** to the current enclosed, we have now arrived at a relationship involving the closed line integral of **H** *per unit area* enclosed and the current *per unit area* enclosed, or current density. We performed a similar analysis in passing from the integral form of Gauss's law, involving flux through a closed surface and charge enclosed, to the point form, relating flux through a closed surface *per unit volume* enclosed and charge *per unit volume* enclosed, or volume charge density. In each case a limit is necessary to produce an equality.

If we choose closed paths which are oriented perpendicularly to each of the remaining two coordinate axes, analogous processes lead to expressions for the y and z components of the current density,

$$\lim_{\Delta y, \Delta z \to 0} \frac{\oint \mathbf{H} \cdot d\mathbf{L}}{\Delta y \Delta z} = \frac{\partial H_z}{\partial y} - \frac{\partial H_y}{\partial z} = J_x \tag{19}$$

and

$$\lim_{\Delta z, \Delta x \to 0} \frac{\oint \mathbf{H} \cdot d\mathbf{L}}{\Delta z \Delta x} = \frac{\partial H_x}{\partial z} - \frac{\partial H_z}{\partial x} = J_y \tag{20}$$

Comparing (18), (19), and (20), we see that a component of the current density is given by the limit of the quotient of the closed line integral of **H** about a small path in a plane normal to that component and of the area enclosed as the path shrinks to zero. This limit has its counterpart in other fields of science and long ago received the name of *curl*. The curl of any vector is a vector, and any component of the curl is given by the limit of the quotient of the closed line integral of the vector about a small path in a plane normal to that component desired and the area enclosed, as the path shrinks to zero. It should be noted that the above definition of curl does not refer specifically to a particular coordinate system. The mathematical form of the definition is

$$(\text{curl } \mathbf{H})_N = \lim_{\Delta S_N \to 0} \frac{\oint \mathbf{H} \cdot d\mathbf{L}}{\Delta S_N} \tag{21}$$

where ΔS_N is the planar area enclosed by the closed line integral. The N subscript indicates that the component of the curl is that component which is *normal* to the surface enclosed by the closed path. It may represent any component in any coordinate system.

In cartesian coordinates the definition (21) shows that the x, y, and z components of the curl of **H** are given by (18), (19), and (20), and therefore

$$\text{curl } \mathbf{H} = \left(\frac{\partial H_z}{\partial y} - \frac{\partial H_y}{\partial z}\right)\mathbf{a}_x + \left(\frac{\partial H_x}{\partial z} - \frac{\partial H_z}{\partial x}\right)\mathbf{a}_y + \left(\frac{\partial H_y}{\partial x} - \frac{\partial H_x}{\partial y}\right)\mathbf{a}_z \tag{22}$$

This result may be written in the form of a determinant,

$$\text{curl } \mathbf{H} = \begin{vmatrix} \mathbf{a}_x & \mathbf{a}_y & \mathbf{a}_z \\ \dfrac{\partial}{\partial x} & \dfrac{\partial}{\partial y} & \dfrac{\partial}{\partial z} \\ H_x & H_y & H_z \end{vmatrix} \tag{23}$$

and may also be written in terms of the vector operator,

$$\text{curl } \mathbf{H} = \nabla \times \mathbf{H} \tag{24}$$

Equation (22) is the result of applying the definition (21) to the cartesian coordinate system. We obtained the z component of this expression by evaluating Ampère's circuital law about an incremental path of sides Δx and Δy, and we could have obtained the other two components just as easily by choosing the appropriate paths. Equation (23) is a neat method of storing the cartesian coordinate expression for curl; the form is symmetrical and easily remembered. Equation (24) is even more concise and leads to (22) upon applying the definitions of the cross product and vector operator.

The expressions for curl **H** in cylindrical and spherical coordinates are derived in the Appendix by applying the definition (21). Although they may be written in determinant form, as explained there, the determinants do not have one row of unit vectors on top and one row of components on the bottom, and they are not easily memorized. For this reason, the curl expansions in cylindrical and spherical coordinates which appear below and inside the back cover are usually referred to whenever necessary.

$$\nabla \times \mathbf{H} = \left(\frac{\partial H_z}{\partial y} - \frac{\partial H_y}{\partial z}\right)\mathbf{a}_x + \left(\frac{\partial H_x}{\partial z} - \frac{\partial H_z}{\partial x}\right)\mathbf{a}_y + \left(\frac{\partial H_y}{\partial x} - \frac{\partial H_x}{\partial y}\right)\mathbf{a}_z \quad \text{(cartesian)} \tag{22}$$

$$\nabla \times \mathbf{H} = \left(\frac{1}{\rho}\frac{\partial H_z}{\partial \phi} - \frac{\partial H_\phi}{\partial z}\right)\mathbf{a}_\rho + \left(\frac{\partial H_\rho}{\partial z} - \frac{\partial H_z}{\partial \rho}\right)\mathbf{a}_\phi + \left[\frac{1}{\rho}\frac{\partial(\rho H_\phi)}{\partial \rho} - \frac{1}{\rho}\frac{\partial H_\rho}{\partial \phi}\right]\mathbf{a}_z \quad \text{(cylindrical)} \tag{25}$$

$$\nabla \times \mathbf{H} = \frac{1}{r \sin\theta}\left[\frac{\partial(H_\phi \sin\theta)}{\partial \theta} - \frac{\partial H_\theta}{\partial \phi}\right]\mathbf{a}_r + \frac{1}{r}\left[\frac{1}{\sin\theta}\frac{\partial H_r}{\partial \phi} - \frac{\partial(rH_\phi)}{\partial r}\right]\mathbf{a}_\theta + \frac{1}{r}\left[\frac{\partial(rH_\theta)}{\partial r} - \frac{\partial H_r}{\partial \theta}\right]\mathbf{a}_\phi \quad \text{(spherical)} \tag{26}$$

Although we have described curl as a line integral per unit area, this does not provide everyone with a satisfactory physical picture of the nature of the curl operation, for the closed line integral itself requires physical interpretation. This integral was first met in the electrostatic field, where we saw that $\oint \mathbf{E} \cdot d\mathbf{L} = 0$. Inasmuch as the integral was zero, we did not belabor the physical picture. More recently we have discussed the closed line integral of **H**, $\oint \mathbf{H} \cdot d\mathbf{L} = I$. Either of these closed line integrals is also known by the name of "circulation," a term obviously borrowed from the field of fluid dynamics.

The circulation of **H**, or $\oint \mathbf{H} \cdot d\mathbf{L}$, is obtained by multiplying the component of **H** parallel to the specified closed path at each point along it by the differential path length and summing the results as the differential lengths approach zero and as their number becomes infinite. We do not require a vanishingly small path. Ampère's circuital law tells us that if **H** does possess circulation about a given path, then current passes through this path. In electrostatics we see that the circulation of **E** is zero about every path, a direct consequence of the fact that zero work is required to carry a charge around a closed path.

We may now describe curl as *circulation per unit area*. The closed path is vanishingly small, and curl is defined at a point. The curl of **E** must be zero, for the circulation is zero. The curl of **H** is not zero, however; the circulation of **H** per unit area is the current density by Ampère's circuital law [or (18), (19), and (20)].

Skilling[5] suggests the use of a very small paddle wheel as a "curl meter." Our vector quantity, then, must be thought of as capable of applying a force to

---

[5] See the Suggested References at the end of the chapter.

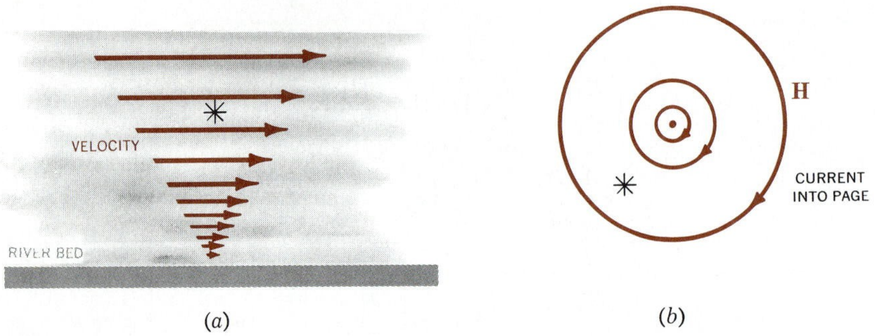

FIGURE 8.14
(*a*) The curl meter shows a component of the curl of the water velocity into the page. (*b*) The curl of the magnetic field intensity about an infinitely long filament is shown.

each blade of the paddle wheel, the force being proportional to the component of the field normal to the surface of that blade. In order to test a field for curl we dip our paddle wheel into the field, with the axis of the paddle wheel lined up with the direction of the component of curl desired, and note the action of the field on the paddle. No rotation means no curl; larger angular velocities mean greater values of the curl; a reversal in the direction of spin means a reversal in the sign of the curl. In order to find the direction of the vector curl and not merely to establish the presence of any particular component, we should place our paddle wheel in the field and hunt around for the orientation which produces the greatest torque. The direction of the curl is then along the axis of the paddle wheel, as given by the right-hand rule.

As an example, consider the flow of water in a river. Figure 8.14*a* shows the longitudinal section of a wide river taken at the middle of the river. The water velocity is zero at the bottom and increases linearly as the surface is approached. A paddle wheel placed in the position shown, with its axis perpendicular to the paper, will turn in a clockwise direction, showing the presence of a component of curl in the direction of an inward normal to the surface of the page. If the velocity of the water does not change as we go up- or downstream and also shows no variation as we go across the river (or even if it decreases in the same fashion toward either bank), then this component is the only component present at the center of the stream, and the curl of the water velocity has a direction into the page.

In Fig. 8.14*b* the streamlines of the magnetic field intensity about an infinitely long filamentary conductor are shown. The curl meter placed in this field of curved lines shows that a larger number of blades have a clockwise force exerted on them but that this force is in general smaller than the counterclockwise force exerted on the smaller number of blades closer to the wire. It seems possible that if the curvature of the streamlines is correct and also if the variation of the field strength is just right, the net torque on the paddle wheel may be zero. Actually, the paddle wheel does not rotate in this case, for since $\mathbf{H} = (I/2\pi\rho)\mathbf{a}_\phi$, we may substitute into

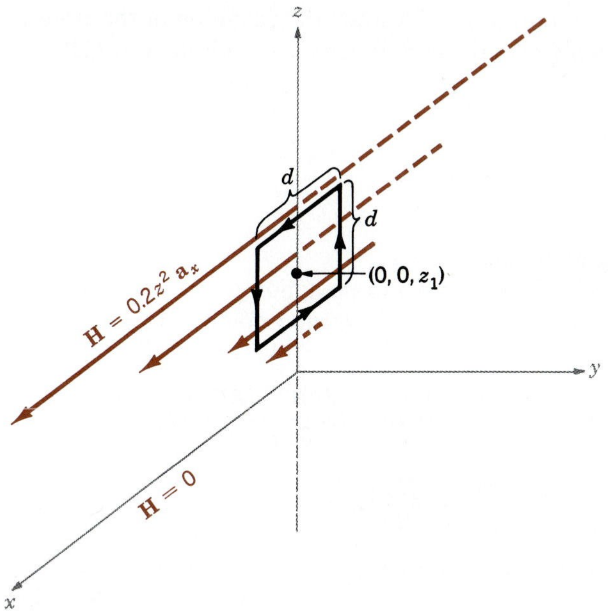

FIGURE 8.15
A square path of side d with its center on the z axis at $z = z_1$ is used to evaluate $\oint \mathbf{H} \cdot d\mathbf{L}$ and find curl $\mathbf{H}$.

(25), obtaining

$$\text{curl } \mathbf{H} = -\frac{\partial H_\phi}{\partial z}\mathbf{a}_\rho + \frac{1}{\rho}\frac{\partial(\rho H_\phi)}{\partial p}\mathbf{a}_z = 0$$

As an example of the evaluation of curl $\mathbf{H}$ from the definition and of the evaluation of another line integral, let us suppose that $\mathbf{H} = 0.2z^2\mathbf{a}_x$ for $z > 0$, and $\mathbf{H} = 0$ elsewhere, as shown in Fig. 8.15. For a square path with side d, centered at $(0,0,z_1)$ in the $y = 0$ plane where $z_1 > \frac{1}{2}d$, we evaluate the line integral of $\mathbf{H}$ along the four segments, beginning at the top:

$$\oint \mathbf{H} \cdot d\mathbf{L} = 0.2(z_1 + \tfrac{1}{2}d)^2 d + 0 - 0.2(z_1 - \tfrac{1}{2}d)^2 d + 0$$
$$= 0.4z_1 d^2$$

In the limit as the area approaches zero, we find

$$(\nabla \times \mathbf{H})_y = \lim_{d \to 0} \frac{\oint \mathbf{H} \cdot d\mathbf{L}}{d^2} = \lim_{d \to 0} \frac{0.4z_1 d^2}{d^2} = 0.4z_1$$

The other components are zero, so $\nabla \times \mathbf{H} = 0.4z_1\mathbf{a}_y$.

To evaluate the curl without trying to illustrate the definition or the evaluation of a line integral, we simply take the partial derivative indicated by (23):

$$\nabla \times \mathbf{H} = \begin{vmatrix} \mathbf{a}_x & \mathbf{a}_y & \mathbf{a}_z \\ \dfrac{\partial}{\partial x} & \dfrac{\partial}{\partial y} & \dfrac{\partial}{\partial z} \\ 0.2z^2 & 0 & 0 \end{vmatrix} = \frac{\partial}{\partial z}(0.2z^2)\mathbf{a}_y = 0.4z\mathbf{a}_y$$

which checks with the result above when $z = z_1$.

Returning now to complete our original examination of the application of Ampère's circuital law to a differential-sized path, we may combine (18), (19), (20), (22), and (24),

$$\text{curl}\,\mathbf{H} = \nabla \times \mathbf{H} = \left(\frac{\partial H_z}{\partial y} - \frac{\partial H_y}{\partial z}\right)\mathbf{a}_x + \left(\frac{\partial H_x}{\partial z} - \frac{\partial H_z}{\partial x}\right)\mathbf{a}_y + \left(\frac{\partial H_y}{\partial x} - \frac{\partial H_x}{\partial y}\right)\mathbf{a}_z = \mathbf{J} \tag{27}$$

and write the *point form of Ampère's circuital law*,

$$\boxed{\nabla \times \mathbf{H} = \mathbf{J}} \tag{28}$$

This is the second of Maxwell's four equations as they apply to non-time-varying conditions. We may also write the third of these equations at this time; it is the point form of $\oint \mathbf{E} \cdot d\mathbf{L} = 0$, or

$$\boxed{\nabla \times \mathbf{E} = 0} \tag{29}$$

The fourth equation appears in Sec. 8.5.

D8.4. (*a*) Evaluate the closed line integral of **H** from $P_1(5,4,1)$ to $P_2(5,6,1)$ to $P_3(0,6,1)$ to $P_4(0,4,1)$ to P_1, using straight line segments, if $\mathbf{H} = 0.1y^3\mathbf{a}_x + 0.4x\mathbf{a}_z$ A/m. (*b*) Determine the quotient of the closed line integral and the area enclosed by the path as an approximation to $(\nabla \times \mathbf{H})_z$. (*c*) Determine $(\nabla \times \mathbf{H})_z$ at the center of the area.

Ans. -76 A; -7.6 A/m^2; -7.5 A/m^2

D8.5. Find $\nabla \times \mathbf{G}$: (*a*) in cartesian coordinates at $P_A(3,2,1)$ if $\mathbf{G} = xyz(\mathbf{a}_x + \mathbf{a}_y)$; (*b*) in cylindrical coordinates at $P_B(2,30°,3)$ if $\mathbf{G} = \rho\phi z\mathbf{a}_z$; (*c*) in spherical coordinates at $P_C(4,30°,45°)$ if $\mathbf{G} = \sin\theta(\mathbf{a}_r + \mathbf{a}_\theta + \mathbf{a}_\phi)$.

Ans. $-6\mathbf{a}_x + 6\mathbf{a}_y - \mathbf{a}_z$; $3\mathbf{a}_\rho - 1.571\mathbf{a}_\phi$; $0.433\mathbf{a}_r - 0.1250\mathbf{a}_\theta - 0.0915\mathbf{a}_\phi$

8.4 STOKES' THEOREM

Although the last section was devoted primarily to a discussion of the curl operation, the contribution to the subject of magnetic fields should not be overlooked.

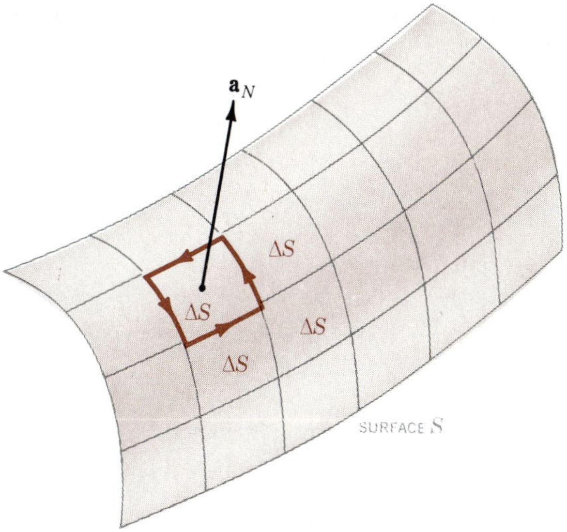

FIGURE 8.16
The sum of the closed line integrals about the perimeter of every ΔS is the same as the closed line integral about the perimeter of S because of cancellation on every interior path.

From Ampère's circuital law we derived one of Maxwell's equations, $\nabla \times \mathbf{H} = \mathbf{J}$. This latter equation should be considered the point form of Ampère's circuital law and applies on a "per-unit-area" basis. In this section we shall again devote a major share of the material to the mathematical theorem known as Stokes' theorem, but in the process we shall show that we may obtain Ampère's circuital law from $\nabla \times \mathbf{H} = \mathbf{J}$. In other words, we are then prepared to obtain the integral form from the point form or to obtain the point form from the integral form.

Consider the surface S of Fig. 8.16 which is broken up into incremental surfaces of area ΔS. If we apply the definition of the curl to one of these incremental surfaces, then

$$\frac{\oint \mathbf{H} \cdot d\mathbf{L}_{\Delta S}}{\Delta S} \doteq (\nabla \times \mathbf{H})_N$$

where the N subscript again indicates the right-hand normal to the surface. The subscript on $d\mathbf{L}_{\Delta S}$ indicates that the closed path is the perimeter of an incremental area ΔS. This result may also be written

$$\frac{\oint \mathbf{H} \cdot d\mathbf{L}_{\Delta S}}{\Delta S} \doteq (\nabla \times \mathbf{H}) \cdot \mathbf{a}_N$$

or

$$\oint \mathbf{H} \cdot d\mathbf{L}_{\Delta S} \doteq (\nabla \times \mathbf{H}) \cdot \mathbf{a}_N \Delta S = (\nabla \times \mathbf{H}) \cdot \Delta \mathbf{S}$$

where $\mathbf{a}_N$ is a unit vector in the direction of the right-hand normal to $\Delta \mathbf{S}$.

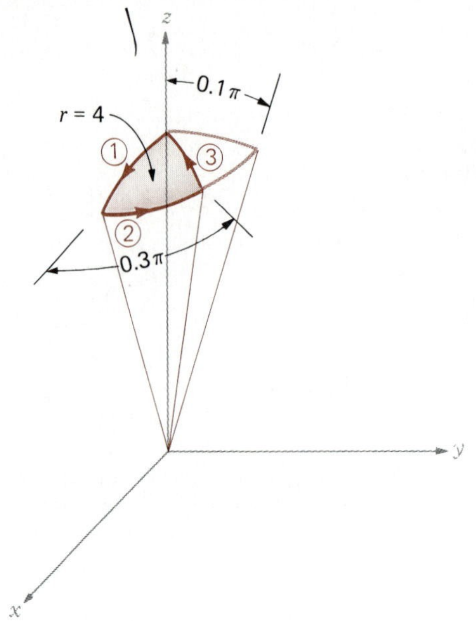

FIGURE 8.17
A portion of a spherical cap is used as a surface and a closed path to illustrate Stokes' theorem.

Now let us determine this circulation for every ΔS comprising S and sum the results. As we evaluate the closed line integral for each ΔS, some cancellation will occur because every *interior* wall is covered once in each direction. The only boundaries on which cancellation cannot occur form the outside boundary, the path enclosing S. Therefore we have

$$\oint \mathbf{H} \cdot d\mathbf{L} \equiv \int_S (\nabla \times \mathbf{H}) \cdot d\mathbf{S} \tag{30}$$

where $d\mathbf{L}$ is taken only on the perimeter of S.

Equation (30) is an identity, holding for any vector field, and is known as *Stokes' theorem.*

A numerical example may help to illustrate the geometry involved in Stokes' theorem. Consider the portion of a sphere shown in Fig. 8.17. The surface is specified by $r = 4, 0 \le \theta \le 0.1\pi, 0 \le \phi \le 0.3\pi$, and the closed path forming its perimeter is composed of three circular arcs. The first path segment is described in spherical coordinates by $r = 4$, $0 \le \theta \le 0.1\pi$, $\phi = 0$; the second one by $r = 4$, $\theta = 0.1\pi$, $0 \le \phi \le 0.3\pi$; and the third by $r = 4$, $0 \le \theta \le 0.1\pi$, $\phi = 0.3\pi$. We are given the field $\mathbf{H} = 6r \sin\phi\, \mathbf{a}_r + 18r \sin\theta \cos\phi\, \mathbf{a}_\phi$ and are asked to evaluate each side of Stokes' theorem.

The differential path element $d\mathbf{L}$ is the vector sum of the three differential lengths of the spherical coordinate system first met in Sec. 1.9,

$$d\mathbf{L} = dr\,\mathbf{a}_r + r\,d\theta\,\mathbf{a}_\theta + r \sin\theta\,d\phi\,\mathbf{a}_\phi$$

The first term is zero on all three segments of the path since $r = 4$ and $dr = 0$, the second is zero on segment 2 since θ is constant, and the third term is zero on both segments 1 and 3. Thus

$$\oint \mathbf{H} \cdot d\mathbf{L} = \int_1 H_\theta r \, d\theta + \int_2 H_\phi r \sin\theta \, d\phi + \int_3 H_\theta r \, d\theta$$

Since $H_\theta = 0$, we have only the second integral to evaluate,

$$\oint \mathbf{H} \cdot d\mathbf{L} = \int_0^{0.3\pi} [18(4) \sin 0.1\pi \cos\phi] 4 \sin 0.1\pi \, d\phi$$

$$= 288 \sin^2 0.1\pi \sin 0.3\pi = 22.2 \quad \text{A}$$

We next attack the surface integral. First, we use (26) to find

$$\nabla \times \mathbf{H} = \frac{1}{r \sin\theta}(36r \sin\theta \cos\theta \cos\phi)\mathbf{a}_r + \frac{1}{r}\left(\frac{1}{\sin\theta} 6r \cos\phi - 36r \sin\theta \cos\phi\right)\mathbf{a}_\theta$$

Since $d\mathbf{S} = r^2 \sin\theta \, d\theta \, d\phi \, \mathbf{a}_r$, the integral is

$$\int_S (\nabla \times \mathbf{H}) \cdot d\mathbf{S} = \int_0^{0.3\pi} \int_0^{0.1\pi} (36 \cos\theta \cos\phi) 16 \sin\theta \, d\theta \, d\phi$$

$$= \int_0^{0.3\pi} 576 (\tfrac{1}{2} \sin^2\theta) \Big|_0^{0.1\pi} \cos\phi \, d\phi$$

$$= 288 \sin^2 0.1\pi \sin 0.3\pi = 22.2 \quad \text{A}$$

Thus, the results check Stokes' theorem, and we note in passing that a current of 22.2 A is flowing upward through this section of a spherical cap.

Next, let us see how easy it is to obtain Ampère's circuital law from $\nabla \times \mathbf{H} = \mathbf{J}$. We merely have to dot each side by $d\mathbf{S}$, integrate each side over the same (open) surface S, and apply Stokes' theorem:

$$\int_S (\nabla \times \mathbf{H}) \cdot d\mathbf{S} = \int_S \mathbf{J} \cdot d\mathbf{S} = \oint \mathbf{H} \cdot d\mathbf{L}$$

The integral of the current density over the surface S is the total current I passing through the surface, and therefore

$$\oint \mathbf{H} \cdot d\mathbf{L} = I$$

This short derivation shows clearly that the current I, described as being "enclosed by the closed path," is also the current passing through any of the infinite number of surfaces which have the closed path as a perimeter.

Stokes' theorem relates a surface integral to a closed line integral. It should be recalled that the divergence theorem relates a volume integral to a closed surface integral. Both theorems find their greatest use in general vector proofs. As an example, let us find another expression for $\nabla \cdot \nabla \times \mathbf{A}$, where $\mathbf{A}$ represents any vector field. The result must be a scalar (why?), and we may let this scalar be T, or

$$\nabla \cdot \nabla \times \mathbf{A} = T$$

Multiplying by dv and integrating throughout any volume v,

$$\int_{\text{vol}} (\nabla \cdot \nabla \times \mathbf{A})\, dv = \int_{\text{vol}} T\, dv$$

we first apply the divergence theorem to the left side, obtaining

$$\oint_S (\nabla \times \mathbf{A}) \cdot d\mathbf{S} = \int_{\text{vol}} T\, dv$$

The left side is the surface integral of the curl of **A** over the *closed* surface surrounding the volume v. Stokes' theorem relates the surface integral of the curl of **A** over the *open* surface enclosed by a given closed path. If we think of the path as the opening of a laundry bag and the open surface as the surface of the bag itself, we see that as we gradually approach a closed surface by pulling on the drawstrings, the closed path becomes smaller and smaller and finally disappears as the surface becomes closed. Hence the application of Stokes' theorem to a *closed* surface produces a zero result, and we have

$$\int_{\text{vol}} T\, dv = 0$$

Since this is true for any volume, it is true for the differential volume dv,

$$T\, dv = 0$$

and therefore

$$T = 0$$

or

$$\boxed{\nabla \cdot \nabla \times \mathbf{A} \equiv 0} \qquad (31)$$

Equation (31) is a useful identity of vector calculus.[6] Of course, it may also be proven easily by direct expansion in cartesian coordinates.

Let us apply the identity to the non-time-varying magnetic field for which

$$\nabla \times \mathbf{H} = \mathbf{J}$$

and then show that

$$\nabla \cdot \mathbf{J} = 0$$

which is the same result we obtained earlier in the chapter by using the continuity equation.

Before introducing several new magnetic field quantities in the following section, we may review our accomplishments at this point. We initially accepted the Biot-Savart law as an experimental result,

$$\mathbf{H} = \oint \frac{I\, d\mathbf{L} \times \mathbf{a}_R}{4\pi R^2}$$

[6] This and other vector identities are tabulated in Appendix A.3.

and tentatively accepted Ampere's circuital law, subject to later proof,

$$\oint \mathbf{H} \cdot d\mathbf{L} = I$$

From Ampère's circuital law the definition of curl led to the point form of this same law,

$$\nabla \times \mathbf{H} = \mathbf{J}$$

We now see that Stokes' theorem enables us to obtain the integral form of Ampère's circuital law from the point form.

D8.6. Given the field $\mathbf{H} = 6r \sin \phi \, \mathbf{a}_r + 18r \sin \theta \cos \phi \, \mathbf{a}_\phi$, evaluate both sides of Stokes' theorem for the portion of the cone $\theta = 0.1\pi$ bounded by $r = 2$, $r = 4$, $\phi = 0$, and $\phi = 0.3\pi$. Let the direction of $d\mathbf{S}$ be $+\mathbf{a}_\theta$.

Ans. 12.44 A; 12.44 A

8.5 MAGNETIC FLUX AND MAGNETIC FLUX DENSITY

In free space, let us define the *magnetic flux density* **B** as

$$\boxed{\mathbf{B} = \mu_0 \mathbf{H}} \qquad \text{(free space only)} \tag{32}$$

where **B** is measured in webers per square meter (Wb/m^2) or in a newer unit adopted in the International System of Units, tesla (T). An older unit that is often used for magnetic flux density is the gauss (G), where 1 T or 1 Wb/m^2 is the same as 10,000 G. The constant μ_0 is not dimensionless and has the *defined* value for free space, in henrys per meter (H/m), of

$$\boxed{\mu_0 = 4\pi \times 10^{-7} \qquad \text{H/m}} \tag{33}$$

The name given to μ_0 is the *permeability* of free space.

We should note that since **H** is measured in amperes per meter, the weber is dimensionally equal to the product of henrys and amperes. Considering the henry as a new unit, the weber is merely a convenient abbreviation for the product of henrys and amperes. When time-varying fields are introduced, it will be shown that a weber is also equivalent to the product of volts and seconds.

The magnetic-flux-density vector **B**, as the name weber per square meter implies, is a member of the flux-density family of vector fields. One of the possible analogies between electric and magnetic fields[7] compares the laws of Biot-Savart

[7] An alternate analogy is presented in Sec. 10.2.

and Coulomb, thus establishing an analogy between **H** and **E**. The relations $\mathbf{B} = \mu_0\mathbf{H}$ and $\mathbf{D} = \epsilon_0\mathbf{E}$ then lead to an analogy between **B** and **D**. If **B** is measured in teslas or webers per square meter, then magnetic flux should be measured in webers. Let us represent magnetic flux by Φ and define Φ as the flux passing through any designated area,

$$\boxed{\Phi = \int_S \mathbf{B} \cdot d\mathbf{S} \qquad \text{Wb}} \tag{34}$$

Our analogy should now remind us of the electric flux Ψ, measured in coulombs, and of Gauss's law, which states that the total flux passing through any *closed* surface is equal to the charge enclosed,

$$\Psi = \oint_S \mathbf{D} \cdot d\mathbf{S} = Q$$

The charge Q is the source of the lines of electric flux and these lines begin and terminate on positive and negative charge, respectively.

No such source has ever been discovered for the lines of magnetic flux. In the example of the infinitely long straight filament carrying a direct current I, the **H** field formed concentric circles about the filament. Since $\mathbf{B} = \mu_0\mathbf{H}$, the **B** field is of the same form. The magnetic flux lines are closed and do not terminate on a "magnetic charge." For this reason Gauss's law for the magnetic field is

$$\boxed{\oint_S \mathbf{B} \cdot d\mathbf{S} = 0} \tag{35}$$

and application of the divergence theorem shows us that

$$\boxed{\nabla \cdot \mathbf{B} = 0} \tag{36}$$

We have not proved (35) or (36) but have only suggested the truth of these statements by considering the single field of the infinite filament. It is possible to show that (35) or (36) follows from the Biot-Savart law and the definition of **B**, $\mathbf{B} = \mu_0\mathbf{H}$, but this is another proof which we shall postpone to Sec. 8.7.

Equation (36) is the last of Maxwell's four equations as they apply to static electric fields and steady magnetic fields. Collecting these equations, we then have for static electric fields and steady magnetic fields

$$\boxed{\begin{aligned} \nabla \cdot \mathbf{D} &= \rho_v \\ \nabla \times \mathbf{E} &= 0 \\ \nabla \times \mathbf{H} &= \mathbf{J} \\ \nabla \cdot \mathbf{B} &= 0 \end{aligned}} \tag{37}$$

To these equations we may add the two expressions relating **D** to **E** and **B** to **H** in free space,

$$\mathbf{D} = \epsilon_0 \mathbf{E} \tag{38}$$

$$\mathbf{B} = \mu_0 \mathbf{H} \tag{39}$$

We have also found it helpful to define an electrostatic potential,

$$\mathbf{E} = -\nabla V \tag{40}$$

and we shall discuss a potential for the steady magnetic field in the following section. In addition, we have extended our coverage of electric fields to include conducting materials and dielectrics, and we have introduced the polarization **P**. A similar treatment will be applied to magnetic fields in the next chapter.

Returning to (37), it may be noted that these four equations specify the divergence and curl of an electric and a magnetic field. The corresponding set of four integral equations that apply to static electric fields and steady magnetic fields is

$$\begin{aligned}
&\oint_S \mathbf{D} \cdot d\mathbf{S} = Q = \int_{\text{vol}} \rho_v \, dv \\
&\oint \mathbf{E} \cdot d\mathbf{L} = 0 \\
&\oint \mathbf{H} \cdot d\mathbf{L} = I = \int_S \mathbf{J} \cdot d\mathbf{S} \\
&\oint_S \mathbf{B} \cdot d\mathbf{S} = 0
\end{aligned} \tag{41}$$

Our study of electric and magnetic fields would have been much simpler if we could have assumed either set of equations, (37) or (41). With a good knowledge of vector analysis, such as we should now have, either set may be readily obtained from the other by applying the divergence theorem or Stokes' theorem. The various experimental laws could have been obtained easily from these equations.

As an example of the use of flux and flux density in magnetic fields, let us find the flux between the conductors of the coaxial line of Fig. 8.8*a*. The magnetic field intensity was found to be

$$H_\phi = \frac{I}{2\pi\rho} \qquad (a < \rho < b)$$

and therefore

$$\mathbf{B} = \mu_0 \mathbf{H} = \frac{\mu_0 I}{2\pi\rho} \mathbf{a}_\phi$$

The magnetic flux contained between the conductors in a length d is the flux crossing any radial plane extending from $\rho = a$ to $\rho = b$ and from, say, $z = 0$ to $z = d$

$$\Phi = \int_S \mathbf{B} \cdot d\mathbf{S} = \int_0^d \int_a^b \frac{\mu_0 I}{2\pi\rho} \mathbf{a}_\phi \cdot d\rho\, dz\, \mathbf{a}_\phi$$

or

$$\Phi = \frac{\mu_0 I d}{2\pi} \ln \frac{b}{a} \tag{42}$$

This expression will be used later to obtain the inductance of the coaxial transmission line.

D8.7. A solid nonmagnetic conductor of circular cross section has its axis on the z axis and carries a uniformly distributed total current of 60 A in the $\mathbf{a}_z$ direction. If the radius is 4 mm, find: (*a*) B_ϕ at $\rho = 5$ mm; (*b*) B_ϕ at $\rho = 3$ mm; (*c*) the total magnetic flux per unit length inside the conductor; (*d*) the total flux per unit length outside the conductor.

Ans. 2.40 mT; 2.25 mT; 6 μWb; ∞

8.6 THE SCALAR AND VECTOR MAGNETIC POTENTIALS

The solution of electrostatic field problems is greatly simplified by the use of the scalar electrostatic potential V. Although this potential possesses a very real physical significance for us, it is mathematically no more than a steppingstone which allows us to solve a problem by several smaller steps. Given a charge configuration, we may first find the potential and then from it the electric field intensity.

We should question whether or not such assistance is available in magnetic fields. Can we define a potential function which may be found from the current distribution and from which the magnetic fields may be easily determined? Can a scalar magnetic potential be defined, similar to the scalar electrostatic potential? We shall show in the next few pages that the answer to the first question is "yes," but the second must be answered "sometimes." Let us attack the last question first by assuming the existence of a scalar magnetic potential, which we designate V_m, whose negative gradient gives the magnetic field intensity

$$\mathbf{H} = -\nabla V_m$$

The selection of the negative gradient will provide us with a closer analogy to the electric potential and to problems which we have already solved.

This definition must not conflict with our previous results for the magnetic field, and therefore

$$\nabla \times \mathbf{H} = \mathbf{J} = \nabla \times (-\nabla V_m)$$

However, the curl of the gradient of any scalar is identically zero, a vector identity the proof of which is left for a leisure moment. Therefore we see that if **H** is to be defined as the gradient of a scalar magnetic potential, then current density must be zero throughout the region in which the scalar magnetic potential is so defined. We then have

$$\boxed{\mathbf{H} = -\nabla V_m \qquad (\mathbf{J} = 0)} \tag{43}$$

Since many magnetic problems involve geometries in which the current-carrying conductors occupy a relatively small fraction of the total region of interest, it is evident that a scalar magnetic potential can be useful. The scalar magnetic potential is also applicable in the case of permanent magnets. The dimensions of V_m are obviously amperes.

This scalar potential also satisfies Laplace's equation. In free space,

$$\nabla \cdot \mathbf{B} = \mu_0 \nabla \cdot \mathbf{H} = 0$$

and hence

$$\mu_0 \nabla \cdot (-\nabla V_m) = 0$$

or

$$\boxed{\nabla^2 V_m = 0 \qquad (\mathbf{J} = 0)} \tag{44}$$

We shall see later that V_m continues to satisfy Laplace's equation in homogeneous magnetic materials; it is not defined in any region in which current density is present.

Although we shall consider the scalar magnetic potential to a much greater extent in the next chapter, when we introduce magnetic materials and discuss the magnetic circuit, one difference between V and V_m should be pointed out now: V_m is not a single-valued function of position. The electric potential V is single-valued; once a zero reference is assigned, there is only one value of V associated with each point in space. Such is not the case with V_m. Consider the cross section of the coaxial line shown in Fig. 8.18. In the region $a < \rho < b$, $\mathbf{J} = 0$, and we may establish a scalar magnetic potential. The value of **H** is

$$\mathbf{H} = \frac{I}{2\pi\rho} \mathbf{a}_\phi$$

where I is the total current flowing in the $\mathbf{a}_z$ direction in the inner conductor. Let us find V_m by integrating the appropriate component of the gradient. Applying (43),

$$\frac{I}{2\pi\rho} = -\nabla V_m\Big|_\phi = -\frac{1}{\rho}\frac{\partial V_m}{\partial \phi}$$

or

$$\frac{\partial V_m}{\partial \phi} = -\frac{I}{2\pi}$$

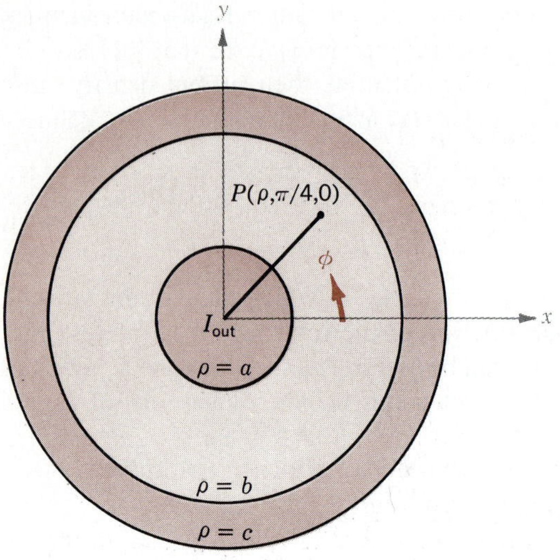

FIGURE 8.18
The scalar magnetic potential V_m is a multivalued function of ϕ in the region $a < \rho < b$. The electrostatic potential is always single-valued.

Thus

$$V_m = -\frac{I}{2\pi}\phi$$

where the constant of integration has been set equal to zero. What value of potential do we associate with point P, where $\phi = \pi/4$? If we let V_m be zero at $\phi = 0$ and proceed counterclockwise around the circle, the magnetic potential goes negative linearly. When we have made one circuit, the potential is $-I$, but that was the point at which we said the potential was zero a moment ago. At P, then, $\phi = \pi/4$, $9\pi/4$, $17\pi/4, \ldots$, or $-7\pi/4$, $-15\pi/4$, $-23\pi/4, \ldots$, or

$$V_{mP} = \frac{I}{2\pi}(2n - \tfrac{1}{4})\pi \qquad (n = 0, \pm 1, \pm 2, \ldots)$$

or

$$V_{mP} = I(n - \tfrac{1}{8}) \qquad (n = 0, \pm 1, \pm 2, \ldots)$$

The reason for this multivaluedness may be shown by a comparison with the electrostatic case. There, we know that

$$\nabla \times \mathbf{E} = 0$$

$$\oint \mathbf{E} \cdot d\mathbf{L} = 0$$

and therefore the line integral

$$V_{ab} = -\int_b^a \mathbf{E} \cdot d\mathbf{L}$$

is independent of the path. In the magnetostatic case, however,

$$\nabla \times \mathbf{H} = 0 \qquad (\text{wherever } \mathbf{J} = 0)$$

but

$$\oint \mathbf{H} \cdot d\mathbf{L} = I$$

even if **J** is zero along the path of integration. Every time we make another complete lap around the current, the result of the integration increases by I. If no current I is enclosed by the path, then a single-valued potential function may be defined. In general, however,

$$V_{m,ab} = -\int_b^a \mathbf{H} \cdot d\mathbf{L} \qquad \text{(specified path)} \tag{45}$$

where a specific path or type of path must be selected. We should remember that the electrostatic potential V is a conservative field; the magnetic scalar potential V_m is not a conservative field. In our coaxial problem let us erect a barrier[8] at $\phi = \pi$; we agree not to select a path which crosses this plane. Therefore we cannot encircle I, and a single-valued potential is possible. The result is seen to be

$$V_m = -\frac{I}{2\pi}\phi \qquad (-\pi < \phi < \pi)$$

and

$$V_{mP} = -\frac{I}{8} \qquad \left(\phi = \frac{\pi}{4}\right)$$

The scalar magnetic potential is evidently the quantity whose equipotential surfaces will form curvilinear squares with the streamlines of **H** in Fig. 8.4. This is one more facet of the analogy between electric and magnetic fields about which we will have more to say in the next chapter.

Let us temporarily leave the scalar magnetic potential now and investigate a vector magnetic potential. This vector field is one which is extremely useful in studying radiation from antennas, from apertures, and radiation leakage from transmission lines, waveguides, and microwave ovens. The vector magnetic potential may be used in regions where the current density is zero or nonzero, and we shall also be able to extend it to the time-varying case later.

Our choice of a vector magnetic potential is indicated by noting that

$$\nabla \cdot \mathbf{B} = 0$$

Next, a vector identity which we proved in Sec. 8.4 shows that the divergence of the curl of any vector field is zero. Therefore we select

$$\mathbf{B} = \nabla \times \mathbf{A} \tag{46}$$

[8] This corresponds to the more precise mathematical term "branch cut."

where **A** signifies a *vector magnetic potential*, and we automatically satisfy the condition that the magnetic flux density shall have zero divergence. The **H** field is

$$\mathbf{H} = \frac{1}{\mu_0} \nabla \times \mathbf{A}$$

and

$$\nabla \times \mathbf{H} = \mathbf{J} = \frac{1}{\mu_0} \nabla \times \nabla \times \mathbf{A}$$

The curl of the curl of a vector field is not zero and is given by a fairly complicated expression,[9] which we need not know now in general form. In specific cases for which the form of **A** is known, the curl operation may be applied twice to determine the current density.

Equation (46) serves as a useful definition of the *vector magnetic potential* **A**. Since the curl operation implies differentiation with respect to a length, the units of **A** are webers per meter.

As yet we have seen only that the definition for **A** does not conflict with any previous results. It still remains to show that this particular definition can help us to determine magnetic fields more easily. We certainly cannot identify **A** with any easily measured quantity or history-making experiment.

We shall show in the following section that, given the Biot-Savart law, the definition of **B**, and the definition of **A**, then **A** may be determined from the differential current elements by

$$\boxed{\mathbf{A} = \oint \frac{\mu_0 I \, d\mathbf{L}}{4\pi R}} \tag{47}$$

The significance of the terms in (47) is the same as in the Biot-Savart law; a direct current I flows along a filamentary conductor of which any differential length $d\mathbf{L}$ is distant R from the point at which **A** is to be found. Since we have defined **A** only through specification of its curl, it is possible to add the gradient of any scalar field to (47) without changing **B** or **H**, for the curl of the gradient is identically zero. In steady magnetic fields, it is customary to set this possible added term equal to zero.

The fact that **A** is a vector magnetic *potential* is more apparent when (47) is compared with the similar expression for the electrostatic potential,

$$V = \int \frac{\rho_L \, dL}{4\pi\epsilon_0 R}$$

[9] $\nabla \times \nabla \times \mathbf{A} \equiv \nabla(\nabla \cdot \mathbf{A}) - \nabla^2\mathbf{A}$. In cartesian coordinates, it may be shown that $\nabla^2\mathbf{A} \equiv \nabla^2 A_x \mathbf{a}_x + \nabla^2 A_y \mathbf{a}_y + \nabla^2 A_z \mathbf{a}_z$. In other coordinate systems, $\nabla^2\mathbf{A}$ may be found by evaluating the second-order partial derivatives in $\nabla^2\mathbf{A} = \nabla(\nabla \cdot \mathbf{A}) - \nabla \times \nabla \times \mathbf{A}$.

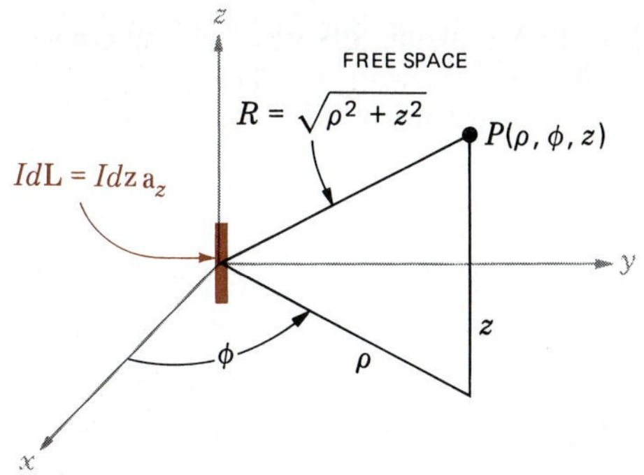

FIGURE 8.19
The differential current element $I\,dz\,\mathbf{a}_z$ at the origin establishes the differential vector magnetic potential field, $d\mathbf{A} = \dfrac{\mu_0 I\,dz\,\mathbf{a}_z}{4\pi\sqrt{\rho^2 + z^2}}$ at $P(\rho,\phi,z)$.

Each expression is the integral along a line source, in one case line charge and in the other case line current; each integrand is inversely proportional to the distance from the source to the point of interest; and each involves a characteristic of the medium (here free space), the permeability or the permittivity.

Equation (47) may be written in differential form,

$$d\mathbf{A} = \frac{\mu_0 I\,d\mathbf{L}}{4\pi R} \tag{48}$$

if we again agree not to attribute any physical significance to any magnetic fields we obtain from (48) until the *entire closed path in which the current flows is considered.*

With this reservation, let us go right ahead and consider the vector magnetic potential field about a differential filament. We locate the filament at the origin in free space, as shown in Fig. 8.19, and allow it to extend in the positive z direction so that $d\mathbf{L} = dz\,\mathbf{a}_z$. We use cylindrical coordinates to find $d\mathbf{A}$ at the point (ρ,ϕ,z):

$$d\mathbf{A} = \frac{\mu_0 I\,dz\,\mathbf{a}_z}{4\pi\sqrt{\rho^2 + z^2}}$$

or

$$dA_z = \frac{\mu_0 I\,dz}{4\pi\sqrt{\rho^2 + z^2}} \qquad dA_\phi = 0 \qquad dA_\rho = 0 \tag{49}$$

We note first that the direction of $d\mathbf{A}$ is the same as that of $I\,d\mathbf{L}$. Each small section of a current-carrying conductor produces a contribution to the total vector magnetic potential which is in the same direction as the current flow in the conductor. The magnitude of the vector magnetic potential varies inversely as the distance to the current element, being strongest in the neighborhood of the current and gradually falling off to zero at distant points. Skilling[10] describes the vector

[10] See the Suggested References at the end of the chapter.

magnetic potential field as "like the current distribution but fuzzy around the edges, or like a picture of the current out of focus."

In order to find the magnetic field intensity, we must take the curl of (49) in cylindrical coordinates, leading to

$$d\mathbf{H} = \frac{1}{\mu_0}\nabla \times d\mathbf{A} = \frac{1}{\mu_0}\left(-\frac{\partial\, dA_z}{\partial \rho}\right)\mathbf{a}_\phi$$

or

$$d\mathbf{H} = \frac{I\,dz}{4\pi}\frac{\rho}{(\rho^2 + z^2)^{3/2}}\mathbf{a}_\phi$$

which is easily shown to be the same as the value given by the Biot-Savart law.

Expressions for the vector magnetic potential **A** can also be obtained for a current source which is distributed. For a current sheet **K**, the differential current element becomes

$$I\,d\mathbf{L} = \mathbf{K}\,dS$$

In the case of current flow throughout a volume with a density **J**, we have

$$I\,d\mathbf{L} = \mathbf{J}\,dv$$

In each of these two expressions the vector character has been given to the current. For the filamentary element it is customary, although not necessary, to use $I\,d\mathbf{L}$ instead of $\mathbf{I}\,dL$. Since the magnitude of the filamentary current is constant, we have chosen the form which allows us to remove one quantity from the integral. The alternative expressions for **A** are then

$$\boxed{\mathbf{A} = \int_S \frac{\mu_0 \mathbf{K}\,dS}{4\pi R}} \tag{50}$$

and

$$\boxed{\mathbf{A} = \int_{\text{vol}} \frac{\mu_0 \mathbf{J}\,dv}{4\pi R}} \tag{51}$$

Equations (47), (50), and (51) express the vector magnetic potential as an integration over all of its sources. From a comparison of the form of these integrals with those which yield the electrostatic potential, it is evident that once again the zero reference for **A** is at infinity, for no finite current element can produce any contribution as $R \to \infty$. We should remember that we very seldom used the similar expressions for V; too often our theoretical problems included charge distributions which extended to infinity and the result would be an infinite potential everywhere. Actually, we calculated very few potential fields until the differential form of the potential equation was obtained, $\nabla^2 V = -\rho/\epsilon$, or better yet, $\nabla^2 V = 0$. We were then at liberty to select our own zero reference.

The analogous expressions for **A** will be derived in the next section, and an example of the calculation of a vector magnetic potential field will be completed.

D8.8. The region, $3.5 < \rho < 4.5$ cm, $0 \le \phi < 2\pi$, $0 < z < 2$ cm, is current-free and contains the field $H_\phi = 70/\rho$ A/m. Find V_m at $\rho = 3.8$ cm, $\phi = 110°$, $z = 0.5$ cm if: (*a*) $V_m = 0$ at $\phi = 0$, and there is a barrier at $\phi = 180°$; (*b*) $V_m = 0$ at $\phi = 0$, and there is a barrier at $\phi = 90°$; (*c*) $V_m = 0$ at $\phi = 180°$, and there is a barrier at $\phi = 0°$; (*d*) $V_m = 0$ at $\phi = 180°$, and there is a barrier at $\phi = 150°$.

Ans. -134.4 A; 305 A; 85.5 A; -354 A

D8.9. The value of **A** within a solid nonmagnetic conductor of radius a carrying a total current I in the $\mathbf{a}_z$ direction may be found easily. Using the known value of **H** or **B** for $\rho < a$, then (46) may be solved for **A**. Select $A = (\mu_0 I \ln 5)/2\pi$ at $\rho = a$ (to correspond with an example in the next section) and find **A** at $\rho =$: (*a*) 0; (*b*) $0.3a$; (*c*) $0.6a$; (*d*) a.

Ans. $0.422I\mathbf{a}_z$ μWb/m; $0.413I\mathbf{a}_z$ μWb/m; $0.386I\mathbf{a}_z$ μWb/m; $0.322I\mathbf{a}_z$ μWb/m

8.7 DERIVATION OF STEADY-MAGNETIC-FIELD LAWS

We shall now carry out our threat to supply the promised proofs of the several relationships between the magnetic field quantities. All these relationships may be obtained from the definitions of **H**,

$$\mathbf{H} = \oint \frac{I\,d\mathbf{L} \times \mathbf{a}_R}{4\pi R^2} \tag{3}$$

of **B** (in free space),

$$\mathbf{B} = \mu_0 \mathbf{H} \tag{32}$$

and of **A**,

$$\mathbf{B} = \nabla \times \mathbf{A} \tag{46}$$

Let us first assume that we may express **A** by the last equation of the preceding section,

$$\mathbf{A} = \int_{\text{vol}} \frac{\mu_0 \mathbf{J}\,dv}{4\pi R} \tag{51}$$

and then demonstrate the correctness of (51) by showing that (3) follows. First we should add subscripts to indicate the point at which the current element is located (x_1,y_1,z_1) and the point at which **A** is given (x_2,y_2,z_2). The differential volume element dv is then written dv_1 and in cartesian coordinates would be $dx_1\,dy_1\,dz_1$. The variables of integration are x_1, y_1, and z_1. Using these subscripts, then,

$$\mathbf{A}_2 = \int_{\text{vol}} \frac{\mu_0 \mathbf{J}_1\,dv_1}{4\pi R_{12}} \tag{52}$$

From (32) and (46) we have

$$\mathbf{H} = \frac{\mathbf{B}}{\mu_0} = \frac{\nabla \times \mathbf{A}}{\mu_0} \tag{53}$$

In order to show that (3) follows from (52), it is necessary to substitute (52) into (53). This step involves taking the curl of $\mathbf{A}_2$, a quantity expressed in terms of the variables x_2, y_2, and z_2, and the curl therefore involves partial derivatives with respect to x_2, y_2, and z_2. We do this, placing a subscript on the del operator to remind us of the variables involved in the partial differentiation process,

$$\mathbf{H}_2 = \frac{\nabla_2 \times \mathbf{A}_2}{\mu_0} = \frac{1}{\mu_0} \nabla_2 \times \int_{\text{vol}} \frac{\mu_0 \mathbf{J}_1 \, dv_1}{4\pi R_{12}}$$

The order of partial differentiation and integration is immaterial, and $\mu_0/4\pi$ is constant, allowing us to write

$$\mathbf{H}_2 = \frac{1}{4\pi} \int_{\text{vol}} \nabla_2 \times \frac{\mathbf{J}_1 \, dv_1}{R_{12}}$$

The curl operation within the integrand represents partial differentiation with respect to x_2, y_2, and z_2. The differential volume element dv_1 is a scalar and a function only of x_1, y_1, and z_1. Consequently, it may be factored out of the curl operation as any other constant, leaving

$$\mathbf{H}_2 = \frac{1}{4\pi} \int_{\text{vol}} \left(\nabla_2 \times \frac{\mathbf{J}_1}{R_{12}} \right) dv_1 \tag{54}$$

The curl of the product of a scalar and a vector is given by an identity which may be checked by expansion in cartesian coordinates or gratefully accepted from Appendix A.3,

$$\nabla \times (S\mathbf{V}) \equiv (\nabla S) \times \mathbf{V} + S(\nabla \times \mathbf{V}) \tag{55}$$

This identity is used to expand the integrand of (54),

$$\mathbf{H}_2 = \frac{1}{4\pi} \int_{\text{vol}} \left[\left(\nabla_2 \frac{1}{R_{12}} \right) \times \mathbf{J}_1 + \frac{1}{R_{12}} (\nabla_2 \times \mathbf{J}_1) \right] dv_1 \tag{56}$$

The second term of this integrand is zero, because $\nabla_2 \times \mathbf{J}_1$ indicates partial derivatives of a function of x_1, y_1 and z_1, taken with respect to the variables x_2, y_2, and z_2; the first set of variables is not a function of the second set, and all partial derivatives are zero.

The first term of the integrand may be determined by expressing R_{12} in terms of the coordinate values,

$$R_{12} = \sqrt{(x_2 - x_1)^2 + (y_2 - y_1)^2 + (z_2 - z_1)^2}$$

and taking the gradient of its reciprocal. Problem 42 shows that the result is

$$\nabla_2 \frac{1}{R_{12}} = -\frac{\mathbf{R}_{12}}{R_{12}^3} = -\frac{\mathbf{a}_{R12}}{R_{12}^2}$$

Substituting this result into (56), we have

$$\mathbf{H}_2 = -\frac{1}{4\pi}\int_{\text{vol}} \frac{\mathbf{a}_{R12} \times \mathbf{J}_1}{R_{12}^2}\, dv_1$$

or

$$\mathbf{H}_2 = \int_{\text{vol}} \frac{\mathbf{J}_1 \times \mathbf{a}_{R12}}{4\pi R_{12}^2}\, dv_1$$

which is the equivalent of (3) in terms of current density. Replacing $\mathbf{J}_1\, dv_1$ by $I_1\, d\mathbf{L}_1$, we may rewrite the volume integral as a closed line integral,

$$\mathbf{H}_2 = \oint \frac{I_1\, d\mathbf{L}_1 \times \mathbf{a}_{R12}}{4\pi R_{12}^2}$$

Equation (51) is therefore correct and agrees with the three definitions (3), (32), and (46).

Next we shall continue with our mathematical orgy and prove Ampère's circuital law in point form,

$$\nabla \times \mathbf{H} = \mathbf{J} \tag{28}$$

Combining (28), (32), and (46), we obtain

$$\nabla \times \mathbf{H} = \nabla \times \frac{\mathbf{B}}{\mu_0} = \frac{1}{\mu_0} \nabla \times \nabla \times \mathbf{A} \tag{57}$$

We now need the expansion in cartesian coordinates for $\nabla \times \nabla \times \mathbf{A}$. Performing the indicated partial differentiations and collecting the resulting terms, we may write the result as

$$\boxed{\nabla \times \nabla \times \mathbf{A} \equiv \nabla(\nabla \cdot \mathbf{A}) - \nabla^2 \mathbf{A}} \tag{58}$$

where

$$\boxed{\nabla^2 \mathbf{A} \equiv \nabla^2 A_x \mathbf{a}_x + \nabla^2 A_y \mathbf{a}_y + \nabla^2 A_z \mathbf{a}_z} \tag{59}$$

Equation (59) is the definition (in cartesian coordinates) of the *Laplacian of a vector*.

Substituting (58) into (57), we have

$$\nabla \times \mathbf{H} = \frac{1}{\mu_0}[\nabla(\nabla \cdot \mathbf{A}) - \nabla^2 \mathbf{A}] \tag{60}$$

and now require expressions for the divergence and the Laplacian of **A**.

We may find the divergence of **A** by applying the divergence operation to (52),

$$\nabla_2 \cdot \mathbf{A}_2 = \frac{\mu_0}{4\pi}\int_{\text{vol}} \nabla_2 \cdot \frac{\mathbf{J}_1}{R_{12}}\, dv_1 \tag{61}$$

and using the vector identity (44) of Sec. 4.8,

$$\nabla \cdot (S\mathbf{V}) \equiv \mathbf{V} \cdot (\nabla S) + S(\nabla \cdot \mathbf{V})$$

Thus,

$$\nabla_2 \cdot \mathbf{A}_2 = \frac{\mu_0}{4\pi} \int_{\text{vol}} \left[\mathbf{J}_1 \cdot \left(\nabla_2 \frac{1}{R_{12}} \right) + \frac{1}{R_{12}} (\nabla_2 \cdot \mathbf{J}_1) \right] dv_1 \tag{62}$$

The second part of the integrand is zero, because $\mathbf{J}_1$ is not a function of x_2, y_2, and z_2.

We have already used the result that $\nabla_2(1/R_{12}) = -\mathbf{R}_{12}/R_{12}^3$, and it is just as easily shown that

$$\nabla_1 \frac{1}{R_{12}} = \frac{\mathbf{R}_{12}}{R_{12}^3}$$

or that

$$\nabla_1 \frac{1}{R_{12}} = -\nabla_2 \frac{1}{R_{12}}$$

Equation (62) can therefore be written as

$$\nabla_2 \cdot \mathbf{A}_2 = \frac{\mu_0}{4\pi} \int_{\text{vol}} \left[-\mathbf{J}_1 \cdot \left(\nabla_1 \frac{1}{R_{12}} \right) \right] dv_1$$

and the vector identity applied again,

$$\nabla_2 \cdot \mathbf{A}_2 = \frac{\mu_0}{4\pi} \int_{\text{vol}} \left[\frac{1}{R_{12}} (\nabla_1 \cdot \mathbf{J}_1) - \nabla_1 \cdot \left(\frac{\mathbf{J}_1}{R_{12}} \right) \right] dv_1 \tag{63}$$

Since we are concerned only with steady magnetic fields, the continuity equation shows that the first term of (63) is zero. Application of the divergence theorem to the second term gives

$$\nabla_2 \cdot \mathbf{A}_2 = -\frac{\mu_0}{4\pi} \oint_{S_1} \frac{\mathbf{J}_1}{R_{12}} \cdot d\mathbf{S}_1$$

where the surface S_1 encloses the volume throughout which we are integrating. This volume must include all the current, for the original integral expression for **A** was an integration such as to include the effect of all the current. Since there is no current outside this volume (otherwise we should have had to increase the volume to include it), we may integrate over a slightly larger volume or a slightly larger enclosing surface without changing **A**. On this larger surface the current density $\mathbf{J}_1$ must be zero, and therefore the closed surface integral is zero, since the integrand is zero. Hence the divergence of **A** is zero.

In order to find the Laplacian of the vector **A** let us compare the x component of (51) with the similar expression for electrostatic potential,

$$A_x = \int_{\text{vol}} \frac{\mu_0 J_x \, dv}{4\pi R} \qquad V = \int_{\text{vol}} \frac{\rho \, dv}{4\pi\epsilon_0 R}$$

We note that one expression can be obtained from the other by a straightforward change of variable, J_x for ρ, μ_0 for $1/\epsilon_0$, and A_x for V. However, we have derived some additional information about the electrostatic potential which we shall not have to repeat now for the x component of the vector magnetic potential. This takes the form of Poisson's equation,

$$\nabla^2 V = -\frac{\rho}{\epsilon_0}$$

which becomes, after the change of variables,

$$\nabla^2 A_x = -\mu_0 J_x$$

Similarly, we have

$$\nabla^2 A_y = -\mu_0 J_y$$

and

$$\nabla^2 A_z = -\mu_0 J_z$$

or

$$\boxed{\nabla^2 \mathbf{A} = -\mu_0 \mathbf{J}} \tag{64}$$

Returning to (60), we can now substitute for the divergence and Laplacian of **A** and obtain the desired answer,

$$\nabla \times \mathbf{H} = \mathbf{J} \tag{28}$$

We have already shown the use of Stokes' theorem in obtaining the integral form of Ampère's circuital law from (28) and need not repeat that labor here.

We thus have succeeded in showing that every result we have essentially pulled from thin air[11] for magnetic fields follows from the basic definitions of **H**, **B**, and **A**. The derivations are not simple, but they should be understandable on a step-by-step basis. It is hoped that the procedure need never be committed to memory.

Finally, let us return to (64) and make use of this formidable second-order vector partial differential equation to find the vector magnetic potential in one simple example. We select the field between conductors of a coaxial cable, with radii of a and b as usual, and current I in the $\mathbf{a}_z$ direction in the inner conductor. Between the conductors, $\mathbf{J} = 0$, and therefore

$$\nabla^2 \mathbf{A} = 0$$

We have already been told (and Prob. 44 gives us the opportunity to check the results for ourselves) that the vector Laplacian may be expanded as the vector sum of the scalar Laplacians of the three components in cartesian coordinates,

$$\nabla^2 \mathbf{A} = \nabla^2 A_x \mathbf{a}_x + \nabla^2 A_y \mathbf{a}_y + \nabla^2 A_z \mathbf{a}_z$$

[11] Free space.

but such a relatively simple result is not possible in other coordinate systems. That is, in cylindrical coordinates, for example,

$$\nabla^2 \mathbf{A} \neq \nabla^2 A_\rho \mathbf{a}_\rho + \nabla^2 A_\phi \mathbf{a}_\phi + \nabla^2 A_z \mathbf{a}_z$$

However, it is not difficult to show for cylindrical coordinates that the z component of the vector Laplacian is the scalar Laplacian of the z component of $\mathbf{A}$, or

$$\nabla^2 \mathbf{A}\Big|_z = \nabla^2 A_z \tag{65}$$

and since the current is entirely in the z direction in this problem, $\mathbf{A}$ has only a z component. Therefore

$$\nabla^2 A_z = 0$$

or

$$\frac{1}{\rho}\frac{\partial}{\partial \rho}\left(\rho \frac{\partial A_z}{\partial \rho}\right) + \frac{1}{\rho^2}\frac{\partial^2 A_z}{\partial \phi^2} + \frac{\partial^2 A_z}{\partial z^2} = 0$$

Thinking symmetrical thoughts about (51) shows us that A_z is a function only of ρ, and thus

$$\frac{1}{\rho}\frac{d}{d\rho}\left(\rho \frac{dA_z}{d\rho}\right) = 0$$

We have solved this equation before, and the result is

$$A_z = C_1 \ln \rho + C_2$$

If we choose a zero reference at $\rho = b$, then

$$A_z = C_1 \ln \frac{\rho}{b}$$

In order to relate C_1 to the sources in our problem, we may take the curl of $\mathbf{A}$,

$$\nabla \times \mathbf{A} = -\frac{\partial A_z}{\partial \rho}\mathbf{a}_\phi = -\frac{C_1}{\rho}\mathbf{a}_\phi = \mathbf{B}$$

obtain $\mathbf{H}$,

$$\mathbf{H} = -\frac{C_1}{\mu_0 \rho}\mathbf{a}_\phi$$

and evaluate the line integral,

$$\oint \mathbf{H} \cdot d\mathbf{L} = I = \int_0^{2\pi} -\frac{C_1}{\mu_0 \rho}\mathbf{a}_\phi \cdot \rho\, d\phi\, \mathbf{a}_\phi = -\frac{2\pi C_1}{\mu_0}$$

Thus

$$C_1 = -\frac{\mu_0 I}{2\pi}$$

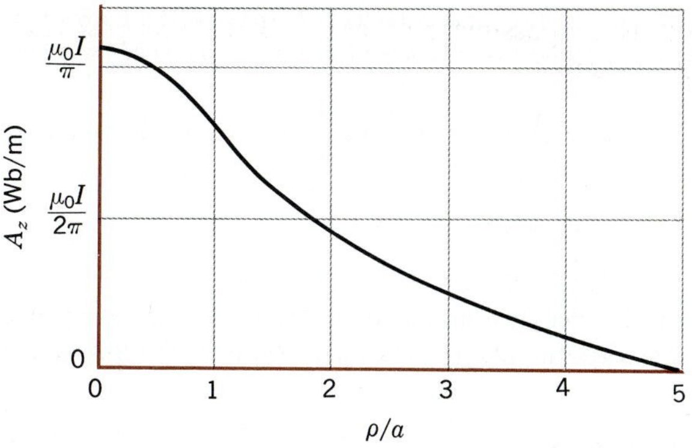

FIGURE 8.20
The vector magnetic potential is shown within the inner conductor and in the region between conductors for a coaxial cable with $b = 5a$ carrying I in the $\mathbf{a}_z$ direction. $A_z = 0$ is arbitrarily selected at $\rho = b$.

or

$$A_z = \frac{\mu_0 I}{2\pi} \ln \frac{b}{\rho} \tag{66}$$

and

$$H_\phi = \frac{I}{2\pi\rho}$$

as before. A plot of A_z versus ρ for $b = 5a$ is shown in Fig. 8.20; the decrease of $|\mathbf{A}|$ with distance from the concentrated current source which the inner conductor represents is evident. The results of Prob. D8.9 have also been added to the graph of Fig. 8.20. The extension of the curve into the outer conductor is left as Prob. 43.

It is also possible to find A_z between conductors by applying a process some of us informally call "uncurling." That is, we know **H** or **B** for the coax, and we may therefore select the ϕ component of $\nabla \times \mathbf{A} = \mathbf{B}$ and integrate to obtain A_z. Try it, you'll like it!

D8.10. Equation (66) is obviously also applicable to the exterior of any conductor of circular cross section carrying a current I in the $\mathbf{a}_z$ direction in free space. The zero reference is arbitrarily set at $\rho = b$. Now consider three conductors, each of 1-cm radius, parallel to the z axis with their axes lying in the $x = 0$ plane. Two have their axes at $y = 4$ cm and $y = -4$ cm, each carrying a current of 12 A in the $\mathbf{a}_z$ direction. The third is located at the z axis and carries -24 A in the $\mathbf{a}_z$ direction. If each current has its zero reference for **A** located 2 cm from its axis, find the combined **A** field at: (*a*) (0,2,0); (*b*) (0,−2,0); (*c*) (2,0,0); (*d*) (3,4,0); (*e*) (∞,0,0).

Ans. $-2.64\mathbf{a}_z$ μWb/m; $-2.64\mathbf{a}_z$ μWb/m; $-3.86\mathbf{a}_z$ μWb/m; $-0.0599\mathbf{a}_z$ μWb/m; 0

SUGGESTED REFERENCES

1. Boast, W. B.: (see Suggested References for Chap. 2). The scalar magnetic potential is defined on p. 220, and its use in mapping magnetic fields is discussed on p. 444.

2. Jordan, E. C., and K. G. Balmain: "Electromagnetic Waves and Radiating Systems," 2d ed., Prentice-Hall, Inc., Englewood Cliffs, N.J., 1968. Vector magnetic potential is discussed on pp. 90–96.

3. Paul, C. R., and S. Y. Nasar: "Introduction to Electromagnetic Fields," McGraw-Hill Book Company, New York, 1982. The scalar magnetic potential is discussed on pp. 147–149, and the vector magnetic potential is presented on pp. 149–154.

4. Skilling, H. H.: (see Suggested References for Chap. 3). The "paddle wheel" is introduced on pp. 23–25.

PROBLEMS

1. A filamentary current of 10 A is directed in from infinity to the origin on the positive x axis, and then back out to infinity along the positive y axis. Use the Biot-Savart law to find $\mathbf{H}$ at $P(0,0,1)$.

2. The parallel filamentary conductors shown in Fig. 8.21 lie in free space. Plot $|\mathbf{H}|$ vs. y, $-5 < y < 5$, along the line $x = 0$, $z = 1$.

3. Find $\mathbf{H}$ at $P(2,3,5)$ in cartesian coordinates if there is an infinitely long current filament passing through the origin and point C. The current of 50 A is directed from the origin to C, where the location of C is: (*a*) $C(0,0,1)$; (*b*) $C(0,1,0)$.

4. Two current filaments are parallel to the z axis. One passes through $(0,0.5,0)$ and carries 100 A in the $\mathbf{a}_z$ direction; the other passes through $(0,-0.5,0)$ and carries 100 A in the $-\mathbf{a}_z$ direction. Plot H_x vs. y, $0 < y < 2$, for points on the y axis.

5. (*a*) A current filament I is formed into a circle, $\rho = a$, in the $z = z'$ plane. Find H_z at $P(0,0,z)$ if I flows in the $\mathbf{a}_\phi$ direction. (*b*) Find H_z at P caused by a uniform surface current density, $\mathbf{K} = K_0\mathbf{a}_\phi$, flowing on the cylindrical surface, $\rho = a$, $0 \le z \le h$. The results of part (*a*) should help.

6. A current element $I\,d\mathbf{L} = 10^{-3}(2\mathbf{a}_x + 4\mathbf{a}_y - \mathbf{a}_z)$ A·m, located at $A(-5,3,-2)$, produces a field $d\mathbf{H}$ at $B(3,-4,3)$. (*a*) Give a unit vector in the direction of $d\mathbf{H}$ at B. (*b*) Find $|d\mathbf{H}|$ at B. (*c*) What direction should $I\,d\mathbf{L}$ have at A so that $d\mathbf{H} = 0$ at B?

7. Given points $A(1,2,4)$, $B(-2,-1,3)$, and $C(3,1,-2)$, let a differential current element with $I = 6$ A and $|d\mathbf{L}| = 10^{-4}$ m be located at A. The direction of $d\mathbf{L}$ is from A to B. Find $d\mathbf{H}$ at C.

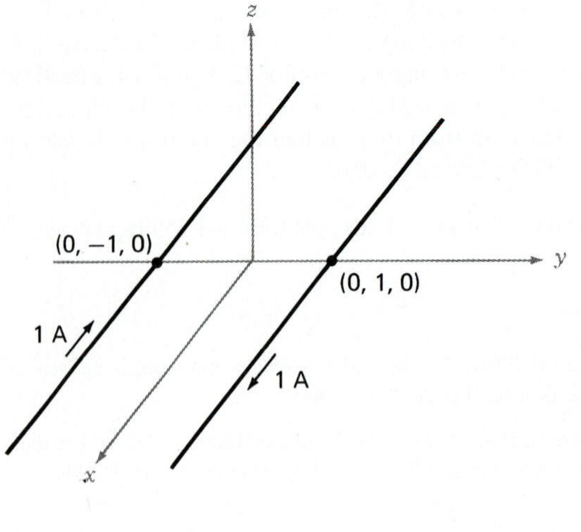

FIGURE 8.21
See Prob. 2.

8. For the finite-length current element on the z axis, as shown in Fig. 8.5, use the Biot-Savart law to derive Eq. (9) of Sec. 8.1.

9. A surface current density, $\mathbf{K} = 20\mathbf{a}_x$ A/m, flows in the $y = 0$ plane throughout the region $-5 < z < 5$ m, $-\infty < x < \infty$. Find $\mathbf{H}$ at $P(0,10,0)$ in free space.

10. (*a*) A filamentary current I flows in the $\mathbf{a}_\phi$ direction in the circular path $\rho = a$, $z = 0$, in free space. Find $\mathbf{H}$ at $P(0,0,z)$. (*b*) Now let identical filamentary currents of 5 A flow in the $\mathbf{a}_\phi$ direction at $\rho = 2$ m, $z = \pm 1$ m. Plot $|\mathbf{H}(0,0,z)|$ vs. z.

11. A surface current, $\mathbf{K} = 10\mathbf{a}_z$ A/m, flows on that portion of the plane $y = 4$ in free space bounded by $x = \pm\infty$, $z = \pm k$. Find $\mathbf{H}$ at the origin if $k =$: (*a*) 3; (*b*) 30; (*c*) ∞.

12. Let 20 A flow in a filament in the $\mathbf{a}_z$ direction on the z axis, let $3\mathbf{a}_z$ A/m flow on the infinite cylinder $\rho = 2$ m, and let $-2.5\mathbf{a}_z$ A/m be at $\rho = 3$. Sketch H_ϕ vs. ρ, $0 < \rho < 5$.

13. The regions, $0 < z < 0.1$ m and $0.3 < z < 0.4$ m, are conducting slabs carrying uniform current densities of 10 A/m^2 in opposite directions, as shown in Fig. 8.22. Find H_x at $z = -0.04$, 0.06, 0.26, 0.36, and 0.46 m.

14. A hollow cylindrical shell of radius a is centered on the z axis and carries a uniform surface current density of $K_a\mathbf{a}_\phi$. (*a*) Show that H is not a function of ϕ or z. (*b*) Show that H_ϕ and H_ρ are everywhere zero. (*c*) Show that $H_z = 0$ for $\rho > a$. (*d*) Show that $H_z = K_a$ for $\rho < a$. (*e*) A second shell, $\rho = b$, carries a current $K_b\mathbf{a}_\phi$. Find $\mathbf{H}$ everywhere.

15. One toroid of rectangular cross section is defined by the cylinders $\rho = 5.5$ cm and $\rho = 10.5$ cm, and the planes $z = \pm 5$ cm. A second toroid is defined by the surfaces $\rho = 6.5$ cm and 9.5 cm, and $z = \pm 1$ cm. Each toroid carries a surface current density of $-20\mathbf{a}_z$ A/m on its outer surface ($\rho = 9.5$ and 10.5 cm). Find $\mathbf{H}$ at: (*a*) $P_A(0,5\text{ cm},0)$; (*b*) $P_B(0,6,0)$; (*c*) $P_C(0,7,0)$; (*d*) $P_D(0,7,2)$; (*e*) $P_E(0,10,0)$.

16. The planes $z = 0$ and $z = 0.1$ m are perfect conductors. A voltage of 2 V is maintained between them, such that $\mathbf{E}$ is in the $\mathbf{a}_z$ direction. For $0 < z < 0.1$ m there is a conducting material for which $\sigma = 1000e^{-100\rho}$ ℧/m. (*a*) Find $\mathbf{E}$ and $\mathbf{J}$. (*b*) Find I_{total} in a cylinder of radius ρ. (*c*) Use Ampère's circuital law to find $\mathbf{H}$. (*d*) Sketch $|\mathbf{H}|$ vs. ρ, $0 < z < 0.05$ m.

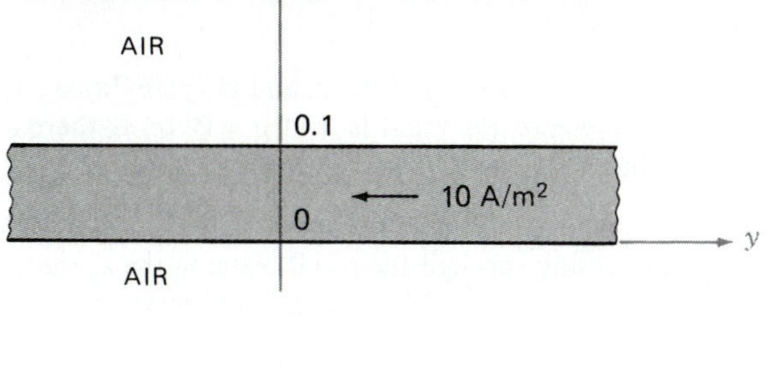

FIGURE 8.22
See Prob. 13.

POINT	H (A/m)
A	$7.56\mathbf{a}_x - 7.80\mathbf{a}_y + 16.80\mathbf{a}_z$
B	$8.37\mathbf{a}_x - 8.20\mathbf{a}_y + 15.20\mathbf{a}_z$
C	$9.20\mathbf{a}_x - 9.23\mathbf{a}_y + 16.00\mathbf{a}_z$
D	$6.80\mathbf{a}_x - 6.83\mathbf{a}_y + 16.00\mathbf{a}_z$
E	$7.20\mathbf{a}_x - 8.00\mathbf{a}_y + 17.64\mathbf{a}_z$
F	$8.80\mathbf{a}_x - 8.00\mathbf{a}_y + 14.44\mathbf{a}_z$

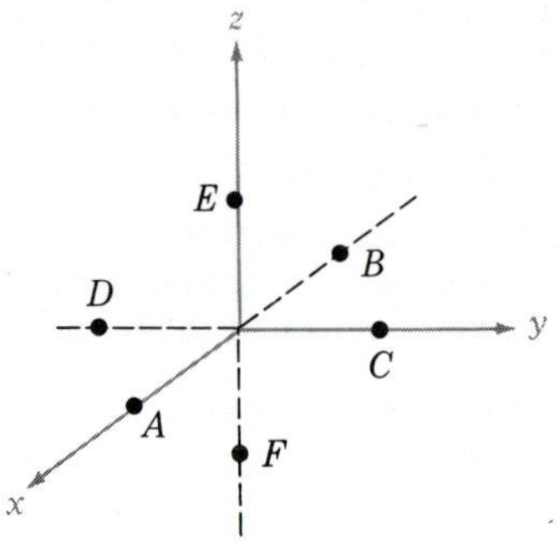

FIGURE 8.23
See Prob. 22.

17. A solid conductor has a circular cross section, radius 1 mm, centered on the z axis, and carries a uniformly distributed total current of 10π A in the $\mathbf{a}_z$ direction. A current sheet $\mathbf{K} = -1000\mathbf{a}_z$ A/m lies at $\rho = 2$ mm. Find $\mathbf{H}$ in the region: (*a*) $0 \le \rho < 1$ mm; (*b*) $1 < \rho < 2$ mm; (*c*) $2 < \rho < 4$ mm. (*d*) What value of current sheet should be located at $\rho = 4$ mm so that $\mathbf{H} = 0$ for $\rho > 4$ mm?

18. Let $\mathbf{J} = 0$ for $|z| > 0.2$ m, and $\mathbf{J} = 20 \sin 5\pi z\, \mathbf{a}_y$ A/m^2 for $-0.2 < z < 0.2$ m. Use Ampère's law and symmetry to find $\mathbf{H}$ everywhere.

19. The cylindrical surface $\rho = 20$ mm carries the current $\mathbf{K} = 100\mathbf{a}_z$ A/m, while the surface $\rho = 40$ mm has the solenoidal current $\mathbf{K} = 80\mathbf{a}_\phi$ A/m. Calculate $|\mathbf{H}|$ at $\rho =$: (*a*) 10 mm; (*b*) 30 mm; (*c*) 50 mm.

20. If $\mathbf{F} = x^2y\mathbf{a}_x - 2z\mathbf{a}_y + (3z^2 + xy)\mathbf{a}_z$, find $\nabla \times [\nabla(\nabla \cdot \mathbf{F})]$.

21. Given $\mathbf{H} = y^2z\mathbf{a}_x + 2(x+1)yz\mathbf{a}_y - (x+1)z^2\mathbf{a}_z$: (*a*) find $\oint \mathbf{H} \cdot d\mathbf{L}$ around the square path going from $P(0,2,0)$ to $A(0, 2+b, 0)$ to $B(0, 2+b, b)$ to $C(0,2,b)$ to P; (*b*) evaluate $\oint \mathbf{H} \cdot d\mathbf{L}$ for $b = 0.1$; (*c*) find $\nabla \times \mathbf{H}$; (*d*) evaluate $(\nabla \times \mathbf{H})_x$ at P; (*e*) show that $(\nabla \times \mathbf{H})_x \doteq [\oint \mathbf{H} \cdot d\mathbf{L}]/\Delta S$.

22. Points A, B, C, D, E, and F are each 1 cm from the origin on the coordinate axis indicated in Fig. 8.23. The value of $\mathbf{H}$ at each point is given. Calculate an approximate value for $\nabla \times \mathbf{H}$ at the origin.

23. A nonmagnetic wire of circular cross section has a radius of 10 mm and a length of 100 m. It carries a total current of 1000 A in the $\mathbf{a}_z$ direction and has a potential difference of 10 V between ends. Let $\mathbf{H} = k\rho^2\mathbf{a}_\phi$, where k is an unknown constant. (*a*) Find k. (*b*) Give σ as a function of the radius ρ.

24. Let $\mathbf{H} = (2/\pi\rho)[1 + (10^7\rho^3/6)]\mathbf{a}_\phi + 8\mathbf{a}_z$ A/m for $0 \le \rho \le 0.01$ m, and $\mathbf{H} = (16/3\pi\rho)\mathbf{a}_\phi + 8\mathbf{a}_z$ A/m for $\rho > 0.01$ m. (*a*) Find $\mathbf{J}$ everywhere. (*b*) What is $\mathbf{J}$ at $\rho = 0$? (*c*) Is there a filamentary current at $\rho = 0$? If so, what is its value? (*d*) What is $|\mathbf{J}|_{max}$ for $\rho > 0$?

25. Let $\mathbf{H} = -y(x^2 + y^2)\mathbf{a}_x + x(x^2 + y^2)\mathbf{a}_y$ A/m in the $z = 0$ plane for $-5 \le x \le 5$ and $-5 \le y \le 5$ m. Find the total current passing through the $z = 0$ plane in the $\mathbf{a}_z$ direction inside the rectangle $-1 < x < 1$ and $-2 < y < 2$ by two different methods.

26. Given the field $\mathbf{G} = \rho^2 \sin^2 \phi \, \mathbf{a}_\rho + \rho^2 \cos^2 \phi \, \mathbf{a}_\phi + 2z^2 \, \mathbf{a}_z$ evaluate both sides of Stokes' theorem for the path formed by the intersection of the cylinder $\rho = 2$ and the plane $z = 1$, and for the surface defined by $\rho = 2$, $1 \le z \le 3$, and $z = 3$, $0 \le \rho \le 2$.

27. In cylindrical coordinates, a magnetic field is given as $\mathbf{H} = (2\rho - \rho^2)\mathbf{a}_\phi$ A/m, $0 \le \rho \le 1$. (*a*) Determine the current density as a function of ρ within this cylinder. (*b*) What total current passes through the surface $z = 0$, $0 \le \rho \le 1$, in the $\mathbf{a}_z$ direction? (*c*) Check (*b*) by another method.

28. The magnetic field intensity is given as $\mathbf{H} = (3/\rho)\mathbf{a}_\rho + 12(z - 5)\mathbf{a}_\phi + 2\rho^2 \sin \phi \, \mathbf{a}_z$ in free space. Determine the value of the surface integral $\int_S (\nabla \times \mathbf{H}) \cdot d\mathbf{S}$ over the paraboloidal surface, $z = \rho^2$, $0 \le z \le 3$.

29. Given $\mathbf{H} = (10r^2/\sin\theta)\mathbf{a}_\theta + 180r \cos\theta \, \mathbf{a}_\phi$ A/m in free space: (*a*) find the current in the $\mathbf{a}_\theta$ direction through the conical surface $\theta = 30°$, $0 \le \phi \le 2\pi$, $0 \le r \le 2$, by using whichever side of Stokes' theorem you like the best. (*b*) Check the result by using the other side of Stokes' theorem.

30. (*a*) Use Ampère's circuital law to find **H** and **B** inside a long, straight, nonmagnetic conductor of radius 8 mm carrying a uniform current density of 100 kA/m^2. (*b*) Show that $\nabla \times \mathbf{H} = \mathbf{J}$. Find the total magnetic flux crossing the surface: (*c*) $z = 0$, for $0.2 < \rho < 8$ mm, $\pi < \phi < 2\pi$; (*d*) $\phi = 0$, $0 < \rho < 8$ mm, $0 < z < 10$ mm; (*e*) $\rho = 0.2$ mm, $0 < \phi < \pi/2$, $0 < z < 10$ mm.

31. A solid nonmagnetic conductor of circular cross section, $\rho = 2$ cm, carries a total current, $I = 60$ A, in the $\mathbf{a}_z$ direction. The conductor is inhomogeneous, having a conductivity that varies with ρ as $\sigma = 10^5(1 + 2.5 \times 10^5\rho^2)$ ℧/m. Find the total flux crossing the radial plane defined by $\phi = 0$, $0 < z < 1$ m, and: (*a*) $0 < \rho < 1$ cm; (*b*) $1 < \rho < 2$ cm.

32. Two solid nonmagnetic cylinders of 0.1-m radius have their axes located in the $x = 0$ plane at $y = \pm 0.3$ m. The one at $y = 0.3$ carries 18 A in the $\mathbf{a}_z$ direction; the other current is 18 A in the $-\mathbf{a}_z$ direction. Find the total magnetic flux crossing the planar surface in the $x = 0$ plane extending from $z = 0$ to $z = 1$ m and: (*a*) $-0.2 < y < 0.2$ m; (*b*) $0.2 < y < 0.4$ m; (*c*) $0.4 < y < \infty$.

33. The region in free space enclosed by planes, $z = 0$ and 3 cm, and by cylinders, $\rho = 5$ and 7 cm, forms a toroid with a rectangular cross section. A surface current, $\mathbf{K} = 600\mathbf{a}_z$ A/m, flows on the inner surface. (*a*) Determine the current on the other three surfaces. (*b*) Find **H** within the toroid. (*c*) Find the total flux within the toroid.

34. A filamentary conductor on the z axis carries a current of 10 A in the $\mathbf{a}_z$ direction, and a coaxial conducting shell at $\rho = 8$ mm carries a total current of 5 A in the $-\mathbf{a}_z$ direction. The region, $3 < \rho < 5$ mm is a solid conducting material carrying a uniformly distributed total current of 5 A in the $-\mathbf{a}_z$ direction. All media are nonmagnetic. (*a*) Find **H**, $0 < \rho < 10$ mm. (*b*) Plot H_ϕ vs. ρ. (*c*) Find the total flux Φ in a 1-m length of the conducting material, $3 < \rho < 5$ mm.

35. A current sheet, $\mathbf{K} = 40\mathbf{a}_z$ A/m, is located in free space at $x = 0.25$ m, and a second sheet, $\mathbf{K} = -40\mathbf{a}_z$ A/m, is at $x = -0.25$ m. (*a*) Let $V_m = 0$ at $P(0.1,0.2,0.3)$ and find $V_m(x,y,z)$ for $-0.25 < x,y,z < 0.25$. (*b*) Let $\mathbf{A} = 0$ at P and determine $\mathbf{A}(x,y,z)$ for $-0.25 < x,y,z < 0.25$.

36. Given the scalar magnetic potential field, $V_m = 30 - 12 \ln \rho$ A: (*a*) sketch the traces (intersections) of the equipotential surfaces $V_m = 0$, 2.5, 5, 7.5, and 10 A with the $z = 0$

plane; (*b*) find **H** and note the direction of **H** in the $z = 0$ plane. The direction lines for **H** are given by ϕ = constant (not Φ = constant). Draw the line for $\phi = 0$. (*c*) Determine ϕ_1 such that the direction line $\phi = \phi_1$ forms curvilinear squares with the equipotentials. Estimate ϕ_1 from your sketch.

37. Find V_m for the region within the toroid of Prob. 33, where $\mathbf{H} = (30/\rho)\mathbf{a}_\phi$ A/m and $V_m = 0$ at $\rho = 6$ cm, $\phi = 0.6$ rad, and $z = 2$ cm. Keep ϕ within the range $0 \leq \phi < 2\pi$.

38. A solenoid in free space is centered on the z axis and has a current of $24\mathbf{a}_\phi$ A/m at $\rho = a = 4$ cm. (*a*) Let $V_m = 0$ at the origin and find $V_m(\rho,\phi,z)$ for all points within the solenoid. (*b*) Let $\mathbf{A} = 0$ at the origin and find $\mathbf{A}(\rho,\phi,z)$ for all points within the solenoid.

39. Let $\mathbf{J} = 8\mathbf{a}_y$ A/m^2 for $-3 < x < 3$ m, and $\mathbf{J} = 0$ elsewhere. (*a*) Obtain expressions for $\mathbf{H}(x,y,z)$ for $|x| < 3$ and $|x| > 3$ m. (*b*) Obtain expressions for $\mathbf{B}(x,y,z)$ for $|x| < 3$ and $|x| > 3$ m if $\mu = \mu_0$. (*c*) Find $V_m(x,y,z)$ for $x > 3$ if $V_m = 0$ at $P(4,-2,5)$. (*d*) Find $\mathbf{A}(x,y,z)$ for $-3 < x < 3$ if $\mathbf{A} = 0$ at $Q(2,-2,4)$.

40. A certain magnetic field intensity is given in free space as $\mathbf{H} = 20(x\mathbf{a}_x + y\mathbf{a}_y)/(x^2 + y^2)$ A/m (*a*) Show that $\nabla \cdot \mathbf{B} = 0$. (*b*) Find the current density **J**. (*c*) Find $V_m(x,y,z)$ if $V_m = 0$ at $P(1,1,1)$. (*d*) Find $\mathbf{A}(x,y,z)$ if $A_x = A_y = 0$ and $A_z = 0$ at $P(1,1,1)$.

41. If $\mathbf{A} = 10\rho^{1.5}\mathbf{a}_z$ Wb/m in free space, find: (*a*) **H**; (*b*) **J**; (*c*) Show that $\oint \mathbf{H} \cdot d\mathbf{L} = I$ for a circular path with $\rho = 1$.

42. Show that $\nabla_2(1/R_{12}) = -\nabla_1(1/R_{12}) = \mathbf{R}_{21}/R_{12}^3$.

43. Compute the vector magnetic potential within the outer conductor for the coaxial line whose vector magnetic potential is shown in Fig. 8.20 if the outer radius of the outer conductor is $7a$. Select the proper zero reference and sketch the results on the figure.

44. By expanding Eq. (58), Sec. 8.7, in cartesian coordinates, show that (59) is correct.

CHAPTER 9

MAGNETIC FORCES, MATERIALS, AND INDUCTANCE

The magnetic field quantities **H**, **B**, Φ, V_m, and **A** introduced in the last chapter have not as yet been given much physical significance. Each of these quantities is merely defined in terms of the distribution of current sources throughout space. If the current distribution is known, we should feel that **H**, **B**, and **A** are determined at every point in space, even though we may not be able to evaluate the defining integrals because of mathematical complexity.

We are now ready to undertake the second half of the magnetic field problem, that of determining the forces and torques exerted by the magnetic field on other charges. The electric field causes a force to be exerted on a charge which may be either stationary or in motion; we shall see that the steady magnetic field is capable of exerting a force only on a *moving* charge. This result appears reasonable; a magnetic field may be produced by moving charges and may exert forces on moving charges; a magnetic field cannot arise from stationary charges and cannot exert any force on a stationary charge.

This chapter initially considers the forces and torques on current-carrying conductors which may either be of a filamentary nature or possess a finite cross section with a known current density distribution. The problems associated with the motion of particles in a vacuum are largely avoided.

With an understanding of the fundamental effects produced by the magnetic field, we may then consider the varied types of magnetic materials, the analysis of elementary magnetic circuits, the forces on magnetic materials, and finally, the important electrical circuit concepts of self-inductance and mutual inductance.

9.1 FORCE ON A MOVING CHARGE

In an electric field the definition of the electric field intensity shows us that the force on a charged particle is

$$\boxed{\mathbf{F} = Q\mathbf{E}} \tag{1}$$

The force is in the same direction as the electric field intensity (for a positive charge) and is directly proportional to both $\mathbf{E}$ and Q. If the charge is in motion, the force at any point in its trajectory is then given by (1).

A charged particle in motion in a magnetic field of flux density $\mathbf{B}$ is found experimentally to experience a force whose magnitude is proportional to the product of the magnitudes of the charge Q, its velocity $\mathbf{v}$, and the flux density $\mathbf{B}$, and to the sine of the angle between the vectors $\mathbf{v}$ and $\mathbf{B}$. The direction of the force is perpendicular to both $\mathbf{v}$ and $\mathbf{B}$ and is given by a unit vector in the direction of $\mathbf{v} \times \mathbf{B}$. The force may therefore be expressed as

$$\boxed{\mathbf{F} = Q\mathbf{v} \times \mathbf{B}} \tag{2}$$

A fundamental difference in the effect of the electric and magnetic fields on charged particles is now apparent, for a force which is always applied in a direction at right angles to the direction in which the particle is proceeding can never change the magnitude of the particle velocity. In other words, the *acceleration* vector is always normal to the velocity vector. The kinetic energy of the particle remains unchanged, and it follows that the steady magnetic field is incapable of transferring energy to the moving charge. The electric field, on the other hand, exerts a force on the particle which is independent of the direction in which the particle is progressing and therefore effects an energy transfer between field and particle in general.

The first two problems at the end of this chapter illustrate the different effects of electric and magnetic fields on the kinetic energy of a charged particle moving in free space.

The force on a moving particle due to combined electric and magnetic fields is obtained easily by superposition,

$$\boxed{\mathbf{F} = Q(\mathbf{E} + \mathbf{v} \times \mathbf{B})} \tag{3}$$

This equation is known as the *Lorentz force equation*, and its solution is required in determining electron orbits in the magnetron, proton paths in the cyclotron, plasma characteristics in a magnetohydrodynamic (MHD) generator, or, in general, charged-particle motion in combined electric and magnetic fields.

D9.1. A negative point charge, $Q = -40$ nC, is moving with a velocity of 6×10^6 m/s in a direction specified by the unit vector $\mathbf{a}_v = -0.48\mathbf{a}_x - 0.6\mathbf{a}_y + 0.64\mathbf{a}_z$. Find the magnitude of the vector force exerted on the moving particle by the field: (*a*) $\mathbf{B} = 2\mathbf{a}_x - 3\mathbf{a}_y + 5\mathbf{a}_z$ mT; (*b*) $\mathbf{E} = 2\mathbf{a}_x - 3\mathbf{a}_y + 5\mathbf{a}_z$ kV/m; (*c*) **B** and **E** acting together.

Ans. 1117 μN; 247 μN; 1144 μN

9.2 FORCE ON A DIFFERENTIAL CURRENT ELEMENT

The force on a charged particle moving through a steady magnetic field may be written as the differential force exerted on a differential element of charge,

$$d\mathbf{F} = dQ\,\mathbf{v} \times \mathbf{B} \tag{4}$$

Physically, the differential element of charge consists of a large number of very small discrete charges occupying a volume which, although small, is much larger than the average separation between the charges. The differential force expressed by (4) is thus merely the sum of the forces on the individual charges. This sum, or resultant force, is not a force applied to a single object. In an analogous way, we might consider the differential gravitational force experienced by a small volume taken in a shower of falling sand. The small volume contains a large number of sand grains, and the differential force is the sum of the forces on the individual grains within the small volume.

If our charges are electrons in motion in a conductor, however, we can show that the force is transferred to the conductor and that the sum of this extremely large number of extremely small forces is of practical importance. Within the conductor, electrons are in motion throughout a region of immobile positive ions which form a crystalline array giving the conductor its solid properties. A magnetic field which exerts forces on the electrons tends to cause them to shift position slightly and produces a small displacement between the centers of "gravity" of the positive and negative charges. The Coulomb forces between electrons and positive ions, however, tend to resist such a displacement. Any attempt to move the electrons, therefore, results in an attractive force between electrons and the positive ions of the crystalline lattice. The magnetic force is thus transferred to the crystalline lattice, or to the conductor itself. The Coulomb forces are so much greater than the magnetic forces in good conductors that the actual displacement of the electrons is almost immeasurable. The charge separation that does result, however, is disclosed by the presence of a slight potential difference across the conductor sample in a direction perpendicular to both the magnetic field and the velocity of the charges. The voltage is known as the *Hall voltage*, and the effect itself is called the *Hall effect*.

Figure 9.1 illustrates the direction of the Hall voltage for both positive and negative charges in motion. In Fig. 9.1*a*, **v** is in the $-\mathbf{a}_x$ direction, $\mathbf{v} \times \mathbf{B}$ is in the $\mathbf{a}_y$ direction, and Q is positive, causing $\mathbf{F}_Q$ to be in the $\mathbf{a}_y$ direction; thus, the positive charges move to the right. In Fig. 9.1*b*, **v** is now in the $+\mathbf{a}_x$ direction, **B** is still

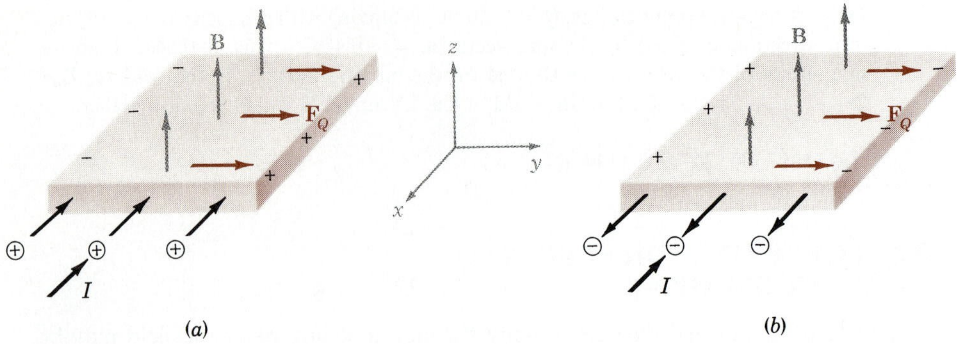

FIGURE 9.1
Equal currents directed into the material are provided by positive charges moving inward in (*a*) and negative charges moving outward in (*b*). The two cases can be distinguished by oppositely directed Hall voltages, as shown.

in the $\mathbf{a}_z$ direction, $\mathbf{v} \times \mathbf{B}$ is in the $-\mathbf{a}_y$ direction, and Q is negative; thus $\mathbf{F}_Q$ is again in the $\mathbf{a}_y$ direction. Hence, the negative charges end up at the right edge. Equal currents provided by holes and electrons in semiconductors can therefore be differentiated by their Hall voltages. This is one method of determining whether a given semiconductor is *n*-type or *p*-type.

Devices employ the Hall effect to measure the magnetic flux density and, in some applications where the current through the device can be made proportional to the magnetic field across it, to serve as electronic wattmeters, squaring elements, and so forth.

Returning to (4), we may therefore say that if we are considering an element of moving charge in an electron beam, the force is merely the sum of the forces on the individual electrons in that small volume element, but if we are considering an element of moving charge within a conductor, the total force is applied to the solid conductor itself. We shall now limit our attention to the forces on current-carrying conductors.

In Chap. 5 we defined convection current density in terms of the velocity of the volume charge density,

$$\boxed{\mathbf{J} = \rho_v \mathbf{v}}$$

The differential element of charge in (4) may also be expressed in terms of volume charge density,[1]

$$dQ = \rho_v \, dv$$

Thus

$$d\mathbf{F} = \rho_v \, dv \, \mathbf{v} \times \mathbf{B}$$

[1] Remember that *dv* is a differential volume element and not a differential increase in velocity.

or

$$\boxed{d\mathbf{F} = \mathbf{J} \times \mathbf{B}\,dv} \tag{5}$$

We saw in the previous chapter that $\mathbf{J}\,dv$ may be interpreted as a differential current element; that is,

$$\mathbf{J}\,dv = \mathbf{K}\,dS = I\,d\mathbf{L}$$

and thus the Lorentz force equation may be applied to surface current density,

$$\boxed{d\mathbf{F} = \mathbf{K} \times \mathbf{B}\,dS} \tag{6}$$

or to a differential current filament,

$$\boxed{d\mathbf{F} = I\,d\mathbf{L} \times \mathbf{B}} \tag{7}$$

Integrating (5), (6), or (7) over a volume, a surface which may be either open or closed (why?), or a closed path, respectively, leads to the integral formulations

$$\mathbf{F} = \int_{\text{vol}} \mathbf{J} \times \mathbf{B}\,dv \tag{8}$$

$$\mathbf{F} = \int_S \mathbf{K} \times \mathbf{B}\,dS \tag{9}$$

and

$$\boxed{\mathbf{F} = \oint I\,d\mathbf{L} \times \mathbf{B} = -I \oint \mathbf{B} \times d\mathbf{L}} \tag{10}$$

One simple result is obtained by applying (7) or (10) to a straight conductor in a uniform magnetic field,

$$\boxed{\mathbf{F} = I\mathbf{L} \times \mathbf{B}} \tag{11}$$

The magnitude of the force is given by the familiar equation

$$F = BIL \sin\theta \tag{12}$$

where θ is the angle between the vectors representing the direction of the current flow and the direction of the magnetic flux density. Equation (11) or (12) applies only to a portion of the closed circuit, and the remainder of the circuit must be considered in any practical problem.

As a numerical example of these equations, consider Fig. 9.2. We have a square loop of wire in the $z = 0$ plane carrying 2 mA in the field of an infinite filament on the y axis, as shown. The field produced in the plane of the loop by

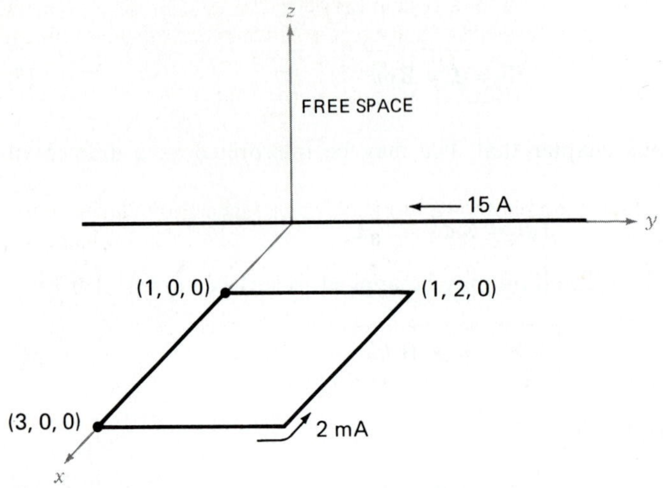

FIGURE 9.2
A square loop of wire in the xy plane carrying 2 mA is subjected to a nonuniform **B** field.

the straight filament is

$$\mathbf{H} = \frac{I}{2\pi x}\mathbf{a}_z = \frac{15}{2\pi x}\mathbf{a}_z \qquad \text{A/m}$$

Therefore,

$$\mathbf{B} = \mu_0 \mathbf{H} = 4\pi \times 10^{-7}\mathbf{H} = \frac{3 \times 10^{-6}}{x}\mathbf{a}_z \qquad \text{T}$$

We use the integral form (10),

$$\mathbf{F} = -I \oint \mathbf{B} \times d\mathbf{L}$$

Let us assume a rigid loop so that the total force is the sum of the forces on the four sides. Beginning with the left side:

$$\begin{aligned}
\mathbf{F} &= -2 \times 10^{-3} \times 3 \times 10^{-6}\left[\int_{x=1}^{3} \frac{\mathbf{a}_z}{x} \times dx\,\mathbf{a}_x + \int_{y=1}^{3} \frac{\mathbf{a}_z}{3} \times dy\,\mathbf{a}_y \right. \\
&\qquad \left. + \int_{x=3}^{1} \frac{\mathbf{a}_z}{x} \times dx\,\mathbf{a}_x + \int_{y=3}^{1} \frac{\mathbf{a}_z}{1} \times dy\,\mathbf{a}_y\right] \\
&= -6 \times 10^{-9}\left[\ln x\Big|_1^3 \mathbf{a}_y + \frac{1}{3} y\Big|_1^3 (-\mathbf{a}_x) + \ln x\Big|_3^1 \mathbf{a}_y + y\Big|_3^1 (-\mathbf{a}_x)\right] \\
&= -6 \times 10^{-9}\left[(\ln 3)\mathbf{a}_y - \frac{2}{3}\mathbf{a}_x + \left(\ln \frac{1}{3}\right)\mathbf{a}_y + 2\mathbf{a}_x\right] \\
&= -8\mathbf{a}_x \qquad \text{pN}
\end{aligned}$$

Thus, the net force on the loop is in the $-\mathbf{a}_x$ direction.

D9.2. Let $V_m = 2x^2 + 4x - 2y^2$ A in a certain region of free space. Find the vector force exerted on a straight wire segment in this region if it extends from the origin to: (*a*) $P_A(1,0,0)$ and carries 5 A in the $\mathbf{a}_x$ direction; (*b*) $P_B(0,0,1)$ and carries 5 A in the $\mathbf{a}_z$ direction; (*c*) $P_C(0.6,0.8,0)$ and carries 5 A away from the origin.

Ans. 0; $-25.1\mathbf{a}_y$ μN; $32.2\mathbf{a}_z$ μN

D9.3. The semiconductor sample shown in Fig. 9.1*a* is *p*-type silicon, having a rectangular cross section of 0.8 mm by 0.9 cm and a length of 1.25 cm. Assume the electron and hole mobilities are 0.14 and 0.05 $m^2/V \cdot s$, respectively, at the operating temperature. With $B = 0.05$ T and an electric field intensity in the direction of the current flow of 750 V/m, find the magnitude of: (*a*) the voltage across the sample length; (*b*) the drift velocity; (*c*) the transverse force per coulomb of moving charge caused by *B*; (*d*) the transverse electric field intensity; (*e*) the Hall voltage.

Ans. 9.38 V; 37.5 m/s; 1.875 N/C; 1.875 V/m; 16.88 mV

9.3 FORCE BETWEEN DIFFERENTIAL CURRENT ELEMENTS

The concept of the magnetic field was introduced to break into two parts the problem of finding the interaction of one current distribution on a second current distribution. It is possible to express the force on one current element directly in terms of a second current element without finding the magnetic field. Since we have claimed that the magnetic-field concept simplifies our work, it then behooves us to show that avoidance of this intermediate step leads to more complicated expressions.

The magnetic field at point 2 due to a current element at point 1 has been found to be

$$d\mathbf{H}_2 = \frac{I_1\, d\mathbf{L}_1 \times \mathbf{a}_{R12}}{4\pi R_{12}^2}$$

Now, the differential force on a differential current element is

$$d\mathbf{F} = I\, d\mathbf{L} \times \mathbf{B}$$

and we apply this to our problem by letting $\mathbf{B}$ be $d\mathbf{B}_2$ (the differential flux density at point 2 caused by current element 1), by identifying $I\,d\mathbf{L}$ as $I_2\, d\mathbf{L}_2$, and by symbolizing the differential amount of our differential force on element 2 as $d(d\mathbf{F}_2)$:

$$d(d\mathbf{F}_2) = I_2\, d\mathbf{L}_2 \times d\mathbf{B}_2$$

Since $d\mathbf{B}_2 = \mu_0\, d\mathbf{H}_2$, we obtain the force between two differential current elements,

$$d(d\mathbf{F}_2) = \mu_0 \frac{I_1 I_2}{4\pi R_{12}^2}\, d\mathbf{L}_2 \times (d\mathbf{L}_1 \times \mathbf{a}_{R12}) \tag{13}$$

To illustrate the use (and misuse) of these results, consider the two differential current elements shown in Fig. 9.3. We have $I_1\, d\mathbf{L}_1 = -3\mathbf{a}_y$ A·m at $P_1(5,2,1)$, and

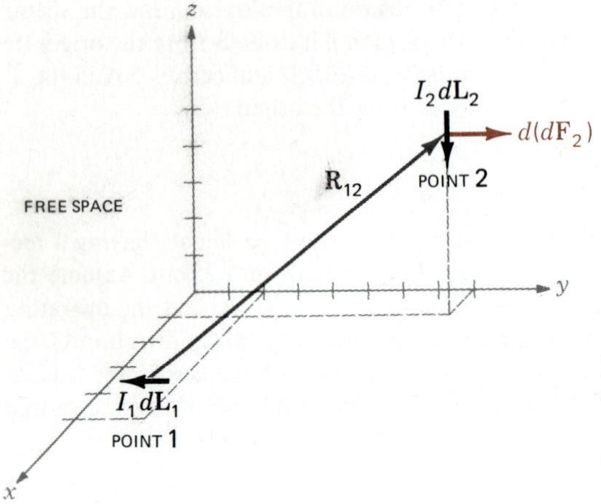

FIGURE 9.3
Given $P_1(5,2,1)$, $P_2(1,8,5)$, $I_1\,d\mathbf{L}_1 = -3\mathbf{a}_y$ A·m, and $I_2\,d\mathbf{L}_2 = -4\mathbf{a}_z$ A·m, the force on $I_2\,d\mathbf{L}_2$ is 8.56 nN in the $\mathbf{a}_y$ direction.

$I_2\,d\mathbf{L}_2 = -4\mathbf{a}_z$ A·m at $P_2(1,8,5)$. Thus, $\mathbf{R}_{12} = -4\mathbf{a}_x + 6\mathbf{a}_y + 4\mathbf{a}_z$, and we may substitute these data into (13),

$$d(d\mathbf{F}_2) = \frac{4\pi 10^{-7}}{4\pi} \frac{(-4\mathbf{a}_z) \times [(-3\mathbf{a}_y) \times (-4\mathbf{a}_x + 6\mathbf{a}_y + 4\mathbf{a}_z)]}{(16 + 36 + 16)^{1.5}}$$

$$= 8.56\mathbf{a}_y \qquad \text{nN}$$

Many chapters ago when we discussed the force exerted by one point charge on another point charge, we found that the force on the first charge was the negative of that on the second. That is, the total force on the system was zero. This is not the case with the differential current elements, and $d(d\mathbf{F}_1) = -12.84\mathbf{a}_z$ nN in the example above. The reason for this different behavior lies with the nonphysical nature of the current element. Whereas point charges may be approximated quite well by small charges, the continuity of current demands that a complete circuit be considered. This we shall now do.

The total force between two filamentary circuits is obtained by integrating twice:

$$\mathbf{F}_2 = \mu_0 \frac{I_1 I_2}{4\pi} \oint \left[d\mathbf{L}_2 \times \oint \frac{d\mathbf{L}_1 \times \mathbf{a}_{R12}}{R_{12}^2} \right]$$

$$= \mu_0 \frac{I_1 I_2}{4\pi} \oint \left[\oint \frac{\mathbf{a}_{R12} \times d\mathbf{L}_1}{R_{12}^2} \right] \times d\mathbf{L}_2 \qquad (14)$$

Equation (14) is quite formidable, but the familiarity gained in the last chapter with the magnetic field should enable us to recognize the inner integral as the integral necessary to find the magnetic field at point 2 due to the current element at point 1.

Although we shall only give the result, it is not very difficult to make use of (14) to find the force of repulsion between two infinitely long, straight, parallel, filamentary conductors with separation d, and carrying equal but opposite currents I, as shown in Fig. 9.4. The integrations are simple, and most errors are made in determining suitable expressions for $\mathbf{a}_{R12}$, $d\mathbf{L}_1$, and $d\mathbf{L}_2$. However, since the

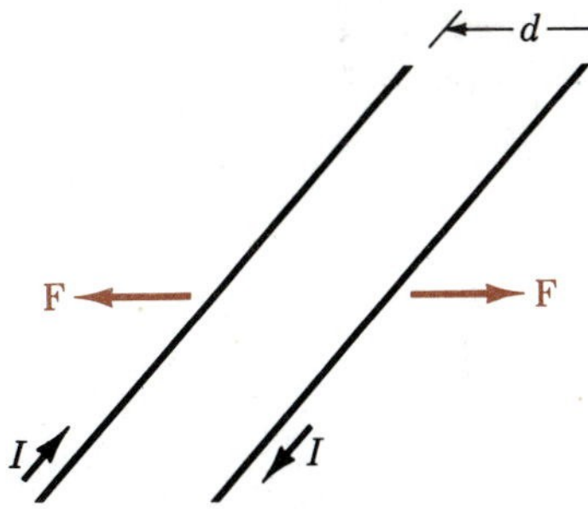

FIGURE 9.4
Two infinite parallel filaments with separation d and equal but opposite currents I experience a repulsive force of $\mu_0 I^2/(2\pi d)$ N/m.

magnetic field intensity at either wire caused by the other is already known to be $I/(2\pi d)$, it is readily apparent that the answer is a force of $\mu_0 I^2/(2\pi d)$ newtons per meter length.

D9.4. A current element, $I_1\,\Delta\mathbf{L}_1 = 10^{-5}\mathbf{a}_z$ A·m, is located at $P_1(1,0,0)$, while a second element, $I_2\,\Delta\mathbf{L}_2 = 10^{-5}(0.6\mathbf{a}_x - 2\mathbf{a}_y + 3\mathbf{a}_z)$ A·m is at $P_2(-1,0,0)$, both in free space. (*a*) Find the vector force exerted on $I_2\,\Delta\mathbf{L}_2$ by $I_1\,\Delta\mathbf{L}_1$. (*b*) Find the vector force exerted on $I_1\,\Delta\mathbf{L}_1$ by $I_2\,\Delta\mathbf{L}_2$.

Ans. $(7.5\mathbf{a}_x - 1.5\mathbf{a}_z)10^{-18}$ N; $-7.5 \times 10^{-18}\mathbf{a}_x$ N

9.4 FORCE AND TORQUE ON A CLOSED CIRCUIT

We have already obtained general expressions for the forces exerted on current systems. One special case is easily disposed of, for if we take our relationship for the force on a filamentary closed circuit, as given by Eq. (10), Sec. 9.2,

$$\mathbf{F} = -I \oint \mathbf{B} \times d\mathbf{L}$$

and assume a *uniform* magnetic flux density, then $\mathbf{B}$ may be removed from the integral:

$$\mathbf{F} = -I\mathbf{B} \times \oint d\mathbf{L}$$

However, we discovered during our investigation of closed line integrals in an electrostatic potential field that $\oint d\mathbf{L} = 0$, and therefore the force on a closed filamentary circuit in a uniform magnetic field is zero.

If the field is not uniform, the total force need not be zero.

This result for uniform fields does not have to be restricted to filamentary circuits only. The circuit may contain surface currents or volume current density as well. If the total current is divided into filaments, the force on each one is zero, as we showed above, and the total force is again zero. Therefore any real closed circuit carrying direct currents experiences a total vector force of zero in a uniform magnetic field.

Although the force is zero, the torque is generally not equal to zero.

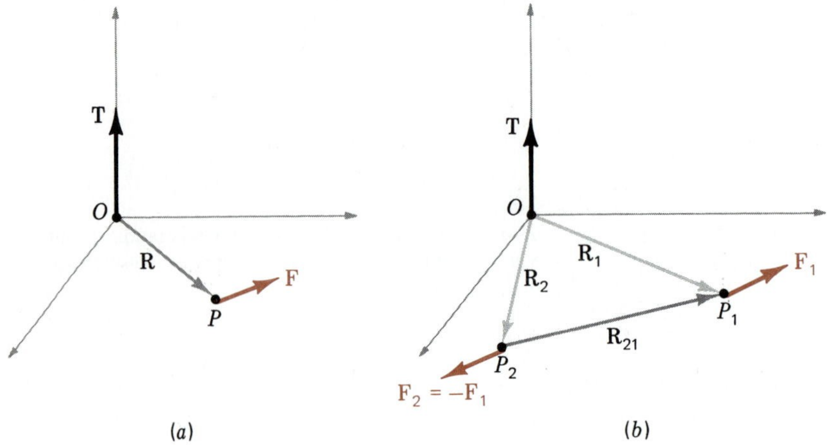

FIGURE 9.5
(*a*) Given a lever arm **R** extending from an origin O to a point P where force **F** is applied, the torque about O is $\mathbf{T} = \mathbf{R} \times \mathbf{F}$. (*b*) If $\mathbf{F}_2 = -\mathbf{F}_1$, then the torque $\mathbf{T} = \mathbf{R}_{21} \times \mathbf{F}_1$ is independent of the choice of origin for $\mathbf{R}_1$ and $\mathbf{R}_2$.

In defining the *torque*, or *moment*, of a force, it is necessary to consider both an origin at or about which the torque is to be calculated, as well as the point at which the force is applied. In Fig. 9.5*a*, we apply a force **F** at point P, and we establish an origin at O with a rigid lever arm **R** extending from O to P. The torque about point O is a vector whose magnitude is the product of the magnitudes of **R**, of **F**, and of the sine of the angle between these two vectors. The direction of the vector torque **T** is normal to both the force **F** and lever arm **R** and is in the direction of progress of a right-handed screw as the lever arm is rotated into the force vector through the smaller angle. The torque is expressible as a cross product,

$$\mathbf{T} = \mathbf{R} \times \mathbf{F}$$

Now let us assume that two forces, $\mathbf{F}_1$ at P_1 and $\mathbf{F}_2$ at P_2, having lever arms $\mathbf{R}_1$ and $\mathbf{R}_2$ extending from a common origin O, as shown in Fig. 9.5*b*, are applied to an object of fixed shape and that the object does not undergo any translation. Then the torque about the origin is

$$\mathbf{T} = \mathbf{R}_1 \times \mathbf{F}_1 + \mathbf{R}_2 \times \mathbf{F}_2$$

where

$$\mathbf{F}_1 + \mathbf{F}_2 = 0$$

and therefore

$$\mathbf{T} = (\mathbf{R}_1 - \mathbf{R}_2) \times \mathbf{F}_1 = \mathbf{R}_{21} \times \mathbf{F}_1$$

The vector $\mathbf{R}_{21} = \mathbf{R}_1 - \mathbf{R}_2$ joins the point of application of $\mathbf{F}_2$ to that of $\mathbf{F}_1$ and is independent of the choice of origin for the two vectors $\mathbf{R}_1$ and $\mathbf{R}_2$. Therefore,

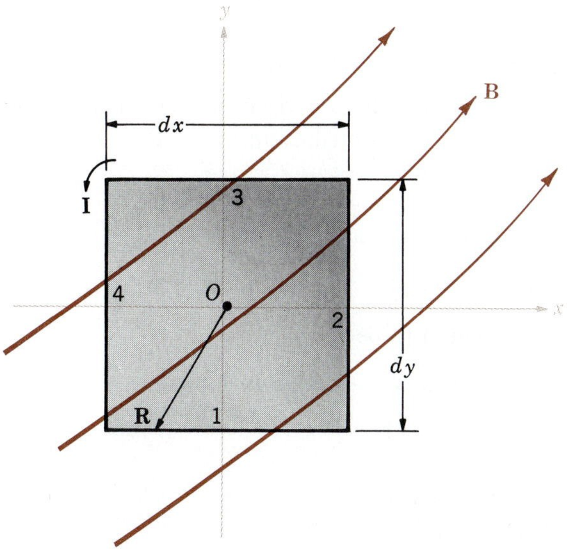

FIGURE 9.6
A differential current loop in a magnetic field **B**. The torque on the loop is $d\mathbf{T} = I(dx\,dy\,\mathbf{a}_z) \times \mathbf{B}_0 = I\,d\mathbf{S} \times \mathbf{B}$.

the torque is also independent of the choice of origin, provided that the total force is zero. This may be extended to any number of forces.

Consider the application of a vertically upward force at the end of a horizontal crank handle on an elderly automobile. This cannot be the only applied force, for if it were, the entire handle would be accelerated in an upward direction. A second force, equal in magnitude to that exerted at the end of the handle, is applied in a downward direction by the bearing surface at the axis of rotation. For a 40-N force on a crank handle 0.3 m in length, the torque is 12 N·m. This figure is obtained regardless of whether the origin is considered to be on the axis of rotation (leading to 12 N·m plus 0 N·m), at the midpoint of the handle (leading to 6 N·m plus 6 N·m), or at some point not even on the handle or an extension of the handle.

We may therefore choose the most convenient origin, and this is usually on the axis of rotation and in the plane containing the applied forces if the several forces are coplanar.

With this introduction to the concept of torque, let us now consider the torque on a differential current loop in a magnetic field **B**. The loop lies in the xy plane (Fig. 9.6); the sides of the loop are parallel to the x and y axes and are of length dx and dy. The value of the magnetic field at the center of the loop is taken as $\mathbf{B}_0$. Since the loop is of differential size, the value of **B** at all points on the loop may be taken as $\mathbf{B}_0$. (Why was this not possible in the discussion of curl and divergence?) The total force on the loop is therefore zero, and we are free to choose the origin for the torque at the center of the loop.

The vector force on side 1 is

$$d\mathbf{F}_1 = I\,dx\,\mathbf{a}_x \times \mathbf{B}_0$$

or

$$d\mathbf{F}_1 = I\,dx(B_{0y}\mathbf{a}_z - B_{0z}\mathbf{a}_y)$$

For this side of the loop the lever arm $\mathbf{R}$ extends from the origin to the midpoint of the side, $\mathbf{R}_1 = -\frac{1}{2}dy\,\mathbf{a}_y$, and the contribution to the total torque is

$$\begin{aligned} d\mathbf{T}_1 &= \mathbf{R}_1 \times d\mathbf{F}_1 \\ &= -\tfrac{1}{2}dy\,\mathbf{a}_y \times I\,dx(B_{0y}\mathbf{a}_z - B_{0z}\mathbf{a}_y) \\ &= -\tfrac{1}{2}dx\,dy\,IB_{0y}\mathbf{a}_x \end{aligned}$$

The torque contribution on side 3 is found to be the same,

$$\begin{aligned} d\mathbf{T}_3 &= \mathbf{R}_3 \times d\mathbf{F}_3 = \tfrac{1}{2}dy\,\mathbf{a}_y \times (-I\,dx\,\mathbf{a}_x \times \mathbf{B}_0) \\ &= -\tfrac{1}{2}dx\,dy\,IB_{0y}\mathbf{a}_x = d\mathbf{T}_1 \end{aligned}$$

and

$$d\mathbf{T}_1 + d\mathbf{T}_3 = -dx\,dy\,IB_{0y}\mathbf{a}_x$$

Evaluating the torque on sides 2 and 4, we find

$$d\mathbf{T}_2 + d\mathbf{T}_4 = dx\,dy\,IB_{0x}\mathbf{a}_y$$

and the total torque is then

$$d\mathbf{T} = I\,dx\,dy(B_{0x}\mathbf{a}_y - B_{0y}\mathbf{a}_x)$$

The quantity within the parentheses may be represented by a cross product,

$$d\mathbf{T} = I\,dx\,dy(\mathbf{a}_z \times \mathbf{B}_0)$$

or

$$\boxed{d\mathbf{T} = I\,d\mathbf{S} \times \mathbf{B}} \qquad (15)$$

where $d\mathbf{S}$ is the vector area of the differential current loop and the subscript on $\mathbf{B}_0$ has been dropped.

We now define the product of the loop current and the vector area of the loop as the differential *magnetic dipole moment* $d\mathbf{m}$, with units of $\text{A}\cdot\text{m}^2$. Thus

$$\boxed{d\mathbf{m} = I\,d\mathbf{S}} \qquad (16)$$

and

$$\boxed{d\mathbf{T} = d\mathbf{m} \times \mathbf{B}} \qquad (17)$$

If we extend the results we obtained in Sec. 4.7 for the differential *electric* dipole by determining the torque produced on it by an *electric* field, we see a similar result,

$$d\mathbf{T} = d\mathbf{p} \times \mathbf{E}$$

Equations (15) and (17) are general results which hold for differential loops of any shape, not just rectangular ones. The torque on a circular or triangular loop is also given in terms of the vector surface or the moment by (15) or (17).

Since we selected a differential current loop so that we might assume **B** was constant throughout it, it follows that the torque on a *planar* loop of any size or shape in a *uniform* magnetic field is given by the same expression,

$$\boxed{\mathbf{T} = I\mathbf{S} \times \mathbf{B} = \mathbf{m} \times \mathbf{B}} \tag{18}$$

We should note that the torque on the current loop always tends to turn the loop so as to align the magnetic field produced by the loop with the applied magnetic field that is causing the torque. This is perhaps the easiest way to determine the direction of the torque.

To illustrate some force and torque calculations, consider the rectangular loop shown in Fig. 9.7. The loop has dimensions of 1 m by 2 m and lies in the uniform field $\mathbf{B}_0 = -0.6\mathbf{a}_y + 0.8\mathbf{a}_z$ T. The loop current is 4 mA, a value that is sufficiently small to avoid causing any magnetic field that might affect $\mathbf{B}_0$.

Let us first calculate the torque by using $\mathbf{T} = I\mathbf{S} \times \mathbf{B}$. We have

$$\mathbf{T} = 4 \times 10^{-3}[(1)(2)\mathbf{a}_z] \times (-0.6\mathbf{a}_y + 0.8\mathbf{a}_z) = 4.8\mathbf{a}_x \qquad \text{mN}\cdot\text{m}$$

Thus, the loop tends to rotate about an axis parallel to the positive x axis. The small magnetic field produced by the 4-mA loop current tends to line up with $\mathbf{B}_0$.

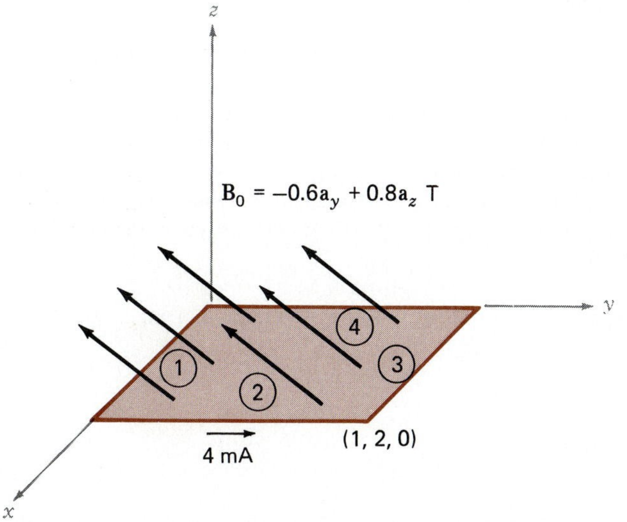

FIGURE 9.7
A rectangular loop is located in a uniform magnetic flux density $\mathbf{B}_0$.

Now let us find the torque once more, this time by calculating the total force and torque contribution for each side. On side 1 we have

$$\mathbf{F}_1 = I\mathbf{L}_1 \times \mathbf{B}_0 = 4 \times 10^{-3}(1\mathbf{a}_x) \times (-0.6\mathbf{a}_y + 0.8\mathbf{a}_z)$$
$$= -3.2\mathbf{a}_y - 2.4\mathbf{a}_z \quad \text{mN}$$

On side 3 we obtain the negative of this result,

$$\mathbf{F}_3 = 3.2\mathbf{a}_y + 2.4\mathbf{a}_z \quad \text{mN}$$

Next we attack side 2:

$$\mathbf{F}_2 = I\mathbf{L}_2 \times \mathbf{B}_0 = 4 \times 10^{-3}(2\mathbf{a}_y) \times (-0.6\mathbf{a}_y + 0.8\mathbf{a}_z)$$
$$= 6.4\mathbf{a}_x \quad \text{mN}$$

with side 4 again providing the negative of this result,

$$\mathbf{F}_4 = -6.4\mathbf{a}_x \quad \text{mN}$$

Since these forces are distributed uniformly along each of the sides, we treat each force as if it were applied at the center of the side. The origin for the torque may be established anywhere since the sum of the forces is zero, and we choose the center of the loop. Thus,

$$\mathbf{T} = \mathbf{T}_1 + \mathbf{T}_2 + \mathbf{T}_3 + \mathbf{T}_4 = \mathbf{R}_1 \times \mathbf{F}_1 + \mathbf{R}_2 \times \mathbf{F}_2 + \mathbf{R}_3 \times \mathbf{F}_3 + \mathbf{R}_4 \times \mathbf{F}_4$$
$$= (-1\mathbf{a}_y) \times (-3.2\mathbf{a}_y - 2.4\mathbf{a}_z) + (0.5\mathbf{a}_x) \times (6.4\mathbf{a}_x)$$
$$+ (1\mathbf{a}_y) \times (3.2\mathbf{a}_y + 2.4\mathbf{a}_z) + (-0.5\mathbf{a}_x) \times (-6.4\mathbf{a}_x)$$
$$= 2.4\mathbf{a}_x + 2.4\mathbf{a}_x = 4.8\mathbf{a}_x \quad \text{mN}\cdot\text{m}$$

Crossing the loop moment with the magnetic flux density is certainly easier.

D9.5. A loop of wire is constructed of three straight segments connecting (0,0,0) to (0.6,0,0) to (0.4,1,0.7) to (0,0,0). A current of 8 mA is in the $\mathbf{a}_x$ direction in the first segment. Given a uniform magnetic field, $\mathbf{B} = 0.2\mathbf{a}_x - 0.1\mathbf{a}_y + 0.2\mathbf{a}_z$ T, find: (*a*) the force on the segment extending from (0,0,0) to (0.6,0,0); (*b*) the total force on the loop; (*c*) the torque on the loop about an origin at (0,0,0); (*d*) the torque on the loop about an origin at (0.45,0.6,0.3).

Ans. $-0.96\mathbf{a}_y - 0.48\mathbf{a}_z$ mN; 0; $-0.096\mathbf{a}_x + 0.48\mathbf{a}_y + 0.336\mathbf{a}_z$ mN·m; $-0.096\mathbf{a}_x + 0.48\mathbf{a}_y + 0.336\mathbf{a}_z$ mN·m

9.5 THE NATURE OF MAGNETIC MATERIALS

We are now in a position to combine our knowledge of the action of a magnetic field on a current loop with a simple model of an atom and obtain some appreciation of the difference in behavior of various types of materials in magnetic fields.

Although accurate quantitative results can only be predicted through the use of quantum theory, the simple atomic model which assumes that there is a central positive nucleus surrounded by electrons in various circular orbits yields reasonable quantitative results and provides a satisfactory qualitative theory. An electron in an orbit is analogous to a small current loop (in which the current is directed oppositely to the direction of electron travel) and as such experiences a torque in an external magnetic field, the torque tending to align the magnetic field produced by the orbiting electron with the external magnetic field. If there were no other magnetic moments to consider, we would then conclude that all the orbiting electrons in the material would shift in such a way as to add their magnetic fields to the applied field, and thus that the resultant magnetic field at any point in the material would be greater than it would be at that point if the material were not present.

A second moment, however, is attributed to *electron spin*. Although it is tempting to model this phenomenon by considering the electron as spinning about its own axis and thus generating a magnetic dipole moment, satisfactory quantitative results are not obtained from such a theory. Instead, it is necessary to digest the mathematics of relativistic quantum theory to show that an electron may have a spin magnetic moment of about $\pm 9 \times 10^{-24}$ A$\cdot$m^2; the plus and minus signs indicate that alignment aiding or opposing an external magnetic field is possible. In an atom with many electrons present, only the spins of those electrons in shells which are not completely filled will contribute to a magnetic moment for the atom.

A third contribution to the moment of an atom is caused by *nuclear spin*. Although this factor provides a negligible effect on the overall magnetic properties of materials, it is the basis of the nuclear magnetic imaging procedure now provided by some of the largest hospitals.

Thus each atom contains many different component moments, and their combination determines the magnetic characteristics of the material and provides its general magnetic classification. We shall describe briefly six different types of material: diamagnetic, paramagnetic, ferromagnetic, antiferromagnetic, ferrimagnetic, and superparamagnetic.

Let us first consider those atoms in which the small magnetic fields produced by the motion of the electrons in their orbits and those produced by the electron spin combine to produce a net field of zero. Note that we are considering here the fields produced by the electron motion itself in the absence of any external magnetic field; we might also describe this material as one in which the permanent magnetic moment $\mathbf{m}_0$ of each atom is zero. Such a material is termed *diamagnetic*. It would seem, therefore, that an external magnetic field would produce no torque on the atom, no realignment of the dipole fields, and consequently an internal magnetic field that is the same as the applied field. With an error that only amounts to about one part in a hundred thousand, this is correct.

Let us select an orbiting electron whose moment $\mathbf{m}$ is in the same direction as the applied field $\mathbf{B}_0$ (Fig. 9.8). The magnetic field produces an outward force on the orbiting electron. Since the orbital radius is quantized and cannot change, the inward Coulomb force of attraction is also unchanged. The force unbalance

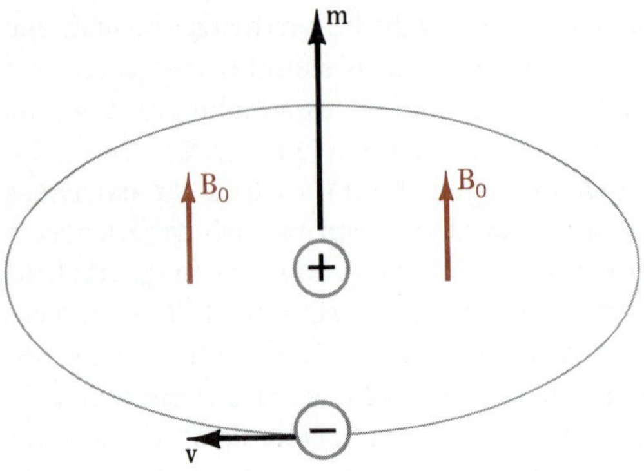

FIGURE 9.8
An orbiting electron is shown having a magnetic moment **m** in the same direction as an applied field $\mathbf{B}_0$.

created by the outward magnetic force must therefore be compensated for by a reduced orbital velocity. Hence, the orbital moment decreases, and a smaller internal field results.

If we had selected an atom for which **m** and $\mathbf{B}_0$ were opposed, the magnetic force would be inward, the velocity would increase, the orbital moment would increase, and greater cancellation of $\mathbf{B}_0$ would occur. Again a smaller internal field would result.

Metallic bismuth shows a greater diamagnetic effect than most other diamagnetic materials, among which are hydrogen, helium, the other "inert" gases, sodium chloride, copper, gold, silicon, germanium, graphite, and sulfur. We should also realize that the diamagnetic effect is present in all materials, because it arises from an interaction of the external magnetic field with every orbiting electron; however, it is overshadowed by other effects in the materials we shall consider next.

Now let us discuss an atom in which the effects of the electron spin and orbital motion do not quite cancel. The atom as a whole has a small magnetic moment, but the random orientation of the atoms in a larger sample produces an *average* magnetic moment of zero. The material shows no magnetic effects in the absence of an external field. When an external field is applied, however, there is a small torque on each atomic moment, and these moments tend to become aligned with the external field. This alignment acts to increase the value of **B** within the material over the external value. However, the diamagnetic effect is still operating on the orbiting electrons and may counteract the above increase. If the net result is a decrease in **B**, the material is still called diamagnetic. However, if there is an increase in **B**, the material is termed *paramagnetic*. Potassium, oxygen, tungsten, and the rare earth elements and many of their salts, such as erbium chloride, neodymium oxide, and yttrium oxide, one of the materials used in masers, are examples of paramagnetic substances.

The remaining four classes of material, ferromagnetic, antiferromagnetic, ferrimagnetic, and superparamagnetic, all have strong atomic moments. Moreover, the interaction of adjacent atoms causes an alignment of the magnetic moments of the atoms in either an aiding or exactly opposing manner.

In *ferromagnetic* materials each atom has a relatively large dipole moment, caused primarily by uncompensated electron spin moments. Interatomic forces cause these moments to line up in a parallel fashion over regions containing a large number of atoms. These regions are called *domains*, and they may have a variety of shapes and sizes ranging from one micrometer to several centimeters, depending on the size, shape, material, and magnetic history of the sample. Virgin ferromagnetic materials will have domains which each have a strong magnetic moment; the domain moments, however, vary in direction from domain to domain. The overall effect is therefore one of cancellation, and the material as a whole has no magnetic moment. Upon application of an external magnetic field, however, those domains which have moments in the direction of the applied field increase their size at the expense of their neighbors, and the internal magnetic field increases greatly over that of the external field alone. When the external field is removed, a completely random domain alignment is not usually attained, and a residual, or remnant, dipole field remains in the macroscopic structure. The fact that the magnetic moment of the material is different after the field has been removed, or that the magnetic state of the material is a function of its magnetic history, is called *hysteresis*, a subject which will be discussed again when magnetic circuits are studied a few pages from now.

Ferromagnetic materials are not isotropic in single crystals, and we shall therefore limit our discussion to polycrystalline materials, except for mentioning that one of the characteristics of anisotropic magnetic materials is magnetostriction, or the change in dimensions of the crystal when a magnetic field is impressed on it.

The only elements that are ferromagnetic at room temperature are iron, nickel, and cobalt, and they lose all their ferromagnetic characteristics above a temperature called the Curie temperature, which is 1043 K (770°C) for iron. Some alloys of these metals with each other and with other metals are also ferromagnetic, as for example alnico, an aluminum-nickel-cobalt alloy with a small amount of copper. At lower temperatures some of the rare earth elements, such as gadolinium and dysprosium, are ferromagnetic. It is also interesting that some alloys of nonferromagnetic metals are ferromagnetic, such as bismuth-manganese and copper-manganese-tin.

In *antiferromagnetic* materials, the forces between adjacent atoms cause the atomic moments to line up in an antiparallel fashion. The net magnetic moment is zero, and antiferromagnetic materials are affected only slightly by the presence of an external magnetic field. This effect was first discovered in manganese oxide, but several hundred antiferromagnetic materials have been identified since then. Many oxides, sulfides, and chlorides are included, such as nickel oxide (NiO), ferrous sulfide (FeS), and cobalt chloride ($CoCl_2$). Antiferromagnetism is only present at relatively low temperatures, often well below room temperature. The effect is not of engineering importance at present.

The *ferrimagnetic* substances also show an antiparallel alignment of adjacent atomic moments, but the moments are not equal. A large response to an external magnetic field therefore occurs, although not as large as that in ferromagnetic

TABLE 9.1
Characteristics of magnetic materials

Classification	Magnetic moments	*B* values	Comments
Diamagnetic	$\mathbf{m}_{orb} + \mathbf{m}_{spin} = 0$	$B_{int} < B_{appl}$	$B_{int} \doteq B_{appl}$
Paramagnetic	$\mathbf{m}_{orb} + \mathbf{m}_{spin} = \text{small}$	$B_{int} > B_{appl}$	$B_{int} \doteq B_{appl}$
Ferromagnetic	$\lvert\mathbf{m}_{spin}\rvert \gg \lvert\mathbf{m}_{orb}\rvert$	$B_{int} \gg B_{appl}$	Domains
Antiferromagnetic	$\lvert\mathbf{m}_{spin}\rvert \gg \lvert\mathbf{m}_{orb}\rvert$	$B_{int} \doteq B_{appl}$	Adjacent moments oppose
Ferrimagnetic	$\lvert\mathbf{m}_{spin}\rvert > \lvert\mathbf{m}_{orb}\rvert$	$B_{int} > B_{appl}$	Unequal adjacent moments oppose; low σ
Superparamagnetic	$\lvert\mathbf{m}_{spin}\rvert \gg \lvert\mathbf{m}_{orb}\rvert$	$B_{int} > B_{appl}$	Nonmagnetic matrix; recording tapes

materials. The most important group of ferrimagnetic materials are the *ferrites*, in which the conductivity is low, several orders of magnitude less than that of semiconductors. The fact that these substances have greater resistance than the ferromagnetic materials results in much smaller induced currents in the material when alternating fields are applied, as for example in transformer cores which operate at the higher frequencies. The reduced currents (eddy currents) lead to lower ohmic losses in the transformer core. The iron oxide magnetite (Fe_3O_4), a nickel-zinc ferrite ($Ni_{1/2}Zn_{1/2}Fe_2O_4$), and a nickel ferrite ($NiFe_2O_4$) are examples of this class of materials. Ferrimagnetism also disappears above the Curie temperature.

Superparamagnetic materials are composed of an assemblage of ferromagnetic particles in a nonferromagnetic matrix. Although domains exist within the individual particles, the domain walls cannot penetrate the intervening matrix material to the adjacent particle. An important example is the magnetic tape used in audio or video tape recorders.

Table 9.1 summarizes the characteristics of the six types of magnetic materials discussed aboved.

9.6 MAGNETIZATION AND PERMEABILITY

In order to place our description of magnetic materials on a more quantitative basis, we shall now devote a page or so to showing how the magnetic dipoles act as a distributed source for the magnetic field. Our result will be an equation that looks very much like Ampère's circuital law, $\oint \mathbf{H} \cdot d\mathbf{L} = I$. The current, however, will be the movement of *bound* charges (orbital electrons, electron spin, and nuclear spin), and the field, which has the dimensions of **H**, will be called the magnetization **M**. The current produced by the bound charges is called a *bound current*, or *Amperian current*.

Let us begin by defining the magnetization **M** in terms of the magnetic dipole moment **m**. The bound current I_b circulates about a path enclosing a differential area $d\mathbf{S}$, establishing a dipole moment,

$$\mathbf{m} = I_b\, d\mathbf{S}$$

If there are n magnetic dipoles per unit volume and we consider a volume Δv, then the total magnetic dipole moment is found by the vector sum,

$$\mathbf{m}_{\text{total}} = \sum_{i=1}^{n\Delta v} \mathbf{m}_i \tag{19}$$

Each of the $\mathbf{m}_i$ may be different. Next, we define the *magnetization* **M** as the *magnetic dipole moment per unit volume*,

$$\mathbf{M} = \lim_{\Delta v \to 0} \frac{1}{\Delta v} \sum_{i=1}^{n\Delta v} \mathbf{m}_i$$

and see that its units must be the same as for **H**, amperes per meter.

Now let us consider the effect of some alignment of the magnetic dipoles as the result of the application of a magnetic field. We shall investigate this alignment along a closed path, a short portion of which is shown in Fig. 9.9. The figure shows several magnetic moments **m** that make an angle θ with the element of path $d\mathbf{L}$; each moment consists of a bound current I_b circulating about an area $d\mathbf{S}$. We are therefore considering a small volume, $dS \cos\theta\, dL$, or $d\mathbf{S} \cdot d\mathbf{L}$, within which there are $n\, d\mathbf{S} \cdot d\mathbf{L}$ magnetic dipoles. In changing from a random orientation to this partial alignment, the bound current crossing the surface enclosed by the path (to our left as we travel in the $\mathbf{a}_L$ direction in Fig. 9.9) has increased by I_b for each of the $n\, d\mathbf{S} \cdot d\mathbf{L}$ dipoles. Thus

$$dI_b = nI_b\, d\mathbf{S} \cdot d\mathbf{L} = \mathbf{M} \cdot d\mathbf{L} \tag{20}$$

and within an entire closed contour,

$$I_b = \oint \mathbf{M} \cdot d\mathbf{L} \tag{21}$$

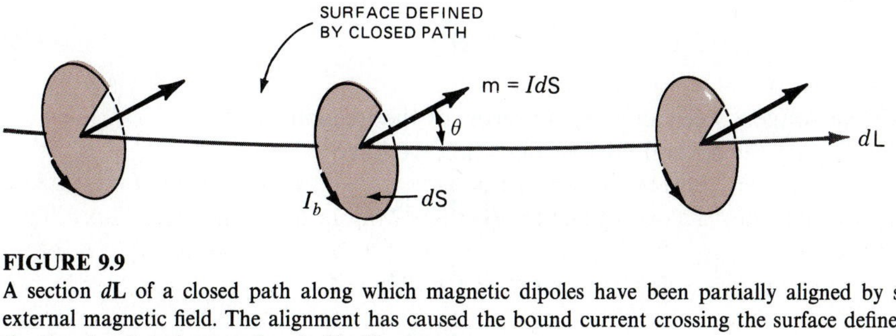

FIGURE 9.9
A section $d\mathbf{L}$ of a closed path along which magnetic dipoles have been partially aligned by some external magnetic field. The alignment has caused the bound current crossing the surface defined by the closed path to increase by $nI_b\, d\mathbf{S} \cdot d\mathbf{L}$ amperes.

Equation (21) merely says that if we go around a closed path and find dipole moments going our way more often than not, there will be a corresponding current composed of, for example, orbiting electrons crossing the interior surface.

This last expression has some resemblance to Ampère's circuital law, and we may now generalize the relationship between **B** and **H** so that it applies to media other than free space. Our present discussion is based on the forces and torques on differential current loops in a **B** field, and we therefore take **B** as our fundamental quantity and seek an improved definition of **H**. We thus write Ampère's circuital law in terms of the *total* current, bound plus free,

$$\oint \frac{\mathbf{B}}{\mu_0} \cdot d\mathbf{L} = I_T \tag{22}$$

where

$$I_T = I_b + I$$

and I is the total *free* current enclosed by the closed path. Note that the free current appears without subscript since it is the most important type of current and will be the only current appearing in Maxwell's equations.

Combining these last three equations, we obtain an expression for the free current enclosed,

$$I = I_T - I_b = \oint \left(\frac{\mathbf{B}}{\mu_0} - \mathbf{M} \right) \cdot d\mathbf{L} \tag{23}$$

We may now define **H** in terms of **B** and **M**,

$$\mathbf{H} = \frac{\mathbf{B}}{\mu_0} - \mathbf{M} \tag{24}$$

and we see that $\mathbf{B} = \mu_0 \mathbf{H}$ in free space where the magnetization is zero. This relationship is usually written in a form that avoids fractions and minus signs:

$$\boxed{\mathbf{B} = \mu_0(\mathbf{H} + \mathbf{M})} \tag{25}$$

We may now use our newly defined **H** field in (23),

$$I = \oint \mathbf{H} \cdot d\mathbf{L} \tag{26}$$

obtaining Ampère's circuital law in terms of the free currents.

Utilizing the several current densities, we have

$$I_b = \oint_S \mathbf{J}_b \cdot d\mathbf{S}$$

$$I_T = \oint_S \mathbf{J}_T \cdot d\mathbf{S}$$

$$I = \oint_S \mathbf{J} \cdot d\mathbf{S}$$

With the help of Stokes' theorem, we may therefore transform (21), (26), and (22) into the equivalent curl relationships:

$$\nabla \times \mathbf{M} = \mathbf{J}_b$$

$$\nabla \times \frac{\mathbf{B}}{\mu_0} = \mathbf{J}_T$$

$$\boxed{\nabla \times \mathbf{H} = \mathbf{J}} \tag{27}$$

We shall emphasize only (26) and (27), the two expressions involving the free charge, in the work that follows.

The relationship between **B**, **H**, and **M** expressed by (25) may be simplified for linear isotropic media where a magnetic susceptibility χ_m can be defined:

$$\boxed{\mathbf{M} = \chi_m \mathbf{H}} \tag{28}$$

Thus we have

$$\mathbf{B} = \mu_0(\mathbf{H} + \chi_m \mathbf{H})$$

$$= \mu_0 \mu_R \mathbf{H}$$

where

$$\mu_R = 1 + \chi_m \tag{29}$$

is defined as the *relative permeability* μ_R. We next define the *permeability* μ:

$$\mu = \mu_0 \mu_R \tag{30}$$

and this enables us to write the simple relationship between **B** and **H**,

$$\mathbf{B} = \mu \mathbf{H} \tag{31}$$

As an example of the use of these several magnetic quantities, let us select a ferrite material for which $\mu_R = 50$ and operate with sufficiently low flux densities that a linear relationship is reasonable. We have

$$\chi_m = \mu_R - 1 = 49$$

and if we take $B = 0.05$ T, then

$$B = \mu_R \mu_0 H$$

and

$$H = \frac{0.05}{50 \times 4\pi \times 10^{-7}} = 796 \qquad \text{A/m}$$

The magnetization is $\chi_m H$, or 39 000 A/m. The alternate ways of relating B and H are, first,

$$B = \mu_0(H + M)$$

or

$$0.05 = 4\pi \times 10^{-7}(796 + 39\,000)$$

and we see that Amperian currents produce 49 times the magnetic field intensity that the free charges do; and second,

$$B = \mu_R \mu_0 H$$

or

$$0.05 = 50 \times 4\pi \times 10^{-7} \times 796$$

where we utilize a relative permeability of 50 and let this quantity account completely for the motion of the bound charges. We shall emphasize the latter interpretation in the chapters that follow.

The first two laws that we investigated for magnetic fields were the Biot-Savart law and Ampère's circuital law. Both were restricted to free space in their application. We may now extend their use to any homogeneous, linear, isotropic magnetic material that may be described in terms of a relative permeability μ_R.

Just as we found for anisotropic dielectric materials, the permeability of an anisotropic magnetic material must be given as a 3×3 matrix, while **B** and **H** are both 3×1 matrices. We have

$$B_x = \mu_{xx}H_x + \mu_{xy}H_y + \mu_{xz}H_z$$

$$B_y = \mu_{yx}H_x + \mu_{yy}H_y + \mu_{yz}H_z$$

$$B_z = \mu_{zx}H_x + \mu_{zy}H_y + \mu_{zz}H_z$$

For anisotropic materials, then, $\mathbf{B} = \mu\mathbf{H}$ is a matrix equation; however $\mathbf{B} = \mu_0(\mathbf{H} + \mathbf{M})$ remains valid, although **B**, **H**, and **M** are no longer parallel in general. The most common anisotropic magnetic material is a single ferromagnetic crystal, although thin magnetic films also exhibit anisotropy. Most applications of ferromagnetic materials, however, involve polycrystalline arrays that are much easier to make.

Our definitions of susceptibility and permeability are also dependent upon the assumption of linearity. Unfortunately, this is true only in the less interesting paramagnetic and diamagnetic materials for which the relative permeability rarely differs from unity by more than one part in a thousand. Some typical values of the susceptibility for diamagnetic materials are hydrogen, -2×10^{-5}; copper, -0.9×10^{-5}; germanium, -0.8×10^{-5}; silicon, -0.3×10^{-5}; and graphite, -12×10^{-5}. Several representative paramagnetic susceptibilities are oxygen, 2×10^{-6}; tungsten, 6.8×10^{-5}; ferric oxide (Fe_2O_3), 1.4×10^{-3}; and yttrium oxide (Y_2O_3), 0.53×10^{-6}. If we simply take the ratio of B to $\mu_0 H$ as the relative permeability of a ferromagnetic material, typical values of μ_R would range from 10 to 100,000. Diamagnetic, paramagnetic, and antiferromagnetic materials are commonly said to be nonmagnetic.

D9.6. Find the magnetic field intensity within a magnetic material where: (*a*) $M = 150$ A/m and $\mu = 1.5 \times 10^{-5}$ H/m; (*b*) $B = 300\ \mu$T and $\chi_m = 15$; (*c*) there are 8.2×10^{28} atoms/m^3, each atom has a dipole moment of 5×10^{-27} A·m^2, and $\mu_R = 30$.

Ans. 13.72 A/m; 14.92 A/m; 14.14 A/m

D9.7. The magnetic flux density in a magnetic material with $\chi_m = 6$ is given in a certain region as $\mathbf{B} = 0.005y^2\mathbf{a}_x$ T. At $y = 0.4$ m, find the magnitude of: (*a*) $\mathbf{J}$; (*b*) $\mathbf{J}_b$; (*c*) $\mathbf{J}_T$.

Ans. 455 A/m^2; 2730 A/m^2; 3180 A/m^2

9.7 MAGNETIC BOUNDARY CONDITIONS

We should have no difficulty in arriving at the proper boundary conditions to apply to **B**, **H**, and **M** at the interface between two different magnetic materials, for we have solved similar problems for both conducting materials and dielectrics. We need no new techniques.

Figure 9.10 shows a boundary between two isotropic homogeneous linear materials with permeabilities μ_1 and μ_2. The boundary condition on the normal components is determined by allowing the surface to cut a small cylindrical gaussian surface. Applying Gauss's law for the magnetic field from Sec. 8.5,

$$\oint_S \mathbf{B} \cdot d\mathbf{S} = 0$$

we find that

$$B_{N1}\,\Delta S - B_{N2}\,\Delta S = 0$$

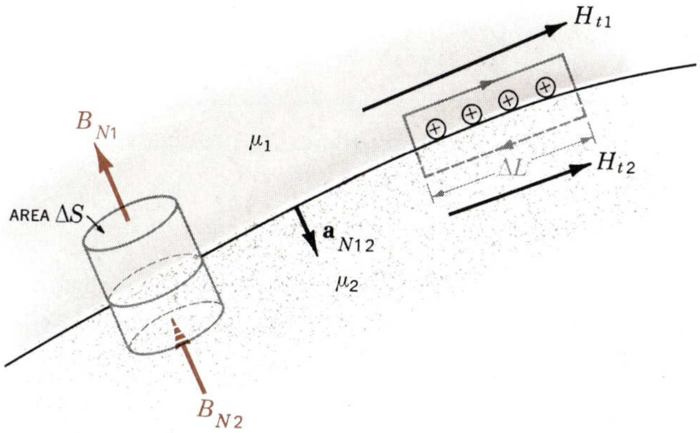

FIGURE 9.10
A gaussian surface and a closed path are constructed at the boundary between media 1 and 2, having permeabilities of μ_1 and μ_2, respectively. From this we determine the boundary conditions $B_{N1} = B_{N2}$ and $H_{t1} - H_{t2} = K$, the component of the surface current density directed into the page.

or

$$\boxed{B_{N2} = B_{N1}} \tag{32}$$

Thus

$$H_{N2} = \frac{\mu_1}{\mu_2} H_{N1} \tag{33}$$

The normal component of **B** is continuous, but the normal component of **H** is discontinuous by the ratio μ_1/μ_2.

The relationship between the normal components of **M**, of course, is fixed once the relationship between the normal components of **H** is known. For linear magnetic materials, the result is written simply as

$$M_{N2} = \frac{\chi_{m2}\mu_1}{\chi_{m1}\mu_2} M_{N1} \tag{34}$$

Next, Ampère's circuital law

$$\oint \mathbf{H} \cdot d\mathbf{L} = I$$

is applied about a small closed path in a plane normal to the boundary surface, as shown to the right in Fig. 9.10. Taking a clockwise trip around the path, we find that

$$H_{t1}\,\Delta L - H_{t2}\,\Delta L = K\,\Delta L$$

where we assume that the boundary may carry a surface current **K** whose component normal to the plane of the closed path is K. Thus

$$\boxed{H_{t1} - H_{t2} = K} \tag{35}$$

The directions are specified more exactly by using the cross product to identify the tangential components,

$$(\mathbf{H}_1 - \mathbf{H}_2) \times \mathbf{a}_{N12} = \mathbf{K}$$

where $\mathbf{a}_{N12}$ is the unit normal at the boundary directed from region 1 to region 2. An equivalent formulation in terms of the vector tangential components may be more convenient for finding **H**:

$$\mathbf{H}_{t1} - \mathbf{H}_{t2} = \mathbf{a}_{N12} \times \mathbf{K}$$

For tangential **B**, we have

$$\frac{B_{t1}}{\mu_1} - \frac{B_{t2}}{\mu_2} = K \tag{36}$$

The boundary condition on the tangential component of the magnetization for linear materials is therefore

$$M_{t2} = \frac{\chi_{m2}}{\chi_{m1}} M_{t1} - \chi_{m2} K \tag{37}$$

The last three boundary conditions on the tangential components are much simpler, of course, if the surface current density is zero. This is a free current density, and it must be zero if neither material is a conductor.

To illustrate these relationships with an example, let us assume that $\mu = \mu_1 = 4\ \mu\text{H/m}$ in region 1 where $z > 0$, while $\mu_2 = 7\ \mu\text{H/m}$ wherever $z < 0$. Moreover, let $\mathbf{K} = 80\mathbf{a}_x$ A/m on the surface $z = 0$. We establish a field, $\mathbf{B}_1 = 2\mathbf{a}_x - 3\mathbf{a}_y + \mathbf{a}_z$ mT, in region 1 and seek the value of $\mathbf{B}_2$.

The normal component of $\mathbf{B}_1$ is

$$\mathbf{B}_{N1} = (\mathbf{B}_1 \cdot \mathbf{a}_{N12})\mathbf{a}_{N12} = [(2\mathbf{a}_x - 3\mathbf{a}_y + \mathbf{a}_z) \cdot (-\mathbf{a}_z)](-\mathbf{a}_z) = \mathbf{a}_z \qquad \text{mT}$$

Thus,

$$\mathbf{B}_{N2} = \mathbf{B}_{N1} = \mathbf{a}_z \qquad \text{mT}$$

We next determine the tangential components:

$$\mathbf{B}_{t1} = \mathbf{B}_1 - \mathbf{B}_{N1} = 2\mathbf{a}_x - 3\mathbf{a}_y \qquad \text{mT}$$

and

$$\mathbf{H}_{t1} = \frac{\mathbf{B}_{t1}}{\mu_1} = \frac{(2\mathbf{a}_x - 3\mathbf{a}_y)10^{-3}}{4 \times 10^{-6}} = 500\mathbf{a}_x - 750\mathbf{a}_y \qquad \text{A/m}$$

Thus,

$$\begin{aligned}\mathbf{H}_{t2} &= \mathbf{H}_{t1} - \mathbf{a}_{N12} \times \mathbf{K} = 500\mathbf{a}_x - 750\mathbf{a}_y - (-\mathbf{a}_z) \times 80\mathbf{a}_x \\ &= 500\mathbf{a}_x - 750\mathbf{a}_y + 80\mathbf{a}_y = 500\mathbf{a}_x - 670\mathbf{a}_y \qquad \text{A/m}\end{aligned}$$

and

$$\mathbf{B}_{t2} = \mu_2 \mathbf{H}_{t2} = 7 \times 10^{-6}(500\mathbf{a}_x - 670\mathbf{a}_y) = 3.5\mathbf{a}_x - 4.69\mathbf{a}_y \qquad \text{mT}$$

Therefore,

$$\mathbf{B}_2 = \mathbf{B}_{N2} + \mathbf{B}_{t2} = 3.5\mathbf{a}_x - 4.69\mathbf{a}_y + \mathbf{a}_z \qquad \text{mT}$$

D9.8. Region 1 is the semi-infinite space in which $2x - 5y > 0$, while region 2 is defined by $2x - 5y < 0$. Let $\mu_{R1} = 3$, $\mu_{R2} = 4$, and $\mathbf{H}_1 = 30\mathbf{a}_x$ A/m. Find: (*a*) $|\mathbf{B}_1|$; (*b*) $|\mathbf{B}_{N1}|$; (*c*) $|\mathbf{H}_{t1}|$; (*d*) $|\mathbf{H}_2|$.

Ans. 113.1 μT; 42.0 μT; 27.9 A/m; 29.1 A/m

9.8 THE MAGNETIC CIRCUIT

In this section we shall digress briefly to discuss the fundamental techniques involved in solving a class of magnetic problems known as magnetic circuits. As we shall see shortly, the name arises from the great similarity to the dc-resistive-circuit analysis with which it is assumed we are all facile. The only important difference lies in the nonlinear nature of the ferromagnetic portions of the magnetic circuit; the methods which must be adopted are similar to those required in nonlinear electric circuits which contain diodes, thermistors, incandescent filaments, and other nonlinear elements.

As a convenient starting point, let us identify those field equations upon which resistive circuit analysis is based. At the same time we shall point out or derive the analogous equations for the magnetic circuit. We begin with the electrostatic potential and its relationship to electric field intensity,

$$\mathbf{E} = -\nabla V \tag{38a}$$

The scalar magnetic potential has already been defined, and its analogous relation to the magnetic field intensity is

$$\boxed{\mathbf{H} = -\nabla V_m} \tag{38b}$$

In dealing with magnetic circuits, it is convenient to call V_m the *magnetomotive force*, or mmf, and we shall acknowledge the analogy to the electromotive force, or emf, by doing so. The units of the mmf are, of course, amperes, but it is customary to recognize that coils with many turns are often employed by using the term "ampere-turns." Remember that no current may flow in any region in which V_m is defined.

The electric potential difference between points A and B may be written as

$$V_{AB} = \int_A^B \mathbf{E} \cdot d\mathbf{L} \tag{39a}$$

and the corresponding relationship between the mmf and the magnetic field intensity,

$$\boxed{V_{mAB} = \int_A^B \mathbf{H} \cdot d\mathbf{L}} \tag{39b}$$

was developed in Chap. 8, where we learned that the path selected must not cross the chosen barrier surface.

Ohm's law for the electric circuit has the point form

$$\mathbf{J} = \sigma \mathbf{E} \tag{40a}$$

and we see that the magnetic flux density will be the analog of the current density,

$$\boxed{\mathbf{B} = \mu \mathbf{H}} \tag{40b}$$

To find the total current, we must integrate:

$$I = \int_S \mathbf{J} \cdot d\mathbf{S} \tag{41a}$$

A corresponding operation is necessary to determine the total magnetic flux flowing through the cross section of a magnetic circuit:

$$\Phi = \int_S \mathbf{B} \cdot d\mathbf{S} \tag{41b}$$

We then defined resistance as the ratio of potential difference and current, or

$$V = IR \tag{42a}$$

and we shall now define *reluctance* as the ratio of the magnetomotive force to the total flux; thus

$$V_m = \Phi \mathscr{R} \tag{42b}$$

where reluctance is measured in ampere-turns per weber (A·t/Wb). In resistors which are made of linear isotropic homogeneous material of conductivity σ and have a uniform cross section of area S and length d, the total resistance is

$$R = \frac{d}{\sigma S} \tag{43a}$$

If we are fortunate enough to have such a linear isotropic homogeneous magnetic material of length d and uniform cross section S, then the total reluctance is

$$\mathscr{R} = \frac{d}{\mu S} \tag{43b}$$

The only such material to which we shall commonly apply this relationship is air.

Finally, let us consider the analog of the source voltage in an electric circuit. We know that the closed line integral of $\mathbf{E}$ is zero,

$$\oint \mathbf{E} \cdot d\mathbf{L} = 0$$

In other words, Kirchhoff's voltage law states that the rise in potential through the source is exactly equal to the fall in potential through the load. The expression for magnetic phenomena takes on a slightly different form,

$$\oint \mathbf{H} \cdot d\mathbf{L} = I_{\text{total}}$$

for the closed line integral is not zero. Since the total current linked by the path is usually obtained by allowing a current I to flow through an N-turn coil, we may express this result as

$$\oint \mathbf{H} \cdot d\mathbf{L} = NI \tag{44}$$

In an electric circuit the voltage source is a part of the closed path; in the magnetic circuit the current-carrying coil will surround or link the magnetic circuit. In tracing a magnetic circuit, we shall not be able to identify a pair of terminals at which the magnetomotive force is applied. The analogy is closer here to a pair of coupled circuits in which induced voltages exist (and in which we shall see in Chap. 10 that the closed line integral of **E** is also not zero).

Let us try out some of these ideas on a simple magnetic circuit. In order to avoid the complications of ferromagnetic materials at this time, we shall assume that we have an air-core toroid with 500 turns, a cross-sectional area of 6 cm^2, a mean radius of 15 cm, and a coil current of 4 A. As we already know, the magnetic field is confined to the interior of the toroid, and if we consider the closed path of our magnetic circuit along the mean radius, we link 2000 A·t,

$$V_{m,\,\text{source}} = 2000 \qquad \text{A}\cdot\text{t}$$

Although the field in the toroid is not quite uniform, we may assume that it is for all practical purposes and calculate the total reluctance of the circuit as

$$\mathscr{R} = \frac{d}{\mu S} = \frac{2\pi 0.15}{4\pi 10^{-7} \times 6 \times 10^{-4}}$$
$$= 1.25 \times 10^{9} \qquad \text{A}\cdot\text{t/Wb}$$

Thus

$$\Phi = \frac{V_{m,S}}{\mathscr{R}} = \frac{2000}{1.25 \times 10^{9}} = 1.6 \times 10^{-6} \qquad \text{Wb}$$

This value of the total flux is in error by less than $\frac{1}{4}$ percent, in comparison with the value obtained when the exact distribution of flux over the cross section is used.

Hence

$$B = \frac{\Phi}{S} = \frac{1.6 \times 10^{-6}}{6 \times 10^{-4}} = 2.67 \times 10^{-3} \qquad \text{T}$$

and finally,

$$H = \frac{B}{\mu} = \frac{2.67 \times 10^{-3}}{4\pi 10^{-7}} = 2120 \qquad \text{A}\cdot\text{t/m}$$

As a check, we may apply Ampère's circuital law directly in this symmetrical problem,

$$H_{\phi} 2\pi r = NI$$

and obtain

$$H_{\phi} = \frac{NI}{2\pi r} = \frac{500 \times 4}{6.28 \times 0.15} = 2120 \qquad \text{A/m}$$

at the mean radius.

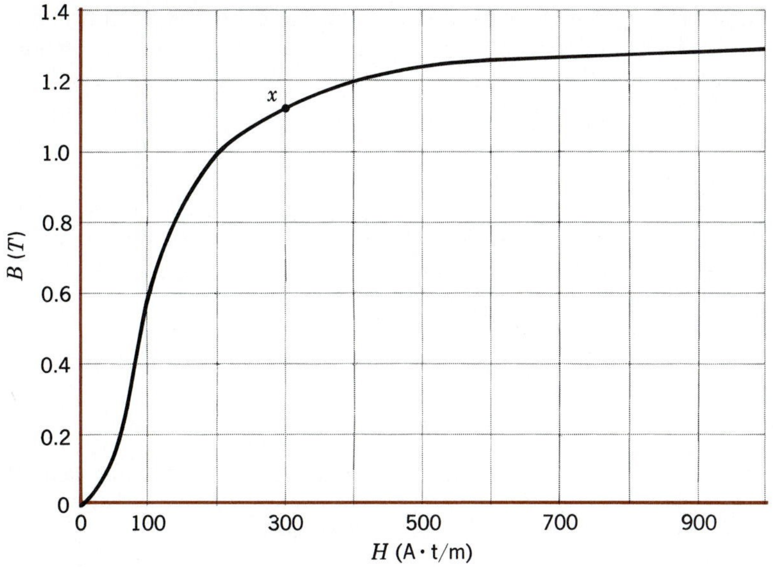

FIGURE 9.11
Magnetization curve of a sample of silicon sheet steel.

Our magnetic circuit in this example does not give us any opportunity to find the mmf across different elements in the circuit, for there is only one type of material. The analogous electric circuit is, of course, a single source and a single resistor. We could make it look just as long as the above analysis, however, if we found the current density, the electric field intensity, the total current, the resistance, and the source voltage.

More interesting and more practical problems arise when ferromagnetic materials are present in the circuit. Let us begin by considering the relationship between B and H in such a material. We may assume that we are establishing a curve of B versus H for a sample of ferromagnetic material which is completely demagnetized; both B and H are zero. As we begin to apply an mmf, the flux density also rises, but not linearly, as the experimental data of Fig. 9.11 show near the origin. After H reaches a value of about 100 A·t/m, the flux density rises more slowly and begins to saturate when H is several hundred A·t/m. Having reached partial saturation, let us now turn to Fig. 9.12, where we may continue our experiment at point x by reducing H. As we do so, the effects of hysteresis begin to show, and we do not retrace our original curve. Even after H is zero, $B = B_r$, the remnant flux density. As H is reversed, then brought back to zero, and the complete cycle traced several times, the hysteresis loop of Fig. 9.12 is obtained. The mmf required to reduce the flux density to zero is identified as H_c, the coercive "force." For smaller maximum values of H smaller hysteresis loops are obtained and the locus of the tips is about the same as the virgin magnetization curve of Fig. 9.11.

Let us make use of the magnetization curve for silicon steel to solve a magnetic circuit problem that is slightly different from our previous example. We shall

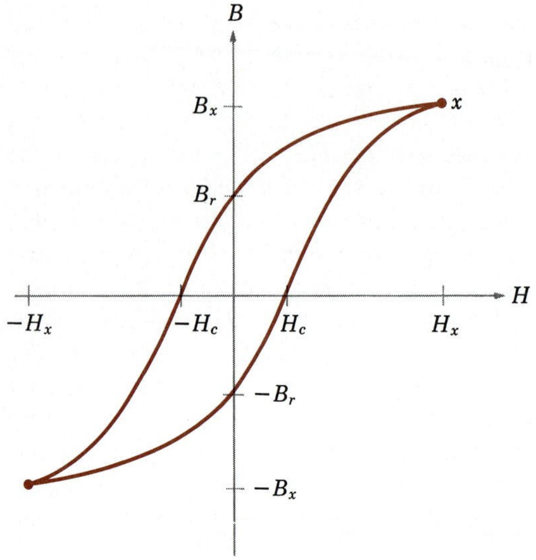

FIGURE 9.12
A hysteresis loop for silicon steel. The coercive force H_c and remnant flux density B_r are indicated.

use a steel core in the toroid, except for an air gap of 2 mm. Magnetic circuits with air gaps occur because gaps are deliberately introduced in some devices, such as inductors which must carry large direct currents, because they are unavoidable in other devices such as rotating machines, or because of unavoidable problems in assembly. There are still 500 turns about the toroid, and we ask what current is required to establish a flux density of 1 T everywhere in the core. This magnetic circuit is analogous to an electric circuit containing a voltage source and two resistors, one of which is nonlinear. Since we are given the "current," it is easy to find the "voltage" across each series element, and hence the total "emf." In the air gap,

$$\mathscr{R}_{\text{air}} = \frac{d_{\text{air}}}{\mu S} = \frac{2 \times 10^{-3}}{4\pi 10^{-7} \times 6 \times 10^{-4}} = 2.65 \times 10^6 \qquad \text{A}\cdot\text{t/Wb}$$

Knowing the total flux,

$$\Phi = BS = 1(6 \times 10^{-4}) = 6 \times 10^{-4} \qquad \text{Wb}$$

which is the same in both steel and air, we may find the mmf required for the gap,

$$V_{m,\,\text{air}} = (6 \times 10^{-4})(2.65 \times 10^6) = 1590 \qquad \text{A}\cdot\text{t}$$

Referring to Fig. 9.11, a magnetic field strength of 200 A·t/m is required to produce a flux density of 1 T in the steel. Thus

$$H_{\text{steel}} = 200 \qquad \text{A}\cdot\text{t}$$

$$V_{m,\,\text{steel}} = H_{\text{steel}} d_{\text{steel}} = 200 \times 0.30\pi$$

$$= 188 \qquad \text{A}\cdot\text{t}$$

The total mmf is therefore 1778 A·t, and a coil current of 3.56 A is required.

We should realize that we have made several approximations in obtaining this answer. We have already mentioned the lack of a completely uniform cross section, or cylindrical symmetry; the path of every flux line is not of the same length. The choice of a "mean" path length can help compensate for this error in problems in which it may be more important than it is in our example. Fringing flux in the air gap is another source of error, and formulas are available by which we may calculate an effective length and cross-sectional area for the gap which will yield more accurate results. There is also a leakage flux between the turns of wire, and in devices containing coils concentrated on one section of the core, a few flux lines bridge the interior of the toroid. Fringing and leakage are problems which seldom arise in the electric circuit because the ratio of the conductivities of air and the conductive or resistive materials used is so high. In contrast, the magnetization curve for silicon steel shows that the ratio of H to B in the steel is about 200 up to the "knee" of the magnetization curve; this compares with a ratio in air of about 800 000. Thus, although flux prefers steel to air by the commanding ratio of 4000 to 1, this is not very close to the ratio of conductivities of, say, 10^{15} for a good conductor and a fair insulator.

As a last example, let us consider the reverse problem. Given a coil current of 4 A in the previous magnetic circuit, what will the flux density be? First let us try to linearize the magnetization curve by a straight line from the origin to $B = 1$, $H = 200$. We then have $B = H/200$ in steel and $B = \mu_0 H$ in air. The two reluctances are found to be 0.314×10^6 for the steel path and 2.65×10^6 for the air gap, or 2.96×10^6 A·t/Wb total. Since V_m is 2000 A·t, the flux is 6.76×10^{-4} Wb, and $B = 1.13$ T. A more accurate solution may be obtained by assuming several values of B and calculating the necessary mmf. Plotting the results enables us to determine the true value of B by interpolation. With this method we obtain $B = 1.10$ T. The good accuracy of the linear model results from the fact that the reluctance of the air gap in a magnetic circuit is often much greater than the reluctance of the ferromagnetic portion of the circuit. A relatively poor approximation for the iron or steel can thus be tolerated.

D9.9. Given the magnetic circuit of Fig. 9.13, assume $B = 0.6$ T at the midpoint of the left leg and find: (*a*) $V_{m,\text{air}}$; (*b*) $V_{m,\text{steel}}$; (*c*) the current required in a 1250-turn coil linking the left leg.

Ans. 4300 A·t; 58 A·t; 3.48 A

D9.10. For values of B below the knee on the magnetization curve for silicon steel, approximate the curve by a straight line with $\mu = 0.005$ H/m. The core shown in Fig. 9.14 has areas of 1.5 cm^2 and lengths of 12 cm in each outer leg, and an area of 2.5 cm^2 and a length of 4 cm in the central leg. A coil of 1250 turns carrying 10 mA is placed around the central leg. Find B in the: (*a*) center leg; (*b*) left leg; (*c*) right leg, if a 0.25-mm air gap is now installed in the center and left legs.

Ans. 0.446 T; 0.372 T; 0.0774 T

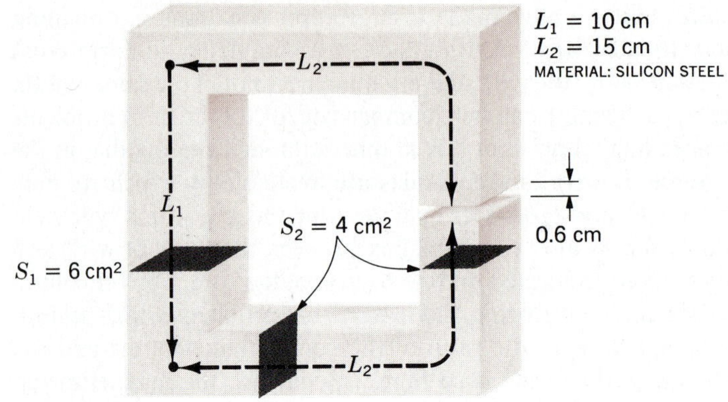

FIGURE 9.13
See Prob. D9.9

9.9 POTENTIAL ENERGY AND FORCES ON MAGNETIC MATERIALS

In the electrostatic field we first introduced the point charge and the experimental law of force between point charges. After defining electric field intensity, electric flux density, and electric potential, we were able to find an expression for the energy in an electrostatic field by establishing the work necessary to bring the prerequisite point charges from infinity to their final resting places. The general expression for energy is

$$W_E = \frac{1}{2}\int_{\text{vol}} \mathbf{D}\cdot\mathbf{E}\,dv \tag{45}$$

where a linear relationship between **D** and **E** is assumed.

This is not as easily done for the steady magnetic field. It would seem that we might assume two simple sources, perhaps two current sheets, find the force on one due to the other, move the sheet a differential distance against this force, and equate the necessary work to the change in energy. If we did, we should be

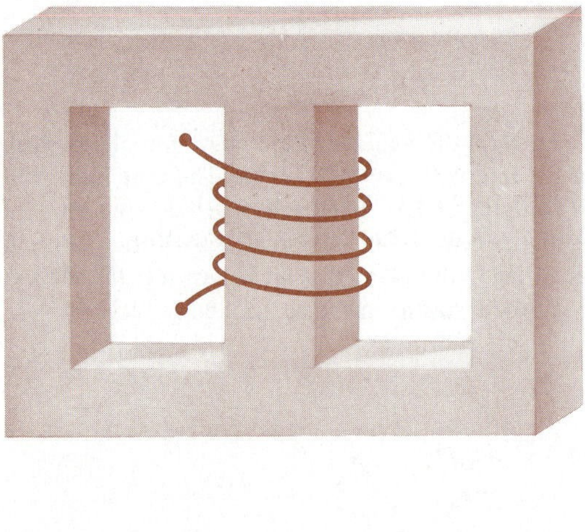

FIGURE 9.14
See Prob. D9.10

wrong, because Faraday's law (coming up in the next chapter) shows that there will be a voltage induced in the moving current sheet against which the current must be maintained. Whatever source is supplying the current sheet turns out to receive half the energy we are putting into the circuit by moving it.

In other words, energy density in the magnetic field may be determined more easily after time-varying fields are discussed. We shall develop the appropriate expression in discussing Poynting's theorem in Chap. 11.

An alternate approach would be possible at this time, however, for we might define a magnetostatic field based on assumed magnetic poles (or "magnetic charges"). Using the scalar magnetic potential, we could then develop an energy expression by methods similar to those used in obtaining the electrostatic energy relationship. These new magnetostatic quantities we should have to introduce would be too great a price to pay for one simple result, and we shall therefore merely present the result at this time and show that the same expression arises in the Poynting theorem later. The total energy stored in a steady magnetic field in which $\mathbf{B}$ is linearly related to $\mathbf{H}$ is

$$\boxed{W_H = \frac{1}{2}\int_{\text{vol}} \mathbf{B}\cdot\mathbf{H}\,dv} \tag{46}$$

Letting $\mathbf{B} = \mu\mathbf{H}$, we have the equivalent formulations

$$W_H = \frac{1}{2}\int_{\text{vol}} \mu H^2\,dv \tag{47}$$

or

$$W_H = \frac{1}{2}\int_{\text{vol}} \frac{B^2}{\mu}\,dv \tag{48}$$

It is again convenient to think of this energy as being distributed throughout the volume with an energy density of $\frac{1}{2}\mathbf{B}\cdot\mathbf{H}$ J/m^3, although we have no mathematical justification for such a statement.

In spite of the fact that these results are valid only for linear media, we may use them to calculate the forces on nonlinear magnetic materials if we focus our attention on the linear media (usually air) which may surround them. For example, suppose that we have a long solenoid with a silicon-steel core. A coil containing n turns/m with a current I surrounds it. The magnetic field intensity in the core is therefore nI A·t/m, and the magnetic flux density can be obtained from the magnetization curve for silicon steel. Let us call this value B_{st}. Suppose that the core is composed of two semi-infinite cylinders[2] which are just touching. We now apply a mechanical force to separate these two sections of the core while keeping the flux density constant. We apply a force F over a distance dL, thus doing work $F\,dL$. Faraday's law does not apply here, for the fields in the core have not changed,

[2] A semi-infinite cylinder is a cylinder of infinite length having one end located in finite space.

and we can therefore use the principle of virtual work to determine that the work we have done in moving one core appears as stored energy in the air gap we have created. By (48) above, this increase is

$$dW_H = F\,dL = \frac{1}{2}\frac{B_{st}^2}{\mu_0} S\,dL$$

where S is the core cross-sectional area. Thus

$$F = \frac{B_{st}^2 S}{2\mu_0}$$

If, for example, the magnetic field intensity is sufficient to produce saturation in the steel, approximately 1.4 T, the force is

$$F = 7.80 \times 10^5 S \qquad \text{N}$$

or about 113 lb_f/in^2.

D9.11. (*a*) What force is being exerted on the pole faces of the circuit described in Prob. D9.9 and Fig. 9.13? (*b*) Is the force trying to open or close the air gap?

Ans. 128.9 N; as Wilhelm Eduard Weber would put it, "schliessen"

9.10 INDUCTANCE AND MUTUAL INDUCTANCE

Inductance is the last of the three familiar parameters from circuit theory which we are defining in more general terms. Resistance was defined in Chap. 5 as the ratio of the potential difference between two equipotential surfaces of a conducting material to the total current crossing either equipotential surface. The resistance is a function of conductor geometry and conductivity only. Capacitance was defined in the same chapter as the ratio of the total charge on either of two equipotential conducting surfaces to the potential difference between the surfaces. Capacitance is a function only of the geometry of the two conducting surfaces and the permittivity of the dielectric medium between or surrounding them. The interpretation of resistance and capacitance as circuit elements will be inspected more closely in Sec. 13.1.

As a prelude to defining inductance, we first need to introduce the concept of flux linkage. Let us consider a toroid of N turns in which a current I produces a total flux Φ. We shall assume first that this flux links or encircles each of the N turns, and we also see that each of the N turns links the total flux Φ. The *flux linkage* $N\Phi$ is defined as the product of the number of turns N and the flux Φ linking each of them.[3] For a coil having a single turn, the flux linkage is equal to the total flux.

---

[3] The symbol λ is commonly used for flux linkages. We shall only occasionally make use of this concept, however, and we will continue to write it as $N\Phi$.

We now define *inductance* (or self-inductance) as the ratio of the total flux linkages to the current which they link,

$$L = \frac{N\Phi}{I} \tag{49}$$

The current I flowing in the N-turn coil produces the total flux Φ and $N\Phi$ flux linkages, where we assume for the moment that the flux Φ links each turn. This definition is applicable only to magnetic media which are linear, so that the flux is proportional to the current. If ferromagnetic materials are present, there is no single definition of inductance which is useful in all cases, and we shall restrict our attention to linear materials.

The unit of inductance is the henry (H), equivalent to one weber-turn per ampere.

Let us apply (49) in a straightforward way to calculate the inductance per meter length of a coaxial cable of inner radius a and outer radius b. We may take the expression for total flux developed as Eq. (42) in Chap. 8,

$$\Phi = \frac{\mu_0 I d}{2\pi} \ln \frac{b}{a}$$

and obtain the inductance rapidly for a length d,

$$L = \frac{\mu_0 d}{2\pi} \ln \frac{b}{a} \qquad \text{H}$$

or, on a per-meter basis,

$$L = \frac{\mu_0}{2\pi} \ln \frac{b}{a} \qquad \text{H/m} \tag{50}$$

In this case, $N = 1$ turn, and all the flux links all the current.

In the problem of a toroidal coil of N turns and a current I, as shown in Fig. 8.12b, we have

$$B_\phi = \frac{\mu_0 N I}{2\pi\rho}$$

If the dimensions of the cross section are small compared with the mean radius of the toroid ρ_0, then the total flux is

$$\Phi = \frac{\mu_0 N I S}{2\pi\rho_0}$$

where S is the cross-sectional area. Multiplying the total flux by N, we have the flux linkages, and dividing by I, we have the inductance

$$L = \frac{\mu_0 N^2 S}{2\pi\rho_0} \tag{51}$$

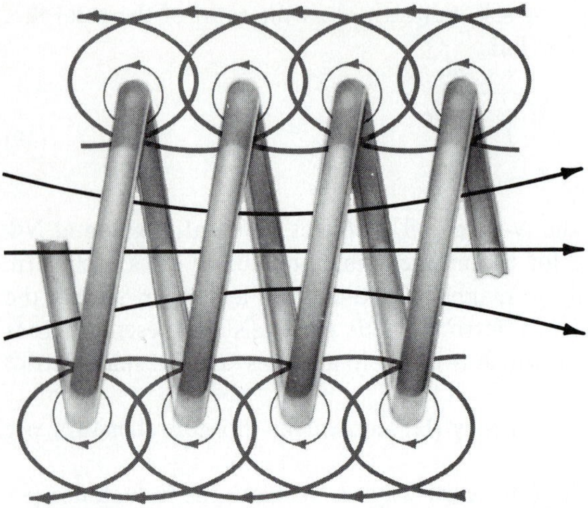

FIGURE 9.15
A portion of a coil showing partial flux linkages. The total flux linkages are obtained by adding the fluxes linking each turn.

Once again we have assumed that all the flux links all the turns, and this is a good assumption for a toroidal coil of many turns packed closely together. Suppose, however, that our toroid has an appreciable spacing between turns, a short part of which might look like Fig. 9.15. The flux linkages are no longer the product of the flux at the mean radius times the total number of turns. In order to obtain the total flux linkages we must look at the coil on a turn-by-turn basis.

$$(N\Phi)_{\text{total}} = \Phi_1 + \Phi_2 + \cdots + \Phi_i + \cdots + \Phi_N$$
$$= \sum_{i=1}^{N} \Phi_i$$

where Φ_i is the flux linking the ith turn. Rather than doing this, we usually rely on experience and empirical quantities called winding factors and pitch factors to adjust the basic formula to apply to the real physical world.

An equivalent definition for inductance may be made using an energy point of view,

$$\boxed{L = \frac{2W_H}{I^2}} \tag{52}$$

where I is the total current flowing in the closed path and W_H is the energy in the magnetic field produced by the current. After using (52) to obtain several other general expressions for inductance, we shall show that it is equivalent to (49). We first express the potential energy W_H in terms of the magnetic fields,

$$L = \frac{\int_{\text{vol}} \mathbf{B} \cdot \mathbf{H}\, dv}{I^2} \tag{53}$$

and then replace **B** by $\nabla \times \mathbf{A}$,

$$L = \frac{1}{I^2} \int_{\text{vol}} \mathbf{H} \cdot (\nabla \times \mathbf{A})\, dv$$

The vector identity

$$\nabla \cdot (\mathbf{A} \times \mathbf{H}) \equiv \mathbf{H} \cdot (\nabla \times \mathbf{A}) - \mathbf{A} \cdot (\nabla \times \mathbf{H}) \tag{54}$$

may be proved by expansion in cartesian coordinates. The inductance is then

$$L = \frac{1}{I^2} \left[\int_{\text{vol}} \nabla \cdot (\mathbf{A} \times \mathbf{H})\, dv + \int_{\text{vol}} \mathbf{A} \cdot (\nabla \times \mathbf{H})\, dv \right] \tag{55}$$

After applying the divergence theorem to the first integral and letting $\nabla \times \mathbf{H} = \mathbf{J}$ in the second integral, we have

$$L = \frac{1}{I^2} \left[\oint_S (\mathbf{A} \times \mathbf{H}) \cdot d\mathbf{S} + \int_{\text{vol}} \mathbf{A} \cdot \mathbf{J}\, dv \right]$$

The surface integral is zero, since the surface encloses the volume containing all the magnetic energy, and this requires that **A** and **H** be zero on the bounding surface. The inductance may therefore be written as

$$L = \frac{1}{I^2} \int_{\text{vol}} \mathbf{A} \cdot \mathbf{J}\, dv \tag{56}$$

Equation (56) expresses the inductance in terms of an integral of the values of **A** and **J** at every point. Since current density exists only within the conductor, the integrand is zero at all points *outside* the conductor and the vector magnetic potential need not be determined there. The vector potential is that which arises from the current **J**, and any other current source contributing a vector potential field in the region of the original current density is to be ignored for the present. Later we shall see that this leads to a *mutual inductance.*

The vector magnetic potential **A** due to **J** is given by Eq. (51), Chap. 8,

$$\mathbf{A} = \int_{\text{vol}} \frac{\mu \mathbf{J}}{4\pi R}\, dv$$

and the inductance may therefore be expressed more basically as a rather formidable double volume integral,

$$L = \frac{1}{I^2} \int_{\text{vol}} \left(\int_{\text{vol}} \frac{\mu \mathbf{J}}{4\pi R}\, dv \right) \cdot \mathbf{J}\, dv \tag{57}$$

A slightly simpler integral expression is obtained by restricting our attention to current filaments of small cross section for which $\mathbf{J}\, dv$ may be replaced by $I\, d\mathbf{L}$ and the volume integral by a closed line integral along the axis of the filament,

$$L = \frac{1}{I^2} \oint \left(\oint \frac{\mu I\, d\mathbf{L}}{4\pi R} \right) \cdot I\, d\mathbf{L}$$

$$= \frac{\mu}{4\pi} \oint \left(\oint \frac{d\mathbf{L}}{R} \right) \cdot d\mathbf{L} \tag{58}$$

Our only present interest in Eqs. (57) and (58) lies in their implication that the inductance is a function of the distribution of the current in space or the geometry of the conductor configuration.

In order to obtain our original definition of inductance (49) let us hypothesize a uniform current distribution in a filamentary conductor of small cross section so that $\mathbf{J}\,dv$ in (56) becomes $I\,d\mathbf{L}$,

$$L = \frac{1}{I} \oint \mathbf{A} \cdot d\mathbf{L} \tag{59}$$

For a small cross section, $d\mathbf{L}$ may be taken along the center of the filament. We now apply Stokes' theorem and obtain

$$L = \frac{1}{I} \int_S (\nabla \times \mathbf{A}) \cdot d\mathbf{S}$$

or

$$L = \frac{1}{I} \int_S \mathbf{B} \cdot d\mathbf{S}$$

or

$$L = \frac{\Phi}{I} \tag{60}$$

Retracing the steps by which (60) has been obtained, we should see that the flux Φ is that portion of the total flux which passes through any and every open surface whose perimeter is the filamentary current path.

If we now let the filament make N identical turns about the total flux, an idealization which may be closely realized in some types of inductors, the closed line integral must consist of N laps about this common path and (60) becomes

$$L = \frac{N\Phi}{I} \tag{61}$$

The flux Φ is now the flux crossing any surface whose perimeter is the path occupied by any *one* of the N turns. The inductance of an N-turn coil may still be obtained from (60), however, if we realize that the flux is that which crosses the complicated surface[4] whose perimeter consists of all N turns.

Use of any of the inductance expressions for a true filamentary conductor (having zero radius) leads to an infinite value of inductance, regardless of the configuration of the filament. Near the conductor Ampère's circuital law shows that the magnetic field intensity varies inversely with the distance from the conductor, and a simple integration soon shows that an infinite amount of energy and an infinite amount of flux are contained within any finite cylinder about the filament. This difficulty is eliminated by specifying a small but finite filamentary radius.

---

[4] Sort of like a spiral ramp.

The interior of any conductor also contains magnetic flux, and this flux links a variable fraction of the total current, depending on its location. These flux linkages lead to an *internal inductance*, which must be combined with the external inductance to obtain the total inductance. The internal inductance of a long straight wire of circular cross section, radius a, and uniform current distribution is

$$\boxed{L_{a,\,\text{int}} = \frac{\mu}{8\pi} \qquad \text{H/m}} \tag{62}$$

a result requested in Prob. 43 at the end of this chapter.

In Chap. 11 it will be seen that the current distribution in a conductor at high frequencies tends to be concentrated near the surface. The internal flux is reduced, and it is usually sufficient to consider only the external inductance. At lower frequencies, however, internal inductance may become an appreciable part of the total inductance.

We conclude by defining the *mutual inductance* between circuits 1 and 2, M_{12}, in terms of mutual flux linkages,

$$\boxed{M_{12} = \frac{N_2 \Phi_{12}}{I_1}} \tag{63}$$

where Φ_{12} signifies the flux produced by I_1 which links the path of the filamentary current I_2, and N_2 is the number of turns in circuit 2. The mutual inductance, therefore, depends upon the magnetic interaction between two currents. With either current alone, the total energy stored in the magnetic field can be found in terms of a single inductance, or self-inductance; with both currents having nonzero values, the total energy is a function of the two self-inductances and the mutual inductance. In terms of a mutual energy, it can be shown that (63) is equivalent to

$$M_{12} = \frac{1}{I_1 I_2} \int_{\text{vol}} (\mathbf{B}_1 \cdot \mathbf{H}_2)\, dv \tag{64}$$

or

$$M_{12} = \frac{1}{I_1 I_2} \int_{\text{vol}} (\mu \mathbf{H}_1 \cdot \mathbf{H}_2)\, dv \tag{65}$$

where $\mathbf{B}_1$ is the field resulting from I_1 (with $I_2 = 0$) and $\mathbf{H}_2$ is the field due to I_2 (with $I_1 = 0$). Interchange of the subscripts does not change the right-hand side of (65), and therefore

$$\boxed{M_{12} = M_{21}} \tag{66}$$

Mutual inductance is also measured in henrys, and we rely on the context to allow us to differentiate it from magnetization, also represented by M.

As an example of the calculation of self- and mutual-inductance, let us assume two coaxial solenoids of radius R_1 and R_2, $R_2 > R_1$, carrying currents I_1 and I_2 with n_1 and n_2 turns/m, respectively. From Eq. (15), Chap. 8, we let $n_1 = N/d$, and obtain

$$\mathbf{H}_1 = n_1 I_1 \mathbf{a}_z \qquad (0 < \rho < R_1)$$
$$= 0 \qquad (\rho > R_1)$$

and

$$\mathbf{H}_2 = n_2 I_2 \mathbf{a}_z \qquad (0 < \rho < R_2)$$
$$= 0 \qquad (\rho > R_2)$$

Thus, for this uniform field

$$\Phi_{12} = \mu_0 n_1 I_1 \pi R_1^2$$

and

$$M_{12} = \mu_0 n_1 n_2 \pi R_1^2$$

Similarly,

$$\Phi_{21} = \mu_0 n_2 I_2 \pi R_1^2$$
$$M_{21} = \mu_0 n_1 n_2 \pi R_1^2 = M_{12}$$

If $n_1 = 50$ turns/cm, $n_2 = 80$ turns/cm, $R_1 = 2$ cm, and $R_2 = 3$ cm, then

$$M_{12} = M_{21} = 63.0 \qquad \text{mH/m}$$

whereas

$$L_1 = 39.4 \qquad \text{mH/m}$$

and

$$L_2 = 227 \qquad \text{mH/m}$$

We see, therefore, that there are many methods available for the calculation of self-inductance and mutual inductance. Unfortunately, even problems possessing a high degree of symmetry present very challenging integrals for evaluation, and only a few problems are available for us to try our skill on.

Inductance will be discussed in circuit terms in Chap. 13.

D9.12. Calculate the inductance of: (*a*) a 10-m length of coaxial cable filled with a material for which $\epsilon_R = 18$, $\sigma = 0$, and $\mu_R = 80$, and having dimensions of $a = 1$ mm, $b = 4$ mm; (*b*) a toroid formed by surfaces $\rho = 3$ cm, $\rho = 5$ cm, $z = 0$, and $z = 1.5$ cm, wrapped with 5000 turns of wire, and filled with a magnetic material for which $\mu_R = 6$; (*c*) a solenoid, 8 cm in length, 2 cm in radius, having $\mu_R = 100$, and carrying 900 turns of wire.

Ans. 222 μH; 0.230 H; 1.599 H

D9.13. Each of the square coils shown in Fig. 9.16 has 1200 turns and a length of 25 cm. The dimensions of the coils are 1.2 × 1.2 cm and 3 × 3 cm. (*a*) With the inner

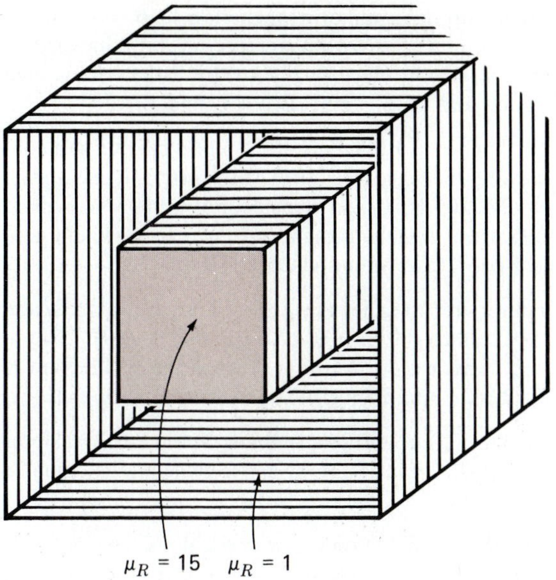

FIGURE 9.16
See Prob. D9.13.

coil and core removed, and a current of 1 mA established in the outer coil, neglect fringing and determine the magnetic field intensity within the outer coil. (*b*) Calculate the self-inductance of this coil. (*c*) With the inner coil and core centered inside the larger coil, as shown in Fig. 9.16, determine the mutual inductance between the coils.

Ans. 4.80 A/m uniform; 6.51 mH; 15.63 mH

SUGGESTED REFERENCES

1. Kraus, J. D.: (see Suggested References for Chap. 3). Examples of the calculation of inductance are given on pp. 164–169.
2. Matsch, L. W.: (see Suggested References for Chap. 5). Chapter 3 is devoted to magnetic circuits and ferromagnetic materials.
3. Paul, C. R., and S. Y. Nasar: (see Suggested References for Chap. 8). Magnetic circuits, including those with permanent magnets, are discussed on pp. 164–173.

PROBLEMS

1. A 0.2-μC point charge, moving through the field $\mathbf{E} = 20\mathbf{a}_z$ V/m, is located at the origin at $t = 0$ with a velocity $\mathbf{v} = 10^5\mathbf{a}_y$ m/s. Let $m = 2 \times 10^{-16}$ kg. Make use of Eq. (1) and Newton's laws to develop the appropriate differential equations and solve them to evaluate at $t = 2$ μs: (*a*) the position $P(x,y,z)$ of the charge; (*b*) its velocity $\mathbf{v}$; and (*c*) its kinetic energy.

2. A 0.2-μC point charge, moving through the field $\mathbf{B} = 20\mathbf{a}_z$ mT, is located at the origin at $t = 0$ with a velocity $\mathbf{v} = 10^5\mathbf{a}_y$ m/s. Let $m = 2 \times 10^{-16}$ kg. Make use of Eq. (2) and Newton's laws to develop the appropriate differential equations and solve them (perhaps with the help of an example given in Sec. 7.5) to evaluate at $t = 2$ μs: (*a*) the position $P(x,y,z)$ of the charge; (*b*) its velocity $\mathbf{v}$; (*c*) and its kinetic energy.

3. A charge, $Q = 5 \times 10^{-18}$ C, is moving through a uniform magnetic field, $\mathbf{B} = -0.4\mathbf{a}_x + 0.2\mathbf{a}_y - 0.1\mathbf{a}_z$ T, with a velocity, $\mathbf{v} = (2\mathbf{a}_x - 3\mathbf{a}_y + 6\mathbf{a}_z)10^5$ m/s, at $t = 0$. (*a*) What electric field is present at $t = 0$ if the net force on the electron is zero? (*b*) If the electric field intensity is entirely in the $\mathbf{a}_x$ direction and $|\mathbf{F}_{\text{net}}| = 2$ pN at $t = 0$, find E_x.

4. An electron ($q_e = -1.602\,18 \times 10^{-19}$ C, $m = 9.109\,39 \times 10^{-31}$ kg) is moving at a constant velocity, $\mathbf{v} = 4 \times 10^7\mathbf{a}_x$ m/s, along the negative x axis. At the origin, it encounters the uniform magnetic field, $\mathbf{B} = -2 \times 10^{-3}\mathbf{a}_y$ T, and remains in it up to $x = 3$ cm. If we assume (with good accuracy) that the electron remains on the x axis while it is in the magnetic field, find its y- and z-coordinate values when $x = 0.4$ m.

5. A current filament carrying 8 A in the $\mathbf{a}_z$ direction lies along the entire z axis in free space. A rectangular loop connecting $A(0,0.2,0)$ to $B(0,0.2,0.3)$ to $C(0,0.7,0.3)$ to $D(0,0.7,0.2)$ to A lies in the $x = 0$ plane. The loop current is 3 mA and it flows in the $\mathbf{a}_z$ direction in the AB segment. (*a*) Find $\mathbf{F}$ on side AB. (*b*) Find $\mathbf{F}$ on side DA. (*c*) Find $\mathbf{F}_{\text{total}}$ on the loop.

6. In a certain region of space, $\mathbf{B}$ is given by $0.1x\mathbf{a}_x + 0.2y\mathbf{a}_y - 0.3z\mathbf{a}_z$ T. Find the total force on the rectangular loop shown in Fig. 9.17 if it lies in the $z = 0$ plane and is bounded by $x = 1$, $x = 3$, $y = 2$, and $y = 5$ m.

7. A filamentary current of 30 A flows in the $\mathbf{a}_z$ direction in free space along the line $x = 0$, $y = 2$ m, and in the $-\mathbf{a}_z$ direction at $x = 0$, $y = -2$ m. A square filamentary loop in the $y = 0$ plane is 2 m on a side and centered on the point $C(2,0,0)$. If the loop current is 2 mA in the $\mathbf{a}_z$ direction in the side nearest the z axis, find: (*a*) the vector force on each side of the loop, and (*b*) the total vector force on the loop.

8. Filamentary currents of $-25\mathbf{a}_z$, $1\mathbf{a}_z$, and $25\mathbf{a}_z$ A are located in the $x = 0$ plane at $y = -1$, $y = k$, and $y = 1$ m, respectively. Find $\mathbf{F}$ on a 1-m length of the 1-A filament and sketch $|\mathbf{F}|$ vs. k if: (*a*) $-1 < k < 1$; (*b*) $k > 1$; (*c*) $k < -1$.

9. A cylindrical conducting shell of radius $\rho = 4$ mm and negligible thickness forms the inner conductor of a coaxial line. The outer conductor is a coaxial conducting cylinder of 20-mm radius and negligible thickness. The cylinders carry equal and opposite total currents of 100 A. What is the magnitude of the force (per unit length) acting to split the outer cylinder apart longitudinally?

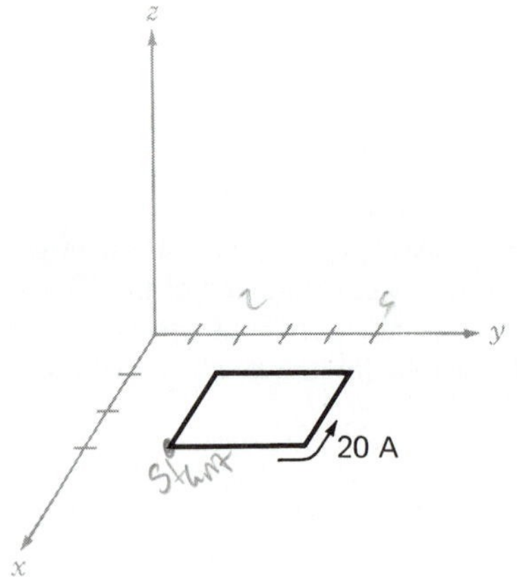

FIGURE 9.17
See Prob. 6.

10. Three infinitely long parallel filaments each carry 50 A in the $\mathbf{a}_z$ direction. If the filaments lie in the plane $x = 0$ with a 2-cm spacing between wires, find the vector force per meter on each filament.

11. (*a*) Use Eq. (14), Sec. 9.3, to show that the force of attraction per unit length between two filamentary conductors in free space with currents $I_1\mathbf{a}_z$ at $x = 0$, $y = d/2$, and $I_2\mathbf{a}_z$ at $x = 0$, $y = -d/2$, is $\mu_0 I_1 I_2/(2\pi d)$. (*b*) Show how a simpler method can be used to check your result.

12. A current filament on the z axis produces the field $\mathbf{H} = -2\mathbf{a}_x/y$ A/m at $P(0,y,z)$ in free space. (*a*) Find the force exerted on a current strip lying between $y = 1$ and 3 m in the $x = 0$ plane carrying a current density $\mathbf{K} = -2\pi\mathbf{a}_z$ A/m. (*b*) What force is exerted on the current filament by the current strip?

13. A straight solid wire segment carrying a current $2\mathbf{a}_y$ A extends from $A(0,1,2)$ to $B(0,4,2)$ in free space. This wire is subjected to the magnetic field of an infinite current filament lying along the z axis and carrying 25 A in the $\mathbf{a}_z$ direction. Find the vector torque on the wire segment about an origin at: (*a*) $P_A(0,0,2)$; (*b*) $P_B(0,0,0)$; (*c*) $P_C(0,2,0)$.

14. The rectangular loop of Prob. 6 is now subjected to the **B** field produced by two current sheets, $\mathbf{K}_1 = 400\mathbf{a}_x$ A/m at $z = 3$ and $\mathbf{K}_2 = -300\mathbf{a}_z$ A/m at $y = 0$, in free space. Find the vector torque on the loop, referred to an origin at: (*a*) (0,0,0); (*b*) the center of the loop.

15. An infinite filament carrying $12\mathbf{a}_z$ A lies on the z axis, and a straight solid conductor segment extends from $P_A(1,-l,0)$ to $P_B(1,l,0)$, carrying $2\mathbf{a}_y$ A. Find the torque on the conductor segment about an origin at $P_0(1,0,0)$ as a function of l.

16. Find the vector torque on the square wire loop shown in Fig. 9.18: (*a*) about the origin by $\mathbf{B} = 0.8\mathbf{a}_x$ T; (*b*) about the origin by $\mathbf{B} = 0.6\mathbf{a}_x - 0.8\mathbf{a}_y$ T; (*c*) about $P(2,3,4)$ by $\mathbf{B} = 0.4\mathbf{a}_x + 0.6\mathbf{a}_y - 0.7\mathbf{a}_z$ T; (*d*) about $P(2,3,4)$ by the field $\mathbf{B} = 0.4\mathbf{a}_x + 0.6\mathbf{a}_y - 0.7\mathbf{a}_z$ T acting only on that side of the loop for which $y = 1$.

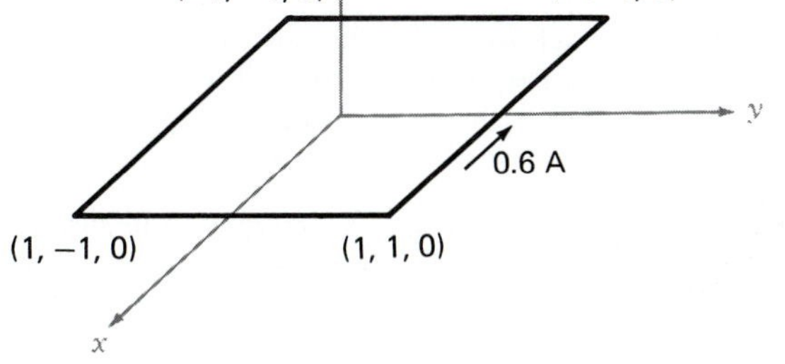

FIGURE 9.18
See Prob. 16.

17. Assume that an electron is describing a circular orbit of radius a about a positively charged nucleus. (*a*) By selecting an appropriate current and area, show that the equivalent orbital dipole moment is $ea^2\omega/2$, where ω is the electron's angular velocity. (*b*) Show that the torque produced by a magnetic field parallel to the plane of the orbit is $ea^2\omega B/2$. (*c*) By equating the Coulomb and centrifugal forces, show that ω is $(4\pi\epsilon_0 ma^3/e^2)^{-1/2}$. (*d*) Find values for the angular velocity, torque, and the orbital magnetic moment for a hydrogen atom, where a is about 6×10^{-11} m; let $B = 1$ T.

18. The hydrogen atom described in Prob. 17 is now subjected to a magnetic field having the same direction as that of the atom. Show that the forces caused by B result in a decrease of the angular velocity by $eB/(2m)$ and a decrease in the orbital moment by $e^2a^2B/(4m)$. What are these decreases for the hydrogen atom in parts per million for the relatively large magnetic flux density of 1 T?

19. Calculate **M** for a material in which: (*a*) $\mu_R = 1.4$ and $\mathbf{H} = 350\mathbf{a}_x$ A/m; (*b*) $\mu_R = 6$ and there are 2.5×10^{29} atoms/m^3, each having a magnetic dipole moment of $2.8 \times 10^{-30}\mathbf{a}_x$ A·m^2; (*c*) $\mathbf{B} = 8 \times 10^{-3}\mathbf{a}_z$ T and $\chi_m = 0.4$; (*d*) bound surface current densities of $20\mathbf{a}_z$ and $-12\mathbf{a}_z$ A/m exist at $\rho = 3$ and 5 m, respectively.

20. Find the magnitude of the magnetic flux density in a material for which: (*a*) the magnetization is 2.8 A/m and the magnetic susceptibility is 0.0025; (*b*) the magnetic field intensity is 1300 A/m and the relative permeability is 1.006; (*c*) there are 8.2×10^{28} atoms per cubic meter, each having a dipole moment of 3×10^{-30} A·m^2 in the same direction, and the magnetic susceptibility is 2×10^{-4}.

21. If $\mathbf{B} = 0.05x\mathbf{a}_y$ T in a material for which $\chi_m = 2.5$, find: (*a*) μ_R; (*b*) μ; (*c*) **H**; (*d*) **M**; (*e*) **J**; (*f*) $\mathbf{J}_b$.

22. A coax with $a = 2$ mm and $b = 9$ mm carries a 15-A current in the $\mathbf{a}_z$ direction in the center conductor. Let $\mu_{R1} = 10$, $2 < \rho < 3$ mm and $8 < \rho < 9$ mm, while $\mu_{R2} = 50$ for $3 < \rho < 8$ mm. Sketch curves of H_ϕ, B_ϕ, and M_ϕ vs. ρ.

23. Current sheets of $50\mathbf{a}_x$, $20\mathbf{a}_x$, and $-30\mathbf{a}_x$ A/m are located at $z = 0$, 2, and 5 m, respectively. If $\mu = \mu_0$ for $z < 1$ and $z > 5$, while $\mu = 8\mu_0$ for $0 < z < 2$ and $\mu = 3\mu_0$ for $2 < z < 5$, find **M** everywhere.

24. Let $\mu_R = 10$ for $\rho < 2$ cm, $\mu_R = 4$ for $2 < \rho < 3$ cm, and $\mu_R = 1$ for $\rho > 3$ cm. If $\mathbf{B} = (80/\rho)\mathbf{a}_\phi - 400\mathbf{a}_z$ μT for $\rho < 2$ cm, find **B** for: (*a*) $2 < \rho < 3$ cm; (*b*) $\rho > 3$ cm.

25. A coaxial transmission line has $a = 5$ mm and $b = 20$ mm. Let the axis of the coax be the z axis and let a direct current of 12 A flow in the $\mathbf{a}_z$ direction in the center conductor. The volume between conductors contains a magnetic material for which $\mu_R = 2.5$, as well as air. Find **H**, **B**, and **M** everywhere between conductors if the magnetic material is located where: (*a*) $a < \rho < 3a$; (*b*) $0 < \phi < \pi$.

26. In region 1, where $\mathbf{r} \cdot (2\mathbf{a}_x - 3\mathbf{a}_y + \mathbf{a}_z) > 5$, let $\mu_{R1} = 10$, and in region 2, where $\mathbf{r} \cdot (2\mathbf{a}_x - 3\mathbf{a}_y + \mathbf{a}_z) < 5$, let $\mu_{R2} = 2$. Let $\mathbf{B}_1 = 10\mathbf{a}_x + 20\mathbf{a}_y + 30\mathbf{a}_z$ mT everywhere in region 1. (*a*) Find the angle θ_1 that $\mathbf{B}_1$ makes with the unit normal $\mathbf{a}_{N21}$. (*b*) Find $\mathbf{B}_2$. (*c*) Find θ_2 between $\mathbf{B}_2$ and $\mathbf{a}_{N21}$.

27. Region 1 where $\mu_{R1} = 2$ and region 2 where $\mu_{R2} = 5$ are separated by the surface $3x - 2y + 5z = 0$. The point $P(2,2,2)$ lies in region 2. If $\mathbf{H}_1 = 4\mathbf{a}_x + 6\mathbf{a}_y - 3\mathbf{a}_z$ A/m, find $\mathbf{H}_2$.

28. In Prob. D9.10*a*, the linear approximation suggested in the statement of the problem leads to a flux density of 0.446 T in the central leg. Using this value of B and the magnetization curve for silicon steel, what current is required in the 1250-turn coil?

29. A magnetic circuit consists of an air gap that is 0.5 mm long with an effective area of 2.2 cm^2 in series with a section of silicon steel that is 10 cm in length and 2 cm^2 in area. An mmf of 400 A·t is applied. Calculate the flux density in the steel if: (*a*) the steel section is assumed to have zero reluctance; (*b*) the steel is assumed to be magnetically linear with $\mu = 0.005$ H/m; (*c*) the magnetization curve for the steel is that given in Fig. 9.11.

30. A toroid has a cross section with an area of 5 cm^2 and an effective radius of 5 cm. The coil contains 5000 turns. The core is composed of two semicircular segments, each with a circular cross section; one is silicon steel and the other is a linear material for which $\mu_R = 250$. If there is a 0.3-mm air gap at each of the two joints, (*a*) what coil current is required to produce a flux density of 1 T in the core? (*b*) What flux density is produced if the current is 0.25 A?

31. Two washerlike ferrite disks each have a 24-mm inner diameter, a 36-mm outer diameter, and a 6-mm thickness. Both materials may be assumed to be magnetically linear, one with a relative permeability of 15, and the other with $\mu_R = 30$. The disks are now laid on top of each other so that they have a common axis, and a coil of 1200 turns links both of them and produces flux in the $\mathbf{a}_\phi$ direction. If a current 10 mA exists in the coil, what is the total flux present in the core?

32. The *B-H* curve of a certain ferrous material is represented by the expression $B = (H/200)(0.2 + e^{-H/400})$, where H is in A/m and B is in T. If a magnetic circuit contains a 10-cm length of this material, as well as a 0.2-mm air gap, assume a uniform cross section of 3 cm^2 and find: (*a*) the total mmf required to produce a flux of 360 μWb; (*b*) the flux if the applied mmf is 500 A·t.

33. Determine the energy density stored in free space by the field: (*a*) $\mathbf{H}_A = 10^3\mathbf{a}_x$ A/m; (*b*) $\mathbf{H}_B = 10^3\mathbf{a}_x + 10^3\mathbf{a}_y$ A/m; (*c*) $\mathbf{H}_C = 10^3\mathbf{a}_x - 10^3\mathbf{a}_y$ A/m; (*d*) $\mathbf{H}_B + \mathbf{H}_C$; (*e*) $10^3\mathbf{a}_x$ A/m + $10^{-3}\mathbf{a}_y$ T.

34. Current sheets are located in free space as follows: $120\mathbf{a}_z$ A/m at $\rho = 5$ mm; $-14\mathbf{a}_z$ A/m at $\rho = 20$ mm; and $-8\mathbf{a}_z$ A/m at $\rho = 40$ mm. (*a*) Calculate, as a function of ρ_0, the total energy stored in the volume $0 < z < 1$ m, $0 < \rho < \rho_0$. (*b*) Make a plot of W_H vs. ρ_0, $0 < \rho_0 < 50$ mm.

35. The cones $\theta = 20°$ and $160°$ are conducting surfaces and carry total currents of 50 A, as shown in Fig. 9.19. The currents return on a spherical conducting surface of 0.2-m radius. (*a*) Find $\mathbf{H}$ in the region $0 < r < 0.2$, $20° < \theta < 160°$, $0 < \phi < 2\pi$. (*b*) How much energy is stored in this region?

36. A toroid having a square cross section is defined by the surfaces $z = 0$, $z = 20$ mm, $\rho = 30$ mm, and $\rho = 50$ mm. The surface at $\rho = 30$ mm carries a current $\mathbf{K} = 10\mathbf{a}_z$ kA/m. (*a*) Find the surface currents on the other three surfaces. (*b*) Find $\mathbf{H}$ within the toroid. (*c*) If $\mu_R = 20$ within the toroid, find the total stored energy.

37. Find the inductance of the cone-sphere configuration described in Prob. 35. The inductance is that offered at the origin between the vertices of the cone.

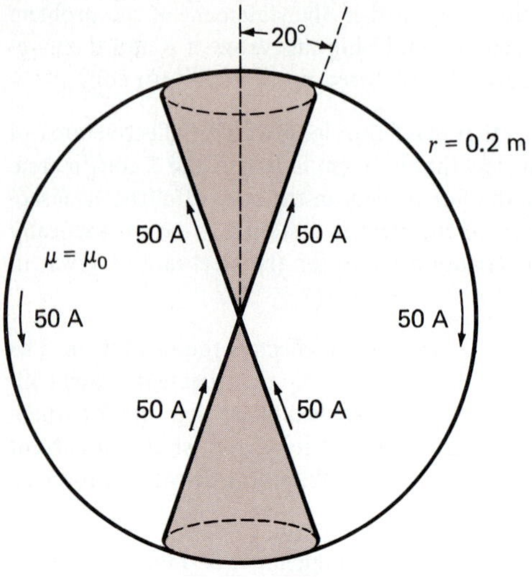

FIGURE 9.19
See Prob. 35.

38. A toroidal core is composed of a material with $\mu_R = 25$. The boundary surfaces are $z = 0$, $z = 0.025$, $\rho = 0.025$, and $\rho = 0.04$ m. The core is wound symmetrically with 12 000 turns of wire such that $\mathbf{H}$ is in the $\mathbf{a}_\phi$ direction. Find the inductance.

39. The radii of the inner and outer conductors of a coax are 0.3 cm and 2 cm. (*a*) Find the volume enclosed between the conductors in a 1-m length of coax. (*b*) Three-quarters of this volume is to have $\mu = \mu_0$, while the remaining 25 percent has $\mu_R = 10$. Show how the magnetic material should be distributed throughout the volume to obtain a maximum inductance per unit length. (*c*) Calculate $L_{\max}$.

40. Conducting planes in air at $z = 0$ and $z = d$ carry surface currents of $\pm K_0\mathbf{a}_x$ A/m. (*a*) Find the energy stored in the magnetic field per unit length $(0 < x < 1)$ in a width w $(0 < y < w)$ and separation d $(0 < z < d)$. (*b*) Calculate the inductance per unit length of this transmission line from $W_H = \frac{1}{2}LI^2$, where I is the total current in a width w in either conductor. (*c*) Calculate the total flux passing through the rectangle $0 < x < 1$, $0 < z < d$, in the plane $y = 0$, and from this result again find the inductance per unit length.

41. An infinite filament lies along the z axis in free space, and a square coil of N turns lies in the plane $y = 0$ with its corners at $(b,0,0)$, $(b + a,0,0)$, $(b + a,0,a)$, and $(b,0,a)$. Find the mutual inductance between the filament and coil in terms of a, b, N, and μ_0.

42. Find the mutual inductance of each of the following conductor systems in free space: (*a*) the solenoid of Fig. 8.11*b* and a filamentary ring of radius b coaxially centered inside the solenoid, for $b < a$; (*b*) two solenoids, each like that of Fig. 8.11*b*, formed by winding two wires simultaneously on the same core (and then removing the core); (*c*) a toroid of N turns with a square cross section of area a^2 and an average radius ρ_0, and an infinite filament on the z axis.

43. (*a*) Use energy relationships to show that the internal inductance of a nonmagnetic cylindrical wire of radius a carrying a uniformly distributed current I is $\mu_0/(8\pi)$ H/m. (*b*) Find the internal inductance if the portion of the conductor for which $\rho < c < a$ is removed.

CHAPTER 10

TIME-VARYING FIELDS AND MAXWELL'S EQUATIONS

The basic relationships of the electrostatic and the steady magnetic field have been obtained in the previous nine chapters, and we are now ready to discuss time-varying fields. The discussion will be short, for vector analysis and vector calculus should now be more familiar tools; some of the relationships are unchanged, and most of the relationships are changed only slightly.

Two new concepts will be introduced: the electric field produced by a changing magnetic field and the magnetic field produced by a changing electric field. The first of these concepts resulted from experimental research by Michael Faraday, and the second from the theoretical efforts of James Clerk Maxwell.

Maxwell actually was inspired by Faraday's experimental work and by the mental picture provided through the "lines of force" that Faraday introduced in developing his theory of electricity and magnetism. He was 40 years younger than Faraday, but they knew each other during the 5 years Maxwell spent in London as a young professor, a few years after Faraday had retired. Maxwell's theory was developed subsequent to this university position and while he was working alone at his home in Scotland. It occupied him for 5 years between the ages of 35 and 40.

The four basic equations of electromagnetic theory presented in this chapter bear his name.

10.1 FARADAY'S LAW

After Oersted[1] demonstrated in 1820 that an electric current affected a compass needle, Faraday professed his belief that if a current could produce a magnetic field, then a magnetic field should be able to produce a current. The concept of the "field" was not available at that time, and Faraday's goal was to show that a current could be produced by "magnetism."

He worked on this problem intermittently over a period of ten years, until he was finally successful in 1831.[2] He wound two separate windings on an iron toroid and placed a galvanometer in one circuit and a battery in the other. Upon closing the battery circuit, he noted a momentary deflection of the galvanometer; a similar deflection in the opposite direction occurred when the battery was disconnected. This, of course, was the first experiment he had made involving a *changing* magnetic field, and he followed it with a demonstration that either a *moving* magnetic field or a moving coil could also produce a galvanometer deflection.

In terms of fields, we now say that a time-varying magnetic field produces an *electromotive force* (emf) which may establish a current in a suitable closed circuit. An electromotive force is merely a voltage that arises from conductors moving in a magnetic field or from changing magnetic fields, and we shall define it below. Faraday's law is customarily stated as

$$\boxed{\text{emf} = -\frac{d\Phi}{dt} \quad \text{V}} \tag{1}$$

Equation (1) implies a closed path, although not necessarily a closed conducting path; the closed path, for example, might include a capacitor, or it might be a purely imaginary line in space. The magnetic flux is that flux which passes through any and every surface whose perimeter is the closed path, and $d\Phi/dt$ is the time rate of change of this flux.

A nonzero value of $d\Phi/dt$ may result from any of the following situations:

1. A time-changing flux linking a stationary closed path
2. Relative motion between a steady flux and a closed path
3. A combination of the two

The minus sign is an indication that the emf is in such a direction as to produce a current whose flux, if added to the original flux, would reduce the magnitude of the emf. This statement that the induced voltage acts to produce an opposing flux is known as *Lenz's law*.[3]

---

[1] Hans Christian Oersted was Professor of Physics at the University of Copenhagen in Denmark.

[2] Joseph Henry produced similar results at Albany Academy in New York at about the same time.

[3] Henri Frederic Emile Lenz was born in Germany, but worked in Russia. He published his law in 1834.

If the closed path is that taken by an N-turn filamentary conductor, it is often sufficiently accurate to consider the turns as coincident and let

$$\text{emf} = -N\frac{d\Phi}{dt} \tag{2}$$

where Φ is now interpreted as the flux passiing through any one of N coincident paths.

We need to define emf as used in (1) or (2). The emf is obviously a scalar, and (perhaps not so obviously) a dimensional check shows that it is measured in volts. We define the emf as

$$\text{emf} = \oint \mathbf{E} \cdot d\mathbf{L} \tag{3}$$

and note that it is the voltage about a specific *closed path.* If any part of the path is changed, the emf in general changes. The departure from static results is clearly shown by (3), for an electric field intensity resulting from a static charge distribution must lead to zero potential difference about a closed path. In electrostatics, the line integral leads to a potential difference; with time-varying fields, the result is an emf or a voltage.

Replacing Φ in (1) by the surface integral of $\mathbf{B}$, we have

$$\text{emf} = \oint \mathbf{E} \cdot d\mathbf{L} = -\frac{d}{dt}\int_S \mathbf{B} \cdot d\mathbf{S} \tag{4}$$

where the fingers of our right hand indicate the direction of the closed path, and our thumb indicates the direction of $d\mathbf{S}$. A flux density $\mathbf{B}$ in the direction of $d\mathbf{S}$ and increasing with time thus produces an average value of $\mathbf{E}$ which is *opposite* to the positive direction about the closed path. The right-handed relationship between the surface integral and the closed line integral in (4) should always be kept in mind during flux integrations and emf determinations.

Let us divide our investigation into two parts by first finding the contribution to the total emf made by a changing field within a stationary path (transformer emf), and then we will consider a moving path within a constant field (motional, or generator, emf).

We first consider a stationary path. The magnetic flux is the only time-varying quantity on the right side of (4), and a partial derivative may be taken under the integral sign,

$$\text{emf} = \oint \mathbf{E} \cdot d\mathbf{L} = -\int_S \frac{\partial \mathbf{B}}{\partial t} \cdot d\mathbf{S} \tag{5}$$

Before we apply this simple result to an example, let us obtain the point form of this integral equation. Applying Stokes' theorem to the closed line integral, we have

$$\int_S (\nabla \times \mathbf{E}) \cdot d\mathbf{S} = -\int_S \frac{\partial \mathbf{B}}{\partial t} \cdot d\mathbf{S}$$

where the surface integrals may be taken over identical surfaces. The surfaces are perfectly general and may be chosen as differentials,

$$(\nabla \times \mathbf{E}) \cdot d\mathbf{S} = -\frac{\partial \mathbf{B}}{\partial t} \cdot d\mathbf{S}$$

and

$$\boxed{\nabla \times \mathbf{E} = -\frac{\partial \mathbf{B}}{\partial t}} \tag{6}$$

This is one of Maxwell's four equations as written in differential, or point, form, the form in which they are most generally used. Equation (5) is the integral form of this equation and is equivalent to Faraday's law as applied to a fixed path. If **B** is not a function of time, (5) and (6) evidently reduce to the electrostatic equations,

$$\oint \mathbf{E} \cdot d\mathbf{L} = 0 \qquad \text{(electrostatics)}$$

and

$$\nabla \times \mathbf{E} = 0 \qquad \text{(electrostatics)}$$

As an example of the interpretation of (5) and (6), let us assume a simple magnetic field which increases exponentially with time within the cylindrical region $\rho < b$,

$$\mathbf{B} = B_0 e^{kt} \mathbf{a}_z \tag{7}$$

where $B_0 = \text{constant}$. Choosing the circular path $\rho = a$, $a < b$, in the $z = 0$ plane, along which E_ϕ must be constant by symmetry, we then have from (5)

$$\text{emf} = 2\pi a E_\phi = -kB_0 e^{kt} \pi a^2$$

The emf around this closed path is $-kB_0 e^{kt} \pi a^2$. It is proportional to a^2, since the magnetic flux density is uniform and the flux passing through the surface at any instant is proportional to the area.

If we now replace a by ρ, $\rho < b$, the electric field intensity at any point is

$$\mathbf{E} = -\tfrac{1}{2} k B_0 e^{kt} \rho \mathbf{a}_\phi \tag{8}$$

Let us now attempt to obtain the same answer from (6), which becomes

$$(\nabla \times \mathbf{E})_z = -kB_0 e^{kt} = \frac{1}{\rho} \frac{\partial (\rho E_\phi)}{\partial \rho}$$

Multiplying by ρ and integrating from 0 to ρ (treating t as a constant, since the derivative is a partial derivative),

$$-\tfrac{1}{2}kB_0e^{kt}\rho^2 = \rho E_\phi$$

or

$$\mathbf{E} = -\tfrac{1}{2}kB_0e^{kt}\rho\mathbf{a}_\phi$$

once again.

If B_0 is considered positive, a filamentary conductor of resistance R would have a current flowing in the negative $\mathbf{a}_\phi$ direction, and this current would establish a flux within the circular loop in the negative $\mathbf{a}_z$ direction. Since E_ϕ increases exponentially with time, the current and flux do also, and thus tend to reduce the time rate of increase of the applied flux and the resultant emf in accordance with Lenz's law.

Before leaving this example, it is well to point out that the given field $\mathbf{B}$ does not satisfy all of Maxwell's equations. Such fields are often assumed (*always* in ac-circuit problems) and cause no difficulty when they are interpreted properly. They occasionally cause surprise, however. This particular field is discussed further in Prob. 20 at the end of the chapter.

Now let us consider the case of a time-constant flux and a moving closed path. Before we derive any special results from Faraday's law (1), let us use the basic law to analyze the specific problem outlined in Fig. 10.1. The closed circuit consists of two parallel conductors which are connected at one end by a high-resistance voltmeter of negligible dimensions and at the other end by a sliding bar moving at a velocity $\mathbf{v}$. The magnetic flux density is constant (in space and time) and is normal to the plane containing the closed path.

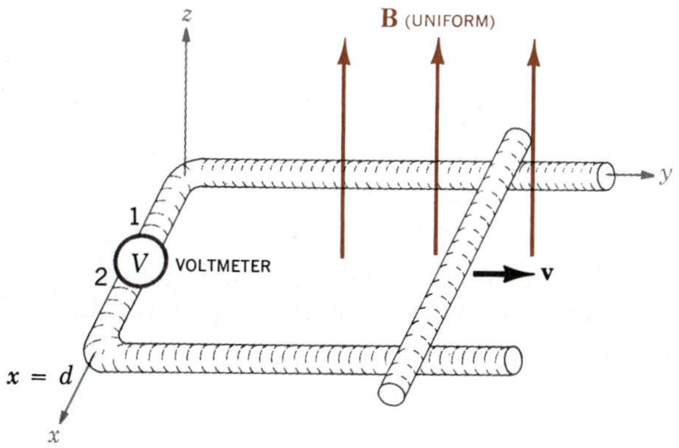

FIGURE 10.1
An example illustrating the application of Faraday's law to the case of a constant magnetic flux density and a moving path. The shorting bar moves to the right with a velocity $\mathbf{v}$, and the circuit is completed through the two rails and an extremely small high-resistance voltmeter. The voltmeter reading is $V_{12} = -Bvd$.

Let the position of the shorting bar be given by y; the flux passing through the surface within the closed path at any time t is then

$$\Phi = Byd$$

From (1), we obtain

$$\text{emf} = -\frac{d\Phi}{dt} = -B\frac{dy}{dt}d = -Bvd \tag{9}$$

Since the emf is defined as $\oint \mathbf{E} \cdot d\mathbf{L}$ and we have a conducting path, we may actually determine $\mathbf{E}$ at every point along the closed path. We found in electrostatics that the tangential component of $\mathbf{E}$ is zero at the surface of a conductor, and we shall show in Sec. 10.4 that the tangential component is zero at the surface of a *perfect* conductor ($\sigma = \infty$) for all time-varying conditions. This is equivalent to saying that a perfect conductor is a "short circuit." The entire closed path in Fig. 10.1 may be considered as a perfect conductor, with the exception of the voltmeter. The actual computation of $\oint \mathbf{E} \cdot d\mathbf{L}$ then must involve no contribution along the entire moving bar, both rails, and the voltmeter leads. Since we are integrating in a counterclockwise direction (keeping the interior of the positive side of the surface on our left as usual), the contribution $E\,\Delta L$ across the voltmeter must be $-Bvd$, showing that the electric field intensity in the instrument is directed from terminal 2 to terminal 1. For an up-scale reading, the positive terminal of the voltmeter should therefore be terminal 2.

The direction of the resultant small current flow may be confirmed by noting that the enclosed flux is reduced by a clockwise current in accordance with Lenz's law. The voltmeter terminal 2 is again seen to be the positive terminal.

Let us now consider this example using the concept of *motional emf*. The force on a charge Q moving at a velocity $\mathbf{v}$ in a magnetic field $\mathbf{B}$ is

$$\boxed{\mathbf{F} = Q\mathbf{v} \times \mathbf{B}}$$

or

$$\frac{\mathbf{F}}{Q} = \mathbf{v} \times \mathbf{B} \tag{10}$$

The sliding conducting bar is composed of positive and negative charges, and each experiences this force. The force per unit charge, as given by (10), is called the *motional* electric field intensity $\mathbf{E}_m$,

$$\boxed{\mathbf{E}_m = \mathbf{v} \times \mathbf{B}} \tag{11}$$

If the moving conductor were lifted off the rails, this electric field intensity would force electrons to one end of the bar (the far end) until the *static field* due to these charges just balanced the field induced by the motion of the bar. The resultant tangential electric field intensity would then be zero along the length of the bar.

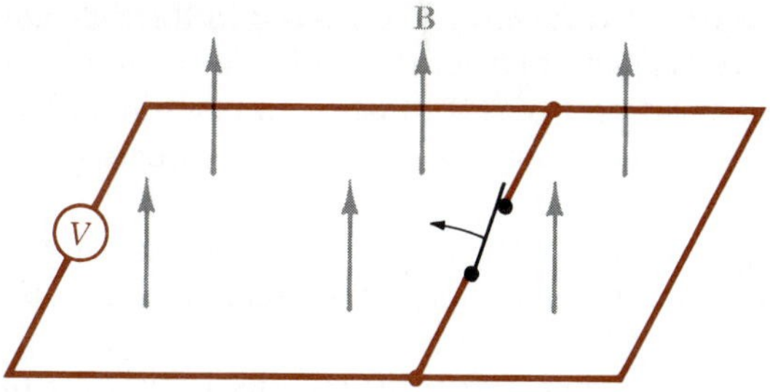

FIGURE 10.2
An apparent increase in flux linkages does not lead to an induced voltage when one part of a circuit is simply substituted for another by opening the switch. No indication will be observed on the voltmeter.

The motional emf produced by the moving conductor is then

$$\text{emf} = \oint \mathbf{E}_m \cdot d\mathbf{L} = \oint (\mathbf{v} \times \mathbf{B}) \cdot d\mathbf{L} \tag{12}$$

where the last integral may have a nonzero value only along that portion of the path which is in motion, or along which $\mathbf{v}$ has some nonzero value. Evaluating the right side of (12), we obtain

$$\oint (\mathbf{v} \times \mathbf{B}) \cdot d\mathbf{L} = \int_d^0 vB\,dx = -Bvd$$

as before. This is the total emf, since $\mathbf{B}$ is not a function of time.

In the case of a conductor moving in a uniform constant magnetic field, we may therefore ascribe a motional electric field intensity $\mathbf{E}_m = \mathbf{v} \times \mathbf{B}$ to every portion of the moving conductor and evaluate the resultant emf by

$$\text{emf} = \oint \mathbf{E} \cdot d\mathbf{L} = \oint \mathbf{E}_m \cdot d\mathbf{L} = \oint (\mathbf{v} \times \mathbf{B}) \cdot d\mathbf{L} \tag{13}$$

If the magnetic flux density is also changing with time, then we must include both contributions, the transformer emf (5) and the motional emf (12),

$$\text{emf} = \oint \mathbf{E} \cdot d\mathbf{L} = -\int_S \frac{\partial \mathbf{B}}{\partial t} \cdot d\mathbf{S} + \oint (\mathbf{v} \times \mathbf{B}) \cdot d\mathbf{L} \tag{14}$$

This expression is equivalent to the simple statement

$$\text{emf} = -\frac{d\Phi}{dt} \tag{1}$$

and either can be used to determine these induced voltages.

Although (1) appears simple, there are a few contrived examples in which its proper application is quite difficult. These usually involve sliding contacts or switches; they always involve the substitution of one part of a circuit by a new part.[4] As an example, consider the simple circuit of Fig. 10.2, containing several perfectly conducting wires, an ideal voltmeter, a uniform constant field $\mathbf{B}$, and a switch. When the switch is opened, there is obviously more flux enclosed in the

---

[4] See Bewley, in Suggested References at the end of the chapter, particularly pp. 12–19.

voltmeter circuit; however, it continues to read zero. The change in flux has not been produced by either a time-changing **B** [first term of (14)] or a conductor moving through a magnetic field [second part of (14)]. Instead, a new circuit has been substituted for the old. Thus it is necessary to use care in evaluating the change in flux linkages.

The separation of the emf into the two parts indicated by (14), one due to the time rate of change of **B** and the other to the motion of the circuit, is somewhat arbitrary in that it is dependent on the relative velocity of the *observer* and the system. A field that is changing with both time and space may look constant to an observer moving with the field. This line of reasoning is developed more fully in applying the special theory of relativity to electromagnetic theory.[5]

D10.1. Given the magnetic flux density, $\mathbf{B} = 6 \cos 10^6 t \sin 0.01x \, \mathbf{a}_z$ mT, find: (*a*) the magnetic flux passing through the surface $z = 0$, $0 < x < 20$ m, $0 < y < 3$ m, at $t = 1\ \mu$s; (*b*) the value of the closed line integral of **E** around the perimeter of the surface specified above, at $t = 1\ \mu$s.

Ans. 19.39 mWb; 30.2 kV

D10.2. Parameters for the device shown in Fig. 10.1 are $d = 5$ cm, $\mathbf{B} = 0.25\mathbf{a}_z$ T, and $\mathbf{v} = 20\sqrt{y}\,\mathbf{a}_y$ m/s. Let $y = 4$ cm at $t = 0$ and find values at $t = 0.06$ s for: (*a*) v_y; (*b*) y; (*c*) V_{12}; (*d*) the current entering terminal 1 of the voltmeter if the meter resistance is 200 kΩ.

Ans. 16 m/s; 64 cm; −0.2 V; −1 μA

10.2 DISPLACEMENT CURRENT

Faraday's experimental law has been used to obtain one of Maxwell's equations in differential form,

$$\nabla \times \mathbf{E} = -\frac{\partial \mathbf{B}}{\partial t} \tag{15}$$

which shows us that a time-changing magnetic field produces an electric field. Remembering the definition of curl, we see that this electric field has the special property of circulation; its line integral about a general closed path is not zero. Now let us turn our attention to the time-changing electric field.

We should first look at the point form of Ampère's circuital law as it applies to steady magnetic fields,

$$\nabla \times \mathbf{H} = \mathbf{J} \tag{16}$$

and show its inadequacy for time-varying conditions by taking the divergence of each side,

$$\nabla \cdot \nabla \times \mathbf{H} \equiv 0 = \nabla \cdot \mathbf{J}$$

---

[5] This is discussed in several of the references listed in the Suggested References at the end of the chapter. See Panofsky and Philips, pp. 142–151; Owen, pp. 231–245; and Harman in several places.

Since the divergence of the curl is identically zero, $\nabla \cdot \mathbf{J}$ is also zero. However, the equation of continuity,

$$\nabla \cdot \mathbf{J} = -\frac{\partial \rho_v}{\partial t}$$

then shows us that (16) can be true only if $\partial \rho_v / \partial t = 0$. This is an unrealistic limitation, and (16) must be amended before we can accept it for time-varying fields. Suppose we add an unknown term $\mathbf{G}$ to (16),

$$\nabla \times \mathbf{H} = \mathbf{J} + \mathbf{G}$$

Again taking the divergence, we have

$$0 = \nabla \cdot \mathbf{J} + \nabla \cdot \mathbf{G}$$

Thus

$$\nabla \cdot \mathbf{G} = \frac{\partial \rho_v}{\partial t}$$

Replacing ρ_v by $\nabla \cdot \mathbf{D}$,

$$\nabla \cdot \mathbf{G} = \frac{\partial}{\partial t}(\nabla \cdot \mathbf{D}) = \nabla \cdot \frac{\partial \mathbf{D}}{\partial t}$$

from which we obtain the simplest solution for $\mathbf{G}$,

$$\mathbf{G} = \frac{\partial \mathbf{D}}{\partial t}$$

Ampère's circuital law in point form therefore becomes

$$\boxed{\nabla \times \mathbf{H} = \mathbf{J} + \frac{\partial \mathbf{D}}{\partial t}} \tag{17}$$

Equation (17) has not been derived. It is merely a form we have obtained which does not disagree with the continuity equation. It is also consistent with all our other results, and we accept it as we did each experimental law and the equations derived from it. We are building a theory, and we have every right to our equations *until they are proved wrong*. This has not yet been done.

We now have a second one of Maxwell's equations and shall investigate its significance. The additional term $\partial \mathbf{D}/\partial t$ has the dimensions of current density, amperes per square meter. Since it results from a time-varying electric flux density (or displacement density), Maxwell termed it a *displacement current density*. We sometimes denote it by $\mathbf{J}_d$:

$$\nabla \times \mathbf{H} = \mathbf{J} + \mathbf{J}_d$$

$$\mathbf{J}_d = \frac{\partial \mathbf{D}}{\partial t}$$

This is the third type of current density we have met. Conduction current density,

$$\mathbf{J} = \sigma \mathbf{E}$$

is the motion of charge (usually electrons) in a region of zero net charge density, and convection current density,

$$\mathbf{J} = \rho_v \mathbf{v}$$

is the motion of volume charge density. Both are represented by **J** in (17). Bound current density is, of course, included in **H**. In a nonconducting medium in which no volume charge density is present, $\mathbf{J} = 0$, and then

$$\nabla \times \mathbf{H} = \frac{\partial \mathbf{D}}{\partial t} \qquad (\text{if } \mathbf{J} = 0) \tag{18}$$

Notice the symmetry between (18) and (15):

$$\nabla \times \mathbf{E} = -\frac{\partial \mathbf{B}}{\partial t} \tag{15}$$

Again the analogy between the intensity vectors **E** and **H** and the flux density vectors **D** and **B** is apparent. Too much faith cannot be placed in this analogy, however, for it fails when we investigate forces on particles. The force on a charge is related to **E** and to **B**, and some good arguments may be presented showing an analogy between **E** and **B** and between **D** and **H**. We shall omit them, however, and merely say that the concept of displacement current was probably suggested to Maxwell by the symmetry first mentioned above.[6]

The total displacement current crossing any given surface is expressed by the surface integral,

$$I_d = \int_S \mathbf{J}_d \cdot d\mathbf{S} = \int_S \frac{\partial \mathbf{D}}{\partial t} \cdot d\mathbf{S}$$

and we may obtain the time-varying version of Ampère's circuital law by integrating (17) over the surface S,

$$\int_S (\nabla \times \mathbf{H}) \cdot d\mathbf{S} = \int_S \mathbf{J} \cdot d\mathbf{S} + \int_S \frac{\partial \mathbf{D}}{\partial t} \cdot d\mathbf{S}$$

and applying Stokes' theorem,

$$\boxed{\oint \mathbf{H} \cdot d\mathbf{L} = I + I_d = I + \int_S \frac{\partial \mathbf{D}}{\partial t} \cdot d\mathbf{S}} \tag{19}$$

---

[6] The analogy that relates **B** to **D** and **H** to **E** is strongly advocated by Fano, Chu, and Adler (see Suggested References for Chap. 5) on pp. 159–160 and 179; the case for comparing **B** to **E** and **D** to **H** is presented in Halliday and Resnick (see Suggested References for this chapter) on pp. 665–668 and 832–836.

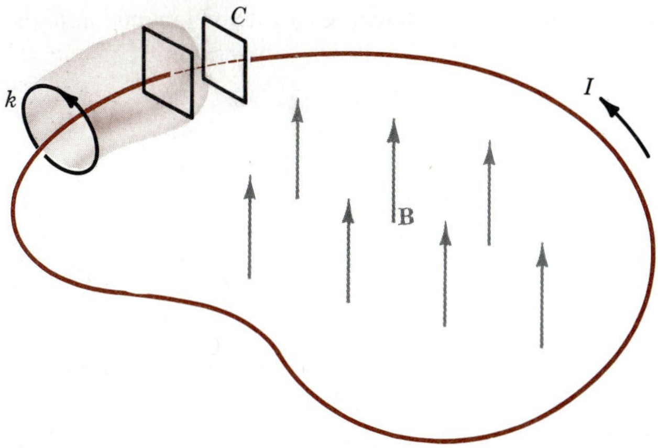

FIGURE 10.3
A filamentary conductor forms a loop connecting the two plates of a parallel-plate capacitor. A time-varying magnetic field inside the closed path produces an emf of $V_0 \cos \omega t$ around the closed path. The conduction current I is equal to the displacement current between the capacitor plates.

What is the nature of displacement current density? Let us study the simple circuit of Fig. 10.3, containing a filamentary loop and a parallel-plate capacitor. Within the loop a magnetic field varying sinusoidally with time is applied to produce an emf about the closed path (the filament plus the dashed portion between the capacitor plates) which we shall take as

$$\text{emf} = V_0 \cos \omega t$$

Using elementary circuit theory and assuming the loop has negligible resistance and inductance, we may obtain the current in the loop as

$$I = -\omega C V_0 \sin \omega t$$

$$= -\omega \frac{\epsilon S}{d} V_0 \sin \omega t$$

where the quantities ϵ, S, and d pertain to the capacitor. Let us apply Ampère's circuital law about the smaller closed circular path k and neglect displacement current for the moment:

$$\oint_k \mathbf{H} \cdot d\mathbf{L} = I_k$$

The path and the value of $\mathbf{H}$ along the path are both definite quantities (although difficult to determine), and $\oint_k \mathbf{H} \cdot d\mathbf{L}$ is a definite quantity. The current I_k is that current through every surface whose perimeter is the path k. If we choose a simple surface punctured by the filament, such as the plane circular surface defined by the circular path k, the current is evidently the conduction current. Suppose now we consider the closed path k as the mouth of a paper bag whose bottom passes between the capacitor plates. The bag is not pierced by the filament, and the conduction current is zero. Now we need to consider displacement current, for within the capacitor

$$D = \epsilon E = \epsilon\left(\frac{V_0}{d}\cos \omega t\right)$$

and therefore

$$I_d = -\omega \frac{\epsilon S}{d} V_0 \sin \omega t$$

This is the same value as that of the conduction current in the filamentary loop. Therefore the application of Ampère's circuital law including displacement current to the path k leads to a definite value for the line integral of **H**. This value must be equal to the total current crossing the chosen surface. For some surfaces the current is almost entirely conduction current, but for those surfaces passing between the capacitor plates, the conduction current is zero, and it is the displacement current which is now equal to the closed line integral of **H**.

Physically, we should note that a capacitor stores charge and that the electric field between the capacitor plates is much greater than the small leakage fields outside. We therefore introduce little error when we neglect displacement current on all those surfaces which do not pass between the plates.

Displacement current is associated with time-varying electric fields and therefore exists in all imperfect conductors carrying a time-varying conduction current. The last part of the drill problem below indicates the reason why this additional current was never discovered experimentally. This comparison is illustrated further in Sec. 11.3.

D10.3. Find the amplitude of the displacement current density: (*a*) in the air near a car antenna, where the field strength of an FM signal is $\mathbf{E} = 80\cos(6.277 \times 10^8 t - 2.092y)\,\mathbf{a}_z$ V/m; (*b*) in an air space within a large power transformer where $\mathbf{H} = 10^6 \cos(377t + 1.2566 \times 10^{-6} z)\,\mathbf{a}_y$ A/m; (*c*) inside a capacitor where $\epsilon_R = 600$ and $\mathbf{D} = 3 \times 10^{-6} \sin(6 \times 10^6 t - 0.3464x)\,\mathbf{a}_z$ C/m^2; (*d*) inside a typical metallic conductor where $f = 1$ kHz, $\sigma = 5 \times 10^7$ ℧/m, $\epsilon_R = 1$; and the conduction current density is $\mathbf{J} = 10^7 \sin(6283t - 444z)\,\mathbf{a}_x$ A/m^2.

Ans. 0.445 A/m^2; 1.257 A/m^2; 18.00 A/m^2; 11.13 nA/m^2

10.3 MAXWELL'S EQUATIONS IN POINT FORM

We have already obtained two of Maxwell's equations for time-varying fields,

$$\boxed{\nabla \times \mathbf{E} = -\frac{\partial \mathbf{B}}{\partial t}} \qquad (20)$$

and

$$\nabla \times \mathbf{H} = \mathbf{J} + \frac{\partial \mathbf{D}}{\partial t} \tag{21}$$

The remaining two equations are unchanged from their non-time-varying form:

$$\nabla \cdot \mathbf{D} = \rho_v \tag{22}$$

$$\nabla \cdot \mathbf{B} = 0 \tag{23}$$

Equation (22) essentially states that charge density is a source (or sink) of electric flux lines. Note that we can no longer say that *all* electric flux begins and terminates on charge, because the point form of Faraday's law (20) shows that **E**, and hence **D**, may have circulation if a changing magnetic field is present. Thus the lines of electric flux may form closed loops. However, the converse is still true, and every coulomb of charge must have one coulomb of electric flux diverging from it.

Equation (23) again acknowledges the fact that "magnetic charges," or poles, are not known to exist. Magnetic flux is always found in closed loops and never diverges from a point source.

These four equations form the basis of all electromagnetic theory. They are partial differential equations and relate the electric and magnetic fields to each other and to their sources, charge and current density. The auxiliary equations relating **D** and **E**,

$$\mathbf{D} = \epsilon \mathbf{E} \tag{24}$$

relating **B** and **H**,

$$\mathbf{B} = \mu \mathbf{H} \tag{25}$$

defining conduction current density,

$$\mathbf{J} = \sigma \mathbf{E} \tag{26}$$

and defining convection current density in terms of the volume charge density ρ_v,

$$\mathbf{J} = \rho_v \mathbf{v} \tag{27}$$

are also required to define and relate the quantities appearing in Maxwell's equations.

The potentials V and **A** have not been included above because they are not strictly necessary, although they are extremely useful. They will be discussed at the end of this chapter.

If we do not have "nice" materials to work with, then we should replace (24) and (25) by the relationships involving the polarization and magnetization fields,

$$\boxed{\mathbf{D} = \epsilon_0\mathbf{E} + \mathbf{P}} \tag{28}$$

$$\boxed{\mathbf{B} = \mu_0(\mathbf{H} + \mathbf{M})} \tag{29}$$

For linear materials we may relate **P** to **E**

$$\mathbf{P} = \chi_e\epsilon_0\mathbf{E} \tag{30}$$

and **M** to **H**

$$\mathbf{M} = \chi_m\mathbf{H} \tag{31}$$

Finally, because of its fundamental importance we should include the Lorentz force equation, written in point form as the force per unit volume,

$$\boxed{\mathbf{f} = \rho_v(\mathbf{E} + \mathbf{v} \times \mathbf{B})} \tag{32}$$

The following chapters are devoted to the application of Maxwell's equations to several simple problems.

D10.4. Select the value of K so that each of the following pairs of fields satisfies Maxwell's equations in a region where $\sigma = 0$ and $\rho_v = 0$: (*a*) $\mathbf{E} = (Kx - 100t)\mathbf{a}_y$ V/m, $\mathbf{H} = (x + 20t)\mathbf{a}_z$ A/m, if $\mu = 0.25$ H/m and $\epsilon = 0.01$ F/m; (*b*) $\mathbf{D} = 5x\mathbf{a}_x - 2y\mathbf{a}_y + Kz\mathbf{a}_z\ \mu\text{C/m}^2$, $\mathbf{B} = 2\mathbf{a}_y$ mT, $\mu = \mu_0$, $\epsilon = \epsilon_0$; (*c*) $\mathbf{E} = 60 \sin 10^6 t \sin 0.01z\ \mathbf{a}_x$ V/m, $\mathbf{H} = 0.6 \cos 10^6 t \cos 0.01z\ \mathbf{a}_y$ A/m, $\mu = K$, $\epsilon = C_1$.

Ans. $-5\ \text{V/m}^2$; $-3\ \mu\text{C/m}^3$; 10^{-6} H/m

10.4 MAXWELL'S EQUATIONS IN INTEGRAL FORM

The integral forms of Maxwell's equations are usually easier to recognize in terms of the experimental laws from which they have been obtained by a generalization process. Experiments must treat physical macroscopic quantities, and their results therefore are expressed in terms of integral relationships. A differential equation always represents a theory. Let us now collect the integral forms of Maxwell's equations of the previous section.

Integrating (20) over a surface and applying Stokes' theorem, we obtain Faraday's law,

$$\boxed{\oint \mathbf{E} \cdot d\mathbf{L} = -\int_S \frac{\partial \mathbf{B}}{\partial t} \cdot d\mathbf{S}} \tag{33}$$

and the same process applied to (21) yields Ampère's circuital law,

$$\boxed{\oint \mathbf{H} \cdot d\mathbf{L} = I + \int_S \frac{\partial \mathbf{D}}{\partial t} \cdot d\mathbf{S}} \tag{34}$$

Gauss's laws for the electric and magnetic fields are obtained by integrating (22) and (23) throughout a volume and using the divergence theorem:

$$\boxed{\oint_S \mathbf{D} \cdot d\mathbf{S} = \int_{\text{vol}} \rho_v \, dv} \tag{35}$$

$$\boxed{\oint_S \mathbf{B} \cdot d\mathbf{S} = 0} \tag{36}$$

These four integral equations enable us to find the boundary conditions on **B**, **D**, **H**, and **E** which are necessary to evaluate the constants obtained in solving Maxwell's equations in partial differential form. These boundary conditions are in general unchanged from their forms for static or steady fields, and the same methods may be used to obtain them. Between any two real physical media (where **K** must be zero on the boundary surface), (33) enables us to relate the tangential **E**-field components,

$$E_{t1} = E_{t2} \tag{37}$$

and from (34),

$$H_{t1} = H_{t2} \tag{38}$$

The surface integrals produce the boundary conditions on the normal components,

$$D_{N1} - D_{N2} = \rho_S \tag{39}$$

and

$$B_{N1} = B_{N2} \tag{40}$$

It is often desirable to idealize a physical problem by assuming a perfect conductor for which σ is infinite but **J** is finite. From Ohm's law, then, in a perfect conductor,

$$\mathbf{E} = 0$$

and it follows from the point form of Faraday's law that

$$\mathbf{H} = 0$$

for time-varying fields. The point form of Ampère's circuital law then shows that the finite value of **J** is

$$\mathbf{J} = 0$$

and current must be carried on the conductor surface as a surface current **K**. Thus, if region 2 is a perfect conductor, (37) to (40) become, respectively,

$$E_{t1} = 0 \tag{41}$$

$$H_{t1} = K \qquad (\mathbf{H}_{t1} = \mathbf{K} \times \mathbf{a}_N) \tag{42}$$

$$D_{N1} = \rho_S \tag{43}$$

$$B_{N1} = 0 \tag{44}$$

where $\mathbf{a}_N$ is an outward normal at the conductor surface.

Note that surface charge density is considered a physical possibility for either dielectrics, perfect conductors, or imperfect conductors, but that surface *current* density is assumed only in conjunction with perfect conductors.

The boundary conditions stated above are a very necessary part of Maxwell's equations. All real physical problems have boundaries and require the solution of Maxwell's equations in two or more regions and the matching of these solutions at the boundaries. In the case of perfect conductors, the solution of the equations within the conductor is trivial (all time-varying fields are zero), but the application of the boundary conditions (41) to (44) may be very difficult.

Certain fundamental properties of wave propagation are evident when Maxwell's equations are solved for an *unbounded* region. This problem is treated in the following chapter. It represents the simplest application of Maxwell's equations, because it is the only problem which does not require the application of any boundary conditions.

D10.5. The unit vector $0.48\mathbf{a}_x - 0.6\mathbf{a}_y + 0.64\mathbf{a}_z$ is directed from region 2 ($\epsilon_{R2} = 2.5$, $\mu_{R2} = 2$, $\sigma_2 = 0$) toward region 1 ($\epsilon_{R1} = 4$, $\mu_{R1} = 10$, $\sigma_1 = 0$). If $\mathbf{H}_1 = (-100\mathbf{a}_x - 50\mathbf{a}_y + 200\mathbf{a}_z) \sin 400t$ A/m at point P in region 1 adjacent to the boundary, find the amplitude at P of: (*a*) $\mathbf{H}_{N1}$; (*b*) $\mathbf{H}_{t1}$; (*c*) $\mathbf{H}_{N2}$; (*d*) $\mathbf{H}_2$.

Ans. 110.0 A/m; 201 A/m; 550 A/m; 586 A/m

D10.6. The surface $x = 0$ is a perfectly conducting plane, while the region $x > 0$ has $\mu_R = 2$, $\epsilon_R = 8$, and $\sigma = 0$. Let $\mathbf{H} = 2 \cos (3 \times 10^8 t - 4z)\mathbf{a}_y$ A/m for $x > 0$, and find: (*a*) $\mathbf{K}$ at $P(0,2,0.1)$ at $t = 5$ ns; (*b*) ρ_S at P at $t = 5$ ns; (*c*) $\mathbf{E}$ at $Q(1,2,3)$ at $t = 4$ ns.

Ans. $0.907\mathbf{a}_z$ A/m; 12.10 nC/m^2; $-73.2\mathbf{a}_x$ V/m

10.5 THE RETARDED POTENTIALS

The time-varying potentials, usually called *retarded* potentials for a reason which we shall see shortly, find their greatest application in radiation problems in which the distribution of the source is known approximately. We should remember that the scalar electric potential V may be expressed in terms of a static charge distribution,

$$V = \int_{\text{vol}} \frac{\rho_v \, dv}{4\pi\epsilon R} \qquad \text{(static)} \tag{45}$$

and the vector magnetic potential may be found from a current distribution which is constant with time,

$$\mathbf{A} = \int_{\text{vol}} \frac{\mu \mathbf{J}\, dv}{4\pi R} \qquad \text{(dc)} \tag{46}$$

The differential equations satisfied by V,

$$\nabla^2 V = -\frac{\rho_v}{\epsilon} \qquad \text{(static)} \tag{47}$$

and $\mathbf{A}$,

$$\nabla^2 \mathbf{A} = -\mu \mathbf{J} \qquad \text{(dc)} \tag{48}$$

may be regarded as the point forms of the integral equations (45) and (46), respectively.

Having found V and $\mathbf{A}$, the fundamental fields are then simply obtained by using the gradient,

$$\mathbf{E} = -\nabla V \qquad \text{(static)} \tag{49}$$

or the curl,

$$\mathbf{B} = \nabla \times \mathbf{A} \qquad \text{(dc)} \tag{50}$$

We now wish to define suitable time-varying potentials which are consistent with the above expressions when only static charges and direct currents are involved.

Equation (50) apparently is still consistent with Maxwell's equations. These equations state that $\nabla \cdot \mathbf{B} = 0$, and the divergence of (50) leads to the divergence of the curl which is identically zero. Let us therefore tentatively accept (50) as satisfactory for time-varying fields and turn our attention to (49).

The inadequacy of (49) is obvious, because application of the curl operation to each side and recognition of the curl of the gradient as being identically zero confront us with $\nabla \times \mathbf{E} = 0$. The point form of Faraday's law states that $\nabla \times \mathbf{E}$ is not generally zero. Let us try to effect an improvement by adding an unknown term to (49),

$$\mathbf{E} = -\nabla V + \mathbf{N}$$

taking the curl,

$$\nabla \times \mathbf{E} = 0 + \nabla \times \mathbf{N}$$

using the point form of Faraday's law,

$$\nabla \times \mathbf{N} = -\frac{\partial \mathbf{B}}{\partial t}$$

and using (50), giving us

$$\nabla \times \mathbf{N} = -\frac{\partial}{\partial t}(\nabla \times \mathbf{A})$$

or

$$\nabla \times \mathbf{N} = -\nabla \times \frac{\partial \mathbf{A}}{\partial t}$$

The simplest solution of this equation is

$$\mathbf{N} = -\frac{\partial \mathbf{A}}{\partial t}$$

and this leads to

$$\mathbf{E} = -\nabla V - \frac{\partial A}{\partial t} \tag{51}$$

We still must check (50) and (51) by substituting them into the remaining two of Maxwell's equations:

$$\nabla \times \mathbf{H} = \mathbf{J} + \frac{\partial \mathbf{D}}{\partial t}$$

$$\nabla \cdot \mathbf{D} = \rho_v$$

Doing this, we obtain the more complicated expressions

$$\frac{1}{\mu} \nabla \times \nabla \times \mathbf{A} = \mathbf{J} + \epsilon \left(-\nabla \frac{\partial V}{\partial t} - \frac{\partial^2 \mathbf{A}}{\partial t^2} \right)$$

and

$$\epsilon \left(-\nabla \cdot \nabla V - \frac{\partial}{\partial t} \nabla \cdot \mathbf{A} \right) = \rho_v$$

or

$$\nabla(\nabla \cdot \mathbf{A}) - \nabla^2 \mathbf{A} = \mu \mathbf{J} - \mu\epsilon \left(\nabla \frac{\partial V}{\partial t} + \frac{\partial^2 \mathbf{A}}{\partial t^2} \right) \tag{52}$$

and

$$\nabla^2 V + \frac{\partial}{\partial t} (\nabla \cdot \mathbf{A}) = -\frac{\rho_v}{\epsilon} \tag{53}$$

There is no apparent inconsistency in (52) and (53). Under static or dc conditions $\nabla \cdot \mathbf{A} = 0$, and (52) and (53) reduce to (48) and (47), respectively. We shall therefore assume that the time-varying potentials may be defined in such a way that **B** and **E** may be obtained from them through (50) and (51). These latter two equations do not serve, however, to define **A** and V *completely*. They represent necessary, but not sufficient, conditions. Our initial assumption was merely that $\mathbf{B} = \nabla \times \mathbf{A}$, and a vector cannot be defined by giving its curl alone. Suppose, for example, that we have a very simple vector potential field in which A_y and A_z are zero. Expansion of (50) leads to

$$B_x = 0$$

$$B_y = \frac{\partial A_x}{\partial z}$$

$$B_z = -\frac{\partial A_x}{\partial y}$$

and we see that no information is available about the manner in which A_x varies with x. This information could be found if we also knew the value of the divergence of **A**, for in our example

$$\nabla \cdot \mathbf{A} = \frac{\partial A_x}{\partial x}$$

Finally, we should note that our information about **A** is given only as partial derivatives and that a space-constant term might be added. In all physical problems in which the region of the solution extends to infinity, this constant term must be zero, for there can be no fields at infinity.

Generalizing from this simple example, we may say that a vector field is defined completely when both its curl and divergence are given and when its value is known at any one point (including infinity). We are therefore at liberty to specify the divergence of **A**, and we do so with an eye on (52) and (53), seeking the simplest expressions. We define

$$\nabla \cdot \mathbf{A} = -\mu\epsilon \frac{\partial V}{\partial t} \tag{54}$$

and (52) and (53) become

$$\nabla^2 \mathbf{A} = -\mu \mathbf{J} + \mu\epsilon \frac{\partial^2 \mathbf{A}}{\partial t^2} \tag{55}$$

and

$$\nabla^2 V = -\frac{\rho_v}{\epsilon} + \mu\epsilon \frac{\partial^2 V}{\partial t^2} \tag{56}$$

These equations are related to the wave equation, which will be discussed in the following chapter. They show considerable symmetry, and we should be highly pleased with our definitions of V and **A**,

$$\mathbf{B} = \nabla \times \mathbf{A} \tag{50}$$

$$\nabla \cdot \mathbf{A} = -\mu\epsilon \frac{\partial V}{\partial t} \tag{54}$$

$$\mathbf{E} = -\nabla V - \frac{\partial \mathbf{A}}{\partial t} \tag{51}$$

The integral equivalents of (45) and (46) for the time-varying potentials follow from the definitions (50), (51), and (54), but we shall merely present the final results and indicate their general nature. In the next chapter a study of the uniform plane wave will introduce the concept of *propagation*, in which any electromagnetic disturbance is found to travel at a velocity

$$v = \frac{1}{\sqrt{\mu\epsilon}}$$

through any homogeneous medium described by μ and ϵ. In the case of free space this velocity turns out to be velocity of light, approximately 3×10^8 m/s. It is logical, then, to suspect that the potential at any point is due not to the value of the charge density at some distant point at the same instant, but to its value at some previous time, since the effect propagates at a finite velocity. Thus (45) becomes

$$\boxed{V = \int_{\text{vol}} \frac{[\rho_v]}{4\pi\epsilon R}\, dv} \tag{57}$$

where $[\rho_v]$ indicates that every t appearing in the expression for ρ_v has been replaced by a *retarded* time,

$$t' = t - \frac{R}{v}$$

Thus, if the charge density throughout space were given by

$$\rho_v = e^{-r} \cos \omega t$$

then

$$[\rho_v] = e^{-r} \cos\left[\omega\left(t - \frac{R}{v}\right)\right]$$

where R is the distance between the differential element of charge being considered and the point at which the potential is to be determined.

The retarded vector magnetic potential is given by

$$\boxed{\mathbf{A} = \int_{\text{vol}} \frac{\mu[\mathbf{J}]}{4\pi R}\, dv} \tag{58}$$

The use of a retarded time has resulted in the time-varying potentials being given the name of retarded potentials. In Chap. 13 we shall apply (58) to the simple situation of a differential current element in which I is a sinusoidal function of time. Other simple applications of (58) are considered in the problems at the end of this chapter.

We may summarize the use of the potentials by stating that a knowledge of the distribution of ρ_v and $\mathbf{J}$ throughout space theoretically enables us to determine V and $\mathbf{A}$ from (57) and (58). The electric and magnetic fields are then obtained by applying (50) and (51). If the charge and current distributions are unknown, or reasonable approximations cannot be made for them, these potentials usually offer no easier path toward the solution than does the direct application of Maxwell's equations.

D10.7. A point charge of $5 \cos 10^7 \pi t$ μC is located at $P_+(0,0,15)$, while $-5 \cos 10^7 \pi t$ μC is at $P_-(0,0,-15)$, both in free space. Find V at $P(r = 3000, \theta, \phi = 0)$ at $t = 150$ ns, where $\theta =$: (*a*) 0°; (*b*) 90°; (*c*) 60°.

Ans. 30.0 V; 0; 21.2 V

SUGGESTED REFERENCES

1. Bewley, L. V.: "Flux Linkages and Electromagnetic Induction," The Macmillan Company, New York, 1952. This little book discusses many of the paradoxical examples involving induced (?) voltages.

2. Faraday, M.: "Experimental Researches in Electricity," B. Quaritch, London, 1839, 1855. Very interesting reading of early scientific research. A more recent and available source is "Great Books of the Western World," vol. 45, Encyclopaedia Britannica, Inc., Chicago, 1952.

3. Halliday, D., and R. Resnick: "Physics," pt. II, 3d ed., John Wiley & Sons, New York, 1978. This text is widely used in the first university-level course in physics. Most of their symbols and equations are the same as ours.

4. Harman, W. W.: "Fundamentals of Electronic Motion," McGraw-Hill Book Company, New York, 1953. Relativistic effects are discussed in a clear and interesting manner.

5. Nussbaum, A.: "Electromagnetic Theory for Engineers and Scientists," Prentice-Hall, Inc., Englewood Cliffs, N.J., 1965. See the rocket-generator example beginning on p. 211.

6. Owen, G. E.: "Electromagnetic Theory," Allyn and Bacon, Inc., Boston, 1963. Faraday's law is discussed in terms of the frame of reference in chap. 8.

7. Panofsky, W. K. H., and M. Phillips: "Classical Electricity and Magnetism," 2d ed., Addison-Wesley Publishing Company, Inc., Reading, Mass., 1962. Relativity is treated at a moderately advanced level in chap. 15.

PROBLEMS

1. In Fig. 10.1, replace the uniform **B** field by $\mathbf{B} = 2e^{-50y}\mathbf{a}_z$ T. Find $V_{12}(t)$ if $d = 4$ cm and $\mathbf{v} = 65\mathbf{a}_y$ mi/h with $y = 0$ at $t = 0$.

2. In Fig. 10.4, let the separation of the rails be 8 cm and let $\mathbf{B} = 0.5e^{0.1x}\mathbf{a}_z$ T. (*a*) Find $V_{ab}(t)$ if the bar moves with a velocity $\mathbf{v} = 5\mathbf{a}_x$ m/s and $x = 0$ at $t = 0$. (*b*) Find $\mathbf{v}(t)$ if $V_{ab}(t) = -0.1t$ V and $x = 0$ at $t = 0$.

3. Let us assume that the sliding bar of Fig. 10.4 is fixed in position at $x = 1$ m with a rail separation of 8 cm. Find $V_{ab}(t)$ if $B_z =$: (*a*) $2 \cos 10^8 t$ μT; (*b*) $2 \cos (10^8 t - x/3)$ μT.

4. A coaxial transmission line has inner and outer radii of a and b, respectively, and has the magnetic field intensity, $H_\phi = (A/\rho) \sin \omega t \cos \beta z$. The conductors are perfect and the dielectric has $\mu_R = 1$, $\epsilon_R = 2.25$, and $\sigma_{\text{diel}} = 0$. The line extends from $z = 0$ to $z = d$. (*a*) Determine the flux $\Phi(t)$ passing through that portion of the plane $\phi = 0$ bounded

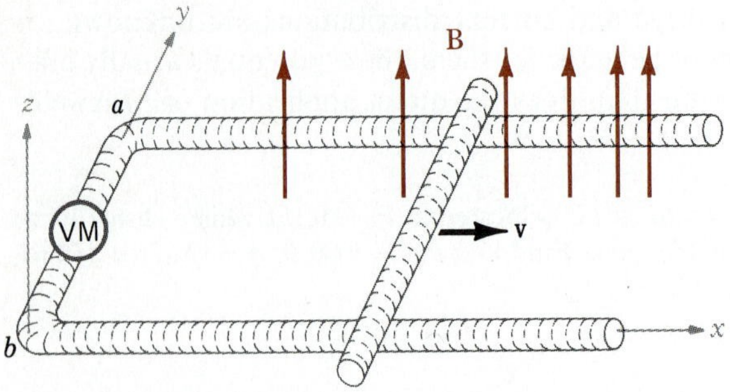

FIGURE 10.4
See Probs. 2,3, and 5.

by $\rho = a$, $\rho = b$, $z = 0$, and $z = d$. (*b*) Find $-d\Phi/dt$. (*c*) If the plane $z = 0$ is perfectly conducting, find $V_{ab}(t)$ at $z = d$.

5. Let $\mathbf{B} = 0.5x\mathbf{a}_z$ T in Fig. 10.4. The position of the sliding bar is given by $x = 4t - 2t^2$. If the separation of the rails is 10 cm, find the voltmeter reading V_{ab} at: (*a*) $t = 0.5$ s; (*b*) $x = 1$ m.

6. A perfectly conducting filament containing a small 500-Ω resistor is formed into a square, as shown in Fig. 10.5. Find $I(t)$ if $\mathbf{B} =$: (*a*) $0.2 \cos 120\pi t\, \mathbf{a}_z$ T; (*b*) $2 \cos [3\pi 10^8(t - x/c)]\, \mathbf{a}_z$ μT, where $c = 3 \times 10^8$ m/s.

7. The rails in Fig. 10.6 each have a resistance of 2.5 Ω/m. The bar moves at a constant speed of 8 m/s in a uniform field of 0.9 T. Find I as a function of time, $0 < t < 2$ s, if the bar leaves the left end at $t = 0$ and: (*a*) a 0.2-Ω resistor is present across the left end with the right end open-circuited; (*b*) a 0.2-Ω resistor is present across each end.

8. The nearer rail in Fig. 10.1 is replaced by a nonstraight conductor such that the separation is given by $d = f(y)$. The sliding bar is sufficiently long to make contact with both rails. Assuming a uniform field $\mathbf{B} = B_0\mathbf{a}_z$, a constant velocity $\mathbf{v} = v_0\mathbf{a}_y$, and a location $y = 0$ at $t = 0$, find $V_{12}(t)$.

9. A square loop of wire has corners at (0,0,0), (0.2,0,0), (0.2,0.2,0), and (0,0.2,0) at $t = 0$. The loop is perfectly conducting except for a small 100-Ω resistor in one side. It is moving through the field $\mathbf{B} = 5[\cos(6 \times 10^8 t - 2x)]\mathbf{a}_z$ μT with a constant velocity of $40\mathbf{a}_y$ m/s. Calculate the power being delivered to the resistor as a function of time.

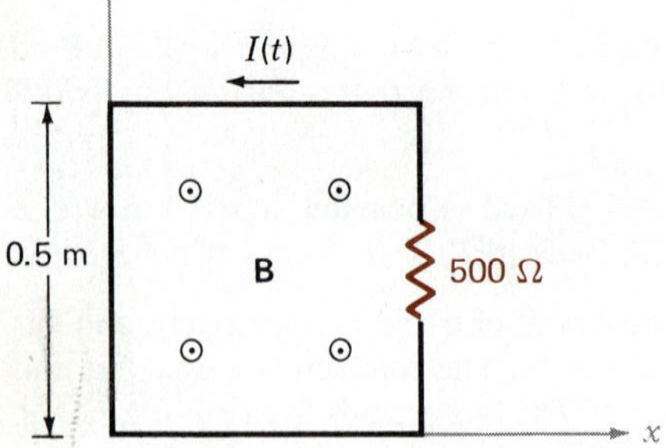

FIGURE 10.5
See Prob. 6.

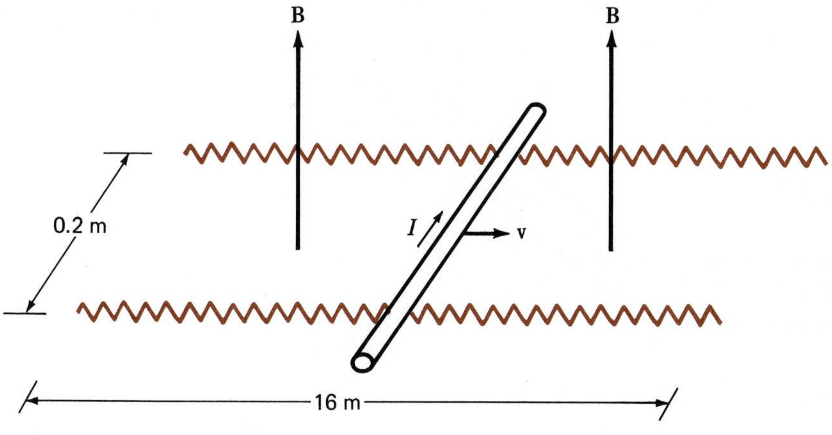

FIGURE 10.6
See Prob. 7.

10. (*a*) Show that the ratio of the amplitudes of the conduction current density and displacement current density is $\sigma/\omega\epsilon$ for the applied field $E = E_m \cos \omega t$. Assume $\mu = \mu_0$. (*b*) What is this amplitude ratio if the applied field is $E = E_m e^{-t/\tau}$, where τ is real?

11. A coaxial transmission line with $a = 1$ cm and $b = 2.718$ cm has a voltage $V_{ab} = 100 \cos (3 \times 10^8 t - z)$ V between conductors. Let $\epsilon_R = \mu_R = 1$. (*a*) If the spatial relationship between the time-varying **E** and V fields is the same as it is for electrostatics, find $\mathbf{E}(\rho,z,t)$. (*b*) Find the displacement current density $\mathbf{J}_d(\rho,z,t)$.

12. A coaxial capacitor has the parameters $a = 5$ mm, $b = 30$ mm, $l = 20$ cm, $\epsilon_R = 8$, and $\sigma = 10^{-6}$ ℧/m. If the conduction current density in the capacitor is $(2/\rho) \sin 10^6 t\, \mathbf{a}_\rho$ A/m^2, find: (*a*) the total conduction current through the capacitor; (*b*) the maximum instantaneous value of the displacement current density; (*c*) the total displacement current; (*d*) the Q $(= \omega CR = I_d/I_\sigma)$ of the capacitor.

13. A voltage source $V_0 \sin \omega t$ is connected between two concentric conducting spheres, $r = a$ and $r = b$, $b > a$, where the region between them is a material for which $\epsilon = \epsilon_R\epsilon_0$, $\mu = \mu_0$, and $\sigma = 0$. Find the total displacement current through the dielectric and compare it with the source current as determined from the capacitance (Sec. 5.10) and circuit-analysis methods.

14. The displacement current density is $5 \cos (2 \times 10^8 t - kz)\, \mathbf{a}_x$ μA/m^2 in a material for which $\sigma = 0$, $\epsilon = 5\epsilon_0$, and $\mu = 4\mu_0$. (*a*) Use the definition of displacement current density to find **D** and **E**. (*b*) Now use the point form of Faraday's law and a time integration to find **B** and **H**. (*c*) Finally, use the point form of Ampère's circuital law to find the displacement current density. What must k be?

15. Assume a homogeneous material of infinite extent with $\epsilon = 2 \times 10^{-10}$ F/m, $\mu = 1.25 \times 10^{-5}$ H/m, and $\sigma = 0$. Let $\mathbf{E} = 400 \cos (10^9 t - kz)\, \mathbf{a}_x$ V/m. If all the fields vary sinusoidally (or cosinusoidaly), use Maxwell's equations to find **D**, **B**, **H**, and k.

16. (*a*) A certain material has $\sigma = 0$ and $\mu_R = 1$. If $\mathbf{E} = 800 \sin (10^6 t - 0.01z)\, \mathbf{a}_y$ V/m, make use of Maxwell's equations to find ϵ_R. (*b*) Find $\mathbf{H}(z,t)$.

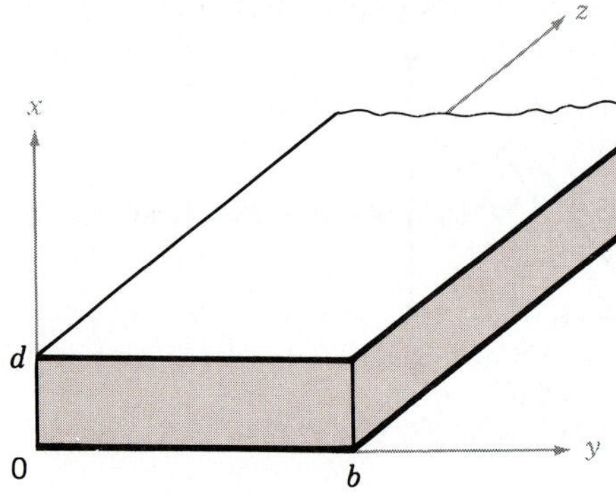

FIGURE 10.7
See Prob. 17.

17. The parallel-strip transmission line shown in Fig. 10.7 has dimensions $b = 5$ cm and $d = 1$ mm, while the medium between the strips is characterized by $\mu_R = 1$, $\epsilon_R = 25$, and $\sigma = 0$. Neglect fields outside the dielectric. Given the field $\mathbf{E} = 10^4 \cos(10^9 t - \beta z)\, \mathbf{a}_x$ V/m, use Maxwell's equations to help find: (*a*) β, if $\beta > 0$; (*b*) the displacement current density at $z = 0$; (*c*) the total displacement current crossing the surface $x = 0.5d$, $0 < y < b$, $0 < z < 0.1$ m in the $\mathbf{a}_x$ direction.

18. The electric field intensity throughout a certain enclosed region of free space is $E_y = A \sin 20x \sin bz \sin(12 \times 10^9 t)$ V/m. Beginning with the $\nabla \times \mathbf{E}$ relationship, use Maxwell's equations to find a numerical value for b, assuming $b > 0$.

19. Let $\mathbf{E} = (-20\mathbf{a}_x + 30\mathbf{a}_y + 70\mathbf{a}_z) \cos 10^6 t$ V/m at point $P(3,-4,1)$ which lies on the surface of a perfect conductor. If the material adjacent to the conductor has $\epsilon_R = 5$, $\mu_R = 2$, and $\sigma = 0$, find: (*a*) a unit vector normal to the conductor surface at P; (*b*) the surface charge density on the conductor surface at P.

20. In the first section of this chapter, Faraday's law was used to show that the field $\mathbf{E} = -\frac{1}{2}kB_0 e^{kt}\mathbf{a}_\phi$ results from the changing magnetic field $\mathbf{B} = B_0 e^{kt}\mathbf{a}_z$. (*a*) Show that these fields do not satisfy Maxwell's other curl equation. (*b*) If we let $B_0 = 1$ T and $k = 10^6$ s^{-1}, we are establishing a fairly large magnetic flux density in 1 μs. Use the $\nabla \times \mathbf{H}$ equation to show that the rate at which B_z should (but does not) change with ρ is only about 5×10^{-6} T per meter in free space at $t = 0$.

21. A section of coaxial cable is bounded by conducting surfaces at $z = 0$ and $z = 0.2$ m. The conductor radii are $a = 2$ mm and $b = 8$ mm, and in the region $a < \rho < b$ we have $\epsilon_R = 6.25$, $\mu_R = 1$, and $\sigma = 0$. If $\mathbf{E} = (100/\rho) \sin 5\pi z \sin \omega t\, \mathbf{a}_\rho$ V/m, find: (*a*) ω with the help of Maxwell's equations; (*b*) the total charge on the inner conductor at $t = \pi/(2\omega)$; (*c*) the surface current density at $\rho = 3$ mm, $\phi = 0$, $z = 0$, $t = 0$.

22. In region 1, $z < 0$, $\epsilon_1 = 10^{-11}$ F/m, $\mu_1 = 4 \times 10^{-6}$ H/m, and $\sigma_1 = 10^{-3}$ ℧/m; in region 2, $z > 0$, $\epsilon_2 = 2\epsilon_1$, $\mu_2 = \mu_1/2$, and $\sigma_2 = 4\sigma_1$. It is known that $\mathbf{E}_1 = (10\mathbf{a}_x + 20\mathbf{a}_y + 30\mathbf{a}_z) \cos 10^9 t$ V/m at $P_1(0,0,0^-)$. (*a*) Find $\mathbf{E}_{N1}$, $\mathbf{E}_{t1}$, $\mathbf{D}_{N1}$, and $\mathbf{D}_{t1}$ at P_1. (*b*) Find $\mathbf{J}_{N1}$ and $\mathbf{J}_{t1}$ at P_1. (*c*) Find $\mathbf{E}_{t2}$, $\mathbf{D}_{t2}$, and $\mathbf{J}_{t2}$ at $P_2(0,0,0^+)$. (*d*) (Harder) Use the continuity equation to help show that $J_{N1} - J_{N2} = \partial D_{N2}/\partial t$, and then determine $\mathbf{D}_{N2}$, $\mathbf{J}_{N2}$, and $\mathbf{E}_{N2}$.

23. Consider a region where $\epsilon = \epsilon_0$, $\mu = \mu_0$, $\sigma = 0$, $\mathbf{J} = 0$, and $\rho_v = 0$. Use Maxwell's equations in cartesian coordinates and assume that $\mathbf{E} = E_x(z,t)\mathbf{a}_x$ and $\mathbf{H} = H_y(z,t)\mathbf{a}_y$. (*a*) Find the second-order partial differential equation that E_x must satisfy. (*b*) Show that $E_x = (200t - az)^2$ is a solution of this equation for a certain value of a, and find that value.

24. A current $\mathbf{I} = 100t\mathbf{a}_y$ A is present in free space on the y axis in the interval $0 < y < 0.2$ m. Find A_y at $P(0,0,2)$ and sketch A_y over the time interval $-0.1 < t < 0.1$ μs.

25. In a region where $\epsilon_R = \mu_R = 1$ and $\sigma = 0$, let $\mathbf{A} = 10^{-3} y \cos 3 \times 10^8 t \cos z\, \mathbf{a}_z$ Wb/m and $V = 3 \times 10^5 y \sin 3 \times 10^8 t \sin z$ V. Find **E** and **H**.

26. Given the retarded potentials, $V = -y(x + ct)$ V and $\mathbf{A} = y(x/c + t)\mathbf{a}_x$ Wb/m, where $c = 1/\sqrt{\mu_0 \epsilon_0}$: (*a*) show that $\nabla \cdot \mathbf{A} = -\mu\epsilon\, \partial V/\partial t$; (*b*) find **B**, **H**, **E**, and **D**. (*c*) Show that these results satisfy Maxwell's equations if $\mathbf{J} = 0$ and $\rho_v = 0$.

CHAPTER 11

THE UNIFORM PLANE WAVE

In this chapter we shall apply Maxwell's equations to introduce the fundamental theory of wave motion. The uniform plane wave represents one of the simplest applications of Maxwell's equations, and yet it illustrates the principles behind the propagation of energy. We shall introduce the velocity of propagation, wavelength, wave impedance, the phase and attenuation constants, and the use of the Poynting theorem in finding the power density. Finally, we shall consider the reflection and transmission of a uniform plane wave at the boundary between two different media. The use of the standing-wave ratio and input impedance will prepare us to consider many of the practical problems of the guided transmission of power and information in the following chapter.

11.1 WAVE MOTION IN FREE SPACE

As we indicated in our discussion of boundary conditions in the previous chapter, the solution of Maxwell's equations without the application of any boundary conditions at all represents a very special type of problem. Although we restrict our attention to a solution in cartesian coordinates, it may seem even then that we are solving several different problems as we consider various special cases in this chapter. Solutions are obtained first for free-space conditions, then for perfect dielectrics, next for lossy dielectrics, and finally for the good conductor. We do this to take advantage of the approximations that are applicable to each special case and to emphasize the special characteristics of wave propagation in these media, but it is not necessary to use a separate treatment; it is possible (and not very difficult) to solve the general problem once and for all. Our discussion of propagation in the lossy dielectric begins with a consideration of the general case, but we shall then simplify the analysis by confining our attention to dielectrics with relatively small losses.

To consider wave motion in free space first, Maxwell's equations may be written in terms of **E** and **H** only as

$$\boxed{\nabla \times \mathbf{H} = \epsilon_0 \frac{\partial \mathbf{E}}{\partial t}} \tag{1}$$

$$\boxed{\nabla \times \mathbf{E} = -\mu_0 \frac{\partial \mathbf{H}}{\partial t}} \tag{2}$$

$$\boxed{\nabla \cdot \mathbf{E} = 0} \tag{3}$$

$$\boxed{\nabla \cdot \mathbf{H} = 0} \tag{4}$$

Now let us see whether wave motion can be inferred from these four equations without actually solving them. The first equation states that if **E** is changing with time at some point, then **H** has curl at that point and thus can be considered as forming a small closed loop linking the changing **E** field. Also, if **E** is changing with time, then **H** will in general also change with time, although not necessarily in the same way. Next, we see from the second equation that this changing **H** produces an electric field which forms small closed loops about the **H** lines. We now have once more a changing electric field, our original hypothesis, but this field is present a small distance away from the point of the original disturbance. We might guess (correctly) that the velocity with which the effect moves away from the original point is the velocity of light, but this must be checked by a more quantitative examination of Maxwell's equations.

Let us first write Maxwell's four equations above for the special case of sinusoidal (more strictly, *cosinusoidal*) variation with time. This is accomplished by complex notation and phasors. Given the vector field

$$\mathbf{E} = E_x \mathbf{a}_x$$

we assume that the component E_x is given as

$$E_x = E_{xyz} \cos(\omega t + \psi) \tag{5}$$

where E_{xyz} is a real function of x, y, z, and perhaps ω, but not of time, and ψ is a phase angle which may also be a function of x, y, z, and ω. Making use of Euler's identity,

$$e^{j\omega t} = \cos \omega t + j \sin \omega t$$

we let

$$E_x = \text{Re } E_{xyz} e^{j(\omega t + \psi)} = \text{Re } E_{xyz} e^{j\psi} e^{j\omega t} \tag{6}$$

where Re signifies that the *real* part of the following quantity is to be taken. If we then simplify the nomenclature by dropping Re and suppressing $e^{j\omega t}$, the field quantity E_x becomes a *phasor*, or a complex quantity, which we identify by use of an s subscript, E_{xs}. Thus

$$E_{xs} = E_{xyz} e^{j\psi} \tag{7}$$

and

$$\mathbf{E}_s = E_{xs} \mathbf{a}_x$$

The s can be thought of as indicating a frequency domain quantity expressed as a function of the complex frequency s, even though we shall consider only those cases in which s is a pure imaginary, $s = j\omega$.

For example, let us take $E_y = 100 \cos (10^8 t - 0.5z + 30°)$ V/m and express it as a phasor. We first go to exponential notation,

$$E_y = \text{Re}\left[100 e^{j(10^8 t - 0.5z + 30°)}\right]$$

and then drop Re and suppress $e^{j10^8 t}$, obtaining the phasor

$$E_{ys} = 100 e^{-j0.5z + j30°}$$

Note that E_y is real, but E_{ys} is in general complex. Note also that a mixed nomenclature is commonly used for the angle. That is, $-0.5z$ is in radians, while 30° is in degrees.

Given a scalar component or a vector expressed as a phasor, we may easily recover the time-domain expression. For example, consider the field intensity vector, identified as a phasor by its subscript s,

$$\mathbf{E}_s = 100\underline{/30°}\,\mathbf{a}_x + 20\underline{/-50°}\,\mathbf{a}_y + 40\underline{/210°}\,\mathbf{a}_z \qquad \text{V/m}$$

Let us assume that the frequency is specified as 1 MHz. We first select exponential notation for mathematical clarity,

$$\mathbf{E}_s = 100 e^{j30°} \mathbf{a}_x + 20 e^{-j50°} \mathbf{a}_y + 40 e^{j210°} \mathbf{a}_z$$

reinsert the $e^{j\omega t}$ factor,

$$\mathbf{E}_s(t) = (100 e^{j30°} \mathbf{a}_x + 20 e^{-j50°} \mathbf{a}_y + 40 e^{j210°} \mathbf{a}_z) e^{j2\pi 10^6 t}$$

$$= 100 e^{j(2\pi 10^6 t + 30°)} \mathbf{a}_x + 20 e^{j(2\pi 10^6 t - 50°)} \mathbf{a}_y + 40 e^{j(2\pi 10^6 t + 210°)} \mathbf{a}_z$$

and take the real part, obtaining the real vector,

$$\mathbf{E}(t) = 100 \cos (2\pi 10^6 t + 30°)\, \mathbf{a}_x + 20 \cos (2\pi 10^6 \text{t} - 50°)\, \mathbf{a}_y$$

$$+ 40 \cos (2\pi 10^6 t + 210°)\, \mathbf{a}_z \qquad \text{V/m}$$

None of the amplitudes or phase angles in this example are expressed as functions of x, y, or z, but, if any are, the same procedure is effective. Thus, if $\mathbf{H}_s =$

$20e^{-(0.1+j20)z}\mathbf{a}_x$ A/m, then

$$\mathbf{H}(t) = \text{Re}\,[20e^{-0.1z}e^{-j20z}e^{j\omega t}]\mathbf{a}_x = 20e^{-0.1z}\cos(\omega t - 20z)\,\mathbf{a}_x \qquad \text{A/m}$$

Now, since

$$\frac{\partial E_x}{\partial t} = \frac{\partial}{\partial t}[E_{xyz}\cos(\omega t + \psi)] = -\omega E_{xyz}\sin(\omega t + \psi)$$

$$= \text{Re}\, j\omega E_{xs}e^{j\omega t}$$

it is evident that taking the partial derivative of any field quantity with respect to time is equivalent to multiplying the corresponding phasor by $j\omega$. As an example, if

$$\frac{\partial E_x}{\partial t} = -\frac{1}{\epsilon_0}\frac{\partial H_y}{\partial z}$$

the corresponding phasor equation is

$$j\omega E_{xs} = -\frac{1}{\epsilon_0}\frac{\partial H_{ys}}{\partial z}$$

where E_{xs} and H_{ys} may both be complex quantities.

We next apply this notation to Maxwell's equations. Thus, given the equation,

$$\nabla \times \mathbf{H} = \epsilon_0 \frac{\partial \mathbf{E}}{\partial t}$$

the corresponding relationship in terms of phasor-vectors is

$$\boxed{\nabla \times \mathbf{H}_s = j\omega\epsilon_0\mathbf{E}_s} \tag{8}$$

Equation (8) and the three equations

$$\boxed{\nabla \times \mathbf{E}_s = -j\omega\mu_0\mathbf{H}_s} \tag{9}$$

$$\boxed{\nabla \cdot \mathbf{E}_s = 0} \tag{10}$$

$$\boxed{\nabla \cdot \mathbf{H}_s = 0} \tag{11}$$

are Maxwell's four equations in phasor notation for sinusoidal time variation in free space. It should be noted that (10) and (11) are no longer independent relationships, for they can be obtained by taking the divergence of (8) and (9), respectively.

Our next step is to obtain the sinusoidal steady-state form of the wave equation, a step we could omit because the simple problem we are going to solve yields easily to simultaneous solution of the four equations above. The wave equation is an important equation, however, and it is a convenient starting point for many other investigations.

The method by which the wave equation is obtained could be accomplished in one line (using four equals signs on a wider sheet of paper):

$$\nabla \times \nabla \times \mathbf{E}_s = \nabla(\nabla \cdot \mathbf{E}_s) - \nabla^2 \mathbf{E}_s = -j\omega\mu_0 \nabla \times \mathbf{H}_s$$
$$= \omega^2 \mu_0 \epsilon_0 \mathbf{E}_s = -\nabla^2 \mathbf{E}_s$$

since $\nabla \cdot \mathbf{E}_s = 0$. Thus

$$\boxed{\nabla^2 \mathbf{E}_s = -\omega^2 \mu_0 \epsilon_0 \mathbf{E}_s} \tag{12}$$

This concise phasor-vector equation is also known as the vector Helmholtz equation.[1] It is fairly formidable when expanded, even in cartesian coordinates, for three scalar phasor equations result, and each has four terms. The x component of (12) becomes, still using the del-operator notation,

$$\nabla^2 E_{xs} = -\omega^2 \mu_0 \epsilon_0 E_{xs} \tag{13}$$

and the expansion of the operator leads to the second-order partial differential equation

$$\frac{\partial^2 E_{xs}}{\partial x^2} + \frac{\partial^2 E_{xs}}{\partial y^2} + \frac{\partial^2 E_{xs}}{\partial z^2} = -\omega^2 \mu_0 \epsilon_0 E_{xs} \tag{14}$$

Let us attempt a solution of (14) by assuming that a simple solution is possible in which E_{xs} does not vary with x or y, so that the two corresponding derivatives are zero, leading to the ordinary differential equation

$$\frac{d^2 E_{xs}}{dz^2} = -\omega^2 \mu_0 \epsilon_0 E_{xs} \tag{15}$$

By inspection, we may write down one solution of (15),

$$E_{xs} = A e^{-j\omega z\sqrt{\mu_0 \epsilon_0}} \tag{16}$$

reinsert the $e^{j\omega t}$ factor, and reduce to trigonometric form by taking the real part,

$$E_x = A \cos\left[\omega(t - z\sqrt{\mu_0 \epsilon_0})\right]$$

where the arbitrary amplitude factor may be replaced by E_{x0}, the value of E_x at $z = 0$, $t = 0$,

$$\boxed{E_x = E_{x0} \cos\left[\omega(t - z\sqrt{\mu_0 \epsilon_0})\right]} \tag{17}$$

---

[1] Hermann Ludwig Ferdinand von Helmholtz (1821–1894) was a professor at Berlin working in the fields of physiology, electrodynamics, and optics. Hertz was one of his students.

Problem 1 at the end of the chapter indicates that

$$E'_x = E'_{x0} \cos [\omega(t + z\sqrt{\mu_0\epsilon_0})] \tag{18}$$

may also be obtained from an alternate solution of the vector Helmholtz equation.

Before we find any other field components, we should understand the physical nature of the single component of the electric field we have obtained in Eq. (17). We see that it is an x component, which we might describe as directed upward at the surface of a plane earth. The radical $\sqrt{\mu_0\epsilon_0}$ has the approximate value $1/(3 \times 10^8)$ s/m, which is the reciprocal of c, the velocity of light in free space,

$$c = \frac{1}{\sqrt{\mu_0\epsilon_0}} = 2.998 \times 10^8 \doteq 3 \times 10^8 \text{ m/s}$$

Let us also allow the z axis to point east and take $z = 0$ in Chicago. In Chicago, then, the field is given by

$$E_x = E_{x0} \cos \omega t$$

which is a simple and familiar variation with time. A free charge (perhaps in a vertical receiving antenna) is accelerated up and down $\omega/2\pi$ times a second. In Cleveland, about 500 km east, we would find

$$E_x = E_{x0} \cos \left[\omega\left(t - \frac{5 \times 10^5}{3 \times 10^8}\right)\right] = E_{x0} \cos [\omega(t - 0.001\ 67)]$$

indicating that the field strength in Cleveland is identical to that which existed in Chicago 0.001 67 s earlier. In general terms, we should then expect the field at any point z meters east of Chicago to lag the reference field by $z\sqrt{\mu_0\epsilon_0}$, or $z/(3 \times 10^8)$ s.

Let us change our viewpoint now and inspect the field everywhere at $t = 0$,

$$E_x = E_{x0} \cos (-\omega z\sqrt{\mu_0\epsilon_0}) = E_{x0} \cos \frac{\omega z}{c}$$

finding a periodic variation with distance. The period of this cosine wave as measured along the z axis is called the *wavelength* λ,

$$\frac{\omega\lambda}{c} = 2\pi$$

$$\lambda = \frac{c}{f} = \frac{3 \times 10^8}{f} \qquad \text{(free space)}$$

At any point, we find a sinusoidal variation with time of period $T = 1/f$; at any time, we find a sinusoidal variation with distance of period λ; at every point and at every instant of time, E_x is directed vertically. Now let us consider the response as both time and location are varied. Certainly we may say that E_x is unchanged if the phase angle $\omega(t - z\sqrt{\mu_0\epsilon_0})$ is unchanged, or

$$\omega(t - z\sqrt{\mu_0\epsilon_0}) = \text{constant}$$

Taking differentials, we have

$$\omega(dt - dz\sqrt{\mu_0\epsilon_0}) = 0$$

Therefore,

$$\boxed{\frac{dz}{dt} = v = \frac{1}{\sqrt{\mu_0\epsilon_0}} = c} \tag{19}$$

This velocity is termed a *phase velocity* since it defines a point of constant phase. Here, the phase velocity is equal to the velocity of light c, and the field is therefore moving in the z direction with the velocity of light c. Whatever the value of the field instantaneously at $z = z_1$, $t = t_1$, it will have the identical value at z_2 at a time $(z_2 - z_1)/c$ later; it will also have the identical value at $t = t_2$, a distance $(t_2 - t_1)c$ farther to the east. The electric field is in motion and is logically termed a *traveling wave*.

Equation (18), which was also a solution to our wave equation, obviously represents a wave traveling in the $-z$ direction, or west. For simplicity, we are considering only the positively traveling wave at this time.

Let us now return to Maxwell's equations, (8) to (11), and determine the form of the **H** field. Given $\mathbf{E}_s$, $\mathbf{H}_s$ is most easily obtained from (9),

$$\nabla \times \mathbf{E}_s = -j\omega\mu_0\mathbf{H}_s \tag{9}$$

which is greatly simplified for a single E_{xs} component varying only with z,

$$\frac{\partial E_{xs}}{\partial z} = -j\omega\mu_0 H_{ys}$$

Using (16) for E_{xs} with $A = E_{x0}$, we have

$$H_{ys} = -\frac{1}{j\omega\mu_0} E_{x0}(-j\omega\sqrt{\mu_0\epsilon_0})e^{-j\omega z/c}$$

and

$$\boxed{H_y = E_{x0}\sqrt{\frac{\epsilon_0}{\mu_0}}\cos\left[\omega\left(t - \frac{z}{c}\right)\right]} \tag{20}$$

We therefore find that this vertical **E** component traveling to the east is accompanied by a horizontal (north-south) magnetic field. Moreover, the ratio of electric to magnetic field intensities, given by the ratio of (17) to (20),

$$\frac{E_x}{H_y} = \sqrt{\frac{\mu_0}{\epsilon_0}} \tag{21}$$

is constant. Using the language of circuit theory, we would say that E_x and H_y are "in phase," but this in-phase relationship refers to space as well as to time. We are accustomed to taking this for granted in a circuit problem in which a current $I_m \cos \omega t$ is assumed to have its maximum amplitude I_m throughout an entire series circuit at $t = 0$. Both (17) and (20) clearly show, however, that the maximum value of either E_x or H_y occurs when $\omega(t - z/c)$ is an integral multiple of 2π rad; neither field is a maximum everywhere at the same instant. It is remarkable, then, that the ratio of these two components, both changing in space and time, should be everywhere a constant.

The square root of the ratio of the permeability to the permittivity is called the *intrinsic impedance* η (eta),

$$\eta = \sqrt{\frac{\mu}{\epsilon}} \tag{22}$$

where η has the dimension of ohms. The intrinsic impedance of free space is

$$\eta_0 = \sqrt{\frac{\mu_0}{\epsilon_0}} = 377 \doteq 120\pi \qquad \Omega$$

This wave is called a *uniform plane wave* because its value is uniform throughout any plane, z = constant. It represents an energy flow in the positive z direction. Both the electric and magnetic fields are perpendicular to the direction of propagation, or both lie in a plane that is transverse to the direction of propagation; the uniform plane wave is a *transverse electromagnetic* wave, or a TEM wave.

Some feeling for the way in which the fields vary in space may be obtained from Fig. 11.1*a* and 11.1*b*. The electric field intensity in Fig. 11.1*a* is shown at $t = 0$, and the instantaneous value of the field is depicted along three lines, the z axis and arbitrary lines parallel to the z axis in the $x = 0$ and $y = 0$ planes. Since the field is uniform in planes perpendicular to the z axis, the variation along all three of the lines is the same. One complete cycle of the variation occurs in a wavelength λ. The values of H_y at the same time and positions are shown in Fig. 11.1*b*.

A uniform plane wave cannot exist physically, for it extends to infinity in two dimensions at least and represents an infinite amount of energy. The distant

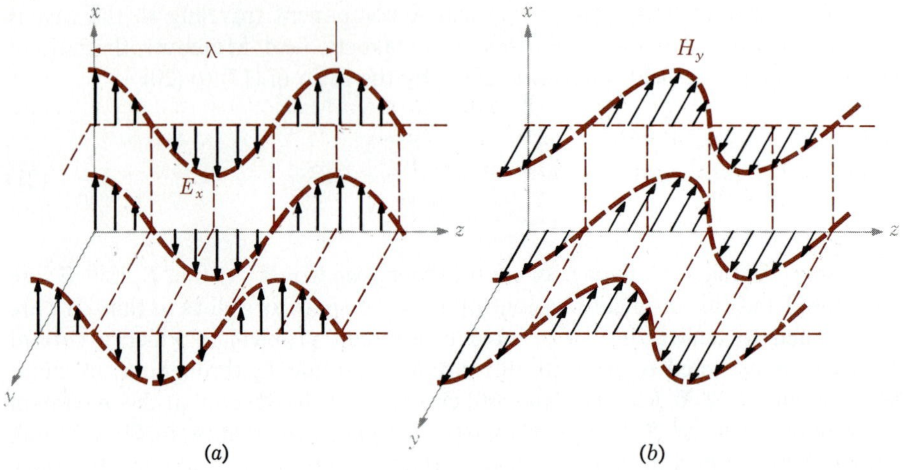

FIGURE 11.1
(*a*) Arrows represent the instantaneous values of $E_{x0} \cos [\omega(t - z/c)]$ at $t = 0$ along the z axis, along an arbitrary line in the $x = 0$ plane parallel to the z axis, and along an arbitrary line in the $y = 0$ plane parallel to the z axis. (*b*) Corresponding values of H_y are indicated. Note that E_x and H_y are in phase at any point at any time.

field of a transmitting antenna, however, is essentially a uniform plane wave in some limited region. A wave reaching a receiving antenna in Cleveland from Chicago is analyzed as a uniform plane wave in the vicinity of the antenna; a radar signal impinging on a distant target is also closely a uniform plane wave.

Although we have considered only a wave varying sinusoidally in time and space, a suitable combination of solutions to the wave equation may be made to achieve a wave of any desired form. The summation of an infinite number of harmonics through the use of a Fourier series can produce a periodic wave of square or triangular shape in both space and time. Nonperiodic waves may be obtained from our basic solution by Fourier integral methods. Finally, waves in other directions may also be included, representing perhaps a wave propagating slightly south of east. These topics are among those considered in the more advanced books on electromagnetic theory.

D11.1. The magnetic field intensity of a uniform plane wave in air is 20 A/m in the $\mathbf{a}_y$ direction. The wave is propagating in the $\mathbf{a}_z$ direction at a frequency of 2 Grad/s (2×10^9 rad/s). Find: (*a*) the wavelength; (*b*) the frequency; (*c*) the period; (*d*) the amplitude of **E**.

Ans. 0.942 m; 318 MHz; 3.14 ns; 7530 V/m

D11.2. Let $\mathbf{E}_s = [500\underline{/-40^\circ}\mathbf{a}_y + (200 - j600)\mathbf{a}_z]e^{-j0.4x}$ V/m in free space. Find: (*a*) ω; (*b*) **E** at (2,3,1), $t = 0$; (*c*) **E** at (2,3,1), $t = 10$ ns; (*d*) **E** at (3,4,2), $t = 20$ ns.

Ans. 119.9 Mrad/s; $36.3\mathbf{a}_y - 291\mathbf{a}_z$ V/m; $478\mathbf{a}_y + 417\mathbf{a}_z$ V/m; $439\mathbf{a}_y + 632\mathbf{a}_z$ V/m

11.2 WAVE MOTION IN PERFECT DIELECTRICS

Let us now extend our analytical treatment of the uniform plane wave to propagation in a perfect (lossless) dielectric of permittivity ϵ and permeability μ. The medium is isotropic and homogeneous, and the wave equation is now

$$\nabla^2 \mathbf{E}_s = -\omega^2 \mu\epsilon \mathbf{E}_s \tag{23}$$

For E_{xs} we have

$$\frac{\partial^2 E_{xs}}{\partial z^2} = -\omega^2 \mu\epsilon E_{xs} \tag{24}$$

Instead of writing down the solution of (24) immediately, let us hypothesize a solution of more general form and use (24) to specify appropriate values for our assumed parameters. We allow for exponential attenuation by assuming

$$E_x = E_{x0} e^{-\alpha z} \cos(\omega t - \beta z)$$

or its equivalent in complex exponential notation,

$$E_{xs} = E_{x0} e^{-\alpha z} e^{-j\beta z}$$

The real exponential factor now allows us to consider cases in which the wave may be attenuated as it propagates in the $+z$ direction; α is called the *attenuation constant.* Since our medium is lossless, we should be able to show that α is zero. If we next note that βz must be measured in radians (assuming β to be real), then it is logical that β be termed the *phase constant.* It is a measure of the radian phase shift per meter. In general, we often combine α and β into the complex *propagation constant* γ (gamma),

$$\boxed{\gamma = \alpha + j\beta} \tag{25}$$

so that we may write

$$E_{xs} = E_{x0} e^{-\gamma z}$$

Now let us substitute back into (24):

$$\gamma^2 E_{x0} e^{-\gamma z} = -\omega^2 \mu\epsilon E_{x0} e^{-\gamma z}$$

and thus we must require

$$\gamma^2 = -\omega^2 \mu\epsilon$$

or

$$\boxed{\gamma = \pm j\omega\sqrt{\mu\epsilon}}$$

Hence

$$\boxed{\alpha = 0}$$

and

$$\boxed{\beta = \omega\sqrt{\mu\epsilon}} \tag{26}$$

where we have selected the root which yields propagation in the positive z direction. Thus

$$E_x = E_{x0} \cos(\omega t - \beta z)$$

and we may interpret this as a wave traveling in the $+z$ direction at a phase velocity v

$$v = \frac{\omega}{\beta} \tag{27}$$

For the uniform plane wave propagating in the perfect dielectric, we find

$$v = \frac{1}{\sqrt{\mu\epsilon}} = \frac{c}{\sqrt{\mu_R\epsilon_R}}$$

The wavelength is the distance required to effect a phase change of 2π radians

$$\beta\lambda = 2\pi$$

which leads to the fundamental definition for wavelength,

$$\boxed{\lambda = \frac{2\pi}{\beta}} \tag{28}$$

It follows that

$$\lambda = \frac{2\pi}{\beta} = \frac{2\pi}{\omega\sqrt{\mu\epsilon}} = \frac{1}{f\sqrt{\mu\epsilon}} = \frac{c}{f\sqrt{\mu_R\epsilon_R}} = \frac{\lambda_0}{\sqrt{\mu_R\epsilon_R}} \tag{29}$$

where λ_0 is the free-space wavelength. Note that $\mu_R\epsilon_R > 1$, and therefore the wavelength is shorter and the velocity is lower in all real media than they are in free space.

Associated with E_x is the magnetic field intensity

$$\boxed{H_y = \frac{E_{x0}}{\eta} \cos(\omega t - \beta z)}$$

where the intrinsic impedance is

$$\boxed{\eta = \sqrt{\frac{\mu}{\epsilon}}} \tag{30}$$

The two fields E_x and H_y are once again perpendicular to each other, perpendicular to the direction of propagation, and in phase with each other everywhere. Note that when **E** is crossed into **H**, the resultant vector is in the direction of propagation. We shall see the reason for this when we discuss the Poynting vector.

Let us apply these results to a 300-MHz wave propagating through fresh water. Although we do not have a lossless medium, we shall neglect the attenuation at this time and assume that $\alpha = 0$. Therefore $\mu_R = 1$, $\epsilon_R = 78$ (at 300 MHz), and

$$\beta = \omega\sqrt{\mu\epsilon} = 2\pi \times 300 \times 10^6\sqrt{78\mu_0\epsilon_0} = 55.5 \qquad \text{rad/m}$$

or 80.8°/in, and

$$v = \frac{\omega}{\beta} = \frac{6\pi \times 10^8}{55.5} = 0.339 \times 10^8 \qquad \text{m/s}$$

$$\lambda = \frac{2\pi}{\beta} = \frac{2\pi}{55.5} = 0.113 \qquad \text{m}$$

whereas the wavelength in air would have been 1 m. Continuing our calculations, we find that

$$\eta = \eta_0\sqrt{\frac{\mu_R}{\epsilon_R}} = \frac{377}{\sqrt{78}} = 42.7 \qquad \Omega$$

If we let the electric field intensity have a maximum amplitude of 0.1 V/m, then

$$E_x = 0.1 \cos(6\pi 10^8 t - 55.5z)$$

$$H_y = \frac{E_x}{\eta} = 2.34 \times 10^{-3} \cos(6\pi 10^8 t - 55.5z)$$

D11.3. A 9375-MHz uniform plane wave is propagating in polystyrene (see Appendix C). If the amplitude of the electric field intensity is 20 V/m and the material is assumed to be lossless, find: (*a*) the phase constant; (*b*) the wavelength (in the polystyrene); (*c*) the velocity of propagation; (*d*) the intrinsic impedance; (*e*) the propagation constant; (*f*) the amplitude of the magnetic field intensity.

Ans. 314 rad/m; 0.0200 m; 1.877×10^8 m/s; 236 Ω; $j314$ m^{-1}; 0.0848 A/m

11.3 PLANE WAVES IN LOSSY DIELECTRICS

All dielectric materials have some conductivity, and while it may be neglected in many cases, even then it is necessary to establish the criteria for doing so. We shall continue to confine our attention to a sinusoidal time variation, and thus Maxwell's curl equations are

$$\nabla \times \mathbf{H}_s = \mathbf{J}_s + j\omega\epsilon\mathbf{E}_s$$

or

$$\nabla \times \mathbf{H}_s = (\sigma + j\omega\epsilon)\mathbf{E}_s$$

and

$$\nabla \times \mathbf{E}_s = -j\omega\mu\mathbf{H}_s \tag{31}$$

The only effect of including the conductivity σ is that the factor $j\omega\epsilon$ has now become $\sigma + j\omega\epsilon$. We may therefore immediately calculate the new value of the propagation constant,

$$\gamma^2 = (\sigma + j\omega\epsilon)j\omega\mu$$

$$\gamma = \pm\sqrt{(\sigma + j\omega\epsilon)j\omega\mu}$$

Dividing throughout by a familiar group of constants, we obtain

$$\gamma = j\omega\sqrt{\mu\epsilon}\sqrt{1 - j\frac{\sigma}{\omega\epsilon}} \tag{32}$$

where we have retained only the plus sign of the radical for a reason which will be evident shortly. This expression differs from the lossless case in the presence of the second radical factor, which becomes unity as σ vanishes. In the general case, values for σ, μ, ϵ, and ω may be inserted into (32), the real and imaginary parts of γ calculated,

$$\gamma = \alpha + j\beta$$

and the x component of the electric field intensity propagating in the $+z$ direction obtained,

$$E_{xs} = E_{x0}e^{-\alpha z}e^{-j\beta z}$$

The use of the plus sign for the radical in (32) leads to positive numerical values for α and β, and hence corresponds to propagation in the $+z$ direction.

Using (31), it is easily shown that H_{ys} is

$$H_{ys} = \frac{E_{x0}}{\eta}e^{-\alpha z}e^{-j\beta z}$$

where the intrinsic impedance is now a complex quantity,

$$\eta = \sqrt{\frac{j\omega\mu}{\sigma + j\omega\epsilon}} = \sqrt{\frac{\mu}{\epsilon}}\frac{1}{\sqrt{1 - j(\sigma/\omega\epsilon)}} \tag{33}$$

The electric and magnetic fields are no longer in time phase.

Before we consider an example illustrating these calculations, the nature of the exponential factor $e^{-\alpha z}$ bears some investigation. For a wave propagating in the $+z$ direction, this factor causes an exponential decrease in amplitude with increasing values of z. The attenuation constant is measured in nepers per meter (Np/m) in order that the exponent of e be measured in the dimensionless units of nepers.[2] Thus, if $\alpha = 0.01$ Np/m, the crest amplitude of the wave at $z = 50$ m will be $e^{-0.5}/e^{-0} = 0.607$ of its value at $z = 0$. In traveling a distance $1/\alpha$ in the $+z$ direction, the amplitude of the wave is reduced by the familiar factor of e^{-1}, or 0.368.

Some numerical results may be calculated for distilled water, which is quite a poor dielectric. At $\omega = 10^{11}$ rad/s, or $f = 15.9$ GHz, a value which falls in the SHF (super-high-frequency) band, representative values of the parameters are $\mu_R = 1$, $\epsilon_R = 50$, and $\sigma = 20$ ℧/m. Working with a little more than our usual accuracy, we have

$$\frac{\sigma}{\omega\epsilon} = \frac{20 \times 10^{12}}{10^{11} \times 50 \times 8.854} = 0.4518$$

and

$$\begin{aligned}\gamma &= j10^{11}\sqrt{4\pi \times 10^{-7} \times 8.854 \times 50 \times 10^{-12}}\sqrt{1 - j0.4518} \\ &= j2359\sqrt{1.0973\underline{/-24.31^\circ}} \\ &= 2471\underline{/77.84^\circ} = 520.3 + j2415 \qquad \text{m}^{-1}\end{aligned}$$

Therefore

$$\alpha = 520.3 \qquad \text{Np/m}$$

and the amplitude of E_x or H_y will attenuate by a factor of 0.368 for every $\frac{1}{520}$ m (about 2 mm) of propagation in the water. The term "propagates" is thus being used quite loosely. The large attenuation indicates the reason why radar is ineffective under water and sonar is used instead; it also suggests that water or rain in the atmosphere may cause propagation problems at high frequencies.

The phase constant is

$$\beta = 2415 \qquad \text{rad/m}$$

and this has been affected only slightly by the nonzero conductivity, for the numerical calculations above show that it would have been 2359 rad/m if σ were zero. The wavelength at this frequency is 1.886 cm in air, and since $\beta = 2\pi/\lambda$, it is 2.601 mm in water.

The intrinsic impedance is found to be

$$\eta = \sqrt{\frac{4\pi 10^{-7+12}}{50 \times 8.854}}\frac{1}{\sqrt{1 - j0.4518}} = 50.86\underline{/12.16^\circ} = 49.72 + j10.71 \qquad \Omega$$

and E_x leads H_y by 12.16° at every point.

Let us now turn our attention to the more practical case of a dielectric material that has some small loss. The criterion by which we should judge whether or not the loss is small is the magnitude of $\sigma/\omega\epsilon$ as compared to unity, as shown by (32)

---

[2] The term *neper* was selected (by some poor speller) to honor John Napier, a Scottish mathematician who first proposed the use of logarithms.

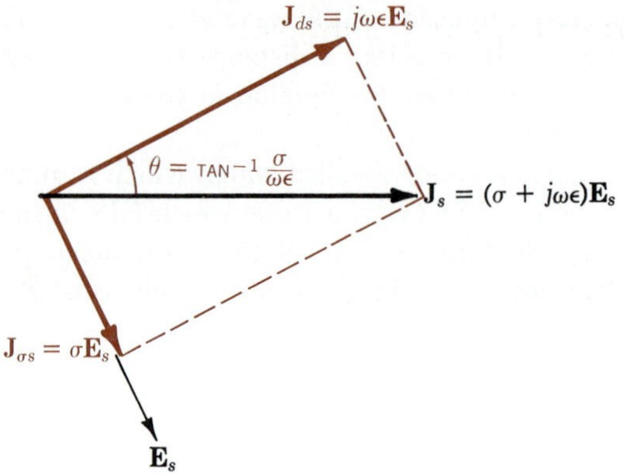

FIGURE 11.2
The time-phase relationship between $\mathbf{J}_{ds}$, $\mathbf{J}_{\sigma s}$, $\mathbf{J}_s$, and $\mathbf{E}_s$. The tangent of θ is equal to $\sigma/\omega\epsilon$, and $90° - \theta$ is the common power-factor angle, or the angle by which $\mathbf{J}_s$ leads $\mathbf{E}_s$.

and (33). The term $\sigma/\omega\epsilon$ is referred to as the *loss tangent* for a reason that becomes clear when we consider Maxwell's curl **H** equation, which formed the starting point for our analysis,

$$\nabla \times \mathbf{H}_s = (\sigma + j\omega\epsilon)\mathbf{E}_s = \mathbf{J}_{\sigma s} + \mathbf{J}_{ds}$$

The ratio of conduction current density to displacement current density is

$$\frac{\mathbf{J}_{\sigma s}}{\mathbf{J}_{ds}} = \frac{\sigma}{j\omega\epsilon}$$

That is, these two vectors point in the same direction in space, but they are 90° out of phase in time. Displacement current density leads conduction current density by 90°, just as the current through a capacitor leads the current through a resistor in parallel with it by 90° in an ordinary electric circuit. This phase relationship is shown in Fig. 11.2. The angle θ (not to be confused with the polar angle in spherical coordinates) may therefore be identified as the angle by which the displacement current density leads the total current density, and

$$\boxed{\tan\theta = \frac{\sigma}{\omega\epsilon}} \tag{34}$$

This relationship has led to the name of "loss tangent" for $\sigma/\omega\epsilon$. Problem 17 at the end of the chapter indicates that the Q of a capacitor (its quality factor, not its charge) which incorporates a lossy dielectric is the reciprocal of the loss tangent.

If the loss tangent is small, then we may obtain useful approximations for the attenuation and phase constants, and the intrinsic impedance. Since

$$\gamma = j\omega\sqrt{\mu\epsilon}\sqrt{1 - j\frac{\sigma}{\omega\epsilon}} \tag{32}$$

we may expand the second radical by the binomial theorem

$$(1+x)^n = 1 + nx + \frac{n(n-1)}{2!}x^2 + \frac{n(n-1)(n-2)}{3!}x^3 + \cdots$$

where $|x| < 1$. We identify x as $-j\sigma/\omega\epsilon$ and n as $\frac{1}{2}$, and thus

$$\gamma = j\omega\sqrt{\mu\epsilon}\left[1 - j\frac{\sigma}{2\omega\epsilon} + \frac{1}{8}\left(\frac{\sigma}{\omega\epsilon}\right)^2 + \cdots\right]$$

Hence

$$\alpha \doteq j\omega\sqrt{\mu\epsilon}\left(-j\frac{\sigma}{2\omega\epsilon}\right) = \frac{\sigma}{2}\sqrt{\frac{\mu}{\epsilon}} \tag{35}$$

and

$$\beta \doteq \omega\sqrt{\mu\epsilon}\left[1 + \frac{1}{8}\left(\frac{\sigma}{\omega\epsilon}\right)^2\right] \tag{36a}$$

or in many cases,

$$\beta \doteq \omega\sqrt{\mu\epsilon} \tag{36b}$$

In a similar way, we find

$$\eta \doteq \sqrt{\frac{\mu}{\epsilon}}\left[1 - \frac{3}{8}\left(\frac{\sigma}{\omega\epsilon}\right)^2 + j\frac{\sigma}{2\omega\epsilon}\right] \tag{37a}$$

or

$$\eta \doteq \sqrt{\frac{\mu}{\epsilon}}\left(1 + j\frac{\sigma}{2\omega\epsilon}\right) \tag{37b}$$

In order to illustrate the accuracy of our approximations, let us recalculate α, β, and η for the distilled-water example discussed earlier, even though the loss tangent has the relatively large value of 0.4518. With $\epsilon_R = 50$, $\sigma = 20$ ℧/m, $\omega = 10^{11}$, and $\mu_R = 1$, we have from (35)

$$\alpha = 532.8 \qquad \text{Np/m}$$

as compared with the exact value of 520.3 Np/m. Using (36*a*), the phase constant is

$$\beta = 2419 \qquad \text{rad/m}$$

which is close to the exact value of 2415 rad/m, while (36*b*) yields the lossless value

$$\beta = 2359 \qquad \text{rad/m}$$

Finally, the intrinsic impedance is found by (37*a*) to be

$$\eta = 50.65\underline{/13.75^\circ} = 49.20 + j12.03 \qquad \Omega$$

as compared to an exact value of $50.86\underline{/12.16^\circ} = 49.72 + j10.71\ \Omega$, while a greater error is introduced by (37*b*),

$$\eta = 54.62\underline{/12.73^\circ} = 53.28 + j12.03 \qquad \Omega$$

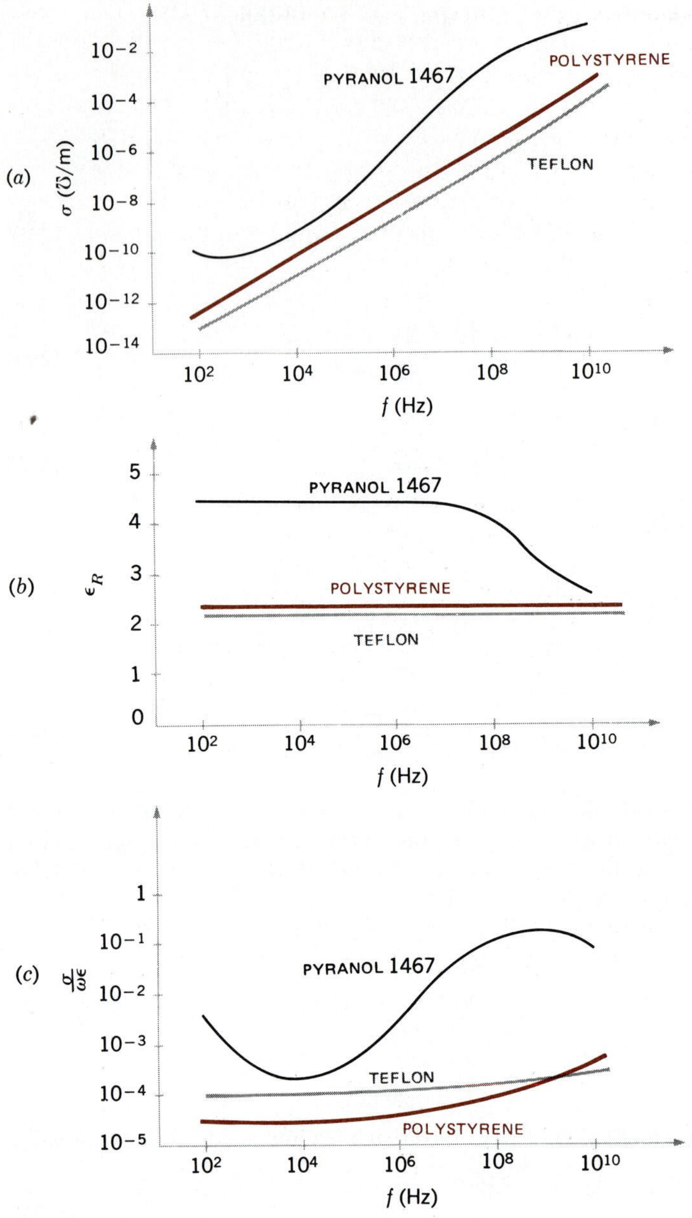

FIGURE 11.3
The variation of σ, ϵ_R, and $\sigma/\omega\epsilon$ is shown in (*a*), (*b*), and (*c*), respectively, for Pyranol 1467, polystyrene, and Teflon. Note the use of logarithmic scales for all axes except ϵ_R.

Although the loss tangent in this example is 0.4518, the errors invloved in using the more approximate formulations are probably unimportant because the conductivity and relative dielectric constant are seldom known to any great accuracy. We shall, however, recommend use of the approximations only when $\sigma/\omega\epsilon < 0.1$; any further approximations should be based on engineering judgment.

In most physical dielectrics the loss tangent is more constant with frequency than is the conductivity. That is, conductivity tends to increase with frequency, although not linearly. There may also be relatively rapid changes of conductivity, permittivity, and loss tangent in the infrared region and again in the ultraviolet region.[3] A sketch of the variation of σ, ϵ_R, and $\sigma/\omega\epsilon$ with frequency for Pyranol 1467, Teflon, and polystyrene is shown in Fig. 11.3 on a logarithmic frequency scale.

D11.4. Given a nonmagnetic material having $\epsilon_R = 2.25$ and $\sigma = 10^{-4}$ ℧/m, find numerical values at 2.5 MHz for: (*a*) the loss tangent; (*b*) the attenuation constant; (*c*) the phase constant; (*d*) the intrinsic impedance.

Ans. 0.320; 0.012 40 Np/m; 0.0796 rad/m; $245\underline{/8.86^\circ}\ \Omega$

D11.5. Consider a material for which $\mu_R = 1$, $\epsilon_R = 4$, and the loss tangent is 0.1. If these three values are constant with frequency for $0.5\text{ MHz} \le f \le 80\text{ MHz}$, calculate: (*a*) σ at 1 and 50 MHz; (*b*) λ at 1 and 50 MHz; (*c*) v at 1 and 50 MHz.

Ans. 22.3 and 1113 μ℧/m; 149.7 and 2.99 m; 1.497×10^8 m/s twice

11.4 THE POYNTING VECTOR AND POWER CONSIDERATIONS

In order to find the power in a uniform plane wave, it is necessary to develop a power theorem for the electromagnetic field known as the Poynting theorem. It was originally postulated in 1884 by an English physicist, John H. Poynting.

Let us begin with Maxwell's equation,

$$\nabla \times \mathbf{H} = \mathbf{J} + \frac{\partial \mathbf{D}}{\partial t}$$

and dot each side of the equation with **E**,

$$\mathbf{E} \cdot \nabla \times \mathbf{H} = \mathbf{J} \cdot \mathbf{E} + \mathbf{E} \cdot \frac{\partial \mathbf{D}}{\partial t}$$

We now make use of the vector identity,

$$\nabla \cdot (\mathbf{E} \times \mathbf{H}) \equiv -\mathbf{E} \cdot \nabla \times \mathbf{H} + \mathbf{H} \cdot \nabla \times \mathbf{E}$$

[3] The variation of all these parameters with frequency is given for many materials in von Hippel. See the Suggested References at the end of the chapter.

which may be proved by expansion in cartesian coordinates. Thus

$$\mathbf{H} \cdot \nabla \times \mathbf{E} - \nabla \cdot (\mathbf{E} \times \mathbf{H}) = \mathbf{J} \cdot \mathbf{E} + \mathbf{E} \cdot \frac{\partial \mathbf{D}}{\partial t}$$

But

$$\nabla \times \mathbf{E} = -\frac{\partial \mathbf{B}}{\partial t}$$

and therefore

$$-\mathbf{H} \cdot \frac{\partial \mathbf{B}}{\partial t} - \nabla \cdot (\mathbf{E} \times \mathbf{H}) = \mathbf{J} \cdot \mathbf{E} + \mathbf{E} \cdot \frac{\partial \mathbf{D}}{\partial t}$$

or

$$-\nabla \cdot (\mathbf{E} \times \mathbf{H}) = \mathbf{J} \cdot \mathbf{E} + \epsilon \mathbf{E} \cdot \frac{\partial \mathbf{E}}{\partial t} + \mu \mathbf{H} \cdot \frac{\partial \mathbf{H}}{\partial t}$$

However,

$$\epsilon \mathbf{E} \cdot \frac{\partial \mathbf{E}}{\partial t} = \frac{\epsilon}{2} \frac{\partial E^2}{\partial t} = \frac{\partial}{\partial t}\left(\frac{\epsilon E^2}{2}\right)$$

and

$$\mu \mathbf{H} \cdot \frac{\partial \mathbf{H}}{\partial t} = \frac{\partial}{\partial t}\left(\frac{\mu H^2}{2}\right)$$

Thus

$$-\nabla \cdot (\mathbf{E} \times \mathbf{H}) = \mathbf{J} \cdot \mathbf{E} + \frac{\partial}{\partial t}\left(\frac{\epsilon E^2}{2} + \frac{\mu H^2}{2}\right)$$

Finally, we integrate throughout a volume,

$$-\int_{\text{vol}} \nabla \cdot (\mathbf{E} \times \mathbf{H})\, dv = \int_{\text{vol}} \mathbf{J} \cdot \mathbf{E}\, dv + \frac{\partial}{\partial t} \int_{\text{vol}} \left(\frac{\epsilon E^2}{2} + \frac{\mu H^2}{2}\right) dv$$

and apply the divergence theorem to obtain

$$\boxed{-\oint_S (\mathbf{E} \times \mathbf{H}) \cdot d\mathbf{S} = \int_{\text{vol}} \mathbf{J} \cdot \mathbf{E}\, dv + \frac{\partial}{\partial t} \int_{\text{vol}} \left(\frac{\epsilon E^2}{2} + \frac{\mu H^2}{2}\right) dv} \qquad (38)$$

If we assume that there are no sources within the volume, then the first integral on the right is the total (but instantaneous) ohmic power dissipated within the volume. If sources are present within the volume, then the result of integrating over the volume of the source will be positive if power is being delivered *to* the source, but it will be negative if power is being delivered *by* the source.

The integral in the second term on the right is the total energy stored in the electric and magnetic fields,[4] and the partial derivatives with respect to time cause this term to be the time rate of increase of energy stored within this volume, or the instantaneous power going to increase the stored energy within this volume. The sum of the expressions on the right must therefore be the total power flowing *into* this volume, and thus the total power flowing *out* of the volume is

$$\oint_S (\mathbf{E} \times \mathbf{H}) \cdot d\mathbf{S}$$

where the integral is over the closed surface surrounding the volume. The cross product $\mathbf{E} \times \mathbf{H}$ is known as the Poynting vector $\mathscr{P}$,

$$\boxed{\mathscr{P} = \mathbf{E} \times \mathbf{H}} \tag{39}$$

which is interpreted as an instantaneous power density, measured in watts per square meter ($\mathrm{W/m^2}$). This interpretation is subject to the same philosophical considerations as was the description of $\frac{1}{2}\mathbf{D} \cdot \mathbf{E}$ or $\frac{1}{2}\mathbf{B} \cdot \mathbf{H}$ as energy densities. We can show rigorously only that the integration of the Poynting vector over a closed surface yields the total power crossing the surface in an outward sense. This interpretation as a power density does not lead us astray, however, especially when applied to sinusoidally varying fields. Problem 18 indicates that strange results may be found when the Poynting vector is applied to time-constant fields.

The direction of the vector $\mathscr{P}$ indicates the direction of the instantaneous power flow at the point, and many of us think of the Poynting vector as a "pointing" vector. This homonym, while accidental, is correct.

Since $\mathscr{P}$ is given by the cross product of $\mathbf{E}$ and $\mathbf{H}$, the direction of power flow at any point is normal to both the $\mathbf{E}$ and $\mathbf{H}$ vectors. This certainly agrees with our experience with the uniform plane wave, for propagation in the $+z$ direction was associated with an E_x and H_y component,

$$E_x\mathbf{a}_x \times H_y\mathbf{a}_y = \mathscr{P}_z\mathbf{a}_z$$

In a perfect dielectric these $\mathbf{E}$ and $\mathbf{H}$ fields are given by

$$E_x = E_{x0} \cos(\omega t - \beta z)$$

$$H_y = \frac{E_{x0}}{\eta} \cos(\omega t - \beta z)$$

and thus

$$\mathscr{P}_z = \frac{E_{x0}^2}{\eta} \cos^2(\omega t - \beta z)$$

[4] This is the expression for magnetic field energy that we have been anticipating since Chap. 9.

To find the time-average power density, we integrate over one cycle and divide by the period $T = 1/f$,

$$\mathscr{P}_{z,\,\text{av}} = f \int_0^{1/f} \frac{E_{x0}^2}{\eta} \cos^2(\omega t - \beta z)\,dt$$

$$= \frac{f}{2}\frac{E_{x0}^2}{\eta} \int_0^{1/f} [1 + \cos(2\omega t - 2\beta z)]\,dt$$

$$= \frac{f}{2}\frac{E_{x0}^2}{\eta} \left[t + \frac{1}{2\omega}\sin(2\omega t - 2\beta z) \right]_0^{1/f}$$

and

$$\boxed{\mathscr{P}_{z,\,\text{av}} = \frac{1}{2}\frac{E_{x0}^2}{\eta} \qquad \text{W/m}^2} \tag{40}$$

If we were using root-mean-square values instead of peak amplitudes, the factor of $\frac{1}{2}$ would not be present.

Finally, the average power flowing through any area S normal to the z axis is[5]

$$P_{z,\,\text{av}} = \frac{1}{2}\frac{E_{x0}^2}{\eta} S \qquad \text{W}$$

In the case of a lossy dielectric, E_x and H_y are not in time phase. We have

$$E_x = E_{x0} e^{-\alpha z} \cos(\omega t - \beta z)$$

If we let

$$\eta = \eta_m \angle \theta_\eta$$

then we may write the magnetic field intensity as

$$H_y = \frac{E_{x0}}{\eta_m} e^{-\alpha z} \cos(\omega t - \beta z - \theta_\eta)$$

Thus,

$$\mathscr{P}_z = E_x H_y = \frac{E_{x0}^2}{\eta_m} e^{-2\alpha z} \cos(\omega t - \beta z) \cos(\omega t - \beta z - \theta_\eta)$$

Now is the time to use the identity $\cos A \cos B \equiv \frac{1}{2}\cos(A + B) + \frac{1}{2}\cos(A - B)$, improving the form of the last equation considerably,

$$\mathscr{P}_z = \frac{1}{2}\frac{E_{x0}^2}{\eta_m} e^{-2\alpha z}[\cos(2\omega t - 2\beta z - \theta_\eta) + \cos\theta_\eta]$$

---

[5] We shall use P for power as well as for polarization. If they both appear in the same equation in this book, it is an error.

We find that the power density has only a second-harmonic component and a dc component. Since the first term has a zero average value over an integral number of periods, the time-average value of the Poynting vector is

$$\mathscr{P}_{z,\,\mathrm{av}} = \frac{1}{2}\frac{E_{x0}^2}{\eta_m} e^{-2\alpha z} \cos \theta_\eta \tag{41}$$

Note that the power density attenuates as $e^{-2\alpha z}$, whereas E_x and H_y fall off as $e^{-\alpha z}$.

Let us now make a few power calculations for the fresh water example considered in the last section. Some of the data are: $f = 300$ MHz, $\epsilon_R = 50$, $\sigma = 20$ ℧/m, $\alpha = 520.3$ Np/m, $\beta = 2415$ rad/m, and $\eta = 50.86\underline{/12.16^\circ}$ Ω. We now select $E_{x0} = 100$ V/m and $z = 1$ mm. Thus, $\alpha z = 0.5203$ and $\beta z = 2.415$ rad or 138.4°. We have

$$E_x = 100e^{-0.5203} \cos(6\pi 10^8 t - 138.4^\circ) = 59.4 \cos(6\pi 10^8 t - 138.4^\circ) \quad \mathrm{V/m}$$

$$H_y = \frac{100}{50.86} e^{-0.5203} \cos(6\pi 10^8 t - 138.4^\circ - 12.16^\circ)$$

or

$$H_y = 1.169 \cos(6\pi 10^8 t - 150.6^\circ) \quad \mathrm{A/m}$$

Thus,

$$\mathscr{P}_z = E_x H_y = 69.4 \cos(6\pi 10^8 t - 138.4^\circ) \cos(6\pi 10^8 t - 150.6^\circ) \quad \mathrm{W/m^2}$$

or

$$\begin{aligned}\mathscr{P}_z &= 34.7[\cos(12\pi 10^8 t - 289.0^\circ) + \cos 12.2^\circ] \\ &= 33.9 + 34.7 \cos(12\pi 10^8 t - 289.0^\circ) \quad \mathrm{W/m^2}\end{aligned}$$

Equation (41) gives us the time-average part of the answer very quickly,

$$\mathscr{P}_{z,\,\mathrm{av}} = \frac{1}{2}\frac{100^2}{50.86} e^{-2\times 0.5203} \cos 12.16^\circ = 33.9 \quad \mathrm{W/m^2}$$

D11.6. At frequencies of 1, 100, and 3000 MHz, the dielectric constant of ice made from pure water has values of 4.15, 3.45, and 3.20, respectively, while the loss tangent is 0.12, 0.035, 0.0009, also respectively. If a uniform plane wave with an amplitude of 100 V/m at $z = 0$ is propagating through such ice, find the time-average power density at $z = 0$ and $z = 10$ m for each frequency.

Ans. 27.1 and 25.7 W/m^2; 24.7 and 6.31 W/m^2; 23.7 and 8.63 W/m^2

11.5 PROPAGATION IN GOOD CONDUCTORS: SKIN EFFECT

As our final example of unbounded propagation we shall investigate the behavior of a good conductor when a uniform plane wave is established in it. Rather than thinking of a source embedded in a block of copper and launching a wave in that material, we should be more interested in a wave that is established by an electromagnetic field existing in some external dielectric that adjoins the conductor surface. We shall see that the primary transmission of energy must take place in the region *outside* the conductor, because all time-varying fields attenuate very quickly *within* a good conductor.

The good conductor has a high conductivity and large conduction currents. The energy represented by the wave traveling through the material therefore decreases as the wave propagates because ohmic losses are continuously present. When we discussed the loss tangent, we saw that the ratio of conduction current density to the displacement current density in a material is given by $\sigma/\omega\epsilon$. Choosing a poor metallic conductor and a very high frequency as a conservative example, this ratio[6] for nichrome ($\sigma \doteq 10^6$) at 100 MHz is about 2×10^8. Thus we have a situation where $\sigma/\omega\epsilon \gg 1$, and we should be able to make several very good approximations to find α, β, and η for a good conductor.

The general expression for the propagation constant is

$$\gamma = j\omega\sqrt{\mu\epsilon}\sqrt{1 - j\frac{\sigma}{\omega\epsilon}}$$

which we immediately simplify to obtain

$$\gamma = j\omega\sqrt{\mu\epsilon}\sqrt{-j\frac{\sigma}{\omega\epsilon}}$$

or

$$\gamma = j\sqrt{-j\omega\mu\sigma}$$

But

$$-j = 1\underline{/-90^\circ}$$

and

$$\sqrt{1\underline{/-90^\circ}} = 1\underline{/-45^\circ} = \frac{1}{\sqrt{2}} - j\frac{1}{\sqrt{2}}$$

Therefore

$$\gamma = j\left(\frac{1}{\sqrt{2}} - j\frac{1}{\sqrt{2}}\right)\sqrt{\omega\mu\sigma}$$

or

$$\gamma = (j1 + 1)\sqrt{\pi f\mu\sigma} \tag{42}$$

Hence

$$\alpha = \beta = \sqrt{\pi f\mu\sigma} \tag{43}$$

[6] It is customary to take $\epsilon = \epsilon_0$ for metallic conductors.

Regardless of the parameters μ and σ of the conductor or of the frequency of the applied field, α and β are equal. If we again assume only an E_x component traveling in the $+z$ direction, then

$$E_x = E_{x0} e^{-z\sqrt{\pi f \mu \sigma}} \cos(\omega t - z\sqrt{\pi f \mu \sigma}) \tag{44}$$

We may tie this field in the conductor to an external field at the conductor surface. We let the region $z > 0$ be the good conductor and the region $z < 0$ be a perfect dielectric. At the boundary surface $z = 0$, (44) becomes

$$E_x = E_{x0} \cos \omega t \qquad (z = 0)$$

This we shall consider as the source field that establishes the fields within the conductor. Since displacement current is negligible,

$$\mathbf{J} = \sigma \mathbf{E}$$

Thus, the conduction current density at any point within the conductor is directly related to **E**:

$$J_x = \sigma E_x = \sigma E_{x0} e^{-z\sqrt{\pi f \mu \sigma}} \cos(\omega t - z\sqrt{\pi f \mu \sigma}) \tag{45}$$

Equations (44) and (45) contain a wealth of information. Considering first the negative exponential term, we find an exponential decrease in the conduction current density and electric field intensity with penetration into the conductor (away from the source). The exponential factor is unity at $z = 0$ and decreases to $e^{-1} = 0.368$ when

$$z = \frac{1}{\sqrt{\pi f \mu \sigma}}$$

This distance is denoted by δ and is termed the *depth of penetration*, or the *skin depth*,

$$\boxed{\delta = \frac{1}{\sqrt{\pi f \mu \sigma}} = \frac{1}{\alpha} = \frac{1}{\beta}} \tag{46}$$

It is an important parameter in describing conductor behavior in electromagnetic fields. To get some idea of the magnitude of the skin depth, let us consider copper, $\sigma = 5.8 \times 10^7$ ℧/m, at several different frequencies. We have

$$\delta_{Cu} = \frac{0.0661}{\sqrt{f}}$$

At a power frequency of 60 Hz, $\delta_{Cu} = 8.53$ mm, or about $\frac{1}{3}$ in. Remembering that the power density carries an exponential term $e^{-2\alpha z}$, we see that the power density is multiplied by a factor of $0.368^2 = 0.135$ for every 8.53 mm of distance into the copper.

At a microwave frequency of 10 000 MHz, δ is 6.61×10^{-4} mm, or about one-eighth the wavelength of visible light.

Thus, all fields in a good conductor such as copper are essentially zero at distances greater than a few skin depths from the surface. Any current density or electric field intensity established at the surface of a good conductor decays rapidly as we progress into the conductor. Electromagnetic energy is not transmitted in the interior of a conductor; it travels in the region surrounding the conductor, while the conductor merely guides the waves. The currents established at the conductor surface propagate into the conductor in a direction perpendicular to the direction of the current density, and they are attenuated by ohmic losses. This power loss is the price exacted by the conductor for acting as a guide. We shall consider guided propagation in more detail in the following chapter.

Suppose we have a copper bus bar in the substation of an electric utility company which we wish to have carry large currents, and we therefore select dimensions of 2 by 4 in. Then much of the copper is wasted, for the fields are greatly reduced in one skin depth, about $\frac{1}{3}$ in.[7] A hollow conductor with a wall thickness of about $\frac{1}{2}$ in would be a much better design. Although we are applying the results of an analysis for an infinite planar conductor to one of finite dimensions, the fields are attenuated in the finite-size conductor in a similar (but not identical) fashion.

The extremly short skin depth at microwave frequencies shows that only the surface coating of the guiding conductor is important. A piece of glass with an evaporated silver surface 0.0001 in thick is an excellent conductor at these frequencies.

Next, let us determine expressions for the velocity and wavelength within a good conductor. From (46), we already have

$$\alpha = \beta = \frac{1}{\delta} = \sqrt{\pi f \mu \sigma}$$

Then, since

$$\beta = \frac{2\pi}{\lambda}$$

we find the wavelength to be

$$\lambda = 2\pi\delta \tag{47}$$

Also, recalling Eq. (27), Sec. 11.2,

$$v = \frac{\omega}{\beta}$$

we have

$$v = \omega\delta \tag{48}$$

For copper at 60 Hz, $\lambda = 5.36$ cm and $v = 3.22$ m/s, or about 7.2 mi/h. A lot of us can run faster than that. In free space, of course, a 60-Hz wave has a wavelength of 3100 mi and travels at the velocity of light.

---

[7] This utility company operates at 60 Hz.

In order to find H_y, we need an expression for the intrinsic impedance of a good conductor. We begin with Eq. (33), Sec. 11.3,

$$\eta = \sqrt{\frac{j\omega\mu}{\sigma + j\omega\epsilon}}$$

Since $\sigma \gg \omega\epsilon$, we have

$$\eta = \sqrt{\frac{j\omega\mu}{\sigma}}$$

which may be written as

$$\eta = \frac{\sqrt{2}\underline{/45^\circ}}{\sigma\delta} = \frac{1}{\sigma\delta} + j\frac{1}{\sigma\delta} \tag{49}$$

Thus, if we rewrite (44) in terms of the skin depth,

$$E_x = E_{x0}e^{-z/\delta}\cos\left(\omega t - \frac{z}{\delta}\right) \tag{50}$$

then

$$H_y = \frac{\sigma\delta E_{x0}}{\sqrt{2}}\, e^{-z/\delta}\cos\left(\omega t - \frac{z}{\delta} - \frac{\pi}{4}\right) \tag{51}$$

and we see that the maximum amplitude of the magnetic field intensity occurs one-eighth of a cycle later than the maximum amplitude of the electric field intensity at every point.

From (50) and (51) we may obtain the time-average Poynting vector by applying (41),

$$\mathscr{P}_{z,\,\text{av}} = \frac{1}{2}\frac{\sigma\delta E_{x0}^2}{\sqrt{2}}\, e^{-2z/\delta}\cos\frac{\pi}{4}$$

or

$$\mathscr{P}_{z,\,\text{av}} = \tfrac{1}{4}\sigma\delta E_{x0}^2 e^{-2z/\delta}$$

We again note that in a distance of one skin depth the power density is only $e^{-2} = 0.135$ of its value at the surface.

The total power loss in a width $0 < y < b$ and length $0 < x < L$ in the direction of the current, as shown in Fig. 11.4, is obtained by finding the power crossing the conductor surface within this area,

$$P_{L,\,\text{av}} = \int_S \mathscr{P}_{z,\,\text{av}}\, dS = \int_0^b \int_0^L \tfrac{1}{4}\sigma\delta E_{x0}^2 e^{-2z/\delta}\Big]_{z=0} dx\, dy$$
$$= \tfrac{1}{4}\sigma\delta b L E_{x0}^2$$

In terms of the current density J_{x0} at the surface,

$$J_{x0} = \sigma E_{x0}$$

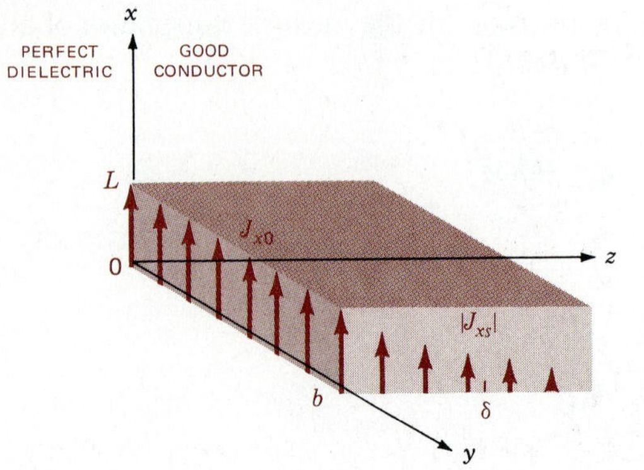

FIGURE 11.4
The current density $J_x = J_{x0}e^{-z/\delta}e^{-jz/\delta}$ decreases in magnitude as the wave propagates into the conductor. The average power loss in the region $0 < x < L$, $0 < y < b$, $z > 0$ is $\delta bLJ_{x0}^2/4\sigma$ watts.

we have

$$P_{L,\,\text{av}} = \frac{1}{4\sigma}\,\delta bLJ_{x0}^2 \tag{52}$$

Now let us see what power loss would result if the *total* current in a width b were distributed *uniformly* in one skin depth. To find the total current, we integrate the current density over the infinite depth of the conductor,

$$I = \int_0^{\infty}\int_0^{b} J_x\,dy\,dz$$

where

$$J_x = J_{x0}e^{-z/\delta}\cos\left(\omega t - \frac{z}{\delta}\right)$$

or in complex exponential notation to simplify the integration,

$$\begin{aligned} J_{xs} &= J_{x0}e^{-z/\delta}e^{-jz/\delta} \\ &= J_{x0}e^{-(1+j1)z/\delta} \end{aligned}$$

Therefore,

$$\begin{aligned} I_s &= \int_0^{\infty}\int_0^{b} J_{x0}e^{-(1+j1)z/\delta}\,dy\,dz \\ &= J_{x0}be^{-(1+j1)z/\delta}\,\frac{-\delta}{1+j1}\Bigg]_0^{\infty} \\ &= \frac{J_{x0}b\delta}{1+j1} \end{aligned}$$

and

$$I = \frac{J_{x0}b\delta}{\sqrt{2}}\cos\left(\omega t - \frac{\pi}{4}\right)$$

If this current is distributed with a uniform density J' throughout the cross section $0 < y < b$, $0 < z < \delta$, then

$$J' = \frac{J_{x0}}{\sqrt{2}} \cos\left(\omega t - \frac{\pi}{4}\right)$$

The ohmic power loss per unit volume is $\mathbf{J} \cdot \mathbf{E}$, and thus the total instantaneous power dissipated in the volume under consideration is

$$P_L = \frac{1}{\sigma}(J')^2 bL\delta = \frac{J_{x0}^2}{2\sigma} bL\delta \cos^2\left(\omega t - \frac{\pi}{4}\right)$$

The time-average power loss is easily obtained, since the average value of the cosine-squared factor is one-half,

$$P_{L,\,\mathrm{av}} = \frac{1}{4\sigma} J_{x0}^2 bL\delta \tag{53}$$

Comparing (52) and (53), we see that they are identical. Thus the average power loss in a conductor with skin effect present may be calculated by assuming that the total current is distributed uniformly in one skin depth. In terms of resistance, we may say that the resistance of a width b and length L of an infinitely thick slab with skin effect is the same as the resistance of a rectangular slab of width b, length L, and thickness δ without skin effect, or with uniform current distribution.

We may apply this to a conductor of circular cross section with little error, provided that the radius a is much greater than the skin depth. The resistance at a high frequency where there is a well-developed skin effect is therefore found by considering a slab of width equal to the circumference $2\pi a$ and thickness δ. Hence

$$R = \frac{L}{\sigma S} = \frac{L}{2\pi a \sigma \delta} \tag{54}$$

A round copper wire of 1-mm radius and 1-km length has a resistance at direct current of

$$R_{\mathrm{dc}} = \frac{10^3}{\pi 10^{-6}(5.8 \times 10^7)} = 5.48 \quad \Omega$$

At 1 MHz the skin depth is 0.0661 mm. Thus $\delta \ll a$, and the resistance at 1 MHz is found by (54),

$$R = \frac{10^3}{2\pi 10^{-3}(5.8 \times 10^7)(0.0661 \times 10^{-3})} = 41.5 \quad \Omega$$

D11.7. A steel pipe is constructed of a material for which we may assume $\mu_R = 200$ and $\sigma = 5 \times 10^6$ ℧/m. The outer and inner radii are 8 and 6 mm, respectively, and the length is 80 m. If the total current carried by the pipe is $2 \cos 10^4 \pi t$ A, find: (*a*) the skin depth; (*b*) the effective resistance; (*c*) the dc resistance; (*d*) the time-average power loss.

Ans. 0.225 mm; 1.414 Ω; 0.1819 Ω; 2.83 W

11.6 REFLECTION OF UNIFORM PLANE WAVES

In order to treat problems of practical interest, we must turn our attention to regions of *finite* size. In this section we shall consider the phenomenon of reflection which occurs when a uniform plane wave is incident on the boundary between regions composed of two different materials. We shall establish expressions for the wave that is reflected from the interface and for that which is transmitted from one region into the other. These results will be directly applicable to impedance-matching problems in ordinary transmission lines as well as to waveguides and other more exotic transmission systems.

We again assume that we have only a single component of the electric field intensity. Let us select region 1 $(\epsilon_1,\mu_1,\sigma_1)$ as $z < 0$ and region 2 $(\epsilon_2,\mu_2,\sigma_2)$ as $z > 0$. Initially we establish the wave traveling in the $+z$ direction in region 1,

$$E_{x1}^{+} = E_{x10}^{+} e^{-\alpha_1 z} \cos(\omega t - \beta_1 z)$$

or

$$E_{xs1}^{+} = E_{x10}^{+} e^{-\gamma_1 z} \tag{55}$$

where we take E_{x10}^{+} as real. The subscript 1 identifies the region and the superscript + indicates a positively traveling wave. Associated with $\mathbf{E}_{s1}^{+}$ is a magnetic field

$$H_{ys1}^{+} = \frac{1}{\eta_1} E_{x10}^{+} e^{-\gamma_1 z} \tag{56}$$

where η_1 is complex unless σ_1 is zero. This uniform plane wave in region 1 which is traveling toward the boundary surface at $z = 0$ is called the *incident* wave. Since the direction of propagation of the incident wave is perpendicular to the boundary plane, we describe it as *normal incidence.*

We now recognize that energy may be transmitted across the boundary surface at $z = 0$ into region 2 by providing a wave moving in the $+z$ direction in that medium,

$$E_{xs2}^{+} = E_{x20}^{+} e^{-\gamma_2 z} \tag{57}$$

$$H_{ys2}^{+} = \frac{1}{\eta_2} E_{x20}^{+} e^{-\gamma_2 z} \tag{58}$$

This wave which moves away from the boundary surface into region 2 is called the *transmitted* wave; note the use of the different propagation constant γ_2 and intrinsic impedance η_2.

Now we must try to satisfy the boundary conditions at $z = 0$ with these assumed fields. E_x is a tangential field; therefore the $\mathbf{E}$ fields in regions 1 and 2 must be equal at $z = 0$. Setting $z = 0$ in (55) and (57) would require that $E_{x10}^{+} = E_{x20}^{+}$. However, H_y is also a tangential field and must be continuous across the boundary (no current sheets are present in real media). When we let $z = 0$ in (56) and (58), however, we find that we must have $E_{x10}^{+}/\eta_1 = E_{x20}^{+}/\eta_2$; since $E_{x10}^{+} = E_{x20}^{+}$, then

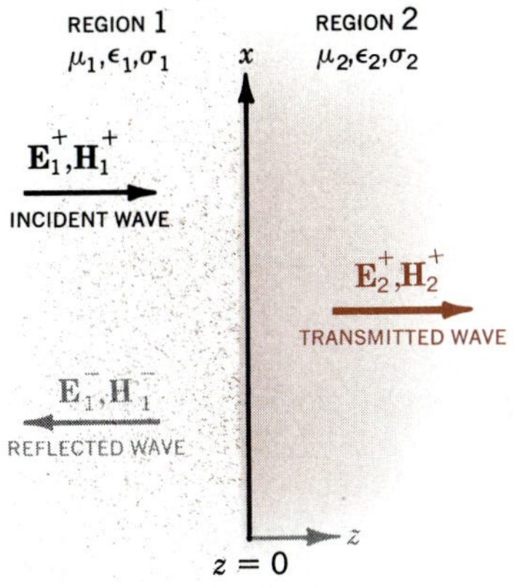

FIGURE 11.5
A wave $\mathbf{E}_1^+$ incident on a plane boundary establishes a reflected wave $\mathbf{E}_1^-$ and a transmitted wave $\mathbf{E}_2^+$.

$\eta_1 = \eta_2$. But this is a very special condition that does not fit the facts in general, and we are therefore unable to satisfy the boundary conditions with only an incident and a transmitted wave. We require a wave traveling away from the boundary in region 1, as shown in Fig. 11.5; this is called a *reflected* wave,

$$E_{xs1}^- = E_{x10}^- e^{\gamma_1 z} \tag{59}$$

$$H_{ys1}^- = -\frac{E_{x10}^-}{\eta_1} e^{\gamma_1 z} \tag{60}$$

where E_{x10}^- may be a complex quantity. Since this field is traveling in the $-z$ direction, $E_{xs1}^- = -\eta_1 H_{ys1}^-$, for the Poynting vector shows that $\mathbf{E}_1^- \times \mathbf{H}_1^-$ must be in the $-\mathbf{a}_z$ direction.

The boundary conditions are now easily satisfied, and in the process the amplitudes of the transmitted and reflected waves may be found in terms of E_{x10}^+. The total electric field intensity is continuous at $z = 0$,

$$E_{xs1} = E_{xs2} \qquad (z = 0)$$

or

$$E_{xs1}^+ + E_{xs1}^- = E_{xs2}^+ \qquad (z = 0)$$

Therefore

$$E_{x10}^+ + E_{x10}^- = E_{x20}^+ \tag{61}$$

Furthermore,

$$H_{ys1} = H_{ys2} \qquad (z = 0)$$

or

$$H_{ys1}^+ + H_{ys1}^- = H_{ys2}^+ \qquad (z = 0)$$

and therefore

$$\frac{E_{x10}^+}{\eta_1} - \frac{E_{x10}^-}{\eta_1} = \frac{E_{x20}^+}{\eta_2} \tag{62}$$

Solving (62) for E_{x20}^{+} and substituting into (61), we find

$$E_{x10}^{+} + E_{x10}^{-} = \frac{\eta_2}{\eta_1} E_{x10}^{+} - \frac{\eta_2}{\eta_1} E_{x10}^{-}$$

or

$$E_{x10}^{-} = E_{x10}^{+} \frac{\eta_2 - \eta_1}{\eta_2 + \eta_1}$$

The ratio of the amplitudes of the reflected and incident electric fields is called the *reflection coefficient* and designated by Γ (gamma),

$$\boxed{\Gamma = \frac{E_{x10}^{-}}{E_{x10}^{+}} = \frac{\eta_2 - \eta_1}{\eta_2 + \eta_1}} \tag{63}$$

The reflection coefficient may be complex, in which case there is a phase shift in the reflected wave.

The relative amplitude of the transmitted electric field intensity is found by combining (63) and (61),

$$\frac{E_{x20}^{+}}{E_{x10}^{+}} = \frac{2\eta_2}{\eta_2 + \eta_1} \tag{64}$$

This is known as the transmission coefficient, but we shall not use it enough to justify designating it by a special symbol.

Let us see how these results may be applied to several special cases. We first let region 1 be a perfect dielectric and region 2 be a perfect conductor. Then, since σ_2 is infinite,

$$\eta_2 = \sqrt{\frac{j\omega\mu_2}{\sigma_2 + j\omega\epsilon_2}} = 0$$

and from (64),

$$E_{x20}^{+} = 0$$

No time-varying fields can exist in the *perfect* conductor. An alternate way of looking at this is to note that the skin depth is zero.

Since $\eta_2 = 0$, then (63) shows that

$$\Gamma = -1$$

and

$$E_{x10}^{-} = -E_{x10}^{+}$$

The reflected wave is equal in amplitude and opposite in sign to the incident wave. All the incident energy is reflected by the perfect conductor, and the total

E field in region 1 is

$$E_{xs1} = E_{xs1}^{+} + E_{xs1}^{-}$$
$$= E_{x10}^{+} e^{-j\beta_1 z} - E_{x10}^{+} e^{j\beta_1 z}$$

where we have let $\gamma_1 = 0 + j\beta_1$ in the perfect dielectric. These terms may be combined and simplified,

$$E_{xs1} = (e^{-j\beta_1 z} - e^{j\beta_1 z})E_{x10}^{+}$$
$$= -j2 \sin \beta_1 z \, E_{x10}^{+}$$

Multiplying by $e^{j\omega t}$ and taking the real part, we may drop the s subscript and obtain the real function of time

$$\boxed{E_{x1} = 2E_{x10}^{+} \sin \beta_1 z \sin \omega t} \tag{65}$$

This total field in region 1 is not a traveling wave, although it was obtained by combining two waves of equal amplitude traveling in opposite directions. Let us compare its form with that of the incident wave,

$$\boxed{E_{x1}^{+} = E_{x10}^{+} \cos (\omega t - \beta_1 z)} \tag{66}$$

Here we see the term $\omega t - \beta_1 z$ or $\omega(t - z/v_1)$, which characterizes a wave traveling in the $+z$ direction at a velocity $v_1 = \omega/\beta_1$. In (65), however, the factors involving time and distance are separate trigonometric terms. At all planes for which $\beta_1 z = n\pi$, E_{x1} is zero for all time. Furthermore, whenever $\omega t = n\pi$, E_{x1} is zero everywhere. A field of the form of (65) is known as a *standing wave*.

The planes on which $E_{x1} = 0$ are located where

$$\beta_1 z = n\pi \qquad (n = 0, \pm 1, \pm 2, \ldots)$$

Thus

$$\frac{2\pi}{\lambda_1} z = n\pi$$

and

$$z = n \frac{\lambda_1}{2}$$

Thus $E_{x1} = 0$ at the boundary $z = 0$ and every half-wavelength from the boundary in region 1, $z < 0$, as illustrated in Fig. 11.6.

Since $E_{xs1}^{+} = H_{ys1}^{+} \eta_1$ and $E_{xs1}^{-} = -H_{ys1}^{-} \eta_1$, the magnetic field is

$$H_{ys1} = \frac{E_{x10}^{+}}{\eta_1} (e^{-j\beta_1 z} + e^{j\beta_1 z})$$

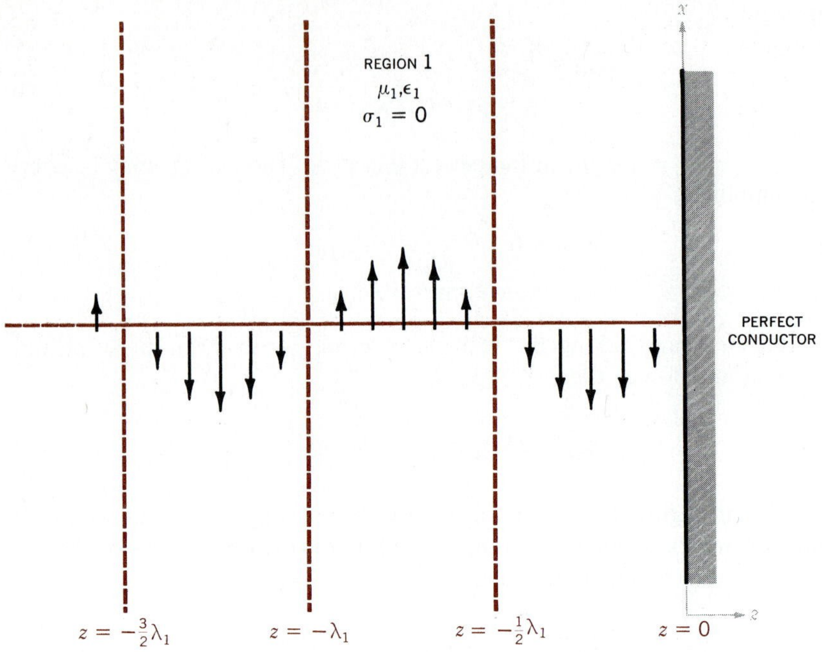

FIGURE 11.6
The instantaneous values of the total field E_{x1} are shown at $t = \pi/2$. $E_{x1} = 0$ for all time at multiples of one half-wavelength from the conducting surface.

or

$$H_{y1} = 2\frac{E_{x10}^+}{\eta_1}\cos\beta_1 z\cos\omega t \tag{67}$$

This is also a standing wave, but it shows a maximum amplitude at the positions where $E_{x1} = 0$. It is also 90° out of time phase with E_{x1} everywhere. Thus no average power is transmitted in either direction.

Let us now consider perfect dielectrics in both regions 1 and 2; η_1 and η_2 are both real positive quantities and $\alpha_1 = \alpha_2 = 0$. Equation (63) enables us to calculate the reflection coefficient and find E_{x1}^- in terms of the incident amplitude E_{x10}^+. Knowing E_{x1}^+ and E_{x1}^-, we then find H_{y1}^+ and H_{y1}^-. In region 2, E_{x2}^+ is found from (64), and this then determines H_{y2}^+. As a numerical example let us select

$$\eta_1 = 300 \quad \Omega$$

$$\eta_2 = 100 \quad \Omega$$

$$E_{x10}^+ = 100 \quad \text{V/m}$$

Then

$$\Gamma = \frac{100 - 300}{100 + 300} = -0.5$$

$$E_{x10}^{-} = -50 \qquad \text{V/m}$$

The magnetic field intensities are

$$H_{y10}^{+} = \frac{100}{300} = 0.333 \qquad \text{A/m}$$

$$H_{y10}^{-} = -\frac{-50}{300} = 0.167 \qquad \text{A/m}$$

The incident average power density is

$$P_{1,\,\text{av}}^{+} = \tfrac{1}{2}E_{x10}^{+}H_{y10}^{+} = 16.67 \qquad \text{W/m}^2$$

while

$$P_{1,\,\text{av}}^{-} = -\tfrac{1}{2}E_{x10}^{-}H_{y10}^{-} = 4.17 \qquad \text{W/m}^2$$

In region 2

$$E_{x20}^{+} = \frac{2\eta_2}{\eta_2 + \eta_1} E_{x10}^{+} = 50 \qquad \text{V/m}$$

and

$$H_{y20}^{+} = \frac{50}{100} = 0.500 \qquad \text{A/m}$$

Thus

$$P_{2,\,\text{av}}^{+} = \tfrac{1}{2}E_{x20}^{+}H_{y20}^{+} = 12.5 \qquad \text{W/m}^2$$

Note that energy is conserved:

$$P_{1,\,\text{av}}^{+} = P_{1,\,\text{av}}^{-} + P_{2,\,\text{av}}^{+}$$

The power relationship between incident, reflected, and transmitted powers should be compared with the equation expressing the continuity of the tangential electric field intensity at the boundary surface,

$$E_{x10}^{+} + E_{x10}^{-} = E_{x20}^{+}$$

The incident and reflected waves in region 1 may be combined to yield the total field there, but this will be reserved for the following section.

D11.8. The plane $x = 0$ is the boundary between two perfect dielectrics. For $x < 0$, $\mu_1 = \mu_0, \epsilon_1 = 3.6\pi$ pF/m, and $\sigma_1 = 0$; for $x > 0, \mu_2 = \mu_0, \epsilon_2 = 14.4\pi$ pF/m, and $\sigma_2 = 0$. If $E_{y1}^{+}(t) = 60 \cos(10^9 t - \beta_1 x)$ V/m, find: *(a)* β_1; *(b)* the incident magnetic field intensity as a function of time; *(c)* $E_{y1}^{-}(t)$; *(d)* β_2; *(e)* $E_{y2}^{+}(t)$.

Ans. 3.77 rad/m; $H_{z1}^{+}(t) = 0.18 \cos(10^9 t - 3.77x)$ A/m; $-20 \cos(10^9 t + 3.77x)$ V/m; 7.54 rad/m; $40 \cos(10^9 t - 7.54x)$ V/m

11.7 STANDING-WAVE RATIO

One of the measurements that is easily made on transmission systems is the *relative* amplitude of the electric or magnetic field intensity through use of a probe. A small coupling loop will give an indication of the amplitude of the magnetic field, while a slightly extended center conductor of a coaxial cable will sample the electric field. Both devices are customarily tuned to the operating frequency to provide increased sensitivity. The output of the probe is rectified and connected directly to a microammeter, or it may be delivered to an electronic voltmeter or a special amplifier. The indication is proportional to the amplitude of the sinusoidal time-varying field in which the probe is immersed.

When a uniform plane wave is traveling through a lossless region, and no reflected wave is present, the probe will indicate the same amplitude at every point. Of course, the instantaneous field which the probe samples will differ in phase by $\beta(z_2 - z_1)$ rad as the probe is moved from $z = z_1$ to $z = z_2$, but the system is insensitive to the phase of the field. The equal-amplitude voltages are characteristic of an unattenuated traveling wave.

When a wave traveling in a lossless medium is reflected by a perfect conductor, the total field is a standing wave and the voltage probe provides no output when it is located an integral number of half-wavelengths from the reflecting surface. As the probe position is changed, its output varies as $|\sin \beta z|$, where z is the distance from the conductor. This sinusoidal amplitude variation is shown in Fig. 11.7, and it characterizes a standing wave.

A more complicated situation arises when the reflected field is neither 0 nor 100 percent of the incident field. Some energy is transmitted into the second region and some is reflected. Region 1 therefore supports a field that is composed of both a traveling wave and a standing wave. It is customary to describe this field as a standing wave even though a traveling wave is also present. We shall see that the field does not have zero amplitude at any point for all time, and the degree to which the field is divided between a traveling wave and a true standing wave is expressed by the ratio of the maximum amplitude found by the probe to the minimum amplitude.

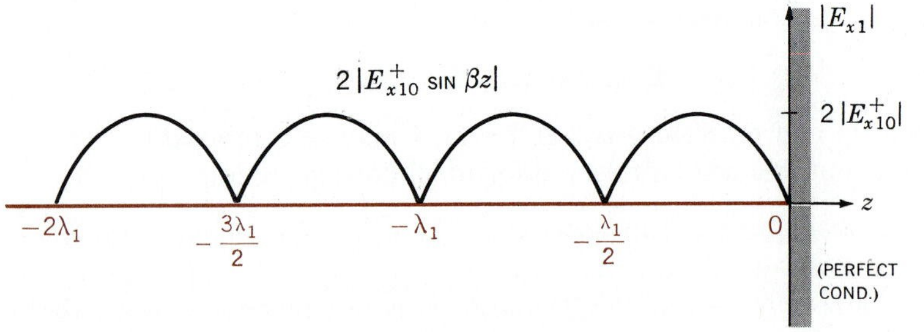

FIGURE 11.7
The standing voltage wave produced in a lossless medium by reflection from a perfect conductor varies as $|\sin \beta z|$.

Using the same fields investigated in the previous section, we combine the incident and reflected electric field intensities,

$$E_{x1} = E_{x1}^{+} + E_{x1}^{-}$$

The field E_{x1} is a sinusoidal function of t (generally with a nonzero phase angle), and it varies with z in a manner as yet unknown. We shall inspect all z to find the maximum and minimum amplitudes, and determine their ratio. We call this ratio the *standing-wave ratio*, and we shall symbolize it by s.

Let us now go through the mechanics of this procedure for the case in which medium 1 is a perfect dielectric, $\alpha_1 = 0$, but region 2 may be any material. We have

$$E_{xs1}^{+} = E_{x10}^{+} e^{-j\beta_1 z}$$

$$E_{xs1}^{-} = \Gamma E_{x10}^{+} e^{j\beta_1 z}$$

where

$$\Gamma = \frac{\eta_2 - \eta_1}{\eta_2 + \eta_1}$$

and η_1 is real and positive but η_2 may be complex. Thus Γ may be complex, and we allow for this possibility by letting

$$\boxed{\Gamma = |\Gamma| e^{j\phi}}$$

If region 2 is a perfect conductor, ϕ is equal to π; if η_2 is real and less than η_1, ϕ is also equal to π; and if η_2 is real and greater than η_1, ϕ is zero. The total field in region 1 is

$$E_{xs1} = (e^{-j\beta_1 z} + |\Gamma| e^{j(\beta_1 z + \phi)}) E_{x10}^{+} \tag{68}$$

We seek the maximum and minimum values of the magnitude of the complex quantity in the larger parentheses. We certainly have a maximum when each term in the larger parentheses has the same phase angle; thus, for E_{x10}^{+} positive and real,

$$E_{xs1,\,\text{max}} = (1 + |\Gamma|) E_{x10}^{+} \tag{69}$$

and this occurs where

$$-\beta_1 z = \beta_1 z + \phi + 2n\pi \qquad (n = 0, \pm 1, \pm 2, \ldots)$$

Thus

$$-\beta_1 z_{\text{max}} = \frac{\phi}{2} + n\pi \tag{70}$$

Note that a voltage maximum is located at the boundary plane ($z = 0$) if $\phi = 0$; moreover, $\phi = 0$ when Γ is real and positive. This occurs for real η_1 and η_2 when $\eta_2 > \eta_1$. Thus there is a voltage maximum at the boundary surface when the intrinsic impedance of region 2 is greater than that of region 1 and both impedances are real.

For the perfect conductor $\phi = \pi$, and these maxima are found at $-\beta_1 z = \pi/2$, $3\pi/2$, and so forth, as we saw before.

The minima must occur where the phase angles of the two terms in the larger parentheses differ by 180°; thus

$$E_{xs1,\,\min} = (1 - |\Gamma|)E_{x10}^{+} \tag{71}$$

and this occurs where

$$-\beta_1 z = \beta_1 z + \phi + \pi + 2n\pi \qquad (n = 0, \pm 1, \pm 2, \ldots)$$

or

$$-\beta_1 z_{\min} = \frac{\phi}{2} + n\pi + \frac{\pi}{2} \tag{72}$$

The minima are separated by multiples of one half-wavelength (as are the maxima), and for the perfect conductor the first minimum occurs when $-\beta_1 z = 0$, or at the conducting surface. In general, a voltage minimum is found at $z = 0$ whenever $\phi = \pi$; this occurs if $\eta_2 < \eta_1$ and both are real.

To illustrate some of these results, let us consider a 100-V/m, 3-GHz wave that is propagating in a material having $\epsilon_{R1} = 4$, $\mu_{R1} = 1$, and $\sigma_1 = 0$. The wave is normally incident on another perfect dielectric in region 2, $z > 0$, where $\epsilon_{R2} = 9$ and $\mu_{R2} = 1$ (Fig. 11.8).

We calculate $\omega = 6\pi 10^9$ rad/s, $\beta_1 = \omega\sqrt{\mu_1\epsilon_1} = 40\pi$ rad/m, and $\beta_2 = \omega\sqrt{\mu_2\epsilon_2} = 60\pi$ rad/m. Although the wavelength would be 10 cm in air, we find here that $\lambda_1 = 2\pi/\beta_1 = 5$ cm, $\lambda_2 = 2\pi/\beta_2 = 3.33$ cm, $\eta_1 = 60\pi\ \Omega$, $\eta_2 = 40\pi\ \Omega$, and $\Gamma = (\eta_2 - \eta_1)/(\eta_2 + \eta_1) = -0.2$. Since Γ is real and negative ($\eta_2 < \eta_1$), there will be a minimum of the electric field at the boundary, and it will be repeated at half-wavelength (2.5 cm) intervals in dielectric 1. From (71), we see that $E_{xs1,\,\min} =$ 80 V/m.

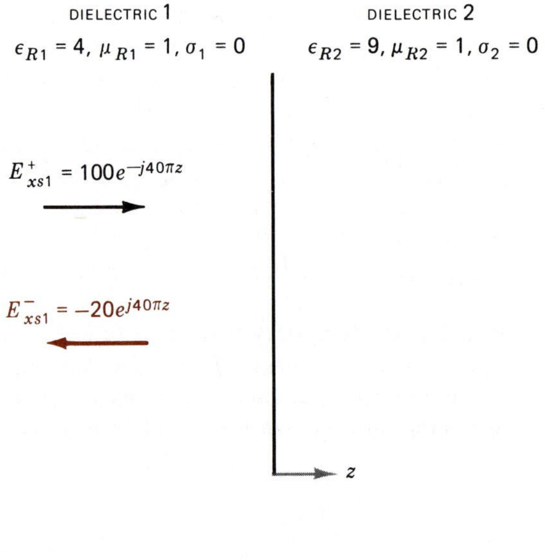

FIGURE 11.8
An incident wave, $E_{xs1}^{+} = 100e^{-j40\pi z}$ V/m is reflected with a reflection coefficient, $\Gamma = -0.2$. Dielectric 2 is infinitely thick.

Maxima of E are found at distances of 1.25, 3.75, 6.25, ... cm from $z = 0$. These maxima all have amplitudes of 120 V/m, as predicted by (69).

There are no maxima or minima in region 2 since there is no reflected wave there.

The ratio of the maximum to minimum amplitudes is called the *standing-wave ratio*:

$$s = \frac{E_{xs1,\max}}{E_{xs1,\min}} = \frac{1 + |\Gamma|}{1 - |\Gamma|} \tag{73}$$

Since $|\Gamma| \leq 1$, s is always positive and greater than or equal to unity. For the example above,

$$s = \frac{1 + |-0.2|}{1 - |-0.2|} = \frac{1.2}{0.8} = 1.5$$

If $|\Gamma| = 1$, the reflected and incident amplitudes are equal, all the incident energy is reflected, and s is infinite. Planes separated by multiples of $\lambda_1/2$ can be found on which E_{x1} is zero at all times. Midway between these planes, E_{x1} has a maximum amplitude twice that of the incident wave.

If $\eta_2 = \eta_1$, then $\Gamma = 0$, no energy is reflected, and $s = 1$; the maximum and minimum amplitudes are equal.

If one-half the incident power is reflected, $|\Gamma|^2 = 0.5$, $|\Gamma| = 0.707$, and $s = 5.83$.

Since the standing-wave ratio is a ratio of amplitudes, the relative amplitudes provided by a probe permit its use to determine s experimentally. For this reason the standing-wave ratio is an important transmission-line parameter, and we shall use it extensively in the following chapter.

Let us now suppose that region 1 is a lossy material for which α_1 is not zero. An incident wave coming in from the left undergoes exponential attenuation as it progresses in the $+z$ direction. The reflected wave attenuates as it propagates in the $-z$ direction. Eventually its amplitude is negligible compared to that of the incident wave. Thus maxima and minima may be discernible near the reflecting surface, but they are smoothed out at greater distances from it. Figure 11.9 shows a sketch of the amplitude of E_{x1} when region 2 is a perfect conductor and region 1 has a propagation constant $\gamma_1 = 1 + j4\pi$. It should be noted that no two maxima have the same amplitude, nor do two successive minima. The standing-wave ratio is a function of z, and its value cannot be specifically defined for a case such as that shown. Unless the location at which the standing-wave ratio is measured is specified, it is much more informative to describe the behavior in terms of the reflection coefficient and the attenuation factor.

Although the case shown in Fig. 11.9 is an extreme one, we must also realize that the true lossless transmission line does not exist in practice, and the standing-wave ratio is always a function of position from the load. The value of s is meaningful only when it does not change appreciably throughout the region in which we are interested.

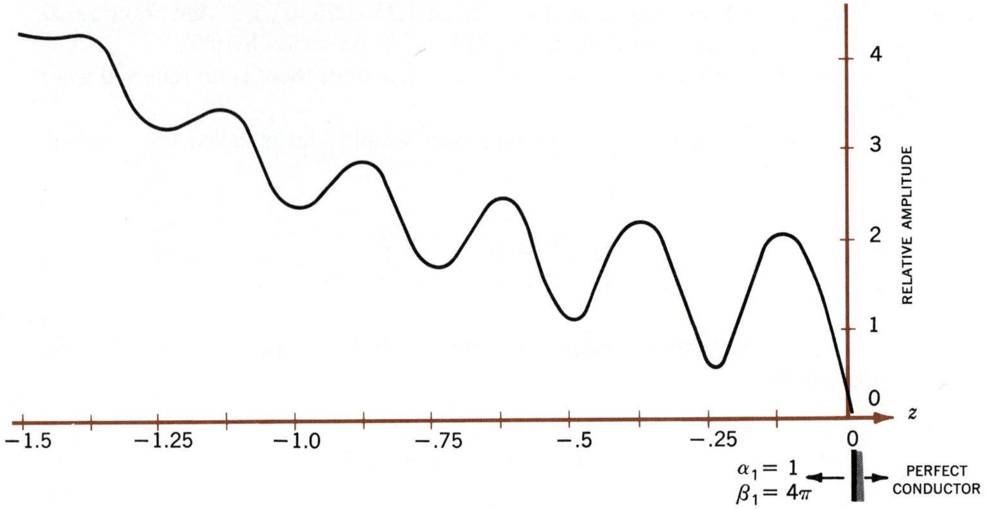

FIGURE 11.9
The effect of a reflected wave disappears in a lossy material.

Again restricting our attention to a lossless medium 1, let us find the ratio of the total electric and magnetic field intensities. For a traveling wave this is $\pm\eta_1$, with the sign depending on the direction of travel. However, reflection from a perfect conductor has shown us that E_{xs1} or H_{ys1} may be zero at certain locations, and their ratio may thus vary from zero to an infinite value. The total fields at $z = -l$ are

$$E_{xs1} = (e^{j\beta_1 l} + \Gamma e^{-j\beta_1 l})E_{x10}^+$$

$$H_{ys1} = (e^{j\beta_1 l} - \Gamma e^{-j\beta_1 l})\frac{E_{x10}^+}{\eta_1}$$

We call the ratio an *input* intrinsic impedance η_{in},

$$\eta_{\text{in}} = \left.\frac{E_{xs1}}{H_{ys1}}\right|_{z=-l} = \eta_1 \frac{e^{j\beta_1 l} + \Gamma e^{-j\beta_1 l}}{e^{j\beta_1 l} - \Gamma e^{-j\beta_1 l}}$$

Letting $\Gamma = (\eta_2 - \eta_1)/(\eta_2 + \eta_1)$ and using Euler's indentity, we have

$$\eta_{\text{in}} = \eta_1 \times \frac{(\eta_2 + \eta_1)(\cos\beta_1 l + j\sin\beta_1 l) + (\eta_2 - \eta_1)(\cos\beta_1 l - j\sin\beta_1 l)}{(\eta_2 + \eta_1)(\cos\beta_1 l + j\sin\beta_1 l) - (\eta_2 - \eta_1)(\cos\beta_1 l - j\sin\beta_1 l)}$$

This is easily simplified to yield

$$\boxed{\eta_{\text{in}} = \eta_1 \frac{\eta_2 + j\eta_1 \tan\beta_1 l}{\eta_1 + j\eta_2 \tan\beta_1 l}} \tag{74}$$

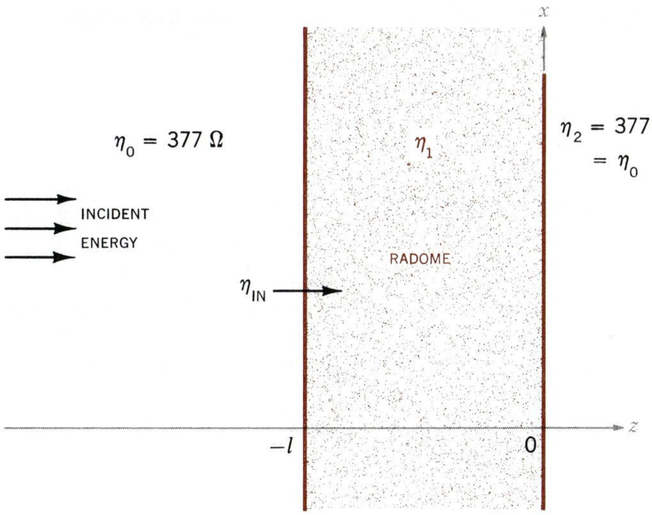

FIGURE 11.10
In order to match the incident wave to free space at the far right, it is necessary that η_{in} be made equal to η_0 by proper choice of η_1 and l.

When η_2 equals η_1, η_{in} also equals η_1; there is no reflection, and we say that the transmission system is *matched.* If $\eta_2 = 0$ (perfect conductor), $\eta_{in} = j\eta_1 \tan \beta_1 l$; the input impedance is zero when $\beta_1 l = n\pi$, and there $E_{xs1} = 0$. Also, η_{in} is infinite at those locations where $H_{ys1} = 0$.

We shall use (74) extensively in the following chapter in a form applicable to transmission lines, but we may conclude this chapter on plane waves by showing how to design a transparent window for a radar antenna. This problem arises because it is necessary to protect these antennas from the weather by appropriate covers or radomes. Let us assume that the antenna is off to the left in free space, $z < -l$, as shown in Fig. 11.10. Region 1 lies between $z = -l$ and 0. We let region 1 be a slab of perfect dielectric, making it as thin as we can, to keep our assumption of zero losses valid. To the right in region 2, $z > 0$, is the free-space region into which the radar signal is to be sent. In order to avoid any reflection of power back into the antenna, or in order to match the antenna to the outside world, we set $\eta_{in} = \eta_0 = 377\ \Omega$. Since $\eta_2 = 377$, we have

$$377 = \eta_1 \frac{377 + j\eta_1 \tan \beta_1 l}{\eta_1 + j377 \tan \beta_1 l}$$

Multiplying out, we find

$$j377^2 \tan \beta_1 l = j\eta_1^2 \tan \beta_1 l$$

Since $\eta_1 < 377$ for all nonmagnetic materials, we can satisfy this equation only by selecting $\beta_1 l = n\pi$. The thinnest radome is obtained when $\beta_1 l = \pi$, or $l = \lambda_1/2$.

Thus, if the operating frequency is 10 000 MHz, we may select a lightweight low-loss plastic for which $\epsilon_{R1} = 2.25$ and use a thickness

$$l = \frac{\lambda_1}{2} = \frac{v_1}{2f_1} = \frac{3 \times 10^8}{2\sqrt{2.25}\,10^{10}} = 10^{-2}\text{ m} \qquad \text{(or 1 cm)}$$

If the radome were 0.5 cm thick, it could be shown that $\eta_{\text{in}} = 167.5\ \Omega$ and 14.8 percent of the incident power would be reflected.

D11.9. Let $\mu_{R1} = 10$, $\epsilon_{R1} = 2.5$, and $\sigma_1 = 0$ in region 1, $z < 0$; for $z > 0$, $\mu_{R2} = 1$, $\epsilon_{R2} = 4$, and $\sigma_2 = 0$. An incident wave, $E_{x1}^+ = 600 \cos(10^9 t - \beta_1 z)$ V/m, is present in region 1. Find: (*a*) s in region 1; (*b*) η_{in} at $z = -\pi$ cm; (*c*) E_{xs} at $z = -\pi$ cm.

Ans. 4; $470\underline{/58.4^\circ}\ \Omega$; $523\underline{/66.6^\circ}$ V/m

SUGGESTED REFERENCES

1. Ginzton, E. L.: "Microwave Measurements," McGraw-Hill Book Company, New York, 1957. This book provides a good description of microwave measurement techniques and theory as of the mid-1950s. Much of it is still valid, and all of it is instructive.
2. International Telephone and Telegraph Co., Inc.: "Reference Data for Radio Engineers," 7th ed., Howard W. Sams & Co., Indianapolis, Ind., 1985. This handbook has some excellent data on the properties of dielectric and insulating materials.
3. Rao, N. Narayana: "Elements of Engineering Electromagnetics," 2d ed., Prentice-Hall, Inc., Englewood Cliffs, N.J., 1987. This undergraduate text emphasizes time-varying fields. The uniform plane wave is the subject of chap. 6.
4. Seshadri, S. R.: "Fundamentals of Transmission Lines and Electromagnetic Fields," Addison-Wesley Publishing Company, Inc., Reading, Mass., 1971. Plane waves are discussed in chaps. 5 and 6.
5. von Hippel, A. R.: "Dielectric Materials and Applications," The Technology Press of the Massachusetts Institute of Technology, Cambridge, Mass., and John Wiley & Sons, Inc., New York, 1954.

PROBLEMS

1. Show that $E_{xs} = Ae^{\pm j(kz+\alpha)}$ is a solution of the vector Helmholtz equation, Sec. 11.1, Eq. (15), for $k = \omega\sqrt{\mu_0\epsilon_0}$ and any α and A.

2. An **E** field in free space is given as $\mathbf{E} = 800 \cos(10^8 t - \beta y)\mathbf{a}_z$ V/m. Find: (*a*) β; (*b*) λ, (*c*) **H** at $P(0.1, 1.5, 0.4)$ at $t = 8$ ns.

3. A uniform plane wave in free space is given by $\mathbf{E}_s = (200\underline{/30^\circ})e^{-j250z}\mathbf{a}_x$ V/m. Find: (*a*) β; (*b*) ω; (*c*) f; (*d*) λ; (*e*) η; (*f*) $\mathbf{H}_s$; (*g*) $|\mathbf{E}|$ at $z = 8$ mm, $t = 6$ ps.

4. A uniform plane wave in free space is propagating in the $-\mathbf{a}_y$ direction at a frequency of 10 MHz. If $\mathbf{E} = 400 \cos \omega t\, \mathbf{a}_z$ V/m at $y = 0$, write expressions for: (*a*) $\mathbf{E}(x,y,z,t)$; (*b*) $\mathbf{E}_s(x,y,z)$; (*c*) $\mathbf{H}_s(x,y,z)$; (*d*) $\mathbf{H}(x,y,z,t)$.

5. A 150-MHz uniform plane wave in free space is traveling in the $\mathbf{a}_x$ direction. The electric field intensity has a maximum amplitude of $200\mathbf{a}_y + 400\mathbf{a}_z$ V/m at $P(10,30,-40)$ at

$t = 0$. Find: (*a*) ω; (*b*) β; (*c*) λ; (*d*) v; (*e*) η; (*f*) $\mathbf{E}(x,y,z,t)$. (*g*) Use one or more of Maxwell's equation to find $\mathbf{H}(x,y,z,t)$.

6. The electric field intensity of a 300-MHz uniform plane wave in free space is given as $\mathbf{E}_s = (20 + j50)(\mathbf{a}_x + 2\mathbf{a}_y)e^{-j\beta z}$ V/m. (*a*) Find ω, λ, v, and β. (*b*) Find $\mathbf{E}$ at $t = 1$ ns, $z = 10$ cm. (*c*) What is $|\mathbf{H}|_{max}$?

7. A wave propagating in a lossless dielectric has the components, $\mathbf{E} = 500 \cos(10^7 t - \beta z)\mathbf{a}_x$ V/m and $\mathbf{H} = 1.1 \cos(10^7 t - \beta z)\mathbf{a}_y$ A/m. If the wave is traveling at $v = 0.5c$, find: (*a*) μ_R; (*b*) ϵ_R; (*c*) β; (*d*) λ; (*e*) η.

8. Let $\mathbf{E}_s = (1000\mathbf{a}_x + 400\mathbf{a}_z)e^{-j10y}$ V/m for a 250-MHz uniform plane wave propagating in a perfect dielectric. If the maximum amplitude of $\mathbf{H}$ is 3 A/m, find β, η, λ, v, ϵ_R, μ_R, and $\mathbf{E}(x,y,z,t)$.

9. A uniform plane wave, $\mathbf{E}_s = (5\mathbf{a}_x + j10\mathbf{a}_y)e^{-j2z}$ V/m, has a frequency of 50 MHz in a lossless dielectric material for which $\mu_R = 1$. (*a*) Find β, ω, v, λ, ϵ_R, and η. (*b*) Calculate $\mathbf{E}$ at the origin for $\omega t = 0$, $\pi/4$, $\pi/2$, $3\pi/4$, and π.

10. A 1-MHz uniform plane wave with an amplitude of 25 V/m is propagating in a material for which $\epsilon_R = 4$, $\mu_R = 9$, and $\sigma = 0$. Propagation is in the $\mathbf{a}_x$ direction. (*a*) Find v, λ, β, and η. (*b*) Find $\mathbf{E}(t)$ if $E_z(x,y,z,t) = 0$, and $E_y = 25$ V/m at $P(10,10,10)$ at 10^{-7} s. (*c*) Find $\mathbf{H}(t)$.

11. Using data given in Fig. 11.3 for Pyranol 1467, determine values of α, β, λ, v, and η at 10^8 Hz for a uniform plane wave.

12. A wave having an amplitude $E_{x0} = 600$ V/m at $(0,0,0, t = 0)$ is propagating in the $\mathbf{a}_z$ direction in a material where $\sigma = 0.1$ ℧/m, $\epsilon = 2 \times 10^{-11}$ F/m, and $\mu = 2 \times 10^{-6}$ H/m. If $\omega = 10^{10}$ rad/s, find: (*a*) E_x at (0,0, 2 cm, 0.1 ns); (*b*) H_y at (0,0, 2 cm, 0.1 ns).

13. A lossy dielectric is characterized by $\epsilon_R = 2.5$, $\mu_R = 4$, and $\sigma = 10^{-3}$ ℧/m at 10 MHz. Let $\mathbf{E}_s = 20e^{-\gamma z}\mathbf{a}_x$ V/m and find: (*a*) α; (*b*) β; (*c*) v; (*d*) λ; (*e*) η; (*f*) $\mathbf{H}_s$; (*g*) $\mathbf{E}$(2,3,4, 10 ns).

14. Parameters for a 50-MHz uniform plane wave propagating in the $\mathbf{a}_z$ direction in a certain lossy material are $\gamma = 0.25 + j2$ m^{-1} and $\eta = 600 + j75$ Ω. (*a*) Find μ_R, ϵ_R, and σ. (*b*) If $\mathbf{E}_s = 100\mathbf{a}_x$ V/m at $z = 0$, find $\mathbf{E}_s$ at $z = 2$ m, and also (*c*) $\mathbf{H}_s$ at $z = 2$ m.

15. A 30-GHz radar signal may be represented as a uniform plane wave in a sufficiently small region. Calculate the wavelength in centimeters and the attenuation in decibels per foot if the wave is propagating in a nonmagnetic material for which: (*a*) $\epsilon_R = 1$ and $\sigma = 0$; (*b*) $\epsilon_R = 1.01$ and $\sigma = 10^{-3}$ ℧/m; (*c*) $\epsilon_R = 2.1$ and $\sigma = 5$ ℧/m.

16. A lossy dielectric is characterized by $\epsilon_R = 1.5$, $\mu_R = 1$, and $\sigma/\omega\epsilon = 2.5 \times 10^{-4}$. At a frequency of 200 MHz, how far can a uniform plane wave propagate in the material before: (*a*) it undergoes an attenuation of 1 Np; (*b*) it undergoes an attenuation of 10 dB; (*c*) its amplitude is halved; (*d*) its phase shifts 180°?

17. The power factor of a capacitor is defined as the cosine of the impedance phase angle, and its Q is ωCR, where R is the parallel resistance. Assume an idealized parallel-plate capacitor having a dielectric characterized by σ, ϵ, and μ_0. Find both the power factor and Q in terms of the loss tangent.

18. An electrostatic field and a steady magnetic field are shown in the same region of space in Fig. 11.11. Determine the direction of the Poynting vector at various points throughout the system and indicate how the Poynting theorem is satisfied.

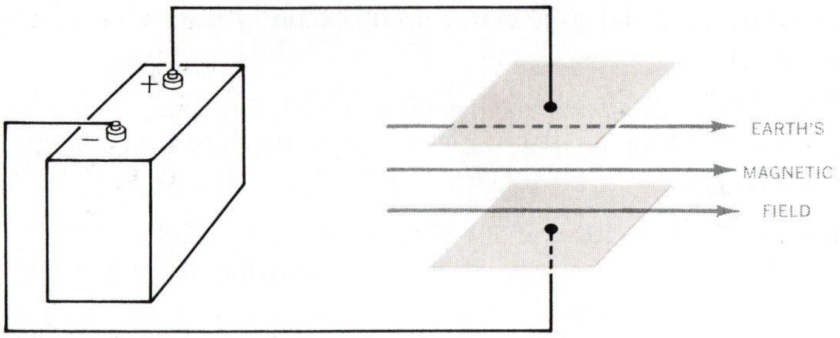

FIGURE 11.11
An electrostatic field and a constant magnetic field can show surprising values of the Poynting vector. See Prob. 18.

19. A uniform plane wave in free space is given by $\mathbf{E}_s = 200e^{-j0.1z}\mathbf{a}_x$ V/m. Find the instantaneous value of the Poynting vector at (*a*) $t = 0$: $z = 0$, 15, 30, and 45 m; (*b*) $z = 0$: $t = 0$, 40, 80, and 120 ns.

20. A uniform plane wave is propagating in the $\mathbf{a}_z$ direction through a lossy material with $\gamma = 0.1 + j1.2\ \text{m}^{-1}$ and $\eta = 300 + j25\ \Omega$. Let $E_s = 100$ V/m at $z = 0$. (*a*) Find $\mathscr{P}_{z,\,\text{av}}$ at $z = 0$ and 1. (*b*) How much average power per cubic meter is being dissipated at $P(2,3,4)$?

21. Certain sinusoidally varying fields in free space are given as $\mathbf{E}_s = 1200\pi[(\sin\theta)/r]e^{-jr}\mathbf{a}_\theta$ V/m and $\mathbf{H}_s = [(\sin\theta)/r]e^{-jr}\mathbf{a}_\phi$ A/m. If $\omega = 3 \times 10^8$ rad/s, find: (*a*) $\mathbf{E}(r,\theta,t)$ and $\mathbf{H}(r,\theta,t)$; (*b*) $\mathscr{P}(r,\theta,t)$; (*c*) the total instantaneous power $P(r,t)$ leaving the surface of a sphere of radius r at time t.

22. An infinitely long coaxial transmission line is air-filled and has dimensions $a = 1$ cm, $b = 2.718$ $(=e)$ cm. Assume an $\mathbf{E}$ field of $(100/\rho)\cos(10^8 t - \beta z)\mathbf{a}_\rho$ V/m. (*a*) Find β. (*b*) Use Maxwell's equations to find $\mathbf{H}(\rho,z,t)$. (*c*) Determine $\mathscr{P}(\rho,z,t)$. (*d*) Find the average power transmitted by the coax.

23. A brass ($\sigma = 10^7$ ℧/m) pipe with inner and outer radii of 1.7 and 2 cm carries a total current of 100 A dc. Find **E**, **H**, and $\mathscr{P}$ within the brass.

24. A two-wire transmission line utilizes circular brass ($\sigma = 1.2 \times 10^7$ ℧/m) tubular conductors, each having an outer diameter of $\frac{3}{16}$ in and a wall thickness of 0.030 in. Find the resistance per meter length of the transmission line at: (*a*) dc; (*b*) 10 MHz; (*c*) 1 GHz.

25. An air-filled coaxial transmission line has brass ($\sigma = 1.4 \times 10^7$ ℧/m) conductors with dimensions $a = 1.5$ mm, $b = 6$ mm. If the line is operated at 500 MHz, determine the resistance per unit length of: (*a*) the inner conductor; (*b*) the outer conductor; (*c*) the coaxial transmission line.

26. (*a*) Most microwave ovens operate at 2.45 GHz. Assume that $\sigma = 1.1 \times 10^6$ ℧/m and $\mu_R = 600$ for the stainless steel interior, and find the depth of penetration. (*b*) Let $\mathbf{E}_s = 10\underline{/0^\circ}$ V/m at the surface of the conductor, and plot a curve of the amplitude of $\mathbf{E}_s$ vs. the angle of $\mathbf{E}_s$ as the field propagates into the stainless steel.

27. A coaxial transmission line with a solid inner conductor of 1-mm radius and an outer conductor of 4-mm inner radius and 4.5-mm outer radius is made of a brass having a conductivity of 1.5×10^7 ℧/m. If the line is operating at 2 GHz: (*a*) find R, the resistance per unit length; (*b*) calculate α, knowing that $\gamma = \alpha + j\beta = \sqrt{j\omega C(R + j\omega L)}$, where C and L are per-unit-length values obtained in Secs. 5.10 and 9.10, respectively, and the coax is air-filled.

28. A 400-MHz wave is propagating along an insulated copper wire. Assume $\delta_{Cu} \ll$ wire radius, let $\epsilon_R = 4$ and $\sigma/\omega\epsilon = 10^{-3}$ for the insulation, and calculate: (*a*) α_{ins}/α_{Cu}; (*b*) $\lambda_{ins}/\lambda_{Cu}$; (*c*) v_{ins}/v_{Cu}.

29. A nonmagnetic good conductor supports a uniform plane wave traveling at a velocity of 2.5×10^5 m/s with a wavelength of 0.25 mm. Find the frequency and the conductivity.

30. In region 1, $z < 0$, $\sigma_1 = 0$, $\epsilon_{R1} = 4$, and $\mu_{R1} = 1$; in region 2, $z > 0$, $\sigma_2 = 0$, $\epsilon_{R2} = 1.44$, and $\mu_{R2} = 6.25$. There is an incident wave in region 1, $\mathbf{E}_{s1}^+ = 400e^{-j20z}\mathbf{a}_x$ V/m. (*a*) Specify f. (*b*) Find the total **E** field in region 1. (*c*) Determine $\mathbf{E}_{s2}^+$.

31. A 10-MHz uniform plane wave in air is incident on a copper sheet that is perpendicular to the direction of propagation. Most of the power is reflected, but a little propagates into the copper sheet. How far from the surface will the wave propagate into the copper before 99 percent of its power is dissipated?

32. Two uniform plane waves, $\mathbf{E}_{s1}^+ = 120e^{-j4\pi z}\mathbf{a}_x$ and $\mathbf{E}_{s1}^- = (30\underline{/50^\circ})e^{j4\pi z}\mathbf{a}_x$ V/m, are both traveling in the same region of space. If this region is free space: (*a*) find the operating frequency. (*b*) What intrinsic impedance in the region $z > 0$ would cause the given reflected wave? (*c*) At what value or values of z, $-40 \le z \le 0$ cm, is the amplitude of the total **E** field a maximum?

33. Region A, $x < 0$, has $\mu_A = 2\ \mu$H/m, $\epsilon_A = 200$ pF/m, and $\sigma_A = 0$; in region B where $x > 0$, $\mu_B = 2\ \mu$H/m, $\epsilon_B = 200$ pF/m, and $\sigma_B = 1.2$ ℧/m. If $\mathbf{E}_A^+ = 1000e^{-\alpha_A x}\cos(10^{10}t - \beta_A x)\mathbf{a}_y$ V/m, find: (*a*) $\mathbf{E}_{sA}$; (*b*) $\mathbf{E}_{sB}^+$; (*c*) $\mathbf{H}_B$.

34. A uniform plane wave in region 1 is incident on the planar boundary separating regions 1 and 2. If $\sigma_1 = \sigma_2 = 0$ and $\mu_{R1} = \mu_{R2} = 1$, find the ratio $\epsilon_{R2}/\epsilon_{R1}$ if 10 percent of the energy in the incident wave is: (*a*) reflected; (*b*) transmitted into region 2.

35. The electric field intensity in a lossless region is given as $\mathbf{E} = 200 \sin 10^9 t \sin 20z\ \mathbf{a}_x$ V/m. If the amplitude of the associated magnetic field intensity is 1 A/m, find: (*a*) ϵ and μ for the medium; (*b*) **H**; (*c*) $\mathbf{E}_s$.

36. A plane interface between two perfect dielectrics is located at $z = 0$. A 4-GHz uniform plane wave traveling in the a_z direction is incident from region 1, $z < 0$, onto region 2, $z > 0$. The wavelengths in the dielectrics are $\lambda_1 = 6$ cm and $\lambda_2 = 4$ cm. Both materials are nonmagnetic. What percentage of the energy incident on the boundary is: (*a*) reflected; (*b*) transmitted? (*c*) What is the standing-wave ratio in region 1?

37. Given region 1, $z < 0$, $\epsilon_1 = 20$ pF/m, and $\mu_1 = 2\ \mu$H/m; region 2, $0 < z < 8$ cm, $\epsilon_2 = 50$ pF/m, and $\mu_2 = 2.5\ \mu$H/m; and region 3, $z > 8$ cm, $\epsilon_3 = \epsilon_1$, and $\mu_3 = \mu_1$; let $\sigma = 0$ everywhere. (*a*) What is the lowest frequency at which a uniform plane wave incident from region 1 on the boundary at $z = 0$ will have no reflection? (*b*) If $f =$ 200 MHz, what will s be in region 1?

38. A lossless dielectric slab for which $\epsilon_R = 9$ is located in the region $0 < z < 0.25$ m. Free space exists where $z < 0$ and $z > 0.25$ m. A 250-MHz uniform plane wave traveling in the $\mathbf{a}_z$ direction is incident on the boundary at $z = 0$. (*a*) Find the standing-wave ratio in each of the three regions. (*b*) Find the location of the maximum $|\mathbf{E}|$ in free space nearest to the slab ($z < 0$).

39. The three regions shown in Fig. 11.12 are all lossless and nonmagnetic. Determine the standing-wave ratio for **E** in the left region when d_2 equals: (*a*) 0; (*b*) 5 cm; (*c*) 8 cm.

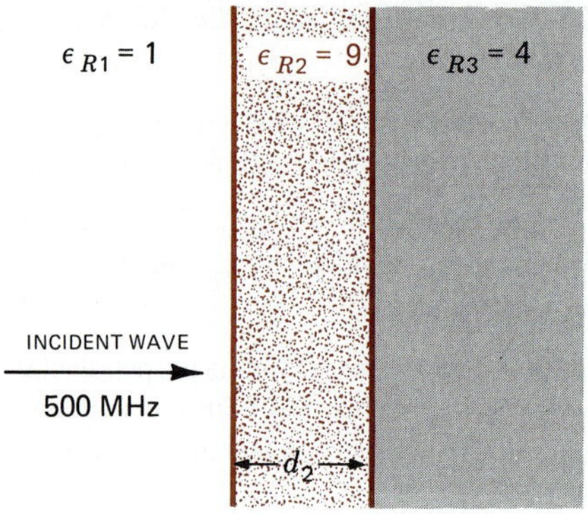

FIGURE 11.12
See Prob. 39.

40. In Fig. 11.10, let $\eta_1 = 0.5\eta_0$. A uniform plane wave is normally incident from the left, as shown. Plot a curve of s in the region to the left: (*a*) as a function of l if $f = 3$ GHz; (*b*) as a function of frequency if $l = 2.5$ cm.

41. In Fig. 11.5, let region 2 be free space, while $\mu_{R1} = 1$, $\sigma_1 = 0$, and ϵ_{R1} is unknown. Find ϵ_{R1} if: (*a*) the amplitude of $\mathbf{E}_1^-$ is one-half that of $\mathbf{E}_1^+$; (*b*) $P_{1,\text{av}}^-$ is one-half of $P_{1,\text{av}}^+$; (*c*) $|\mathbf{E}_1|_{\text{min}}$ is one-half of $|\mathbf{E}_1|_{\text{max}}$.

CHAPTER 12

TRANSMISSION LINES

Transmission lines are used to transmit electric energy and signals from one point to another. The basic transmission line connects a source to a load. This may be a transmitter and an antenna, a shift register and the memory core in a digital computer, a hydroelectric generating plant and a substation several hundred miles away, a television antenna and a receiver, or one channel of a stereo turntable and one input of the preamplifier. Several different types of transmission lines are involved in these various uses, and we shall study their characteristics in the second section of this chapter.

First, however, we need to show that there is a direct analogy between the uniform transmission line and the uniform plane wave. We shall find that the effort devoted to the uniform plane wave in the previous chapter makes it possible to develop analogous results for the uniform transmission line easily and rapidly. The field distributions for the uniform plane wave and for the uniform transmission line are both known as transverse electromagnetic (TEM) waves because **E** and **H** are both perpendicular to the direction of propagation, or both lie in the transverse plane. The great similarity in results is a direct consequence of the fact that we are dealing with TEM waves in each case. In the transmission line, however, it is possible and customary to define a voltage and a current. These quantities are the ones for which we shall write equations, obtain solutions, and find propagation constants, reflection coefficients, and input impedances. We shall also consider power instead of power density.

An important addition to our analytical and design tools will be the use of a graphical technique for solving reflection and matching problems. This method is also applicable to the uniform plane wave.

12.1 THE TRANSMISSION-LINE EQUATIONS

We shall first obtain the differential equations which the voltage or current must satisfy on a uniform transmission line. This may be done by any of several methods. For example, an obvious method would be to solve Maxwell's equations subject to the boundary conditions imposed by the particular transmission line we are considering. We could then define a voltage and a current, thus obtaining our desired equations. It is also possible to solve the general TEM-wave problem once and for all for any two-conductor transmission line having lossless conductors. Instead, we shall construct a circuit model for an incremental length of line, write two circuit equations, and show that the resultant equations are analogous to the fundamental equations from which the wave equation was developed in the previous chapter. By these means we shall begin to tie field theory and circuit theory together, a task which we shall take up again in the following chapter.

Our circuit model will contain the inductance, capacitance, shunt conductance, and series resistance associated with an incremental length of line. Let us do our thinking in terms of a coaxial transmission line containing a dielectric of permeability μ (usually μ_0), permittivity ϵ, and conductivity σ. The inner and outer conductors have a high conductivity σ_c. Knowing the operating frequency and the dimensions, we can then determine the values of R, G, L, and C on a per-unit-length basis by using formulas developed in earlier chapters. We shall review these expressions and collect the information on several different types of lines in the following section.

Let us again assume propagation in the $\mathbf{a}_z$ direction. We therefore cut out a section of length Δz containing a resistance $R\,\Delta z$, an inductance $L\,\Delta z$, a conductance $G\,\Delta z$, and a capacitance $C\,\Delta z$, as shown in Fig. 12.1. Since the section of the line looks the same from either end, we divide the series elements in half to produce a symmetrical network. We could equally well have placed half the conductance and half the capacitance at each end.

Since we are already familiar with the basic characteristics of wave propagation, let us turn immediately to the case of sinusoidal time variation, and use the

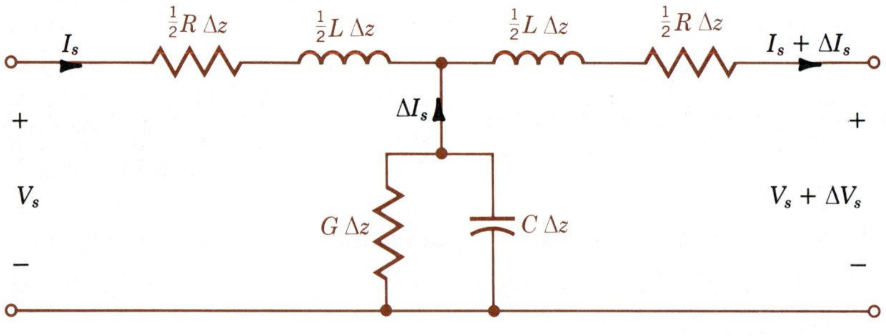

FIGURE 12.1
An incremental length of a uniform transmission line. R, G, L, and C are functions of the transmission-line configuration and materials.

notation for complex quantities we developed in the last chapter. The voltage V between conductors is in general a function of z and t, as, for example,

$$V = V_0 \cos(\omega t - \beta z + \psi)$$

We may use Euler's identity to express this in complex notation,

$$V = \text{Re}\, V_0 e^{j(\omega t - \beta z + \psi)} = \text{Re}\, V_0 e^{j\psi} e^{-j\beta z} e^{j\omega t}$$

By dropping Re and suppressing $e^{j\omega t}$, we transform the voltage to a phasor, which we indicate by an s subscript,

$$V_s = V_0 e^{j\psi} e^{-j\beta z}$$

We may now write the voltage equation around the perimeter of the circuit of Fig. 12.1,

$$V_s = (\tfrac{1}{2}R\,\Delta z + j\tfrac{1}{2}\omega L\,\Delta z)I_s + (\tfrac{1}{2}R\,\Delta z + j\tfrac{1}{2}\omega L\,\Delta z)(I_s + \Delta I_s) + V_s + \Delta V_s$$

or

$$\frac{\Delta V_s}{\Delta z} = -(R + j\omega L)I_s - (\tfrac{1}{2}R + j\tfrac{1}{2}\omega L)\,\Delta I_s$$

As we let Δz approach zero, ΔI_s also approaches zero, and the second term on the right vanishes. In the limit,

$$\boxed{\frac{dV_s}{dz} = -(R + j\omega L)I_s} \tag{1}$$

Neglecting second-order effects, we approximate the voltage across the central branch as V_s and obtain a second equation,

$$\frac{\Delta I_s}{\Delta z} \doteq -(G + j\omega C)V_s$$

or

$$\boxed{\frac{dI_s}{dz} = -(G + j\omega C)V_s} \tag{2}$$

Instead of solving these equations, let us save some time by comparing them with the equations which arise from Maxwell's curl equations for the uniform plane wave. From

$$\nabla \times \mathbf{E}_s = -j\omega\mu \mathbf{H}_s$$

we set $\mathbf{E}_s = E_{xs}\mathbf{a}_x$ and $\mathbf{H}_s = H_{ys}\mathbf{a}_y$, where E_{xs} and H_{ys} are functions of z only, and obtain the single scalar equation

$$\frac{dE_{xs}}{dz} = -j\omega\mu H_{ys} \qquad \left[\frac{dV_s}{dz} = -(R + j\omega L)I_s \qquad (1)\right]$$

with Eq. (1) shown for easy comparison. Similarly, from

$$\nabla \times \mathbf{H}_s = (\sigma + j\omega\epsilon)\mathbf{E}_s$$

we have

$$\frac{dH_{ys}}{dz} = -(\sigma + j\omega\epsilon)E_{xs} \qquad \left[\frac{dI_s}{dz} = -(G + j\omega C)V_s \qquad (2)\right]$$

Careful comparison of the two equations in the last line shows a direct analogy between the following pairs of quantities: I_s and H_{ys}, G and σ, C and ϵ, and V_s and E_{xs}. Replacing the variables in one equation by the corresponding quantities produces the other equation. The analogy is particularly strong in this pair of equations, for the corresponding quantities are measured in almost the same units.

Carrying this same analogy over to the first pair of equations, we see that it continues to hold and provides one additional analogous pair, L and μ. However, there is also a surprise, for the transmission-line equation is more complicated than the field equation. There is no analog for the conductor resistance per unit length R. Although it would be good salesmanship to say that this shows that field theory is simpler than circuit theory, let us be fair in determining the reason for this omission. Conductor resistance must be determined by obtaining a separate solution to Maxwell's equations *within* the conductors and forcing the two solutions to satisfy the necessary boundary conditions at the interface. We considered steady current fields in conductors back in Chap. 5, and we have just considered the high-frequency case under the guise of "skin effect"; however, we have looked only briefly at the problem of matching two solutions at the boundary. Thus the term that is omitted in the field equation represents the problem of the fields within the conductors, and the solution of this problem enables us to obtain a value for R in the circuit equation. We maintain the analogy by agreeing to replace $j\omega\mu$ by $R + j\omega L$.[1]

The boundary conditions on V_s and E_{xs} are the same, as are those for I_s and H_{ys}, and thus the solution of our two circuit equations may be obtained from a knowledge of the solution of the two field equations, as obtained in the last chapter. From

$$E_{xs} = E_{x0}e^{-\gamma z}$$

we obtain the voltage wave

$$V_s = V_0 e^{-\gamma z} \qquad (3)$$

which propagates in the $+z$ direction with an amplitude $V_s = V_0$ at $z = 0$ (and $V = V_0$ at $z = 0$, $t = 0$ for $\psi = 0$). The propagation constant for the uniform plane wave,

$$\gamma = \sqrt{j\omega\mu(\sigma + j\omega\epsilon)}$$

[1] When ferrite materials enter the field problem, a complex permeability $\mu = \mu' - j\mu''$ is often used to include the effect of nonohmic losses in that material. Under these special conditions $\omega\mu''$ is analogous to R.

becomes

$$\boxed{\gamma = \alpha + j\beta = \sqrt{(R + j\omega L)(G + j\omega C)}} \tag{4}$$

The wavelength is still defined as the distance that provides a phase shift of 2π rad; therefore,

$$\boxed{\lambda = \frac{2\pi}{\beta}} \tag{5}$$

Also, the phase velocity has been defined as

$$\boxed{v = \frac{\omega}{\beta}} \tag{6}$$

and this expression is valid both for the uniform plane wave and transmission lines.

For a lossless line ($R = G = 0$) we see that

$$\gamma = j\beta = j\omega\sqrt{LC}$$

Hence

$$\boxed{v = \frac{1}{\sqrt{LC}}} \tag{7}$$

From the expression for the magnetic field intensity

$$H_{ys} = \frac{E_{x0}}{\eta} e^{-\gamma z}$$

we see that the positively traveling current wave

$$\boxed{I_s = \frac{V_0}{Z_0} e^{-\gamma z}} \tag{8}$$

is related to the positively traveling voltage wave by a *characteristic impedance* Z_0 that is analogous to η. Since

$$\eta = \sqrt{\frac{j\omega\mu}{\sigma + j\omega\epsilon}}$$

we have

$$\boxed{Z_0 = \sqrt{\frac{R + j\omega L}{G + j\omega C}}} \tag{9}$$

When a uniform plane wave in medium 1 strikes the interface with medium 2, the fraction of the incident wave that is reflected is called the reflection coefficient Γ,

$$\Gamma = \frac{E_{x0}^-}{E_{x0}^+} = \frac{\eta_2 - \eta_1}{\eta_2 + \eta_1}$$

Thus the fraction of the incident voltage wave that is reflected by a line with a different characteristic impedance, say Z_{02}, is

$$\boxed{\Gamma = |\Gamma| e^{j\phi} = \frac{V_0^-}{V_0^+} = \frac{Z_{02} - Z_{01}}{Z_{02} + Z_{01}}} \tag{10}$$

Knowing the reflection coefficient, we may find the standing-wave ratio,

$$\boxed{s = \frac{1 + |\Gamma|}{1 - |\Gamma|}} \tag{11}$$

Finally, when $\eta = \eta_2$ for $z > 0$, the ratio of E_{xs} to H_{ys} at $z = -l$ is

$$\eta_{\text{in}} = \eta_1 \frac{\eta_2 + j\eta_1 \tan \beta_1 l}{\eta_1 + j\eta_2 \tan \beta_1 l}$$

and therefore the input impedance

$$\boxed{Z_{\text{in}} = Z_{01} \frac{Z_{02} + jZ_{01} \tan \beta_1 l}{Z_{01} + jZ_{02} \tan \beta_1 l}} \tag{12}$$

is the ratio of V_s to I_s at $z = -l$ when $Z_0 = Z_{02}$ for $z > 0$. We often terminate a transmission line at $z = 0$ with a load impedance Z_L which may represent an antenna, the input circuit of a television receiver, or an amplifier on a telephone line. The input impedance at $z = -l$ is then written simply as

$$\boxed{Z_{\text{in}} = Z_0 \frac{Z_L + jZ_0 \tan \beta l}{Z_0 + jZ_L \tan \beta l}} \tag{13}$$

We shall use these equations to gain some familiarity with transmission lines as soon as we are able to determine values for the pertinent parameters, R, G, L, and C.

D12.1. At a frequency of 100 MHz, the following values are appropriate for a certain transmission line: $L = 0.25\ \mu$H/m, $C = 80$ pF/m, $R = 0.15\ \Omega$/m, and $G = 8\ \mu$℧/m. Calculate values for: (*a*) α; (*b*) β; (*c*) λ; (*d*) v; (*e*) Z_0.

Ans. 1.565×10^{-3} Np/m; 2.81 rad/m; 2.24 m; 2.24×10^8 m/s; $55.9\underline{/-0.0228^\circ}\ \Omega$

12.2 TRANSMISSION-LINE PARAMETERS

Let us use this section to collect previous results and develop new ones where necessary, so that values for R, G, L, and C are available for the simpler types of transmission lines.

Coaxial (High Frequencies)

We begin by seeing how many of the necessary expressions we already have for a coaxial cable in which the dielectric has an inner radius a and outer radius b (Fig. 12.2). The capacitance per unit length, obtained as Eq. (46) of Sec. 5.10, is

$$\boxed{C = \frac{2\pi\epsilon}{\ln (b/a)}} \tag{14}$$

The value of permittivity used should be appropriate for the range of operating frequencies considered.

The conductance per unit length may be determined easily from the capacitance expression above by use of the current analogy described in Sec. 6.3. Thus,

$$\boxed{G = \frac{2\pi\sigma}{\ln (b/a)}} \tag{15}$$

where σ is the conductivity of the dielectric between the conductors at the operating frequency.

The inductance per unit length was computed for the coaxial cable as Eq. (50) in Sec. 9.10,

$$\boxed{L_{\text{ext}} = \frac{\mu}{2\pi} \ln (b/a)} \tag{16}$$

where μ is the permeability of the dielectric between conductors, usually μ_0. This is an *external* inductance, for its calculation does not take into account any flux

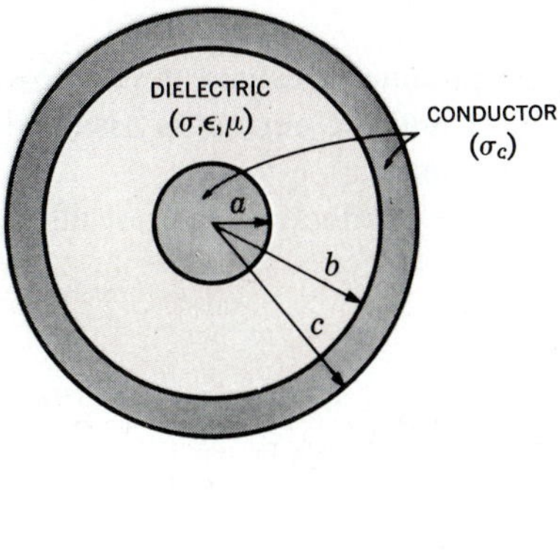

FIGURE 12.2
The geometry of the coaxial transmission line. A homogeneous dielectric is assumed.

within either conductor. Equation (16) is usually an excellent approximation to the total inductance of a high-frequency transmission line, however, for the skin depth is so small at typical operating frequencies that there is negligible flux within either conductor and negligible internal inductance. Note that $L_{ext}C = \mu\epsilon = 1/v^2$, and we are therefore able to evaluate the external inductance for any transmission line for which we know the capacitance and insulator characteristics.

The last of the four parameters that we need is the resistance R per unit length. If the frequency is very high and the skin depth δ is very small, then we obtain an appropriate expression for R by distributing the total current uniformly throughout a depth δ. For a circular conductor of radius a and conductivity σ_c, we let Eq. (54) of Sec. 11.5 apply to a unit length, obtaining

$$R_{\text{inner}} = \frac{1}{2\pi a\,\delta\sigma_c}$$

There is also a resistance for the outer conductor, which has an inner radius b. We assume the same conductivity σ_c and the same value of skin depth δ, leading to

$$R_{\text{outer}} = \frac{1}{2\pi b\,\delta\sigma_c}$$

Since the line current flows through these two resistances in series, the total resistance is the sum:

$$\boxed{R = \frac{1}{2\pi\,\delta\sigma_c}\left(\frac{1}{a} + \frac{1}{b}\right)} \tag{17}$$

It is convenient to include the common expression for the characteristic impedance of a coax here with the parameter formulas. Thus

$$\boxed{Z_0 = \sqrt{\frac{L_{ext}}{C}} = \frac{1}{2\pi}\sqrt{\frac{\mu}{\epsilon}}\ln\frac{b}{a}} \tag{18}$$

If necessary, a more accurate value may be obtained from (9).

Coaxial (Low Frequencies)

Now let us spend a few paragraphs obtaining the parameter values at very low frequencies where there is no appreciable skin effect and the current is assumed to be distributed uniformly throughout the cross section.

We first note that the current distribution in the conductor does not affect either the capacitance or conductance per unit length. Hence

$$\boxed{C = \frac{2\pi\epsilon}{\ln(b/a)}} \tag{14}$$

and

$$G = \frac{2\pi\sigma}{\ln(b/a)} \tag{15}$$

The resistance per unit length may be calculated by dc methods, $R = L/(\sigma_c S)$, where $L = 1$ m and σ_c is the conductivity of the outer and inner conductors. The area of the center conductor is πa^2 and that of the outer is $\pi(c^2 - b^2)$. Adding the two resistance values, we have

$$R = \frac{1}{\sigma_c \pi}\left(\frac{1}{a^2} + \frac{1}{c^2 - b^2}\right) \tag{19}$$

Only one of the four parameter values remains to be found, the inductance per unit length. The external inductance that we calculated at high frequencies is the greatest part of the total inductance. To it, however, must be added smaller terms representing the internal inductances of the inner and outer conductors.

At very low frequencies where the current distribution is uniform, the internal inductance of the center conductor is the subject of Prob. 43 in Chap. 9; the relationship is also given as Eq. (62) in Sec. 9.10:

$$L_{a,\,\text{int}} = \frac{\mu}{8\pi} \qquad \text{H/m} \tag{20}$$

The determination of the internal inductance of the outer shell is a more difficult problem, and most of the work is requested in Prob. 7 at the end of this chapter. There, we find that the energy stored per unit length in an outer cylindrical shell of inner radius b and outer radius c with uniform current distribution is

$$W_H = \frac{\mu I^2}{16\pi(c^2 - b^2)}\left(b^2 - 3c^2 + \frac{4c^4}{c^2 - b^2}\ln\frac{c}{b}\right)$$

Thus the internal inductance of the outer conductor at very low frequencies is

$$L_{bc,\,\text{int}} = \frac{\mu}{8\pi(c^2 - b^2)}\left(b^2 - 3c^2 + \frac{4c^4}{c^2 - b^2}\ln\frac{c}{b}\right) \tag{21}$$

At low frequencies the total inductance is obtained by combining (16), (20), and (21):

$$L = \frac{\mu}{2\pi}\left[\ln\frac{b}{a} + \frac{1}{4} + \frac{1}{4(c^2 - b^2)}\left(b^2 - 3c^2 + \frac{4c^4}{c^2 - b^2}\ln\frac{c}{b}\right)\right] \tag{22}$$

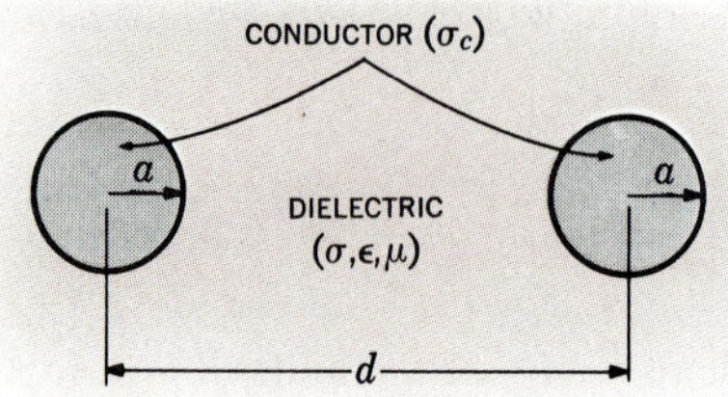

FIGURE 12.3
The geometry of the two-wire transmission line.

Coaxial (Intermediate Frequencies)

There still remains the frequency interval where the skin depth is neither very much larger than nor very much smaller than the radius. In this case, the current distribution is governed by Bessel functions, and both the resistance and internal inductance are complicated expressions. Values are tabulated in the handbooks, and it is necessary to use them for very small conductor sizes at high frequencies and for larger conductor sizes used in power transmission at low frequencies.[2]

Two-Wire (High Frequencies)

For the two-wire transmission line of Fig. 12.3 with conductors of radius a and conductivity σ_c with center-to-center separation d in a medium of permeability μ, permittivity ϵ, and conductivity σ, the capacitance was found in Sec. 5.11 to be

$$\boxed{C = \frac{\pi\epsilon}{\cosh^{-1}(d/2a)}} \tag{23}$$

or

$$C \doteq \frac{\pi\epsilon}{\ln(d/a)} \qquad (a \ll d)$$

The external inductance may be found from $L_{\text{ext}}C = \mu\epsilon$. It is

$$\boxed{L_{\text{ext}} = \frac{\mu}{\pi}\cosh^{-1}(d/2a)} \tag{24}$$

or

$$L_{\text{ext}} \doteq \frac{\mu}{\pi}\ln(d/a) \qquad (a \ll d)$$

---

[2] The current distribution, internal inductance, and internal resistance of round wires is discussed (with numerical examples) in Weeks, pp. 35–44. See the Suggested References at the end of this chapter.

The conductance per unit length may be written immediately from an inspection of the capacitance expression,

$$G = \frac{\pi\sigma}{\cosh^{-1}(d/2a)} \tag{25}$$

The resistance per unit length is twice that of the center conductor of the coax,

$$R = \frac{1}{\pi a\,\delta\sigma_c} \tag{26}$$

Finally, using the capacitance and the external inductance expressions, we obtain a value for the characteristic impedance,

$$Z_0 = \sqrt{\frac{L_{\text{ext}}}{C}} = \frac{1}{\pi}\sqrt{\frac{\mu}{\epsilon}}\cosh^{-1}(d/2a) \tag{27}$$

Two-Wire (Low Frequencies)

At low frequencies where a uniform current distribution may be assumed, we again must modify the L and R expressions. We therefore have the same relationships for C and G:

$$C = \frac{\pi\epsilon}{\cosh^{-1}(d/2a)} \tag{23}$$

$$G = \frac{\pi\sigma}{\cosh^{-1}(d/2a)} \tag{25}$$

but the inductance per unit length must be increased by twice the internal inductance of a straight round wire,

$$L = \frac{\mu}{\pi}\left[\frac{1}{4} + \cosh^{-1}(d/2a)\right] \tag{28}$$

and the resistance becomes twice the dc resistance of a wire of radius a, conductivity σ_c, and unit length:

$$R = \frac{2}{\sigma_c \pi a^2} \tag{29}$$

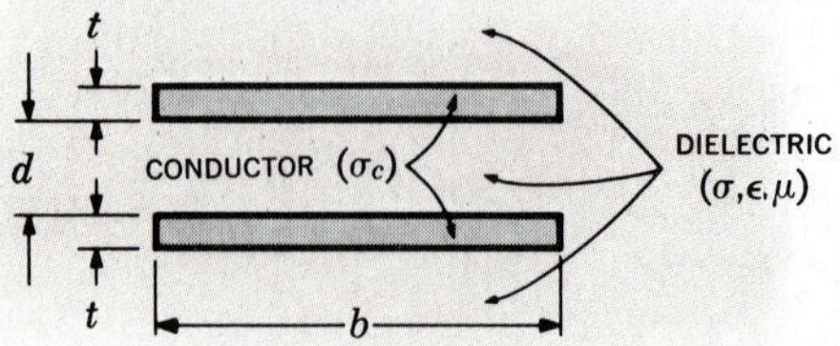

FIGURE 12.4
The geometry of the planar transmission line.

Planar (High Frequencies)

If we have the parallel-plane or planar transmission line of Fig. 12.4, with two conducting planes of conductivity σ_c, thickness t, and separation d, and a dielectric with parameters ϵ, μ, and σ, then we may easily determine the circuit parameters per unit length for a width b. It is necessary to assume either that $b \gg d$ or that we are considering a width b of a much wider guiding system. We have

$$C = \frac{\epsilon b}{d} \tag{30}$$

$$G = \frac{\sigma b}{d} \tag{31}$$

$$L_{\text{ext}} = \frac{\mu d}{b} \tag{32}$$

and

$$R = \frac{2}{\sigma_c \delta b} \tag{33}$$

Here we have assumed a well-developed skin effect such that $\delta \ll t$, the thickness of either plane.

Finally,

$$Z_0 = \sqrt{\frac{L_{\text{ext}}}{C}} = \sqrt{\frac{\mu}{\epsilon}\frac{d}{b}} \tag{34}$$

Low-frequency parameter values are unimportant since the planar transmission line has little use there.

D12.2. A coaxial line has the dimensions $a = 0.5$ cm, $b = 2$ cm, and $c = 2.4$ cm. Let $\mu = \mu_0$, $\epsilon_R = 2.5$, and $\sigma/\omega\epsilon = 0.02$ for the dielectric, while $\sigma_c = 10^7$ ℧/m for the inner and outer conductors. Assuming that the loss tangent is constant with frequency, calculate values for: (*a*) L, C, G, R, and Z_0 at 10^8 Hz, (*b*) L and R at 60 Hz.

Ans. 0.277 μH/m, 100.3 pF/m, 1.261 m℧/m, 0.250 Ω/m, 52.6 Ω; 0.341 μH/m, 1.454 mΩ/m

D12.3. A two-wire transmission line consists of two 3-mm-radius conductors ($\sigma_c = 10^7$ ℧/m), separated a center-to-center distance of 1.5 cm in a medium for which $\epsilon_R = 4$, $\mu_R = 1$, and $\sigma = 2 \times 10^{-9}$ ℧/m at the operating frequency of 60 Hz. Find: (*a*) δ; (*b*) C; (*c*) G; (*d*) L; (*e*) R.

Ans. 2.05 cm; 71.0 pF/m; 4.01 n℧/m; 0.727 μH/m; 7.07 mΩ/m

D12.4. A planar transmission line has the dimensions $b = 5$ mm, $d = 0.2$ mm, and $t = 20\ \mu$m. If the material constants are $\sigma_c = 5 \times 10^7$ ℧/m, $\epsilon = 20$ pF/m, $\mu = \mu_0$, and $\sigma/\omega\epsilon = 0.02$ at a frequency of 500 MHz, calculate: (*a*) α; (*b*) β; (*c*) v; (*d*) Z_0.

Ans. 0.283 Np/m; 15.75 rad/m; 1.995 × 10⁸ m/s; $10.03\underline{/0.1171^\circ}\ \Omega$

12.3 SOME TRANSMISSION-LINE EXAMPLES

In this section we shall apply many of the results that we have obtained in the previous two sections to several typical transmission-line problems. We shall simplify our work by restricting our attention to the lossless line.

Let us begin by assuming a two-wire 300-Ω line ($Z_0 = 300\ \Omega$), such as the lead-in wire from the antenna to a television or FM receiver. The circuit is shown in Fig. 12.5. The line is 2 m long and the dielectric constant is such that the velocity on the line is 2.5×10^8 m/s. We shall terminate the line with a receiver having an input resistance of 300 Ω and represent the antenna by its Thévenin equivalent, $Z_{\text{Th}} = 300\ \Omega$ in series with $V_{s,\text{Th}} = 60$ V at 100 MHz. This antenna voltage is larger by a factor of about 10^5 than it would be in a practical case, but it also provides simpler values to work with; in order to think practical thoughts, divide currents or voltages by 10^5, divide powers by 10^{10}, and leave impedances alone.

Since the load impedance is equal to the characteristic impedance, the line is matched; the reflection coefficient is zero, and the standing-wave ratio is unity. For the given velocity and frequency, the wavelength on the line is $v/f = 2.5$ m,

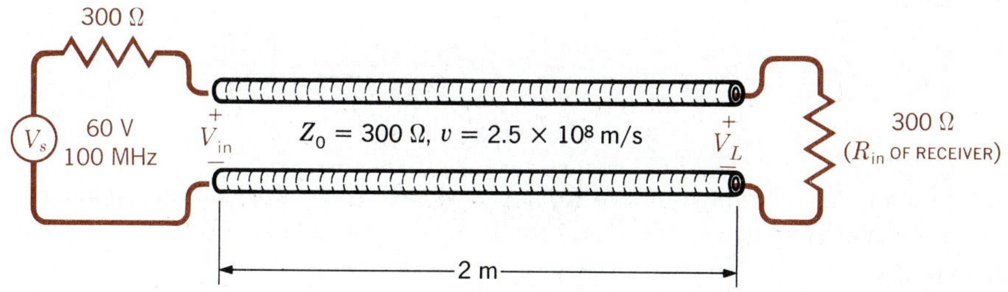

FIGURE 12.5
A transmission line that is matched at each end produces no reflections and thus delivers maximum power to the load.

and the phase constant is $2\pi/\lambda = 0.8\pi$ rad/m; the attenuation constant is zero. The electrical length of the line is $\beta l = (0.8\pi)2$, or 1.6π rad. This length may also be expressed as 288°, or 0.8 wavelength.

The input impedance offered to the voltage source is 300 Ω, and since the internal impedance of the source is 300 Ω, the voltage at the input to the line is half of 60 V, or 30 V. The source is matched to the line and delivers the maximum available power to the line. Since there is no reflection and no attenuation, the voltage at the load is 30 V, but it is delayed in phase by 1.6π rad. Thus

$$V_{in} = 30 \cos 2\pi 10^8 t \qquad \text{V}$$

whereas

$$V_L = 30 \cos (2\pi 10^8 t - 1.6\pi) \qquad \text{V}$$

The input current is

$$I_{in} = \frac{V_{in}}{300} = 0.1 \cos 2\pi 10^8 t \qquad \text{A}$$

while the load current is

$$I_L = 0.1 \cos (2\pi 10^8 t - 1.6\pi) \qquad \text{A}$$

The average power delivered to the input of the line by the source must all be delivered to the load by the line,

$$P_{in} = P_L = \tfrac{1}{2} \times 30 \times 0.1 = 1.5 \qquad \text{W}$$

Now let us connect a second receiver, also having an input resistance of 300 Ω, across the line in parallel with the first receiver. The load impedance is now 150 Ω, the reflection coefficient is

$$\Gamma = \frac{150 - 300}{150 + 300} = -\frac{1}{3}$$

and the standing-wave ratio on the line is

$$s = \frac{1 + \frac{1}{3}}{1 - \frac{1}{3}} = 2$$

The input impedance is no longer 300 Ω, but is now

$$Z_{in} = Z_0 \frac{Z_L + jZ_0 \tan \beta l}{Z_0 + jZ_L \tan \beta l} = 300 \frac{150 + j300 \tan 288^\circ}{300 + j150 \tan 288^\circ}$$

$$= 510\underline{/-23.8^\circ} = 466 - j206 \qquad \Omega$$

which is a capacitive impedance. Physically, this means that this length of line stores more energy in its electric field than in its magnetic field. The input current phasor is thus

$$I_{s,\,in} = \frac{60}{300 + 466 - j206} = 0.0756\underline{/15.0^\circ} \qquad \text{A}$$

and the power supplied to the line by the source is

$$P_{in} = \tfrac{1}{2} \times (0.0756)^2 \times 466 = 1.333 \qquad \text{W}$$

Since there are no losses in the line, 1.333 W must also be delivered to the load. Note that this is less than the 1.50 W which we were able to deliver to a matched load; moreover, this power must divide equally between two receivers, and thus each receiver now receives only 0.667 W. Since the input impedance of each receiver is 300 Ω, the voltage across the receiver is easily found as

$$0.667 = \frac{1}{2}\frac{|V_{s,L}|^2}{300}$$

$$|V_{s,L}| = 20 \qquad \text{V}$$

in comparison with the 30 V obtained across the single load.

Before we leave this example, let us ask ourselves several questions about the voltages on the transmission line. Where is the voltage a maximum and a minimum, and what are these values? Does the phase of the load voltage still differ from the input voltage by 288°? Presumably, if we can answer these questions for the voltage, we could do the same for the current.

We answered questions of this nature for the uniform plane wave in the last chapter, and our analogy should therefore provide us with the corresponding information for the transmission line. In Sec. 11.7, Eq. (70) serves to locate the voltage maxima at

$$-\beta z_{max} = \frac{\phi}{2} + n\pi \qquad (n = 0, \pm 1, \pm 2, \ldots)$$

where $\Gamma = |\Gamma|e^{j\phi}$. Thus, with $\beta = 0.8\pi$ and $\phi = \pi$, we find

$$z_{max} = -0.625 \text{ and } -1.875 \qquad \text{m}$$

while the minima are $\lambda/4$ distant from the maxima,

$$z_{min} = 0 \text{ and } -1.25 \qquad \text{m}$$

and we find that the load voltage (at $z = 0$) is a voltage minimum. This, of course, verifies the general conclusion we reached in the last chapter: a voltage minimum occurs at the load if $Z_L < Z_0$, and a voltage maximum occurs if $Z_L > Z_0$, where both impedances are pure resistances.

The minimum voltage on the line is thus the load voltage, 20 V; the maximum voltage must be 40 V, since the standing-wave ratio is 2. The voltage at the input end of the line is

$$V_{s,\,in} = I_{s,\,in}Z_{in} = (0.0756\underline{/15.0^\circ})(510\underline{/-23.8^\circ}) = 38.5\underline{/-8.8^\circ}$$

The input voltage is almost as large as the maximum voltage anywhere on the line because the line is about three-quarters wavelength long, a length which would place the voltage maximum at the input when $Z_L < Z_0$.

The final question we posed ourselves deals with the relative phase of the input and load voltages. Although we have found each of these voltages, we do not know the phase angle of the load voltage. From Sec. 11.7, Eq. (68), the voltage at any point on the line is

$$\boxed{V_s = (e^{-j\beta z} + \Gamma e^{j\beta z})V_0^+} \tag{35}$$

We may use this expression to determine the voltage at any point on the line in terms of the voltage at any other point. Since we know the voltage at the input to the line, we let $z = -l$,

$$V_{s,\,\text{in}} = (e^{j\beta l} + \Gamma e^{-j\beta l})V_0^+ \tag{36}$$

and solve for V_0^+,

$$V_0^+ = \frac{V_{s,\,\text{in}}}{e^{j\beta l} + \Gamma e^{-j\beta l}} = \frac{38.5\underline{/-8.8^\circ}}{e^{j1.6\pi} - \frac{1}{3}e^{-j1.6\pi}} = 30.0\underline{/72.0^\circ} \qquad \text{V}$$

We may now let $z = 0$ in (35) to find the load voltage,

$$V_{s,L} = (1 + \Gamma)V_0^+ = 20\underline{/72^\circ} = 20\underline{/-288^\circ}$$

The amplitude agrees with our previous value. The presence of the reflected wave causes $V_{s,\,\text{in}}$ and $V_{s,L}$ to differ in phase by about -279° instead of -288°.

In order to provide a slightly more complicated example, let us now place a purely capacitive impedance of $-j300\ \Omega$ in parallel with the two 300-Ω receivers. The load impedance is therefore 150 Ω in parallel with $-j300\ \Omega$, or

$$\mathbf{Z}_L = \frac{150(-j300)}{150 - j300} = \frac{-j300}{1 - j2} = 120 - j60 \qquad \Omega$$

We first calculate the reflection coefficient and the standing-wave ratio:

$$\Gamma = \frac{120 - j60 - 300}{120 - j60 + 300} = \frac{-180 - j60}{420 - j60} = 0.447\underline{/-153.4^\circ}$$

$$s = \frac{1 + 0.447}{1 - 0.447} = 2.62$$

Thus, the standing-wave ratio is higher and the mismatch is therefore worse.

Let us next calculate the input impedance. The electrical length of the line is still 288°, so that

$$Z_{\text{in}} = 300\,\frac{(120 - j60) + j300 \tan 288^\circ}{300 + j(120 - j60) \tan 288^\circ} = 755 - j138.5 \qquad \Omega$$

This leads to a source current of

$$I_{s,\,\text{in}} = \frac{V_{\text{Th}}}{Z_{\text{Th}} + Z_{\text{in}}} = \frac{60}{300 + 755 - j138.5} = 0.0564\underline{/7.47^\circ} \qquad \text{A}$$

Therefore, the average power delivered to the input of the line is $P_{\text{in}} = \frac{1}{2}(0.0564)^2 755 = 1.200$ W. Since the line is lossless, it follows that $P_L = 1.200$ W, and each receiver gets only 0.6 W.

As a final example let us terminate our line with a purely capacitive impedance, $Z_L = -j300\ \Omega$. Obviously, we cannot deliver any average power to the load. As a consequence, the reflection coefficient is

$$\Gamma = \frac{-j300 - 300}{-j300 + 300} = -j1 = 1\underline{/-90^\circ}$$

and the reflected wave is equal in amplitude to the incident wave. Hence it should not surprise us to see that the standing-wave ratio is

$$s = \frac{1 + |-j1|}{1 - |-j1|} = \infty$$

and the input impedance is a pure reactance,

$$Z_{\text{in}} = 300 \frac{-j300 + j300 \tan 288^\circ}{300 + j(-j300) \tan 288^\circ} = j589$$

to which no average power can be delivered.

Although we could continue to find numerous other facts and figures for these examples, much of the work may be done more easily for problems of this type by using graphical techniques. We shall encounter these in the following section.

D12.5. A lossless line has $Z_0 = 100\ \Omega$, $\beta l = 0.9\pi$, and $v = 0.8c$. Let $Z_L = 125 + j0\ \Omega$. If $V_{s,\text{Th}} = 200\underline{/0^\circ}$ V at 100 MHz, in series with $Z_{\text{Th}} = 50 + j0\ \Omega$, is connected to the input of the line, find: (*a*) the length of the line, l; (*b*) Γ; (*c*) s; (*d*) the input impedance at $z = -l$.

Ans. 1.08 m; 0.1111; 1.250; $118.6 + j15.69\ \Omega$

D12.6. For the line of Prob. D12.5, also find: (*e*) $V_{s,\text{in}}$; (*f*) $V_{s,L}$; (*g*) z_{max}; (*h*) $|V_s(z)|_{\text{max}}$.

Ans. $141.3\underline{/2.22^\circ}$ V; $143.8\underline{/-163.2^\circ}$ V; 0, −1.2, −2.4, . . . , *n*; 143.8 V

12.4 GRAPHICAL METHODS

Transmission-line problems often involve manipulations with complex numbers, making the time and effort required for a solution several times greater than that needed for a similar sequence of operations on real numbers. One means of reducing the labor without seriously affecting the accuracy is by using transmission-line charts. Probably the most widely used one is the Smith chart.[3]

[3] P. H. Smith, Transmission-Line Calculator, *Electronics*, vol. 12, pp. 29–31, January 1939.

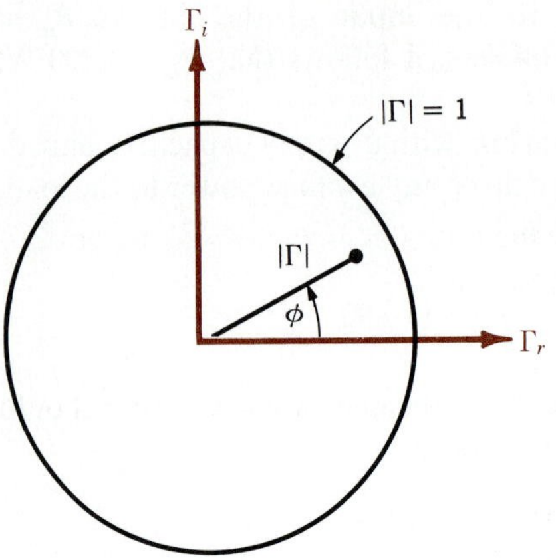

FIGURE 12.6
The polar coordinates of the Smith chart are the magnitude and phase angle of the reflection coefficient; the cartesian coordinates are the real and imaginary parts of the reflection coefficient. The entire chart lies within the unit circle $|\Gamma| = 1$.

Basically, this diagram shows curves of constant resistance and constant reactance; these may represent either an input impedance or a load impedance. The latter, of course, is the input impedance of a zero-length line. An indication of location along the line is also provided, usually in terms of the fraction of a wavelength from a voltage maximum or minimum. Although they are not specifically shown on the chart, the standing-wave ratio and the magnitude and angle of the reflection coefficient are very quickly determined. As a matter of fact, the diagram is constructed within a circle of unit radius, using polar coordinates, with radius variable $|\Gamma|$ and counterclockwise angle variable ϕ, where $\Gamma = |\Gamma|e^{j\phi}$. Figure 12.6 shows this circle. Since $|\Gamma| \leq 1$, all our information must lie on or within the unit circle. Peculiarly enough, the reflection coefficient itself will not be plotted on the final chart, for these additional contours would make the chart very difficult to read.

The basic relationship upon which the chart is constructed is

$$\boxed{\Gamma = \frac{Z_L - Z_0}{Z_L + Z_0}} \tag{37}$$

The impedances which we plot on the chart will be *normalized* with respect to the characteristic impedance. Let us identify the normalized load impedance as z_L,

$$\boxed{z_L = r + jx = \frac{Z_L}{Z_0} = \frac{R_L + jX_L}{Z_0}}$$

and thus

$$\boxed{\Gamma = \frac{z_L - 1}{z_L + 1}}$$

or

$$z_L = \frac{1+\Gamma}{1-\Gamma} \tag{38}$$

In polar form, we have used $|\Gamma|$ and ϕ as the magnitude and angle of Γ; let us now select Γ_r and Γ_i as the real and imaginary parts of Γ,

$$\Gamma = \Gamma_r + j\Gamma_i \tag{39}$$

Thus

$$r + jx = \frac{1 + \Gamma_r + j\Gamma_i}{1 - \Gamma_r - j\Gamma_i} \tag{40}$$

The real and imaginary parts of this equation are

$$r = \frac{1 - \Gamma_r^2 - \Gamma_i^2}{(1 - \Gamma_r)^2 + \Gamma_i^2} \tag{41}$$

$$x = \frac{2\Gamma_i}{(1 - \Gamma_r)^2 + \Gamma_i^2} \tag{42}$$

After several lines of elementary algebra, we may write (41) and (42) in forms which readily display the nature of the curves on Γ_r, Γ_i axes,

$$\left(\Gamma_r - \frac{r}{1+r}\right)^2 + \Gamma_i^2 = \left(\frac{1}{1+r}\right)^2 \tag{43}$$

$$(\Gamma_r - 1)^2 + \left(\Gamma_i - \frac{1}{x}\right)^2 = \left(\frac{1}{x}\right)^2 \tag{44}$$

The first equation describes a family of circles, where each circle is associated with a specific value of resistance r. For example, if $r = 0$ the radius of this zero-resistance circle is seen to be unity, and it is centered at $\Gamma_r = 0$, $\Gamma_i = 0$, the origin. This checks, for a pure reactance termination leads to a reflection coefficient of unity magnitude. On the other hand, if $r = \infty$, then $z_L = \infty$ and we have $\Gamma = 1 + j0$. The circle described by (43) is centered at $\Gamma_r = 1$, $\Gamma_i = 0$ and has zero radius. It is therefore the point $\Gamma = 1 + j0$, as we decided it should be. As another example, the circle for $r = 1$ is centered at $\Gamma_r = 0.5$, $\Gamma_i = 0$ and has a radius of 0.5. This circle is shown on Fig. 12.7, along with circles for $r = 0.5$ and $r = 2$. All circles are centered on the Γ_r axis and pass through the point $\Gamma = 1 + j0$.

Equation (44) also represents a family of circles, but each of these circles is defined by a particular value of x, rather than r. If $x = \infty$, then $z_L = \infty$, and $\Gamma = 1 + j0$ again. The circle described by (44) is centered at $\Gamma = 1 + j0$ and has zero radius; it is therefore the point $\Gamma = 1 + j0$. If $x = +1$, then the circle is centered at $\Gamma = 1 + j1$ and has unit radius. Only one-quarter of this circle lies within the

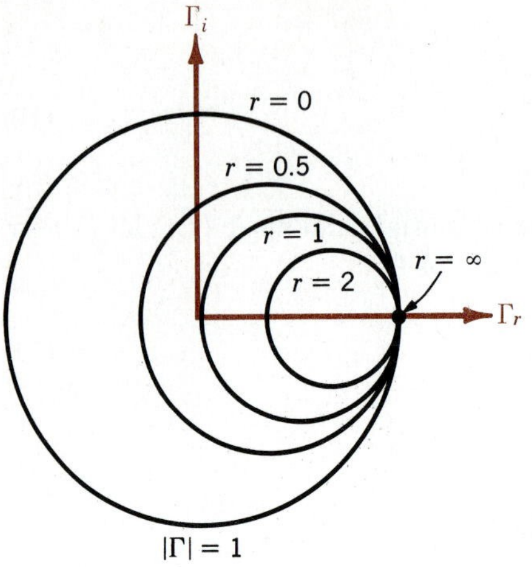

FIGURE 12.7
Constant-r circles are shown on the Γ_r, Γ_i plane. The radius of any circle is $1/(1 + r)$.

boundary curve $|\Gamma| = 1$, as shown in Fig. 12.8. A similar quarter-circle appears below the Γ_r axis for $x = -1$. The portions of other circles for $x = 0.5, -0.5, 2,$ and -2 are also shown. The "circle" representing $x = 0$ is the Γ_r axis; this is also labeled on Fig. 12.8.

The two families of circles both appear on the Smith chart, as shown in Fig. 12.9. It is now evident that if we are given Z_L, we may divide by Z_0 to obtain z_L, locate the appropriate r and x circles (interpolating as necessary), and determine Γ by the intersection of the two circles. Since the chart does not have concentric circles showing the values of $|\Gamma|$, it is necessary to measure the radial distance from the origin to the intersection with dividers or compass and use an auxiliary scale to find $|\Gamma|$. The graduated line segment below the chart in Fig. 12.9 serves this purpose. The angle of Γ is ϕ, and it is the counterclockwise angle from the Γ_r axis. Again, radial lines showing the angle would clutter up the chart badly, so the angle is indicated on the circumference of the circle. A straight line from the origin

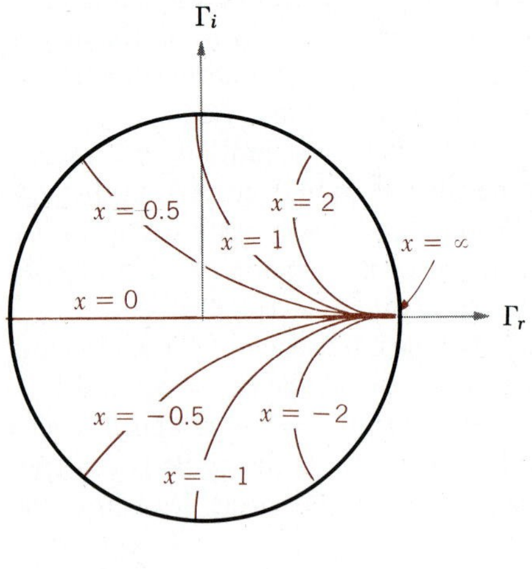

FIGURE 12.8
The portions of the circles of constant x lying within $|\Gamma| = 1$ are shown on the Γ_r, Γ_i axes. The radius of a given circle is $1/|x|$.

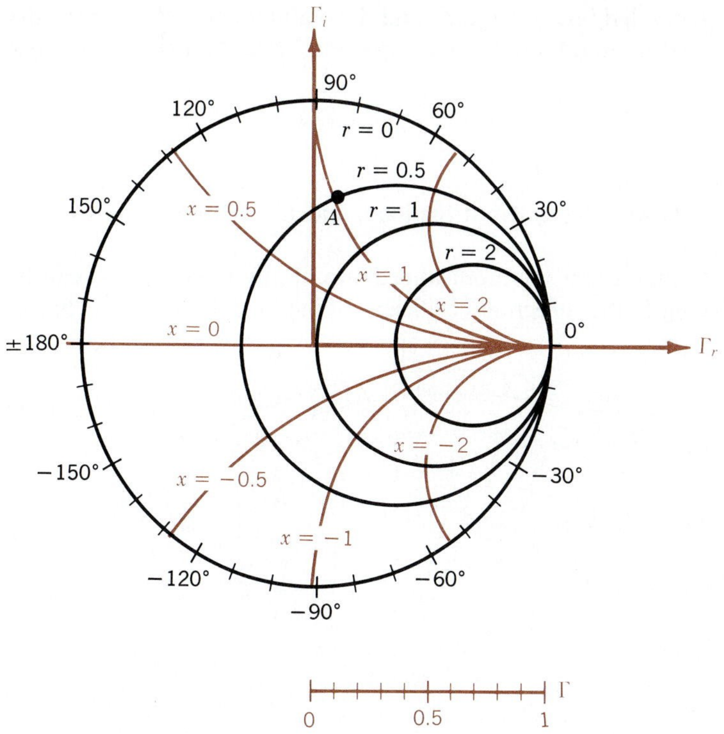

FIGURE 12.9
The Smith chart contains the constant-r circles and constant-x circles, an auxiliary radial scale to determine $|\Gamma|$, and an angular scale on the circumference for measuring ϕ.

through the intersection may be extended to the perimeter of the chart. As an example, if $Z_L = 25 + j50\ \Omega$ on a 50-Ω line, $z_L = 0.5 + j1$, and point A on Fig. 12.9 shows the intersection of the $r = 0.5$ and $x = 1$ circles. The reflection coefficient is approximately 0.62 at an angle ϕ of 83°.

The Smith chart is completed by adding a second scale on the circumference by which distance along the line may be computed. This scale is in wavelength units, but the values placed on it are not obvious. To obtain them, we first divide the voltage at any point along the line,

$$V_s = V_0^+(e^{-j\beta z} + \Gamma e^{j\beta z})$$

by the current

$$I_s = \frac{V_0^+}{Z_0}(e^{-j\beta z} - \Gamma e^{j\beta z})$$

obtaining the normalized input impedance

$$z_{\text{in}} = \frac{V_s}{Z_0 I_s} = \frac{e^{-j\beta z} + \Gamma e^{j\beta z}}{e^{-j\beta z} - \Gamma e^{j\beta z}}$$

Replacing z by $-l$ and dividing numerator and denominator by $e^{j\beta l}$, we have the general equation relating normalized *input* impedance, reflection coefficient, and line length,

$$z_{\text{in}} = \frac{1 + \Gamma e^{-j2\beta l}}{1 - \Gamma e^{-j2\beta l}} = \frac{1 + |\Gamma| e^{j(\phi - 2\beta l)}}{1 - |\Gamma| e^{j(\phi - 2\beta l)}} \tag{45}$$

Note that when $l = 0$, we are located at the load, and $z_{\text{in}} = (1 + \Gamma)/(1 - \Gamma) = z_L$, as shown by (38).

Equation (45) shows that the input impedance at any point $z = -l$ can be obtained by replacing Γ, the reflection coefficient of the load, by $\Gamma e^{-j2\beta l}$. That is,

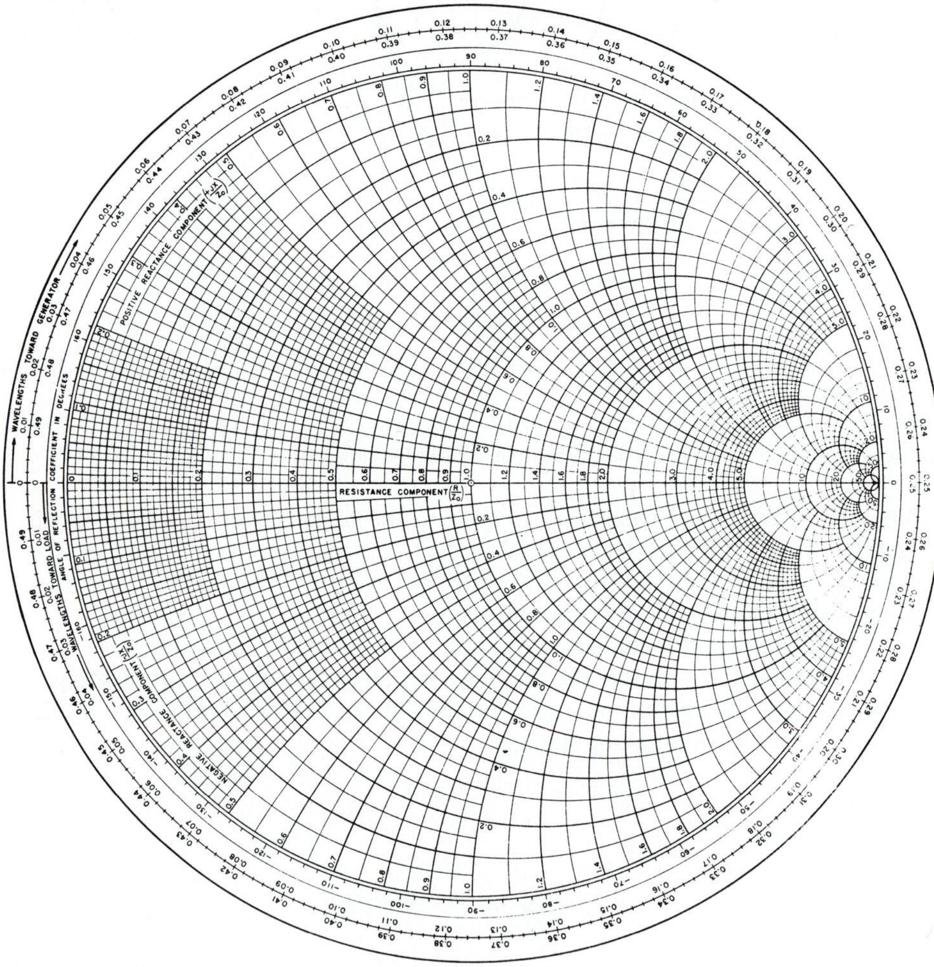

FIGURE 12.10
A photographic reduction of one version of a useful Smith chart (*courtesy of the Emeloid Company, Hillside, N.J.*). For accurate work, larger charts are available wherever fine technical books are sold.

we decrease the angle of Γ by $2\beta l$ radians as we move from the load to to the line input. Only the angle of Γ is changed; the magnitude remains constant.

Thus, as we proceed from the load z_L to the input impedance z_{in}, we move *toward* the generator a distance l on the transmission line, but we move through a *clockwise* angle of $2\beta l$ on the Smith chart. Since the magnitude of Γ stays constant, the movement toward the source is made along a constant-radius circle. One lap around the chart is accomplished whenever βl changes by π rad, or when l changes by one-half wavelength. This agrees with our earlier discovery that the input impedance of a half-wavelength lossless line is equal to the load impedance.

The Smith chart is thus completed by the addition of a scale showing a change of 0.5λ for one circumnavigation of the unit circle. For convenience, two scales are usually given, one showing an increase in distance for clockwise movement and the other an increase for counterclockwise travel. These two scales are shown in Fig. 12.10. Note that the one marked "wavelength toward generator" (wtg) shows increasing values of l/λ for clockwise travel, as described above. The zero point of the wtg scale is rather arbitrarily located to the left. This corresponds to input impedances having phase angles of 0° and $R_L < Z_0$. We have also seen that voltage minima are always located here.

The use of the transmission-line chart is best shown by example. Let us again consider the load impedance $Z_L = 25 + j50\ \Omega$, terminating a 50-Ω line. Thus $z_L = 0.5 + j1$, as marked at A on Fig. 12.11, $\Gamma = 0.62\underline{/83^\circ}$. By drawing a straight line from the origin through A to the circumference, we note a reading of 0.135 on the wtg scale. If the line is 60 cm long and the operating frequency is such that the wavelength on the line is 2 m, then $l/\lambda = 0.3$, and it is therefore 0.3λ from the load to the input. We therefore find z_{in} on the $|\Gamma| = 0.62$ circle opposite a wtg reading of $0.135 + 0.300 = 0.435$. This construction is shown in Fig. 12.11, and the point locating the input impedance is marked B. The normalized input impedance is read as $0.28 - j0.40$, and thus $Z_{in} = 14 - j20$. A more accurate analytical calculation gives $Z_{in} = 13.7 - j20.2$.

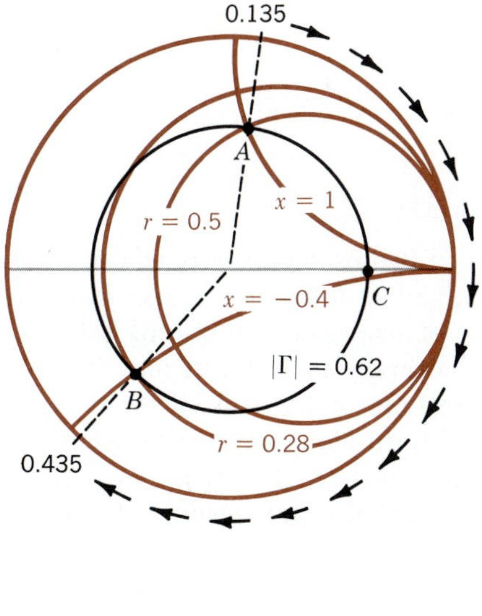

FIGURE 12.11
The normalized input impedance produced by a normalized load impedance $z_L = 0.5 + j1$ on a line 0.3λ long is $z_{in} = 0.28 - j0.40$.

Information concerning the location of the voltage maxima and minima is also readily obtained on the Smith chart. We already know that a maximum or minimum must occur at the load when Z_L is a pure resistance; if $R_L > Z_0$ there is a maximum at the load, and if $R_L < Z_0$ there is a minimum. We may extend this result now by noting that we could cut off the load end of a transmission line at a point where the input impedance is a pure resistance and replace that section with a resistance R_{in}; there would be no changes on the generator portion of the line. It follows, then, that the location of voltage maxima and minima must be at those points where Z_{in} is a pure resistance. Purely resistive input impedances must occur on the $x = 0$ line (the Γ_r axis) of the Smith chart. Voltage maxima or current minima occur when $r > 1$, or at wtg $= 0.25$, and voltage minima or current maxima occur when $r < 1$, or at wtg $= 0$. In the example above, then, the maximum at wtg $= 0.250$ must occur $0.250 - 0.135 = 0.115$ wavelength toward the generator from the load. This is a distance of 0.115×200, or 23 cm from the load.

We should also note that since the standing-wave ratio produced by a resistive load R_L is either R_L/R_0 or R_0/R_L, whichever is greater than unity, the value of s may be read directly as the value of r at the intersection of the $|\Gamma|$ circle and the r axis, $r > 1$. In our example this intersection is marked point C, and $r = 4.2$; thus, $s = 4.2$.

Transmission-line charts may also be used for normalized admittances, although there are several slight differences in such use. We let $y_L = Y_L/Y_0 = g + jb$ and use the r circles as g circles and the x circles as b circles. The two differences, then, are, first, that $g > 1$, $b = 0$ corresponds to a voltage *minimum*, and, second, that 180° must be added to the angle of Γ as read from the perimeter of the chart. We shall use the Smith chart in this way in the following section.

Special charts are also available for nonnormalized lines, particularly 50-Ω charts and 20-m℧ charts.

D12.7. A normalized load, $z_L = 0.4 + j0.6$, is located 0.45λ from the input of a lossless transmission line. (*a*) Use the Smith chart to find z_{in}. (*b*) How many wavelengths from the load is the nearest voltage maximum? (*c*) How far is it from the input to the nearest current maximum? (*d*) Find s.

Ans. $0.31 + j0.26$; 0.155λ; 0.045λ; 3.5

12.5 SEVERAL PRACTICAL PROBLEMS

In this section we shall direct our attention to two examples of practical transmission-line problems. The first is the determination of load impedance from experimental data, and the second is the design of a single-stub matching network.

Let us assume that we have made experimental measurements on a 50-Ω air line which show that there is a standing-wave ratio of 2.5. This has been determined by moving a sliding carriage back and forth along the line to determine maximum and minimum readings. A scale provided on the track along which the carriage moves indicates that a *minimum* occurs at a scale reading of 47.0 cm,

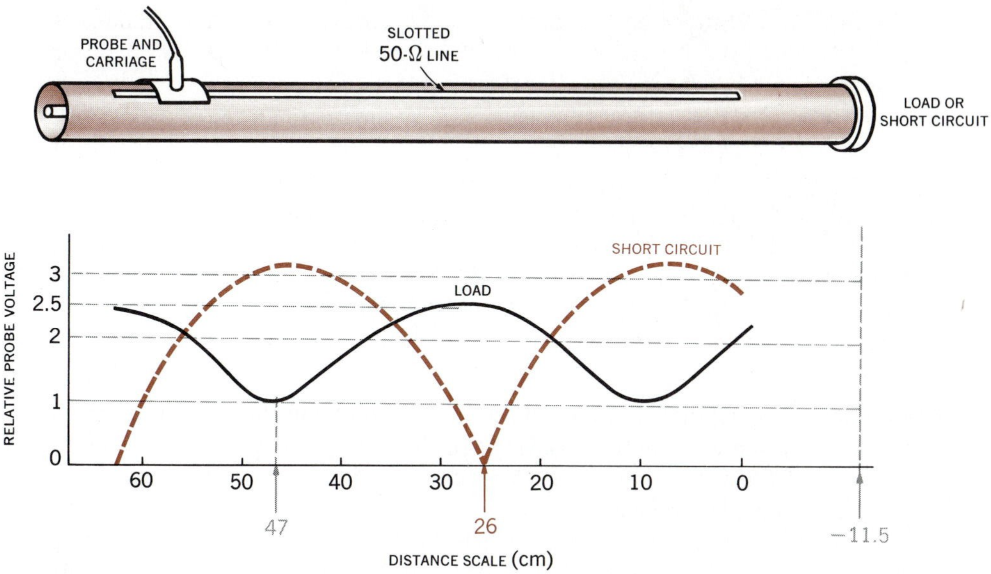

FIGURE 12.12
A sketch of a coaxial slotted line. The distance scale is on the slotted line. With the load in place, $s = 2.5$, and the minimum occurs at a scale reading of 47 cm; for a short circuit the minimum is located at a scale reading of 26 cm. The wavelength is 75 cm.

as shown in Fig. 12.12. The zero point of the scale is arbitrary and does not correspond to the location of the load. The location of the minimum is usually specified instead of the maximum because it can be determined more accurately than that of the maximum; think of the sharper minima on a rectified sine wave. The frequency of operation is 400 MHz, so the wavelength is 75 cm. In order to pinpoint the location of the load, we remove it and replace it with a short circuit; the position of the minimum is then determined as 26.0 cm.

We know that the short circuit must be located an integral number of half-wavelengths from the minimum; let us arbitrarily locate it one half-wavelength away at $26.0 - 37.5 = -11.5$ cm on the scale. Since the short circuit has replaced the load, the load is also located at -11.5 cm. Our data thus show that the minimum is $47.0 - (-11.5) = 58.5$ cm from the load, or subtracting one-half wavelength, a minimum is 21.0 cm from the load. The voltage *maximum* is thus $21.0 - (37.5/2) = 2.25$ cm from the load, or $2.25/75 = 0.030$ wavelength from the load.

With this information, we can now turn to the Smith chart. At a voltage maximum the input impedance is a pure resistance equal to sR_0; on a normalized basis, $z_{in} = 2.5$. We therefore enter the chart at $z_{in} = 2.5$ and read 0.250 on the wtg scale. Subtracting 0.030 wavelength to reach the load, we find that the intersection of the $s = 2.5$ (or $|\Gamma| = 0.429$) circle and the radial line to 0.220 wavelength is at $z_L = 2.1 + j0.8$. The construction is sketched on the Smith chart of Fig. 12.13. Thus $Z_L = 105 + j40\ \Omega$, a value which assumes its location at a scale reading of -11.5 cm, or an integral number of half-wavelengths from that position. Of

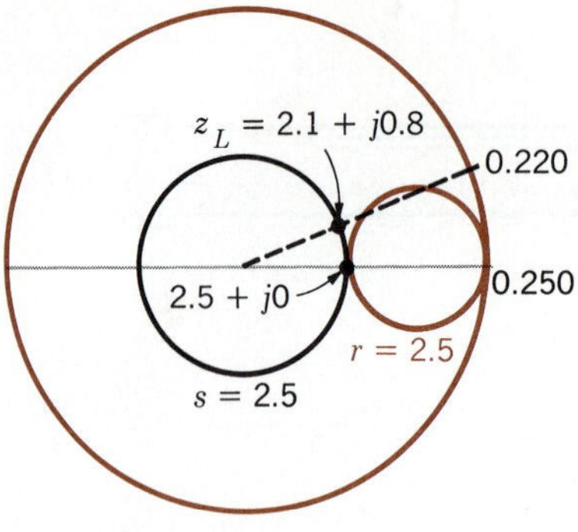

FIGURE 12.13
If $z_{in} = 2.5 + j0$ on a line 0.03 wavelength long, then $z_L = 2.1 + j0.8$.

course, we may select the "location" of our load at will by placing the short circuit at that point which we wish to consider as the load location. Since load locations are not well defined, it is important to specify the point (or plane) at which the load impedance is determined.

As a final example, let us try to match this load to the 50-Ω line by placing a short-circuited stub of length d_1 a distance d from the load (see Fig. 12.14). The stub line has the same characteristic impedance as the main line. The lengths d and d_1 are to be determined.

The input impedance to the stub is a pure reactance; when combined in parallel with the input impedance of the length d containing the load, the resultant input impedance must be $1 + j0$. Since it is much easier to combine admittances in parallel than impedances, let us rephrase our goal in admittance language: the input admittance of the length d containing the load must be $1 + jb_{in}$ for the addition of the input admittance of the stub jb_{stub} to produce a total admittance of $1 + j0$. Hence the stub admittance is $-jb_{in}$. We shall therefore use the Smith chart as an admittance chart instead of an impedance chart.

The impedance of the load is $2.1 + j0.8$, and its location is at -11.5 cm. The admittance of the load is therefore $1/(2.1 + j0.8)$, and this value may be determined

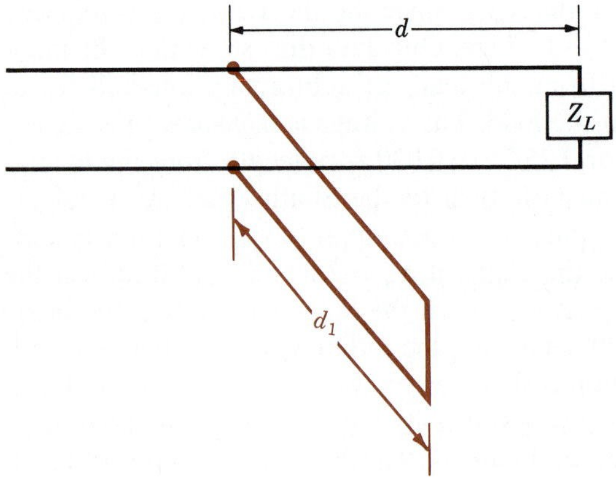

FIGURE 12.14
A short-circuited stub of length d_1, located a distance d from a load Z_L, is used to provide a matched load to the left of the stub.

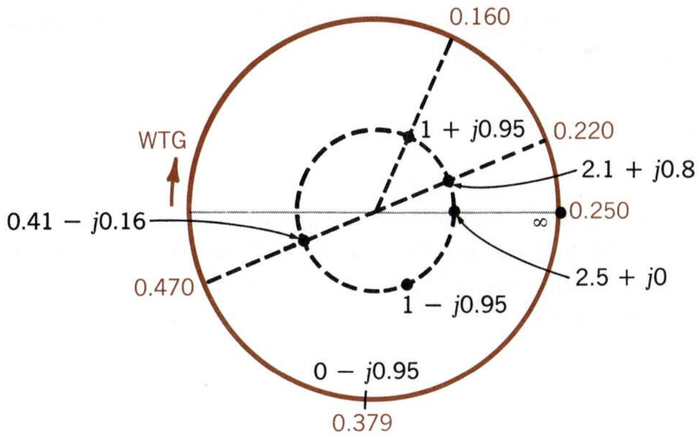

FIGURE 12.15
A normalized load $z_L = 2.1 + j0.8$ is matched by placing a 0.129-wavelength short-circuited stub 0.19 wavelength from the load.

by adding one-quarter wavelength on the Smith chart, since Z_{in} for a quarter-wavelength line is R_0^2/Z_L, or $z_{\text{in}} = 1/z_L$, or $y_{\text{in}} = z_L$. Entering the chart (Fig. 12.15) at $z_L = 2.1 + j0.8$, we read 0.220 on the wtg scale; we add (or subtract) 0.250 and find the admittance $0.41 - j0.16$ corresponding to this impedance. This point is still located on the $s = 2.5$ circle. Now, at what point or points on this circle is the real part of the admittance equal to unity? There are two answers, $1 + j0.95$ at wtg = 0.16, and $1 - j0.95$ at wtg = 0.34, as shown in Fig. 12.15. Let us select the former value since this leads to the shorter stub. Hence $y_{\text{stub}} = -j0.95$, and the stub location corresponds to wtg = 0.16. Since the load admittance was found at wtg = 0.470, then we must move $(0.5 - 0.47) + 0.16 = 0.19$ wavelength to get to the stub location.

Finally, we may use the chart to determine the necessary length of the short-circuited stub. The input conductance is zero for any length of short-circuited stub, so we are restricted to the perimeter of the chart. At the short circuit, $y = \infty$ and wtg = 0.250. We find that $b_{\text{in}} = -0.95$ is achieved at wtg = 0.379, as shown in Fig. 12.15. The stub is therefore $0.379 - 0.250 = 0.129$ wavelength, or 9.67 cm long.

Although we shall conclude our introduction to transmission lines at this point, there are many important and interesting problems that we have not discussed. Perhaps the most important involve the effects of the unavoidable losses. Others include the use of transmission lines as circuit elements, resonant circuits, and delay lines; the transient behavior of the lines; transmission-line equivalent circuits; nonuniform lines; other matching techniques; and special approximations that are useful in treating low-frequency high-voltage transmission lines. Most of these topics are discussed in most of the references listed below.

D12.8. A lossless 100-Ω air line is operating with a load Z_L that produces a standing-wave ratio of 2.8 with one minimum located at a scale reading of 29 cm. When the load is replaced by a short circuit, two adjacent minima are found at scale readings of 21 and 41 cm. Find: (*a*) λ; (*b*) f; (*c*) Z_L; (*d*) Γ.

Ans. 0.4 m; 750 MHz; $169.4 - j121.6\ \Omega$; $0.474\angle -36^\circ$

D12.9. A load $Z_L = 90 + j120\ \Omega$ terminates a 60-Ω lossless line. If the wavelength on the line is 18 cm: (*a*) how far from the load should the shortest possible short-circuited, 60-Ω, lossless stub be located? (*b*) What is the stub length? What is the standing-wave ratio: (*c*) on the line between stub and load? (*d*) on the line between stub and source? (*e*) on the stub itself?

Ans. 4.19 cm; 1.53 cm; 4.62; 1.00; ∞

SUGGESTED REFERENCES

1. Brown, R. G., R. A. Sharpe, W. L. Hughes, and R. E. Post: "Lines, Waves, and Antennas," 2d ed., The Ronald Press Company, New York, 1973. Transmission lines are covered in the first six chapters, with numerous examples.
2. Paris, D. T., and F. K. Hurd: "Basic Electromagnetic Theory," McGraw-Hill Book Company, New York, 1969. Plane waves are covered in chaps. 7 and 8, and in chap. 9 transmission lines and waveguides are discussed.
3. Seshadri, S. R.: (See Suggested References for Chap. 11).
4. Weeks, W. L.: "Transmission and Distribution of Electrical Energy," Harper and Row, Publishers, New York, 1981. Line parameters for various configurations of power transmission and distribution systems are discussed in chap. 2, along with typical parameter values.

PROBLEMS

1. A transmission line operating at 500 Mrad/s has $L = 0.5\ \mu$H/m, $C = 32$ pF/m, $G = 100\ \mu\mho$/m, and $R = 25\ \Omega$/m. (*a*) Calculate values for γ, α, β, v, λ, and Z_0. (*b*) What distance down the line can a voltage wave travel before it is reduced to 10 percent of its initial amplitude? (*c*) What distance must it travel to undergo a 90° phase shift?

2. The velocity of propagation on a certain lossless transmission line is 250 m/μs. If $C = 30$ pF/m, find (*a*) L; (*b*) Z_0; (*c*) β at 100 MHz. (*d*) The line is terminated by a 50-Ω resistor. Find Γ.

3. At a frequency of 80 MHz, a lossless transmission line has a characteristic impedance of 300 Ω and a wavelength of 2.5 m. (*a*) Find L. (*b*) Find C. (*c*) If the line is terminated with the parallel combination of 200 Ω and 5 pF, determine Γ and s.

4. A 50-Ω lossless transmission line is operating at 40 MHz with a velocity on the line of 2×10^8 m/s. If a load $Z_L = 60 + j10\ \Omega$ is located at $z = 0$, find Z_{in} at $z =$: (*a*) -2.5 m; (*b*) -5 m; (*c*) -1 m.

5. Parameter values on a certain transmission line operating at 10 MHz are $\gamma = 0 + j0.3\ \text{m}^{-1}$ and $Z_0 = 150\ \Omega$. (*a*) Find v, λ, L, and C for the line. The line is terminated at $z = 0$ with a short circuit. (*b*) Find Γ. (*c*) Find Z_{in} at $z = -5$ m.

6. A 50-Ω lossless transmission line has an air dielectric and a phase constant of 0.2π rad/m. A load is located at $z = 0$. The input impedance is $Z_{in} = 90 + j0\ \Omega$ at a point 4 m from the load. (*a*) Find Z_L. (*b*) Find R_{in} at some other location, where $Z_{in} = R_{in} + j0$, and $R_{in} < 50\ \Omega$.

7. The dimensions of the outer conductor of a coaxial cable are b and c, $c > b$. Assume $\sigma = \sigma_c$ and let $\mu = \mu_0$. Find the magnetic energy stored in the region $b < \rho < c$ for a uniformly distributed total current I flowing in the opposite directions in the inner and outer conductors.

8. A two-wire transmission line utilizes #20 AWG copper conductors (31.96 mils in diameter) embedded in a dielectric for which $\epsilon_R = 1.5$ and $\sigma = 10^{-4}$ ℧/m. (*a*) What is the spacing between wire centers if $Z_0 = 300\ \Omega$? (*b*) Find R at 100 MHz. (*c*) Find G. (*d*) What is the attenuation of the line in dB/ft at 100 MHz? (1 Np is equivalent to 8.686 dB.)

9. Specify the dimensions of a transmission line containing an air dielectric that is for use at high frequencies, if the maximum dimension of the cross section is 1 cm and the line is: (*a*) a 50-Ω coax with a zero-thickness outer conductor; (*b*) a 300-Ω two-wire line; (*c*) a 20-Ω parallel-strip line constructed of zero-thickness conductors.

10. Both conductors of a two-wire transmission line are steel wires (1 cm diameter, $\sigma_{st} = 5 \times 10^6$ ℧/m, $\mu \doteq 10^{-4}$ H/m), with a 1-mm-thick copper coating. If the wires have a center-to-center separation of 5 cm in air, find C, L, G, and R at $\omega = 10^8$ rad/s.

11. A transmission line consists of a very thin flat strip that is 2 cm wide, parallel to and 1.25 mm from an infinite, perfectly conducting plane. The strip has $\mu = \mu_0$ and $\sigma_c = 4 \times 10^7$ ℧/m, and the surrounding medium has $\epsilon = 2 \times 10^{-11}$ F/m and $\sigma = 10^{-5}$ ℧/m. At a frequency of 150 MHz, find: (*a*) C; (*b*) G; (*c*) L; (*d*) R; (*e*) Z_0.

12. A coaxial transmission line utilizes a central conductor which is a 1-mm-diameter copper wire embedded in a dielectric for which $\epsilon_R = 1.8$ and $\sigma = 10^{-4}$ ℧/m. Assume low-frequency operation. (*a*) What is the inner diameter of the outer conductor if $Z_0 = 50\ \Omega$? Assume $Z_0 = \sqrt{\mu/\epsilon}(1/2\pi)\ln(b/a)$. (*b*) What is the resistance of the coax per meter length if the outer conductor is also copper and has a thickness of 0.5 mm? (*c*) Find the inductance and capacitance per meter length. (*d*) Calculate the conductance per unit length. (*e*) Determine α at 100 Hz.

13. Two lossless coaxial lines operating at 200 MHz both utilize a dielectric with $\mu = \mu_0$, $\sigma = 0$, and $\epsilon_R = 2.25$. For both outer conductors, the inner radius b is 4 mm. Line 2 has $a_2 = 1$ mm and is terminated with $R_L = Z_{02}$. Line 1 has $a_1 = 0.8$ mm and is terminated by line 2. (*a*) What is s on line 1? (*b*) If a 1-m length of line 1 is inserted in front of the load at the load end of line 2, what is s on the input section of line 2?

14. A lossless coaxial line has $v = 2.5 \times 10^8$ m/s and $Z_0 = 60\ \Omega$. The incident phasor current wave is $I_s^+ = 2.5e^{j20°}e^{-j\pi z/4}$ A. (*a*) What is f? (*b*) Find $V_s^+(z)$. If the line for $z > 0$ is now replaced by a load $Z_L = 30 - j40\ \Omega$ at $z = 0$, find: (*c*) Γ_L; (*d*) $V_s^-(z)$; (*e*) V_s at $z = -1.2$ m.

15. For the transmission line represented in Fig. 12.16, let $f = 400$ kHz and find: (*a*) $V_{s,\,in}$; (*b*) $V_{s,\,out}$.

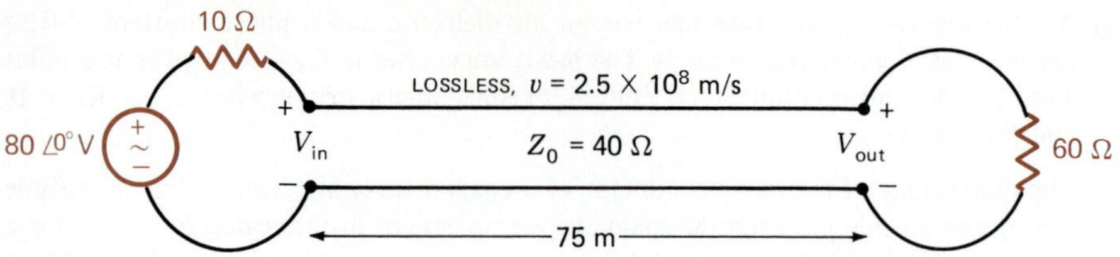

FIGURE 12.16
See Prob. 15.

16. A lossless transmission line with $Z_0 = 300\ \Omega$ is excited so that the wavelength on the line is 1.2 m. The line is 0.44 m long and short-circuited at the load end. If the amplitude of the line voltage is 4 V at the midpoint, find the: (*a*) input voltage amplitude; (*b*) input current amplitude; (*c*) current amplitude at the midpoint.

17. Determine the average power absorbed by each resistor in Fig. 12.17.

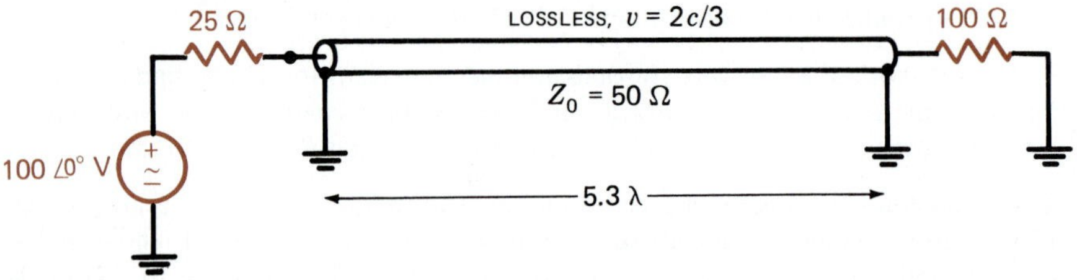

FIGURE 12.17
See Prob. 17.

18. All line segments shown in Fig. 12.18 are lossless and have $Z_0 = 50\ \Omega$. Find s on each of the three sections of line.

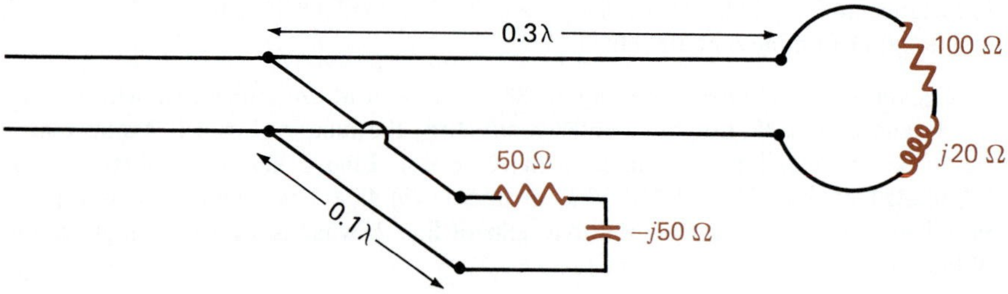

FIGURE 12.18
See Prob. 18.

19. Let $L = 0.4\ \mu\text{H/m}$, $C = 40\ \text{pF/m}$, and $f = 10^8$ Hz for a lossless transmission line that is 60 cm long and terminated in an open circuit. If the amplitude of the line voltage is 25 V at a point 20 cm from the open circuit: (*a*) what is the voltage amplitude at the input of the line, and (*b*) what is the current amplitude at the input of the line?

20. (*a*) Determine s on the transmission line of Fig. 12.19. Note that the dielectric is air. (*b*) Find the input impedance. (*c*) If $1/\omega C = 10\ \Omega$, find I_s. (*d*) What value of C will produce a maximum value for $|I_s|$ at $\omega = 10^9$ rad/s? For this value of C, calculate the average power: (*e*) supplied by the source; (*f*) delivered to $Z_L = 40 - j30\ \Omega$.

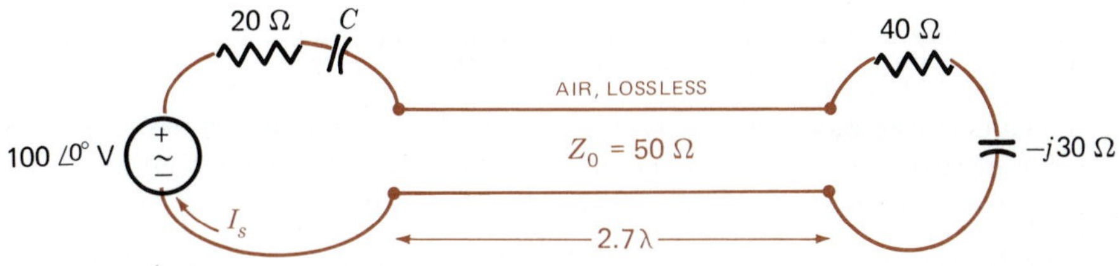

FIGURE 12.19
See Prob. 20.

21. A lossless 300-Ω line operating at $\omega = 5 \times 10^8$ rad/s is terminated in a 300-Ω load in parallel with a 10-pF capacitor. Use the Smith chart to determine: (*a*) the number of wavelengths from the load to the point at which the first voltage maximum occurs; (*b*) the voltage standing-wave ratio on the line; (*c*) Z_{in}, if the line is 80 ft long and $v = 2.4 \times 10^8$ m/s.

22. The input impedance of a lossless air line operating at 500 MHz is $(0.8 - j1.25)Z_0$. Use analytic methods or the Smith chart to find: (*a*) s; (*b*) Z_L/Z_0 if the load is located at the voltage minimum nearest to the input; (*c*) the distance from the input to the load.

23. The normalized load on a lossless transmission line is $z_L = 2 - j1$. Let $\lambda = 10$ m and make use of the Smith chart to answer the following questions: (*a*) What is the shortest distance from the load to a point at which $z_{\text{in}} = r_{\text{in}} + j0$, where $r_{\text{in}} > 1$? (*b*) What is z_{in}? (*c*) The line is cut at this point and the portion containing z_L is thrown away. A resistor $r_L = r_{\text{in}}$ of part (*a*) is connected across the line. What is s on the remainder of the line? (*d*) What is the shortest distance from this resistor to a point at which $z_{\text{in}} = 2 - j1$?

24. With the aid of the Smith chart, plot a curve of s on the left 60-Ω line in Fig. 12.20 as a function of l/λ, $0 \le l/\lambda \le 0.25$.

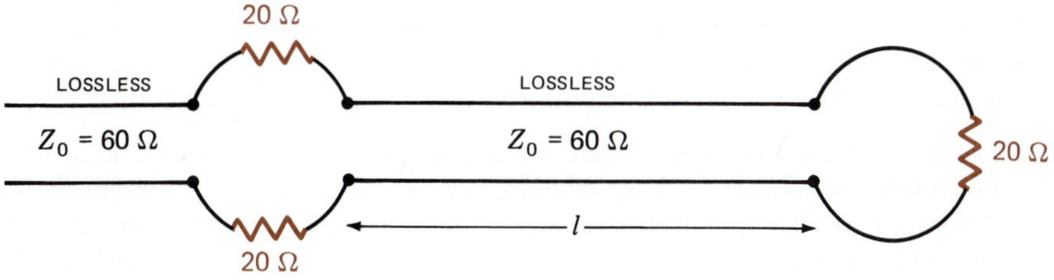

FIGURE 12.20
See Prob. 24.

25. A 50-Ω lossless transmission line has $v = 0.8c$ and a load $Z_L = 80 - j40\ \Omega$ located at $z = 0$. Use the Smith chart to determine: (*a*) s; (*b*) the distance from the load to the nearest voltage minimum if $f = 250$ MHz; (*c*) the distance from the load to the nearest voltage maximum if $f = 400$ MHz; (*d*) the input impedance looking toward Z_L at $z = -147$ cm if $f = 300$ MHz.

26. When a short circuit is placed on a 50-Ω line, two adjacent voltage minima are found to have a separation of 40 cm. With the load Z_L at $z = 0$, use the Smith chart to find Z_L if readings of total voltage on the line are: (*a*) $|V|_{max} = 13$ V at $z = -28$ cm and $|V|_{min} = 0$; (*b*) $|V|_{max} = 60$ V at $z = -16$ cm and $|V|_{min} = 25$ V.

27. A lossless transmission line is operating with a wavelength of 8 cm. Let the normalized input impedance be $z_{in} = 2 - j1$, and use the Smith chart to determine: (*a*) s; (*b*) z_L, if the line is 11 cm in length; (*c*) $z_L = 1 + jx_L$, where $x_L < 0$.

28. A lossless line is terminated in a normalized load admittance, $y_L = 0.6 - j1$. Use the Smith chart to find: (*a*) y_{in}, if $l = 0.2\lambda$; (*b*) the distance in wavelengths from y_L to the nearest voltage maximum; (*c*) Γ_L.

29. A two-wire air line constructed of lossless wires of circular cross section is gradually flared into a coupling loop that looks like an egg beater. At the point X, indicated by the arrows in Fig. 12.21, a short circuit is placed across the line. A probe is moved along the line and indicates that the first voltage minimum to the left of X is 20 cm from X. With the short circuit removed, a voltage minimum is found 5 cm to the left of X, and a voltage maximum is located that is 2.6 times the voltage of the minimum. Use the Smith chart to determine: (*a*) f; (*b*) s; (*c*) the normalized input impedance of the egg beater as seen looking to the right at point X.

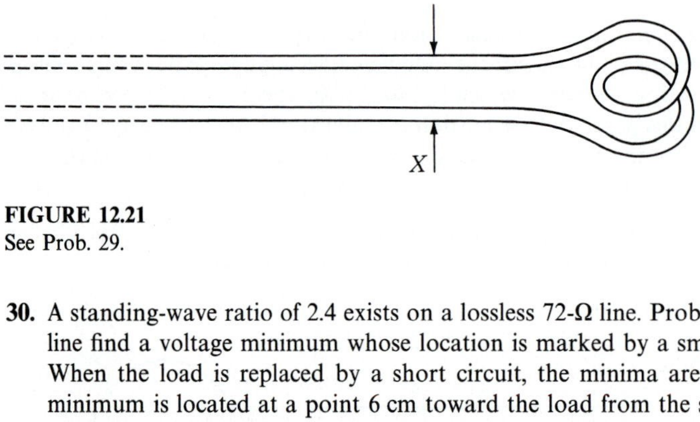

FIGURE 12.21
See Prob. 29.

30. A standing-wave ratio of 2.4 exists on a lossless 72-Ω line. Probe measurements on the line find a voltage minimum whose location is marked by a small scratch on the line. When the load is replaced by a short circuit, the minima are 20 cm apart, and one minimum is located at a point 6 cm toward the load from the scratch. Find Z_L.

31. A lossless 50-Ω line is operating with $v = c$ at $f = 5$ MHz. With the load connected, it is found that $s = 1.6$ and a voltage minimum occurs 8 m to the left of a point on the line marked by a small spot of chartreuse paint. If the load is replaced by a short circuit, the minimum moves to a spot 3 m to the right of the chartreuse dot. (*a*) Find Z_L. (*b*) What load would produce $s = 1.6$ with $|V|_{min}$ at the paint spot?

32. In order to compare the shapes of the maxima and minima of a voltage standing wave, let $z_L = 3 + j0$ be located at $z = 0$. Plot $|V|$ vs. z/λ, $-0.6\lambda < z < -0.1\lambda$; let $|V|_{min} = 1$.

33. In Fig. 12.14, let $Z_L = 40 + j10\ \Omega$, $Z_0 = 50\ \Omega$, $f = 750$ MHz, and $v = c$. (*a*) Find the shortest length d_1 of a short-circuited stub, and the shortest distance d that it may be located from the load, in order to provide a perfect match on the main line to the left of the stub. (*b*) Repeat for an open-circuited stub.

34. The lossless line shown in Fig. 12.22 is operating with $\lambda = 100$ cm. If $d_1 = 10$ cm, $d = 25$ cm, and the line is matched to the left of the stub, what is Z_L?

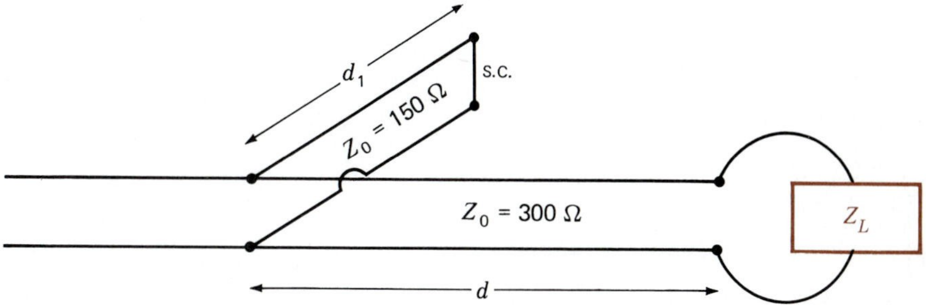

FIGURE 12.22
See Prob. 34.

35. A load, $Z_L = 80 - j60\ \Omega$, is located at $z = 0$ on a lossless line having $Z_0 = 40\ \Omega$ and $v = c$. (*a*) If $f = 150$ MHz, find the smallest distance d ($z = -d$) at which the input impedance has a real part equal to Z_0 and a positive reactance. (*b*) Find the value of the series capacitance C_s that may be inserted (say, in series with the center conductor of a coax) at $z = -d$ to provide a matched line.

36. The two-wire lines shown in Fig. 12.23 are all lossless and have $Z_0 = 250\ \Omega$. Find d and the shortest possible value for d_1 to provide a matched load if $\lambda = 100$ cm.

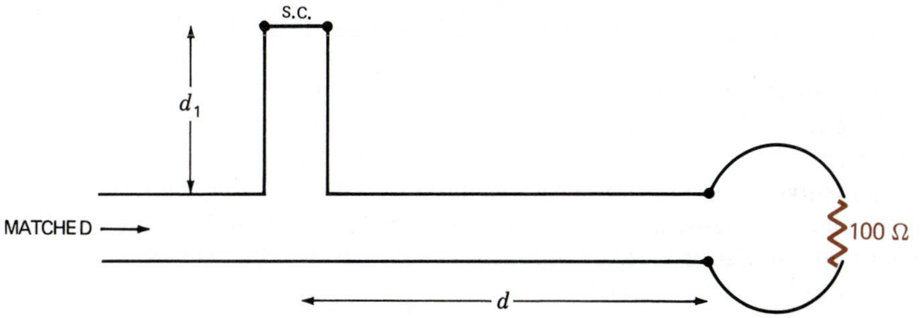

FIGURE 12.23
See Prob. 36.

CHAPTER 13

SEVERAL OTHER APPLICATIONS OF MAXWELL'S EQUATIONS

We shall close our introduction to electromagnetic theory by considering several other important applications of Maxwell's equations—circuit theory, the resonant cavity, and radiation from an antenna. The coverage of these subjects not only indicates the application and usefulness of the fundamental equations presented in the preceding chapters, but it should also provide us with some important concepts about the approximations inherent in circuit theory, resonance phenomena at high frequencies, and the mechanism whereby the energy in a transmission line can be launched into space. The references listed at the end of the chapter should all be very readable by now, and those who have particular interests in any of these applications should be able to continue their studies with enjoyment.

13.1 THE LAWS OF CIRCUIT THEORY

To indicate how Maxwell's equations, the potential definitions, and the concepts of resistance, capacitance, and inductance combine to produce the common expressions of circuit analysis, consider the configuration shown in Fig. 13.1. Between the two points 0 and 1 an external electric field is applied. These terminals are very close together, and a sinusoidal electric field may be assumed. Perhaps we could visualize a microscopic transistor oscillator, a rotating machine the size of a pinhead (complete with prime mover), or even a cooperative flea moving a mouthful of charge alternately toward point 1 and point 0. Whatever the nature

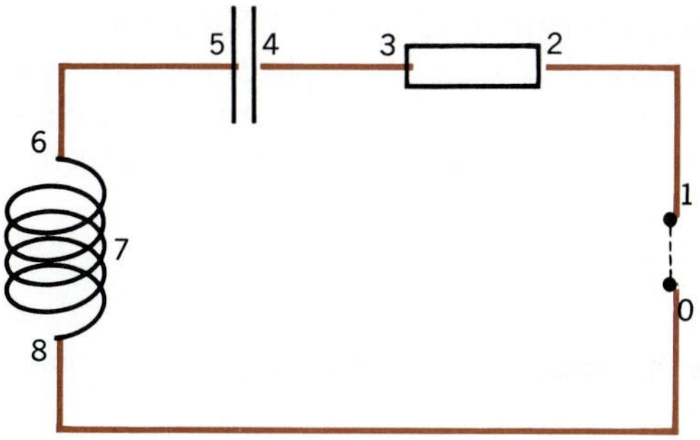

FIGURE 13.1
The field version of an *RLC* circuit. The configuration and materials are arranged so that a resistor, a capacitor, an inductor, and a voltage source may be identified.

of the source, it continues to produce an electric field between these two terminals that is independent of any currents that may consequently flow. Between points 2 and 3 there is a region of lossy material of cross-sectional area S_R, length d_R, and conductivity σ. At points 4 and 5 are located two capacitor plates, each of area S_C, with separation d_C and dielectric of permittivity ϵ. These several points are connected as shown by a filamentary perfect conductor of negligible cross section. Between points 6 and 8 the filament is wound into a helix of N turns having a very fine pitch.

It is probably obvious that we are now going to develop the familiar circuit equation

$$V_{10} = IR + L\frac{dI}{dt} + \frac{1}{C}\int_{-\infty}^{t} I\,dt \tag{1}$$

from Maxwell's equations. As we do so, it is interesting to watch how each of these terms arises and what assumptions we have to make in the process. Our beginning point is the integral form of Faraday's law,

$$\oint \mathbf{E}\cdot d\mathbf{L} = -\frac{\partial}{\partial t}\int_S \mathbf{B}\cdot d\mathbf{S} \tag{2}$$

We shall see that the right side of this equation provides us with only one term in (1), that involving the inductance. The other three terms all arise from the closed line integral.

Let us consider the surface integral on the right side of (2). Since the configuration of the circuit does not change with time, the partial derivative may be replaced with the ordinary derivative. Also, the filament between points 6 and 8, an N-turn helix, serves to produce a much larger magnetic field within the helix than in any other region along the filament. If we assume that a total magnetic flux Φ links all N turns, the surface integral becomes $-d\Phi/dt$, or $-L(dI/dt)$ by the definition of inductance, and

$$\oint \mathbf{E}\cdot d\mathbf{L} = -L\frac{dI}{dt}$$

where I is the filamentary current in each turn of the helix.

The closed line integral is applied to the path 0-1-2-3-4-5-6-7-8-0 of Fig. 13.1. It is taken along the filament, directly between the capacitor plates and points 0 and 1, as indicated by the dashed line. The contribution from the perfectly conducting filament is zero, because tangential **E** must be zero there; this includes the helix, surprising as that may be. We therefore have

$$\oint \mathbf{E} \cdot d\mathbf{L} = \int_0^1 + \int_2^3 + \int_4^5$$

The first integral on the right is the negative of the voltage between points 1 and 0,

$$\int_0^1 \mathbf{E} \cdot d\mathbf{L} = -V_{10}$$

This integral is a function only of the external source and does not depend on the configuration shown in Fig. 13.1. The path is directly between the adjacent terminals, and since we are more used to considering an external source as a voltage than as an electric field intensity, we usually call V_{10} an *applied voltage*.

The second integral is taken across the lossy material, and we apply Ohm's law in point form and the definition of total resistance R,

$$\int_2^3 \mathbf{E} \cdot d\mathbf{L} = \int_2^3 \frac{\mathbf{J}}{\sigma} \cdot d\mathbf{L} = \int_2^3 \frac{J\,dL}{\sigma} = \frac{J d_R}{\sigma} = \frac{I d_R}{\sigma S_R} = IR$$

The same total current I is assumed, and this is justified only when two conditions are met. There can be no displacement currents flowing from one portion of the filament to another (such as from point 3 to point 8), because we require the continuity of conduction plus displacement current density to be satisfied by conduction current alone. In other language, we are assuming that stray capacitances are neglected. Also the dimensions of the filamentary path must be small compared to a wavelength. This will be amplified in the final section of this chapter, but our study of wave motion and transmission lines should indicate the complete reversal which may occur in a field over a distance of one half-wavelength. Here we wish to avoid radiation, but later in the chapter it will provide our main item of interest.

The third integral is evaluated across the region between the capacitor plates where conduction current is zero but displacement current is equal to the current I, as we assumed earlier. Here we may represent the integral by

$$\int_4^5 \mathbf{E} \cdot d\mathbf{L} = \int_4^5 \frac{D}{\epsilon}\,dL = \frac{D d_C}{\epsilon} = \frac{Q d_C}{\epsilon S_C} = \frac{Q}{C}$$

or

$$\int_4^5 \mathbf{E} \cdot d\mathbf{L} = \frac{1}{C}\int_{-\infty}^{t} I\,dt$$

where we assume that there is no charge on the capacitor at $t = -\infty$.

Combining these results, we have

$$-V_{10} + IR + \frac{1}{C}\int_{-\infty}^{t} I\,dt = -L\frac{dI}{dt}$$

or

$$V_{10} = IR + L\frac{dI}{dt} + \frac{1}{C}\int_{-\infty}^{t} I\,dt$$

which is the familiar equation for an *RLC* series circuit that we hoped we would obtain.

The assumptions upon which this equality is based, implicitly assumed in most circuit problems, are as follows:

1. A filamentary conductor defines the closed path or circuit.
2. The maximum dimensions of the circuit are small compared to a wavelength.
3. Displacement current is confined to capacitors.
4. Magnetic flux is confined to inductors.
5. Imperfect conductivity is confined to resistors.

The first assumption simply defines what we mean by a "circuit." The second implies that any circuit ceases to obey the laws of circuit theory if the frequency is sufficiently high. The last three assumptions restrict the circuit relationships to ideal elements. If our capacitors have resistance, or our inductors have capacitance, we cannot apply the circuit equations until we have replaced the lossy capacitor or the inductor plus stray capacitance by some networks containing only ideal elements. The selection of an appropriate combination of ideal elements is made possible by our knowledge of the behavior of the electromagnetic fields in and around the devices; experience also comes in handy.

It is also possible to use a "circuits approach" to many devices which do not satisfy the conditions above, such as transmission lines. Waveguides, resonant cavities, and many antenna problems may also be treated as circuits in certain respects. When this is done, the detailed description of the electric and magnetic fields must be of secondary importance; the device is described in terms of an equivalent voltage and current at each input or output. The resonant cavity discussed in the following section is one such example.

D13.1. A lossless 20-cm length of coax is terminated with a 40-pF capacitance. If $v/c = \frac{2}{3}$ and $Z_0 = 50\ \Omega$ for the coax, let $Z_{in} = 1/j\omega C_{eq}$ and calculate C_{eq} at $f =$: (*a*) 1 kHz; (*b*) 20 MHz; (*c*) 75 MHz.

Ans. 60.0 pF; 62.1 pF; 118.6 pF

13.2 THE RESONANT COAXIAL CAVITY

We found in the last section that the foundations of circuit theory are based on Maxwell's equations, but that we must also have a closed path (not necessarily a conducting path), circuit dimensions that are small compared with a wavelength, and identifiable circuit elements. Now let us consider a device for which most of these requirements are not met, but that yet provides many analytical results that may be interpreted in circuit-theory terms.

We may think of a closed laboratory into which a single coaxial conductor passes. Within the laboratory there is a distributed circuit, that is, one having dimensions comparable to a wavelength with its resistive, inductive, and capacitive properties distributed throughout this region. Circuit relationships are therefore not applicable in the laboratory. At the point at which the coaxial conductor enters the laboratory, however, we are able to define a voltage between the outer and inner conductors and the current in each. This is possible because the radial dimensions of the coax are assumed to be a small fraction of a wavelength even though its length is not. The problem we wish to consider is the description of the device in the laboratory in terms of an equivalent circuit. If we are successful, then it should be impossible to determine whether the laboratory contains an equivalent circuit that satisfies all the requirements of circuit theory or a distributed circuit that does not. We shall see that within a sufficiently narrow frequency band, this is indeed the case.

The device which we shall use as an example in developing an equivalent circuit is a coaxial resonant cavity. Such resonators are very selective with respect to frequency, and they thus find use in frequency meters, tuned amplifiers, and oscillators; they may also be used to determine the conductivity and permittivity of dielectric materials, perhaps to control some manufacturing process or to identify soil characteristics.

Let us consider the coaxial cavity shown in Fig. 13.2. We assume that t, a, and b are much smaller than a wavelength; l may be any length, although we shall specifically select one-quarter wavelength. Since the constants of the dielectric and conductor are known, we could calculate R, G, L, and C for the line, and from these, Z_0 and γ. We could then spend a few minutes obtaining an expression for the

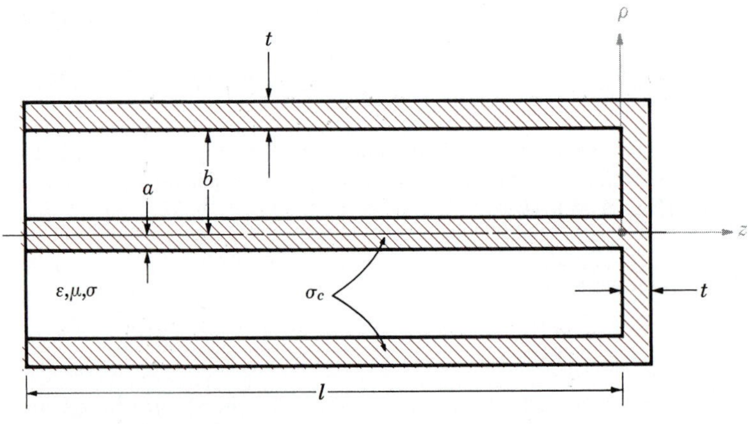

FIGURE 13.2
A coaxial cavity for which $b \ll \lambda$. This device may be represented by a parallel resonant circuit in the neighborhood of the resonant frequency.

input impedance of a *lossy* transmission line, and thus determine the input impedance of the line terminated in a low-impedance load, the short-circuiting plate. Having the input impedance, we would then be able to determine the form of an equivalent circuit providing the same input impedance, at least within some desired frequency range about the resonant frequency and to some specified accuracy.

Let us select a more general procedure, however, which illustrates the principles involved in determining an equivalent circuit for any microwave cavity near resonance. We shall begin by finding the fields that would exist in a *lossless* cavity. We then evaluate these fields at the conductor boundaries and calculate the losses that would occur if these fields were present. Although we could then use the results of the loss analysis to recalculate the cavity fields and use these values to improve our loss calculations, this iterative procedure is not necessary in low-loss cavities; the first approximation provides excellent accuracy.

For the lossless coaxial line terminated in a short circuit at $z = 0$, where the z axis is directed to the right, the phasor voltage at any point along the line is the sum of the incident and reflected waves,

$$V_s = V_0 e^{-j\beta z} - V_0 e^{j\beta z} \tag{3}$$

or

$$V_s = -j2V_0 \sin \beta z$$

and the current is

$$I_s = \frac{V_0}{Z_0} e^{-j\beta z} + \frac{V_0}{Z_0} e^{j\beta z}$$

or

$$I_s = \frac{2V_0}{Z_0} \cos \beta z \tag{4}$$

where

$$\beta = \omega\sqrt{LC} = \omega\sqrt{\mu\epsilon}$$

and

$$Z_0 = \sqrt{\frac{L}{C}} = \frac{1}{2\pi}\sqrt{\frac{\mu}{\epsilon}} \ln \frac{b}{a}$$

The magnetic field intensity is related directly to the current,

$$H_{\phi s} = \frac{I_s}{2\pi\rho}$$

and thus

$$H_{\phi s} = \frac{V_0}{\pi\rho Z_0} \cos \beta z \tag{5}$$

We then recall[1] that the relationship of the electric field intensity to the voltage between conductors of a coaxial line is

$$E_{\rho s} = \frac{V_s}{\rho \ln (b/a)} = \frac{-j2V_0}{\rho \ln (b/a)} \sin \beta z \tag{6}$$

where the voltage of the outer conductor is zero and that of the inner is V_s.

These last two equations give us the interior fields for the lossless cavity.

Power losses in the cavity arise in several different ways: in an imperfect dielectric which might fill the cavity, and in the conductors forming the center conductor, the outer conductor, and the single end plate. Let us first assume that the power loss in the dielectric is much greater than that in the walls, as would be the case if the cavity were filled with beer, flour, or some other healthful food whose production we are monitoring electrically. The loss in the dielectric is determined by integrating the ohmic power density throughout the cavity. We begin by finding the conduction current density,

$$J_{\rho s} = \sigma E_{\rho s} = \frac{-j2\sigma V_0}{\rho \ln (b/a)} \sin \beta z$$

or as a real quantity,

$$J_\rho = \frac{2\sigma V_0}{\rho \ln (b/a)} \sin \beta z \sin \omega t$$

The instantaneous power being dissipated in the dielectric is

$$P_d = \int_{\text{vol}} \frac{1}{\sigma} J_\rho^2 \, dv$$

$$= \int_{-l}^{0} \int_{0}^{2\pi} \int_{a}^{b} \frac{4\sigma V_0^2}{\rho [\ln (b/a)]^2} \sin^2 \beta z \sin^2 \omega t \, d\rho \, d\phi \, dz$$

or

$$P_d = \frac{4\pi\sigma V_0^2}{\ln (b/a)} \left(l - \frac{\sin 2\beta l}{2\beta} \right) \sin^2 \omega t$$

and the time-average power loss is

$$P_{d,\,\text{av}} = \frac{2\pi\sigma V_0^2}{\ln (b/a)} \left(l - \frac{\sin 2\beta l}{2\beta} \right)$$

In order to provide a *resonant* cavity, let us now select the length l as one-quarter wavelength. If the cavity were truly lossless, the current at the input would be zero and the input impedance would be infinite. Instead, we shall find that the power lost in the dielectric and conducting walls causes a small input current and an input impedance which is a high resistance at resonance. If we identify the

[1] Readers who do not have total recall may refer to Example 2 in Sec. 7.3.

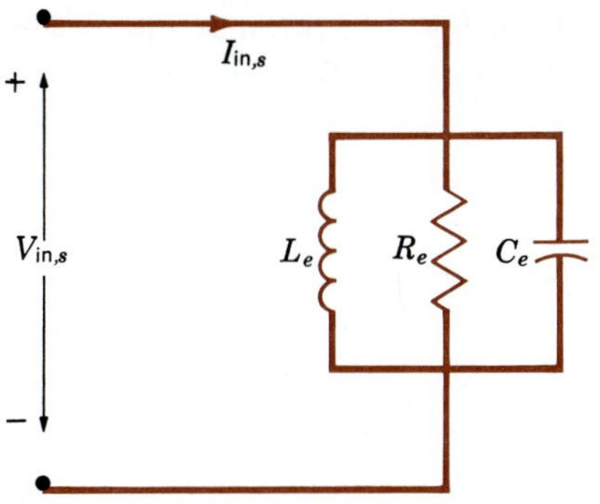

FIGURE 13.3
Suitable selections of R_e, L_e, and C_e cause this circuit to be equivalent to the resonant coaxial cavity of Fig. 13.2 near $f_0 = 1/(2\pi\sqrt{L_eC_e}) = 1/(4l\sqrt{\mu\epsilon})$, where l is the length of the cavity.

resonant frequency as f_0, then

$$l = \frac{\lambda_0}{4} = \frac{1}{4f_0\sqrt{\mu\epsilon}}$$

and

$$\beta_0 = \frac{2\pi}{\lambda_0} = \frac{\pi}{2l}$$

Thus the average power loss in the dielectric at the resonant frequency is

$$P_{d0,\,\mathrm{av}} = \frac{2\pi\sigma V_0^2 l}{\ln(b/a)} \tag{7}$$

We now try to fit our results to an equivalent circuit of the form shown in Fig. 13.3, a parallel resonant circuit. At resonance the input impedance is a pure resistance R_e (by the definition of resonance). From (3) we have the voltage at any point at any time,

$$V = 2V_0 \sin \beta z \sin \omega t$$

and thus the input ($z = -l$) voltage is

$$V_{\mathrm{in}} = -2V_0 \sin \beta l \sin \omega t$$

At resonance, $\beta_0 l = \pi/2$ and

$$V_{0,\,\mathrm{in}} = -2V_0 \sin \omega t$$

Since the crest amplitude is $2V_0$, the average power loss

$$P_{d0,\,\mathrm{av}} = \frac{1}{2}\frac{(-2V_0)^2}{R_e} = \frac{2\pi\sigma V_0^2 l}{\ln(b/a)}$$

enables us to find the equivalent resistance

$$R_e = \frac{\ln(b/a)}{\pi\sigma l} \tag{8}$$

This expression, of course, assumes no losses in the conductor boundaries.

In order to find expressions for C_e and L_e in Fig. 13.3, we may establish a value for the energy stored in the resonator at resonance. We can then use the general definition of Q,

$$Q = 2\pi \frac{\text{energy stored}}{\text{energy lost per cycle}}$$

or

$$Q = \omega \frac{\text{energy stored}}{\text{average power loss}} \tag{9}$$

to determine Q. If we know Q, familiar circuit equations will enable us to find L_e and C_e.

The energy stored in the cavity is the sum of the energies stored in the electric and magnetic fields. It can be shown, however, that this *total* energy is the same as the *maximum* energy stored in either the electric or magnetic fields. That is, the total energy is constant, and when the magnetic energy is zero, the energy in the electric field is a maximum, and vice versa. Let us select the maximum value of the magnetic field (5),

$$H_{\phi,\,\max} = \frac{V_0}{\pi \rho Z_0} \cos \beta z$$

and thus

$$\begin{aligned} W_{H,\,\max} &= \int_{\text{vol}} \tfrac{1}{2} \mu H^2_{\max}\, dv \\ &= \frac{\mu V_0^2}{2\pi^2 Z_0^2} \int_{-l}^{0} \int_0^{2\pi} \int_a^b \frac{1}{\rho} \cos^2 \beta z\, d\rho\, d\phi\, dz \\ &= \frac{\mu V_0^2 \ln (b/a)}{2\pi Z_0^2} \left(l + \frac{\sin 2\beta l}{2\beta} \right) \end{aligned}$$

At resonance

$$W_{H0,\,\max} = W_{0,\,\max} = \frac{\mu l V_0^2 \ln (b/a)}{2\pi Z_0^2} = \frac{2\pi \epsilon l V_0^2}{\ln (b/a)} \tag{10}$$

A determination of the maximum energy stored in the electric field leads to the same result.

Combining our expressions for the average power loss in the dielectric (7), the definition of Q (9), and the total energy (10), all evaluated at the resonant frequency, we find the Q produced by the dielectric at resonance.

$$Q_{d0} = \omega_0 \frac{2\pi \epsilon l V_0^2 / \ln (b/a)}{2\pi \sigma l V_0^2 / \ln (b/a)}$$

or

$$Q_{d0} = \frac{\omega_0 \epsilon}{\sigma}$$

which, interestingly enough, turns out to be the reciprocal of the loss tangent, the value it would have for a coaxial capacitor at lower frequencies.

The equivalent capacitance C_e may be found either from the expression for the energy in a capacitor,

$$W_{0,\,\max} = \tfrac{1}{2}C_e(2V_0)^2$$

or the Q of a parallel resonant circuit,

$$Q_{d0} = \omega_0 C_e R_e$$

It is

$$C_e = \frac{\pi \epsilon l}{\ln (b/a)} \tag{11}$$

Knowing the capacitance, we find the inductance by

$$\omega_0^2 = \frac{1}{L_e C_e}$$

or from the Q,

$$Q_{d0} = \frac{R_e}{\omega_0 L_e}$$

It is

$$L_e = \frac{4\mu l}{\pi^3} \ln \frac{b}{a} \tag{12}$$

For a cavity in which $b/a = 2.72$, or $\ln (b/a) = 1$, $a = 1$ cm, $f_0 = 100$ MHz, $\sigma/\omega\epsilon = 0.001$, $\epsilon_R = 4$, and $\mu_R = 1$, we have

$$l = 37.5 \quad \text{cm}$$

$$Q_{d0} = 1000$$

$$R_e = 38\,100 \quad \Omega$$

$$C_e = 41.7 \quad \text{pF}$$

$$L_e = 0.0607 \quad \mu\text{H}$$

The cavity bandwidth is therefore $f_0/Q_{d0} = 0.1$ MHz. Also, if the dielectric constant of the material filling the cavity decreased 0.25 percent from 4 to 3.99, the resonant frequency would increase 0.125 percent, or 125 kHz. Such a change could be recognized easily, and thus measurement of the resonant frequency for a sample in the cavity might be used to control the dryness of flour, for example.

Now let us consider the losses in the cavity walls. The value of H_ϕ at the outer cylinder $\rho = b$ is

$$H_{\phi b} = \frac{V_0}{\pi b Z_0} \cos \beta z \cos \omega t$$

If the conductor were *perfect*, a surface current

$$K_z = -H_{\phi b} = \frac{-V_0}{\pi b Z_0} \cos \beta z \cos \omega t$$

would be present. With a *finite* conductivity, this total current per unit width is distributed throughout a thin layer near the surface. As we saw in Chap. 11, the total power loss may be found by assuming a uniform current density throughout a region one skin depth thick. This uniform current density is therefore

$$J_z = \frac{K_z}{\delta} = \frac{-V_0}{\pi b \delta Z_0} \cos \beta z \cos \omega t$$

The ohmic power loss in the outer cylinder (of conductivity σ_c) is thus

$$\begin{aligned} P_b &= \int_{\text{vol}} \frac{1}{\sigma_c} J_z^2 \, dv \\ &= \int_{-l}^{0} \frac{1}{\sigma_c} J_z^2 \delta 2\pi b \, dz \\ &= \frac{V_0^2}{\sigma_c \pi b \delta Z_0^2} \left(l + \frac{\sin 2\beta l}{2\beta} \right) \cos^2 \omega t \end{aligned}$$

Therefore the average power loss at resonance is

$$P_{b0,\,\text{av}} = \frac{V_0^2 l}{2\pi \sigma_c b \delta Z_0^2} \tag{13}$$

A similar calculation for the inner conductor gives

$$P_{a0,\,\text{av}} = \frac{V_0^2 l}{2\pi \sigma_c a \delta Z_0^2} \tag{14}$$

The only remaining surface is the end plate at $z = 0$. The magnetic field intensity at $z = 0$ is

$$H_\phi \Big|_{z=0} = \frac{V_0}{\pi \rho Z_0} \cos \omega t$$

leading to a surface current

$$K_\rho = \frac{V_0}{\pi \rho Z_0} \cos \omega t$$

and a current density that is independent of z in the first skin depth,

$$J_\rho = \frac{K_\rho}{\delta} = \frac{V_0}{\pi\rho\delta Z_0} \cos \omega t$$

Thus the power loss in the end plate is

$$P_{\text{end}} = \int_a^b \int_0^{2\pi} \frac{1}{\sigma_c} \frac{V_0^2}{\pi^2\rho^2\delta^2 Z_0^2} \delta \cos^2 \omega t \, \rho \, d\phi \, d\rho$$

$$= \frac{2V_0^2}{\pi\sigma_c\delta Z_0^2} \ln \frac{b}{a} \cos^2 \omega t$$

and therefore

$$P_{\text{end, 0, av}} = \frac{V_0^2}{\pi\sigma_c\delta Z_0^2} \ln \frac{b}{a} \tag{15}$$

Combining (13), (14), and (15), the total power loss in the conducting walls at resonance is

$$P_{\text{av}} = \frac{V_0^2 l}{2\pi\sigma_c\delta Z_0^2} \left(\frac{1}{a} + \frac{1}{b} + \frac{2}{l} \ln \frac{b}{a} \right)$$

Since we know the total energy stored, we may calculate the Q of the coaxial cavity having wall losses only,

$$Q_{w0} = \frac{(2/\delta) \ln (b/a)}{(1/a) + (1/b) + (2/l) \ln (b/a)}$$

The elements of the equivalent circuit are now readily obtained. Since the total energy stored has not changed, C_e is unchanged. L_e may again be found from C_e and the resonant frequency ω_0; thus, it is also unchanged. Hence the changed Q affects only the equivalent resistance R_e,

$$R_e = \frac{Q_{w0}}{\omega_0 C_e} = \frac{4\eta \, [\ln (b/a)]^2}{\pi^2\delta \, [(1/a) + (1/b) + (2/l) \ln (b/a)]} \tag{16}$$

where

$$\eta = \sqrt{\frac{\mu}{\epsilon}}$$

Using the same cavity as an example, let us assume that it is silver-plated ($\sigma_c = 6.18 \times 10^7$ ℧/m):

$$\delta = \frac{1}{\sqrt{\pi f \mu \sigma_c}} = 6.40 \qquad \mu\text{m}$$

and thus

$$Q_{w0} = 2200$$

while

$$R_e = 83\,900 \qquad \Omega$$

When both losses are present, the total power loss is their sum, and the equivalent resistance turns out to be the parallel combination of the resistances obtained by considering each form of loss separately. The Q of the cavity at resonance is

$$Q_0 = \frac{1}{(1/Q_{d0}) + (1/Q_{w0})}$$

Concluding our example, the Q at resonance of the silver-plated cavity filled with the lossy dielectric is

$$Q_0 = \frac{1}{1/1000 + 1/2200} = 687$$

and the equivalent resistance is reduced to 26 200 Ω.

If $V_0 = 1$ V ($|V_{\text{in},s}| = 2$ V), the total power loss is 76.3 μW. This is distributed among the several parts of the cavity as follows: dielectric, 52.4 μW; center conductor, 16.8 μW; outer conductor, 6.2 μW; and end plate, 0.9 μW.

By using the element values of the equivalent circuit, the input impedance of the coaxial cavity may be determined at frequencies near resonance. This is a typical circuit problem, and we shall not try to develop or use these formulas. It is worth our while, however, to ask how far from resonance the equivalence holds. Two facts must be kept in mind. First, the coaxial cavity will show a series resonance when its length is one-half wavelength, or at 200 MHz; in contrast, the equivalent circuit provides no other resonances. Second, the cavity parameters, such as the conductivity and permittivity of the dielectric, vary with frequency, while no provisions have been made in the equivalent circuit for this phenomenon.

These differing characteristics cause the analogy between the circuit and the cavity to be limited to a frequency range of perhaps 20 percent of the resonant frequency. Since our interest in the cavity behavior is probably limited to a narrow frequency band near resonance, one which contains all side bands of appreciable amplitude for a modulated signal, for example, our equivalent circuit proves to be very useful.

D13.2. The parameters of a lossless quarter-wavelength resonant cavity are $l = 20$ cm, $a = 0.02\lambda_0$, $b = 0.05\lambda_0$, $\epsilon_R = 5$, and $\mu_R = 1$. Find: (*a*) the volume of the cavity; (*b*) the internal surface area, assuming one end is open; (*c*) the resonant frequency.

Ans. 844 cm^3; 746 cm^2; 167.6 MHz

D13.3. Data for a quarter-wavelength resonant cavity are: $a = 2.5$ cm, $b = 5$ cm, $l = 40$ cm, $\mu_R = 1$, $\epsilon_R = 5$, $\sigma = 10^{-4}$ ℧/m, and $\sigma_c = 5 \times 10^7$ ℧/m. Let the voltage amplitude across the open end of the cavity be 80 V at resonance and calculate the average power loss in the: (*a*) dielectric; (*b*) outer conductor; (*c*) inner conductor; (*d*) end plate.

Ans. 580 mW; 15.17 mW; 30.3 mW; 2.63 mW

13.3 RADIATION

In this last example of the application of Maxwell's equations we shall find the electromagnetic field which results from a given current distribution. For the first time, therefore, we shall have the specific field which results from a specific time-varying source. In the discussion of the uniform plane wave only the wave motion in free space was investigated, and the source of the field was not considered. The current distribution in a conductor was a similar problem, although we did at least relate the current to an assumed electric field intensity at the conductor surface. This might be considered as a source, but it is not a very practical one for it is infinite in extent.

We now assume a current filament as the source. It is taken as a differential length, but we shall be able to extend the results easily to a filament which is short compared to a wavelength, specifically less than about one-quarter of a wavelength overall. The differential filament is shown at the origin and is oriented along the z axis in Fig. 13.4. The positive sense of the current is taken in the $\mathbf{a}_z$ direction. We assume a uniform current $I_0 \cos \omega t$ in this short length d and do not concern ourselves at present with the apparent discontinuity at each end.

We shall not attempt at this time to discover the "source of the source," but shall merely assume that the current distribution cannot be changed by any field which it produces.

The first step is the application of the retarded vector magnetic potential expression, Sec. 10.5,

$$\mathbf{A} = \int \frac{\mu[I]\,d\mathbf{L}}{4\pi R}$$

where $[I]$ is a function of the retarded time $t - R/v$. Since no integration is required for the very short filament assumed, we have

$$\mathbf{A} = \frac{\mu[I]d}{4\pi R}\,\mathbf{a}_z$$

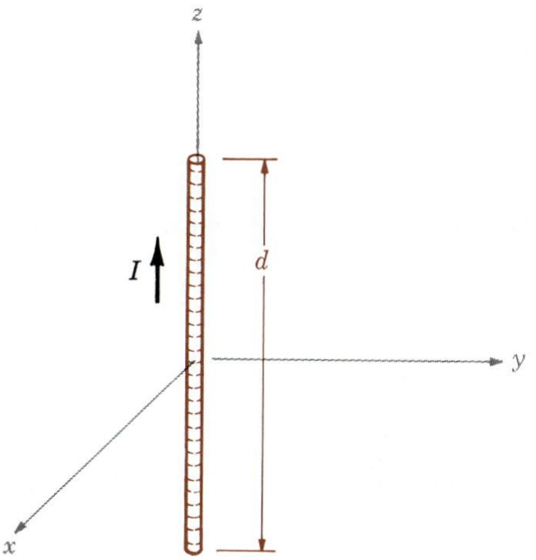

FIGURE 13.4
A differential current filament of length d carries a current $I = I_0 \cos \omega t$.

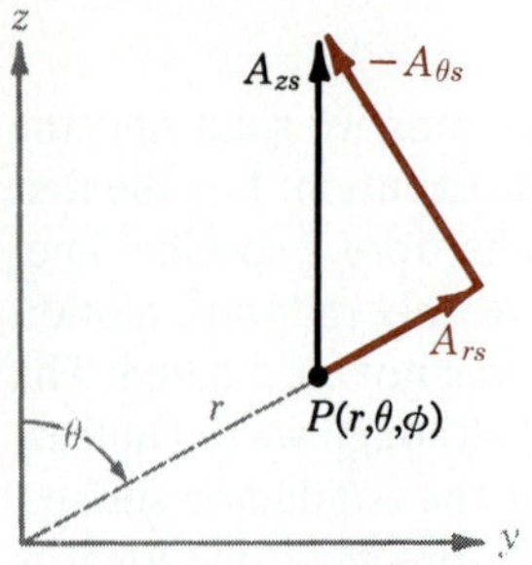

FIGURE 13.5
The resolution of A_{zs} at $P(r,\theta,\phi)$ into the two spherical components A_{rs} and $A_{\theta s}$. The sketch is arbitrarily drawn in the $\phi = 90°$ plane.

Only the z component of **A** is present, for the current is only in the $\mathbf{a}_z$ direction. At any point P, distant R from the origin, the current is retarded by R/v and

$$I = I_0 \cos \omega t$$

becomes

$$[I] = I_0 \cos\left[\omega\left(t - \frac{R}{v}\right)\right]$$

$$[I_s] = I_0 e^{-j\omega R/v}$$

Thus

$$A_{zs} = \frac{\mu I_0 d}{4\pi R} e^{-j\omega R/v}$$

Using a mixed coordinate system for the moment, let us replace R by the small r of the spherical coordinate system and then determine which spherical components are represented by A_{zs}. Figure 13.5 helps to determine that

$$A_{rs} = A_{zs} \cos \theta$$

$$A_{\theta s} = -A_{zs} \sin \theta$$

and therefore

$$A_{rs} = \frac{\mu I_0 d}{4\pi r} \cos \theta \, e^{-j\omega r/v}$$

$$A_{\theta s} = -\frac{\mu I_0 d}{4\pi r} \sin \theta \, e^{-j\omega r/v}$$

From these two components of the vector magnetic potential at P we may find $\mathbf{B}_s$ or $\mathbf{H}_s$ from the definition of $\mathbf{A}_s$,

$$\mathbf{B}_s = \mu \mathbf{H}_s = \nabla \times \mathbf{A}_s$$

by merely taking the indicated partial derivatives. Thus

$$H_{\phi s} = \frac{1}{\mu r}\frac{\partial}{\partial r}(r A_{\theta s}) - \frac{1}{\mu r}\frac{\partial A_{rs}}{\partial \theta}$$

$$H_{rs} = H_{\theta s} = 0$$

and

$$H_{\phi s} = \frac{I_0 d}{4\pi} \sin\theta \, e^{-j\omega r/v} \left(j\frac{\omega}{vr} + \frac{1}{r^2} \right)$$

The components of the electric field which must be associated with this magnetic field are found from the point form of Ampère's circuital law as it applies to a region in which conduction and convection current are absent,

$$\nabla \times \mathbf{H} = \frac{\partial \mathbf{D}}{\partial t}$$

or in complex notation,

$$\nabla \times \mathbf{H}_s = j\omega\epsilon \mathbf{E}_s$$

Expansion of the curl in spherical coordinates leads to

$$E_{rs} = \frac{1}{j\omega\epsilon} \frac{1}{r \sin\theta} \frac{\partial}{\partial\theta} (H_{\phi s} \sin\theta)$$

$$E_{\theta s} = \frac{1}{j\omega\epsilon} \left(-\frac{1}{r} \right) \frac{\partial}{\partial r} (rH_{\phi s})$$

or

$$E_{rs} = \frac{I_0 d}{2\pi} \cos\theta \, e^{-j\omega r/v} \left(\frac{1}{\epsilon v r^2} + \frac{1}{j\omega\epsilon r^3} \right)$$

$$E_{\theta s} = \frac{I_0 d}{4\pi} \sin\theta \, e^{-j\omega r/v} \left(\frac{j\omega}{\epsilon v^2 r} + \frac{1}{\epsilon v r^2} + \frac{1}{j\omega\epsilon r^3} \right)$$

In order to simplify the interpretation of the terms enclosed in parentheses above, we make the substitutions $\omega = 2\pi f$, $f\lambda = v$, $v = 1/\sqrt{\mu\epsilon}$, and $\eta = \sqrt{\mu/\epsilon}$, producing

$$\boxed{H_{\phi s} = \frac{I_0 d}{4\pi} \sin\theta \, e^{-j2\pi r/\lambda} \left(j\frac{2\pi}{\lambda r} + \frac{1}{r^2} \right)} \tag{17}$$

$$\boxed{E_{rs} = \frac{I_0 d\eta}{2\pi} \cos\theta \, e^{-j2\pi r/\lambda} \left(\frac{1}{r^2} + \frac{\lambda}{j2\pi r^3} \right)} \tag{18}$$

$$\boxed{E_{\theta s} = \frac{I_0 d\eta}{4\pi} \sin\theta \, e^{-j2\pi r/\lambda} \left(j\frac{2\pi}{\lambda r} + \frac{1}{r^2} + \frac{\lambda}{j2\pi r^3} \right)} \tag{19}$$

These three equations are indicative of the reason that so many problems involving antennas are solved by experimental rather than theoretical methods. They have resulted from three general steps: an integration (atypically trivial) and two differentiations. These steps are sufficient to cause the simple current element and its simple current expression to "blow up" into the complicated field described by (17) to (19). In spite of this complexity, several interesting observations are possible.

We might notice first the $e^{-j2\pi r/\lambda}$ factor appearing with each component. This indicates propagation outward from the origin in the positive r direction with a phase factor $\beta = 2\pi/\lambda$; thus the wavelength is λ and the velocity $v = 1/\sqrt{\mu\epsilon}$. We use the term "wavelength" now in a somewhat broader sense than the original definition, which identified the wavelength of a uniform plane wave with the distance between two points, measured in the direction of propagation, at which the wave has identical instantaneous values. Here there are additional complications caused by the terms enclosed in parentheses, which are complex functions of r. These variations must now be neglected in determining the wavelength. This is equivalent to a determination of the wavelength at a large distance from the origin, and we may demonstrate this by sketching the H_ϕ component as a function of r under the following conditions:

$$I_0 d = 4\pi \qquad \theta = 90^\circ \qquad t = 0 \qquad f = 300 \text{ MHz}$$

$$v = 3 \times 10^8 \text{ m/s (free space)} \qquad \lambda = 1 \text{ m}$$

Therefore

$$H_{\phi s} = \left(j\frac{2\pi}{r} + \frac{1}{r^2} \right) e^{-j2\pi r}$$

and the real part may be determined at $t = 0$,

$$H_\phi = \sqrt{\left(\frac{2\pi}{r}\right)^2 + \frac{1}{r^4}} \cos\left[(\tan^{-1} 2\pi r) - 2\pi r\right]$$

Knowing that $\cos(a - b) = \cos a \cos b + \sin a \sin b$ and that $\cos(\tan^{-1} x) = 1/\sqrt{1 + x^2}$, we may simplify this result to

$$H_\phi = \frac{1}{r^2}(\cos 2\pi r + 2\pi r \sin 2\pi r)$$

Values obtained from this last equation are plotted against r in the range $1 \le r \le 2$ in Fig. 13.6*a*; the curve is noticeably nonsinusoidal. At $r = 1$, $H_\phi = 1$, while at $r = 2$, one wavelength greater, $H_\phi = 0.25$. Moreover, the curve crosses the axis (with positive slope) at $r = 1 - 0.0258$ and $r = 2 - 0.0127$, again a distance not equal to a wavelength. If a similar sketch is made in the range $101 \le r \le 102$, shown in Fig. 13.6*b* on a different amplitude scale, an essentially sinusoidal wave is obtained and the instantaneous values of H_ϕ at $r = 101$ and $r = 102$ are 0.000 099 8 and 0.000 099 6. The maximum amplitudes of the positive and negative portions of the waveform differ by less than 1 percent, and we may say that for all practical purposes the wave in this region is a uniform plane wave having a sinusoidal variation with distance (and time, of course) and a well-defined wave-

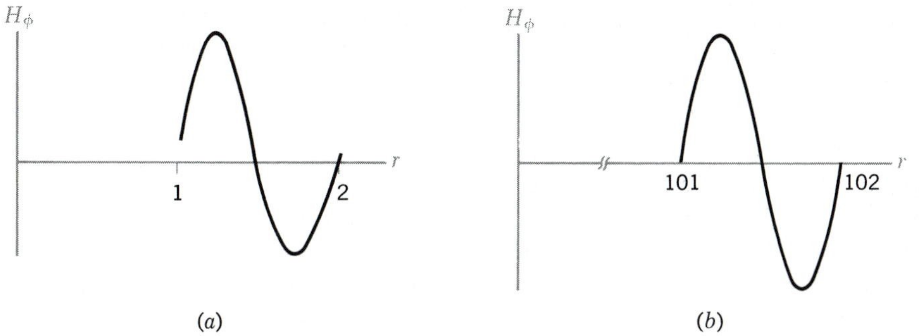

FIGURE 13.6
The instantaneous amplitude of H_ϕ for the special case of a current element having $I_0 d = 4\pi$ and $\lambda = 1$ is plotted at $\theta = 90°$ and $t = 0$ (*a*) in the region $1 \leq r \leq 2$ near the antenna, and (*b*) in the region $101 \leq r \leq 102$ distant from the antenna. The left curve is noticeably nonsinusoidal.

length. This wave evidently carries energy away from the differential antenna, and we shall calculate this power shortly.

Continuing the investigation of (17) to (19), let us now take a more careful look at the expressions containing terms varying as $1/r^3$, $1/r^2$, and $1/r$. At points very close to the current element the $1/r^3$ term must be dominant. In the numerical example we have used, the relative values of the terms in $1/r^3$, $1/r^2$, and $1/r$ in the $E_{\theta s}$ expression are about 250, 16, and 1, respectively, when r is 1 cm. The variation of an electric field as $1/r^3$ should remind us of the *electrostatic* field of the dipole (Chap. 4). This term represents energy stored in a reactive (capacitive) field, and it does not contribute to the radiated power. The inverse-square term in the $H_{\phi s}$ expression is similarly important only in the region very near to the current element and corresponds to the *induction* field of the dc element given by the Biot-Savart law. At distances corresponding to 10 or more wavelengths from the oscillating current element, all terms except the inverse-distance ($1/r$) term may be neglected and the *distant* or *radiation* fields become

$$E_{rs} = 0$$

$$E_{\theta s} = j\frac{I_0\, d\eta}{2\lambda r} \sin\theta\, e^{-j2\pi r/\lambda} \qquad (20)$$

$$H_{\phi s} = j\frac{I_0 d}{2\lambda r} \sin\theta\, e^{-j2\pi r/\lambda} \qquad (21)$$

or

$$E_{\theta s} = \eta H_{\phi s}$$

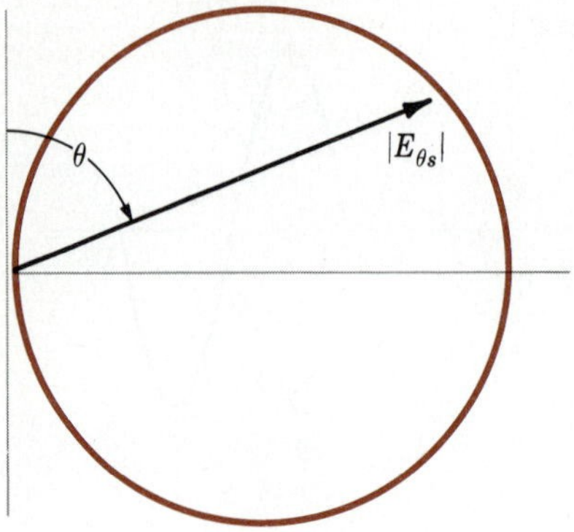

FIGURE 13.7
The polar plot of the vertical pattern of a vertical current element. The crest amplitude of $E_{\theta s}$ is plotted as a function of the polar angle θ at a constant distance r. The locus is a circle.

The relationship between $E_{\theta s}$ and $H_{\phi s}$ is thus seen to be that between the electric and magnetic fields of the uniform plane wave, thus substantiating the conclusion we reached when investigating the wavelength.

The variation of both radiation fields with the polar angle θ is the same; the fields are maximum in the equatorial plane of the current element and vanish along either extension of the element. The variation with angle may be shown by plotting a *vertical pattern* (assuming a vertical orientation of the current element) in which the relative magnitude of $E_{\theta s}$ is plotted against θ for a constant r. The pattern is usually shown on polar coordinates, as in Fig. 13.7. A *horizontal pattern* may also be plotted for more complicated antenna systems and shows the variation of field intensity with ϕ. The horizontal pattern of the current element is a circle centered at the origin since the field is not a function of the azimuth angle.

In order to obtain a quantitative expression for the power radiated, we need to apply the Poynting vector $\mathscr{P} = \mathbf{E} \times \mathbf{H}$ developed in Sec. 11.4. The instantaneous expressions for the radiation components of the electric and magnetic field intensities are

$$E_\theta = \eta H_\phi$$

$$H_\phi = -\frac{I_0 d}{2\lambda r} \sin\theta \sin\left(\omega t - \frac{2\pi r}{\lambda}\right)$$

and thus

$$\mathscr{P}_r = E_\theta H_\phi = \left(\frac{I_0 d}{2\lambda r}\right)^2 \eta \sin^2\theta \sin^2\left(\omega t - \frac{2\pi r}{\lambda}\right)$$

The total (in space) instantaneous (in time) power crossing the surface of a sphere of radius r_0 is then

$$P = \int_{\phi=0}^{2\pi} \int_{\theta=0}^{\pi} \mathscr{P}_r r_0^2 \sin\theta \, d\theta \, d\phi$$

$$= \left(\frac{I_0 d}{\lambda}\right)^2 \eta \frac{2\pi}{3} \sin^2\left(\omega t - \frac{2\pi r_0}{\lambda}\right)$$

and the time-average power is given by one-half the maximum amplitude,

$$P_{av} = \left(\frac{I_0 d}{\lambda}\right)^2 \eta \frac{\pi}{3} = 40\pi^2 \left(\frac{I_0 d}{\lambda}\right)^2$$

where $\eta = 120\pi\ \Omega$ in free space.

This is the same power as that which would be dissipated in a resistance R_{rad} by the current I_0 in the absence of any radiation, where

$$\boxed{P_{av} = \tfrac{1}{2} I_0^2 R_{rad}}$$

$$\boxed{R_{rad} = \frac{2P_{av}}{I_0^2} = 80\pi^2 \left(\frac{d}{\lambda}\right)^2} \tag{22}$$

If we assume the differential length is 0.01λ, then R_{rad} is about $0.08\ \Omega$. This small resistance is probably comparable to the *ohmic* resistance of a practical antenna, and thus the efficiency of the antenna might be unsatisfactorily low. Effective matching to the source also becomes very difficult to achieve, for the input reactance of a short antenna is much greater in magnitude than the input resistance R_{rad}. This is the basis for the statement that an effective antenna should be an appreciable fraction of a wavelength long.

The actual current distribution on a thin linear antenna is very nearly sinusoidal, even for antennas that may be several wavelengths long. Note that if the conductors of an open-circuited two-wire transmission line are folded back 90°, the standing-wave distribution on the line is sinusoidal. The current is zero at each end and a maximum one-quarter wavelength from each end, and the current continues to vary in this manner toward the center. The current at the center, therefore, will be very small for an antenna whose length is an integral number of wavelengths, but it will be equal to the maximum found at any point on the antenna if the antenna length is $\lambda/2$, $3\lambda/2$, $5\lambda/2$, and so forth.

On a short antenna, then, we see only the first portion of the sine wave; the amplitude of the current is zero at each end and increases approximately in a linear manner to a maximum value of I_0 at the center. This is suggested by the sketch of Fig. 13.8. Note that this antenna has identical currents in the two halves and it may be fed conveniently by a two-wire line, where the currents in the two conductors are equal in amplitude but opposite in direction. The gap at the feed point is small and has negligible effects. A symmetrical antenna of this type is called a *dipole*. The linear current variation with distance is a reasonable assumption for antennas having an overall length less than about one-quarter wavelength.

It is possible to extend the analysis of the differential current element to the short dipole if we assume that the length is short enough that retardation effects may be neglected. That is, we consider that signals arriving at any field point P from the two ends of the antenna are in phase. The average current along the antenna is $I_0/2$, where I_0 is the input current at the center terminals. Thus, the

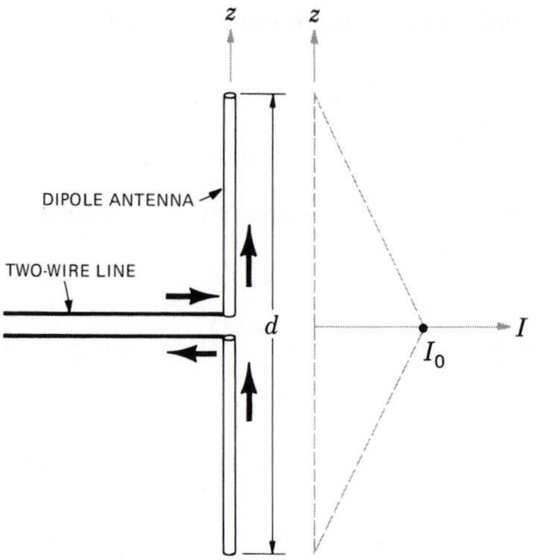

FIGURE 13.8
A short antenna ($d < \lambda/4$) has a linear current distribution and may be driven by a two-wire line.

electric and magnetic field intensities will be one-half the values given in (20) and (21), and there are no changes in the vertical and horizontal patterns. The power will be one-quarter of its previous value, and thus the radiation resistance will also be one-quarter of the value given by (22).

If we try to improve our results by assuming a sinusoidal variation of current amplitude with distance along the antenna, and if the effects of retardation are included in the analysis, then the integrations required to find **A** and the power radiated become enormously more difficult. Since we want to hurry along toward the last page, let us merely note that for the world's most popular antenna, the half-wave dipole ($d = \lambda/2$), the following results are eventually obtained:

$$E_{\theta s} = \frac{I_0 \eta}{2\pi r} \frac{\cos\left(\dfrac{\pi}{2}\cos\theta\right)}{\sin\theta} \tag{23}$$

$$H_{\phi s} = \frac{E_{\theta s}}{\eta} \tag{24}$$

$$R_{\text{rad}} = 30\left[\frac{(2\pi)^2}{2 \cdot 2!} - \frac{(2\pi)^4}{4 \cdot 4!} + \frac{(2\pi)^6}{6 \cdot 6!} - \frac{(2\pi)^8}{8 \cdot 8!} + \cdots\right] = 73.1 \quad \Omega \tag{25}$$

Let us compare this accurate value with results obtained by more approximate means. Suppose we first attempt to find the radiation resistance by assuming a uniform current distribution and neglecting the effects of retardation. The result is obtained from (22) with $d/\lambda = \frac{1}{2}$; $R_{\text{rad}} = 20\pi^2 = 197.4\ \Omega$. This is much greater than 73.1 Ω, but we have also assumed a much greater current on the antenna than is actually present.

The result may be improved by considering a linear current distribution while still ignoring retardation. The average current is half the maximum value, the power is one-quarter, and the radiation resistance drops to $5\pi^2$ or 49.3 Ω. Now the result is too small, primarily because the average value of a triangular wave is less than the average value of a sine wave.

Finally, if we assume a sinusoidal current distribution, we have an average value of $2/\pi$ times the maximum, and the radiation resistance comes to $(2/\pi)^2(20\pi^2)$, or 80 Ω. This is reasonably close to the true value, and the discrepancy lies in neglecting retardation. In a linear antenna, the effect of retardation is always one of cancellation, and therefore its consideration must always lead to smaller values of radiation resistance. This decrease is of relatively small magnitude here (from 80 to 73.1 Ω) because the current elements tending to cancel each other are those at the ends of the dipole, and these are of small amplitude; moreover, the cancellation is greatest in a direction along the antenna axis where all radiation fields are zero for a linear antenna.

Familiar antennas that fall into the dipole classification are the elements used in the common TV and FM receiving antennas.

As a final example of a practical antenna, let us collect a few facts about the *monopole* antenna. This is one-half a dipole plus a perfectly conducting plane, as shown in Fig. 13.9*a*. The image principle discussed in Sec. 5.5 provides the image shown in Fig. 13.9*b* and assures us that the fields above the plane are the same for the monopole and the dipole. Hence, the expressions of (20) and (21) are equally valid for the monopole. The Poynting vector is therefore also the same above the plane, but the integration to find the total power radiated is extended through but one-half the volume. Thus the power radiated and the radiation resistance for the monopole are half the corresponding values for the dipole. As an example, a monopole with an assumed uniform current distribution has $R_{rad} = 40\pi^2(d/\lambda)^2$; a triangular current leads to $R_{rad} = 10\pi^2(d/\lambda)^2$; and the sinusoidal current distribution of a $\lambda/4$ monopole leads to $R_{rad} = 36.5\ \Omega$.

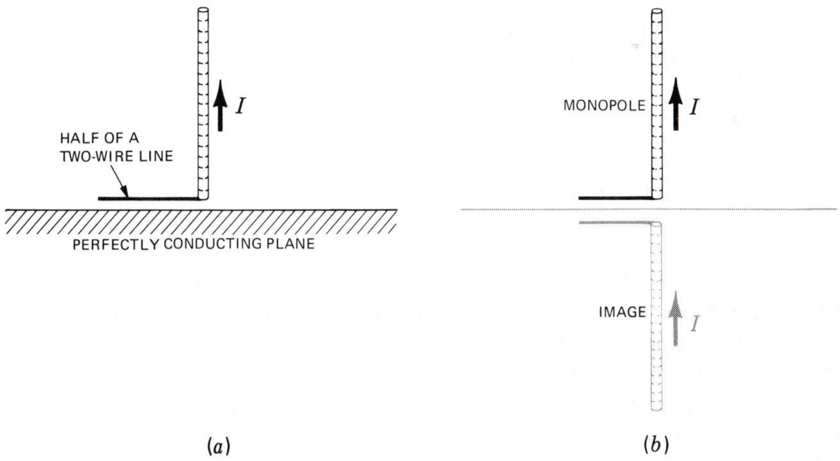

FIGURE 13.9
(*a*) The ideal monopole antenna is always associated with a perfectly conducting plane. (*b*) The monopole plus its image form a dipole.

Monopole antennas may be driven by a coaxial cable below the plane, having its center conductor connected to the antenna through a small hole, and having its outer conductor connected to the plane. If the region below the plane is inaccessible or inconvenient, the coax may be laid on top of the plane and its outer conductor connected to it. Examples of this type of antenna include AM broadcasting towers and CB antennas.

D13.4. Calculate values for the curve shown in Fig. 13.6*a* at $r = 1, 1.2, 1.4, 1.6, 1.8$, and 2.

Ans. 1.000; 5.19; 2.23; −2.62; −3.22; 0.250

D13.5. A short antenna with a uniform current distribution in air has $I_0 d = 3 \times 10^{-4}$ A·m and $\lambda = 10$ cm. Find $|E_{\theta s}|$ at $\theta = 90°$, $\phi = 0°$, and $r =$: (*a*) 2 cm; (*b*) 20 cm; (*c*) 200 cm.

Ans. 24.8 V/m; 2.82 V/m; 0.283 V/m

D13.6. The monopole antenna of Fig. 13.9*a* has a length $\frac{1}{2}d = 0.08$ m and may be assumed to carry a triangular current distribution for which the feed current I_0 is 16 A at a frequency of 375 MHz in free space. At point $P(r = 400 \text{ m}, \theta = 60°, \phi = 45°)$ find: (*a*) $H_{\phi s}$; (*b*) $E_{\theta s}$; (*c*) the amplitude of $\mathscr{P}_r$.

Ans. $j1.732$ mA/m; $j0.653$ V/m; 1.131 mW/m^2

SUGGESTED REFERENCES

1. American Radio Relay League: "The A.R.R.L. Antenna Book," The American Radio Relay League, Inc., Newington, Conn., 1970. This publication contains a wealth of practical and descriptive information on antennas and transmission lines. It also costs very little.
2. Jordan, E. C., and K. G. Balmain: (see Suggested References for Chap. 8). Each of the topics covered in this chapter is thoroughly discussed.
3. Ramo, S., J. R. Whinnery, and T. Van Duzer: (see Suggested References for Chap. 6). Each of the subjects discussed in this chapter is treated in more detail.
4. Weeks, W. L.: "Antenna Engineering," McGraw-Hill Book Company, New York, 1968. This excellent text probably contains more about antennas than you want to know.

PROBLEMS

1. The voltage between the inner and outer conductors of an infinitely long coaxial cable is $100 \cos 2\pi 10^8 t$ V at $z = 0$. Let $a = 4$ mm and $b = 14$ mm, with $\epsilon_R = 1.96$, $\mu_R = 1$, and $\sigma = 0$ for the dielectric. Assume propagation in the $\mathbf{a}_z$ direction. If the conductors are perfect: (*a*) find $E_\rho(\rho,z,t)$; (*b*) find $H_\phi(\rho,z,t)$; (*c*) show that Eq. (2) of Sec. 13.1 yields an equality for the rectangular path ($\rho = 4$ mm, $\phi = 0$, $z = 0$) to (14 mm, 0,0) to (14 mm, 0, 30 cm) to (4 mm, 0, 30 cm) to (4 mm, 0,0).

2. A toroidal coil has a square cross section with an inner radius of 3 cm and an outer radius of 5 cm. It carries 800 turns of #36 enameled copper wire (diameter of copper is 0.005 inch, diameter of wire is 0.0055 inch). The turns compose a single tightly packed layer. (*a*) Find the total length of the wire. (*b*) What fraction of a wavelength is this at 1 MHz, assuming $\mu_R = \epsilon_R = 1$ everywhere? (*c*) Calculate the low-frequency inductance of this toroid. (*d*) Also find the low-frequency resistance.

3. The external inductance of a straight piece of nonmagnetic wire of radius a and length d in air is given in Jordan and Balmain[2] as $L_{\text{ext}} = (\mu_0 d/2\pi) \ln(3 \times 10^8/\omega a)$. The internal inductance may be neglected at high frequencies. If an 8-pF capacitor has leads that are each 4 cm long and constructed of relatively small diameter wire, $a = 0.2$ mm, what will be the series resonant frequency of the L_{ext}-C combination?

4. Two straight wires of 0.793-mm radius are parallel and have a center-to-center separation of 1.6 cm. They are used to connect an ideal voltage source, $170 \cos 2\pi ft$ V, to a 720-Ω resistor. If the source and resistor are 93.7 cm apart in air, calculate the resistor voltage if $f =$: (*a*) 60 Hz; (*b*) 80 MHz. So whatever happened to Kirchhoff's voltage law?

5. An *RLC* parallel circuit is surrounded by a closed surface, as suggested in Fig. 13.10. Apply Poynting's theorem, Eq. (38) in Sec. 11.4, to the closed surface and show that it leads to the circuit-theory result that the instantaneous power being supplied by the generator between points 0 and 1 is equal to the sum of the powers being delivered to the three circuit elements.

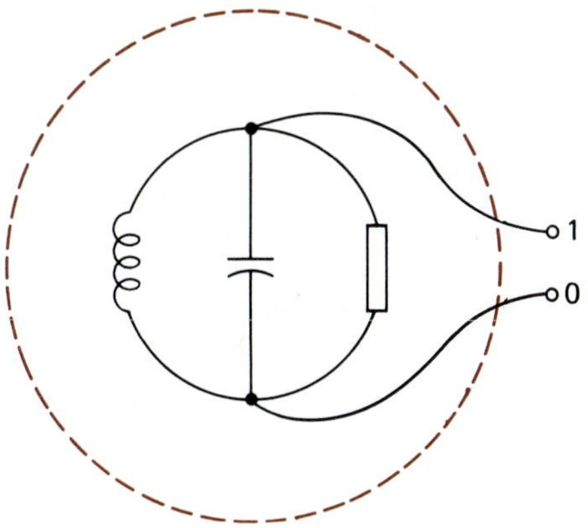

FIGURE 13.10
See Prob. 5.

6. A parallel-strip transmission line utilizes conductors that are 0.6 cm wide and separated 0.4 mm by a lossless dielectric for which $\epsilon = 10^{-10}$ F/m. Assume perfect conductors. If the line is 5 cm long and short-circuited at one end, make a sketch of $|Z_{\text{in}}|$ vs. f, $0 < f < 700$ MHz.

7. Three conducting paths are shown in Fig. 13.11. External magnetic fields are applied in a and b to achieve total fluxes of $2 \sin 3t$ Wb within the loops. No external magnetic field is applied in c. Loop a contains a lumped 2-Ω resistor, loop b has a uniformly distributed total resistance of 2 Ω, and loop c contains a lumped 2-Ω resistor and an ac voltage source, $6 \cos 3t$ V. All connecting wires are perfectly conducting. (*a*) Compute I for each loop, neglecting any flux that might be produced by I itself. (*b*) Evaluate the counterclockwise

[2] See Suggested References for Chap. 8

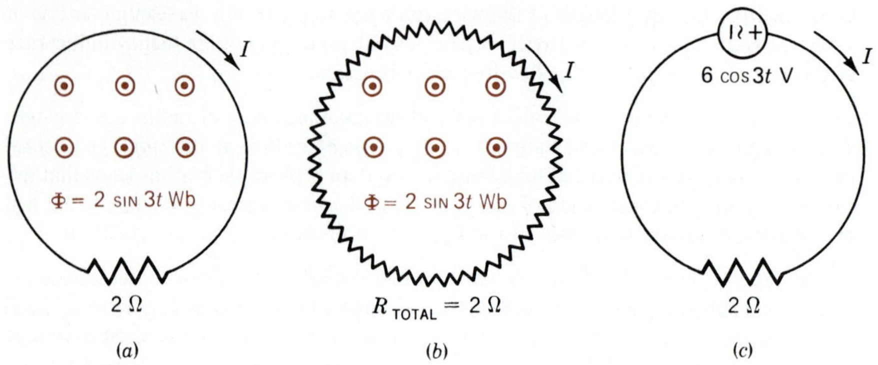

FIGURE 13.11
See Prob. 7.

closed line integral of the electric field intensity about each loop, giving the total result and the contribution from each part of the path.

8. The resonant coaxial cavity used as an example in Sec. 13.2 leads to an equivalent parallel-resonant circuit containing the element values: $R_e = 26.2\ \text{k}\Omega$, $C_e = 41.7$ pF, and $L_e = 0.0607\ \mu$H. Determine corresponding values for a similar cavity that is: (*a*) twice as long, short-circuited at both ends, and fed at the midpoint, (*b*) three times as long, short-circuited at one end, and fed at the other end.

9. A parallel-strip line utilizes conductors with $b = 6$ mm and $d = 0.4$ mm. It is 8 cm long, short-circuited at one end, and filled with a dielectric for which $\epsilon_R = 5$. Neglect fringing and external fields. (*a*) Find the parallel-resonant frequency. Then determine element values for an equivalent parallel-resonant circuit if: (*b*) $\sigma_c = \infty$ and $\sigma = 10^{-4}$ ℧/m; (*c*) $\sigma_c = 2 \times 10^7$ ℧/m and $\sigma = 0$; (*d*) $\sigma_c = 2 \times 10^7$ ℧/m and $\sigma = 10^{-4}$ ℧/m.

10. A lossless coaxial transmission line has $a = 1$ mm, $b = 5$ mm, and a length of 10 cm with one end open-circuited and the other short-circuited. Find the lowest resonant frequency if: (*a*) the line is air-filled; (*b*) the line is filled with a dielectric for which $\epsilon_R = 2.5$; (*c*) half (5 cm) the line is air and the other half has $\epsilon_R = 2.5$.

11. A cubical cavity is defined by the perfectly conducting surfaces $x = 0$ and 5 cm, $y = 0$ and 5 cm, and $z = 0$ and 5 cm. The interior region contains the field $\mathbf{E} = 5000 \sin 20\pi x \sin 20\pi y \cos 4 \times 10^9 \pi t\, \mathbf{a}_z$ V/m. Let $\mu_R = 1$, $\epsilon_R = 4.5$, and $\sigma/\omega\epsilon = 10^{-5}$ inside the cavity, and calculate the average power lost in the dielectric.

12. Two element values in the equivalent circuit of a resonant cavity are $L_e = 0.1\ \mu$H and $R_e = 50\ \text{k}\Omega$. If $\omega_0 = 10^9$ rad/s, what is $I_{\text{in}}(t)$ if $V_{\text{in}}(t) =$: (*a*) $100 \cos \omega_0 t$ V; (*b*) $100 \cos 1.005\omega_0 t$ V?

13. The magnetic field intensity adjacent to a good conductor planar surface is

$$H_{x0} \sin \frac{\pi x}{a} \cos \frac{\pi z}{d} \cos \omega t\, \mathbf{a}_x + H_{z0} \cos \frac{\pi x}{a} \sin \frac{\pi z}{d} \cos \omega t\, \mathbf{a}_z \qquad \text{A/m}$$

Find the average power loss in one side of a conductor surface at $y = 0$ for which $\sigma = \sigma_c$ and $\mu = \mu_0$ over the region $0 < x < a$, $0 < z < d$.

14. A quarter-wavelength coaxial cavity similar to that shown in Fig. 13.2 has dimensions $a = 1.2$ cm, $b = 8$ cm, and $l = 40$ cm. All materials are nonmagnetic. Determine values for C_e, L_e, and R_e if: (*a*) $\sigma_c = \infty$, $\sigma = 2 \times 10^{-4}$ ℧/m, $\epsilon_R = 3$, (*b*) $\sigma_c = 4 \times 10^7$ ℧/m, $\sigma = 0$, $\epsilon_R = 3$; (*c*) $\sigma_c = 4 \times 10^7$ ℧/m, $\sigma = 2 \times 10^{-4}$ ℧/m, $\epsilon_R = 3$.

15. A quarter-wavelength coaxial cavity is constructed of a material for which $\mu_R = 1$ and $\sigma_c = \infty$. Its dimensions are $a = 0.8$ cm, $b = 3.2$ cm, and $l = 50$ cm. The interior contains two dielectrics with different conductivities. Forty percent of the length has $\epsilon_R = 2.5$ and $\sigma = 10^{-5}$ ℧/m, and the remaining 30 cm has $\epsilon_R = 2.5$ and $\sigma = 5 \times 10^{-6}$ ℧/m. Find (*a*) f_0; (*b*) C_e; (*c*) L_e; (*d*) R_e if the more lossy material is at the short-circuited end; (*e*) Q_0.

16. A lossless coaxial line with $Z_0 = 100\ \Omega$ is 50 cm in length and short-circuited at the load end. It is air-filled. (*a*) Calculate values for f_0, L_e, and C_e. (*b*) On the same axes, plot curves of the magnitude of the input impedance of the coaxial line and the equivalent parallel L_e-C_e circuit vs. f, $0 < f < 300$ MHz.

17. A short dipole carrying a current $I_0 \cos \omega t$ in the $\mathbf{a}_z$ direction is located at the origin in free space. (*a*) If $\beta = 1$ rad/m, $r = 2$ m, $\theta = 45°$, $\phi = 0$, and $t = 0$, give a unit vector in rectangular components that shows the instantaneous direction of **E**. (*b*) What fraction of the total average power is radiated in the belt, $80° < \theta < 100°$?

18. Prepare a curve, r vs. θ in polar coordinates, showing the locus in the $\phi = 0$ plane where: (*a*) the radiation field $|E_{\theta s}|$ is one-half of its value at $r = 10^4$ m, $\theta = \pi/2$; (*b*) the average radiated power density $\mathscr{P}_{r,\,av}$ is one-half of its value at $r = 10^4$ m, $\theta = \pi/2$.

19. Two short antennas at the origin in free space carry identical currents of $5 \cos \omega t$ A, one in the $\mathbf{a}_z$ direction and one in the $\mathbf{a}_y$ direction. Let $\lambda = 2\pi$ m and $d = 0.1$ m. Find $\mathbf{E}_s$ at the distant point: (*a*) $(x = 0, y = 1000, z = 0)$; (*b*) (0,0,1000); (*c*) (1000,0,0). (*d*) Find **E** at (1000,0,0) at $t = 0$. (*e*) Find $|\mathbf{E}|$ at (1000,0,0) at $t = 0$.

20. A short current element has a length $d = 0.03\lambda$. Calculate the radiation resistance for each of the following current distributions: (*a*) uniform, I_0; (*b*) linear, $I(z) = I_0(0.5d - |z|)/0.5d$; (*c*) step, I_0 for $0 < |z| < 0.25d$ and $0.5I_0$ for $0.25d < |z| < 0.5d$.

21. A dipole antenna in free space has a linear current distribution. If the length d is 0.02λ, what value of I_0 is required to: (*a*) provide a radiation-field amplitude of 100 mV/m at a distance of 1 mi, at $\theta = 90°$; (*b*) radiate a total power of 1 W?

22. A monopole antenna in free space, extending vertically over a perfectly conducting plane, has a linear current distribution. If the length of the antenna is 0.01λ, what value of I_0 is required to: (*a*) provide a radiation-field amplitude of 100 mV/m at a distance of 1 mi, at $\theta = 90°$; (*b*) radiate a total power of 1 W?

23. The radiation field of a certain short vertical current element is $E_{\theta s} = (20/r) \sin \theta\, e^{-j10\pi r}$ V/m if it is located at the origin in free space. (*a*) Find $E_{\theta s}$ at $P(r = 100, \theta = 90°, \phi = 30°)$. (*b*) Find $E_{\theta s}$ at $P(100, 90°, 30°)$ if the vertical element is located at $A(0.1, 90°, 90°)$. (*c*) Find $E_{\theta s}$ at $P(100, 90°, 30°)$ if identical vertical elements are located at $A(0.1, 90°, 90°)$ and $B(0.1, 90°, 270°)$.

APPENDIX A

VECTOR ANALYSIS

A.1 GENERAL CURVILINEAR COORDINATES

Let us consider a general orthogonal coordinate system in which a point is located by the intersection of three mutually perpendicular surfaces (of unspecified form or shape),

$$u = \text{constant}$$

$$v = \text{constant}$$

$$w = \text{constant}$$

where u, v, and w are the variables of the coordinate system. If each variable is increased by a differential amount and three more mutually perpendicular surfaces are drawn corresponding to these new values, a differential volume is formed which is closely a rectangular parallelepiped. Since u, v, and w need not be measures of length, such as, for example, the angle variables of the cylindrical and spherical coordinate systems, each must be multiplied by a general function of u, v, and w in order to obtain the differential sides of the parallelepiped. Thus we define the scale factors h_1, h_2, and h_3 each as a function of the three variables u, v, and w and write the lengths of the sides of the differential volume as

$$dL_1 = h_1\, du$$

$$dL_2 = h_2\, dv$$

$$dL_3 = h_3\, dw$$

In the three coordinate systems discussed in Chap. 1, it is apparent that the variables and scale factors are

$$\begin{array}{llll}
\text{Cartesian:} & u = x & v = y & w = z \\
& h_1 = 1 & h_2 = 1 & h_3 = 1 \\
\text{Cylindrical:} & u = \rho & v = \phi & w = z \\
& h_1 = 1 & h_2 = \rho & h_3 = 1 \\
\text{Spherical:} & u = r & v = \theta & w = \phi \\
& h_1 = 1 & h_2 = r & h_3 = r \sin \theta
\end{array} \tag{1}$$

The choice of u, v, and w above has been made so that $\mathbf{a}_u \times \mathbf{a}_v = \mathbf{a}_w$ in all cases. More involved expressions for h_1, h_2, and h_3 are to be expected in other less familiar coordinate systems.[1]

A.2 DIVERGENCE, GRADIENT, AND CURL IN GENERAL CURVILINEAR COORDINATES

If the method used to develop divergence in Secs. 3.4 and 3.5 is applied to the general curvilinear coordinate system, the flux of the vector $\mathbf{D}$ passing through the surface of the parallelepiped whose unit normal is $\mathbf{a}_u$ is

$$D_{u0}\, dL_2\, dL_3 + \frac{1}{2}\frac{\partial}{\partial u}(D_u\, dL_2\, dL_3)\, du$$

or

$$D_{u0} h_2 h_3\, dv\, dw + \frac{1}{2}\frac{\partial}{\partial u}(D_u h_2 h_3\, dv\, dw)\, du$$

and for the opposite face it is

$$-D_{u0} h_2 h_3\, dv\, dw + \frac{1}{2}\frac{\partial}{\partial u}(D_u h_2 h_3\, dv\, dw)\, du$$

giving a total for these two faces of

$$\frac{\partial}{\partial u}(D_u h_2 h_3\, dv\, dw)\, du$$

Since u, v, and w are independent variables, this last expression may be written as

$$\frac{\partial}{\partial u}(h_2 h_3 D_u)\, du\, dv\, dw$$

[1] The variables and scale factors are given for nine orthogonal coordinate systems on pp. 50–59 in J. A. Stratton, "Electromagnetic Theory," McGraw-Hill Book Company, New York, 1941. Each system is also described briefly.

and the other two corresponding expressions obtained by a simple permutation of the subscripts and of u, v, and w. Thus the total flux leaving the differential volume is

$$\left[\frac{\partial}{\partial u}(h_2 h_3 D_u) + \frac{\partial}{\partial v}(h_3 h_1 D_v) + \frac{\partial}{\partial w}(h_1 h_2 D_w)\right] du\, dv\, dw$$

and the divergence of **D** is found by dividing by the differential volume

$$\nabla \cdot \mathbf{D} = \frac{1}{h_1 h_2 h_3}\left[\frac{\partial}{\partial u}(h_2 h_3 D_u) + \frac{\partial}{\partial v}(h_3 h_1 D_v) + \frac{\partial}{\partial w}(h_1\, h_2\, D_w)\right] \tag{2}$$

The components of the gradient of a scalar V may be obtained (following the methods of Sec. 4.6) by expressing the total differential of V,

$$dV = \frac{\partial V}{\partial u} du + \frac{\partial V}{\partial v} dv + \frac{\partial V}{\partial w} dw$$

in terms of the component differential lengths, $h_1\, du$, $h_2\, dv$, and $h_3\, dw$,

$$dV = \frac{1}{h_1}\frac{\partial V}{\partial u} h_1\, du + \frac{1}{h_2}\frac{\partial V}{\partial v} h_2\, dv + \frac{1}{h_3}\frac{\partial V}{\partial w} h_3\, dw$$

Then, since

$$d\mathbf{L} = h_1\, du\, \mathbf{a}_u + h_2\, dv\, \mathbf{a}_v + h_3\, dw\, \mathbf{a}_w \qquad \text{and} \qquad dV = \nabla V \cdot d\mathbf{L}$$

we see that

$$\nabla V = \frac{1}{h_1}\frac{\partial V}{\partial u}\mathbf{a}_u + \frac{1}{h_2}\frac{\partial V}{\partial v}\mathbf{a}_v + \frac{1}{h_3}\frac{\partial V}{\partial w}\mathbf{a}_w \tag{3}$$

The components of the curl of a vector **H** are obtained by considering a differential path first in a $u =$ constant surface and finding the circulation of **H** about that path, as discussed for cartesian coordinates in Sec. 8.3. The contribution along the segment in the $\mathbf{a}_v$ direction is

$$H_{v0} h_2\, dv - \frac{1}{2}\frac{\partial}{\partial w}(H_v h_2\, dv)\, dw$$

and that from the oppositely directed segment is

$$-H_{v0} h_2\, dv - \frac{1}{2}\frac{\partial}{\partial w}(H_v h_2\, dv)\, dw$$

The sum of these two parts is

$$-\frac{\partial}{\partial w}(H_v h_2\, dv)\, dw$$

or

$$-\frac{\partial}{\partial w}(h_2 H_v)\, dv\, dw$$

and the sum of the contributions from the other two sides of the path is

$$\frac{\partial}{\partial v}(h_3 H_w)\, dv\, dw$$

Adding these two terms and dividing the sum by the enclosed area, $h_2 h_3\, dv\, dw$, we see that the $\mathbf{a}_u$ component of curl $\mathbf{H}$ is

$$(\nabla \times \mathbf{H})_u = \frac{1}{h_2 h_3}\left[\frac{\partial}{\partial v}(h_3 H_w) - \frac{\partial}{\partial w}(h_2 H_v)\right]$$

and the other two components may be obtained by cyclic permutation. The result is expressible as a determinant,

$$\nabla \times \mathbf{H} = \begin{vmatrix} \dfrac{\mathbf{a}_u}{h_2 h_3} & \dfrac{\mathbf{a}_v}{h_3 h_1} & \dfrac{\mathbf{a}_w}{h_1 h_2} \\ \dfrac{\partial}{\partial u} & \dfrac{\partial}{\partial v} & \dfrac{\partial}{\partial w} \\ h_1 H_u & h_2 H_v & h_3 H_w \end{vmatrix} \tag{4}$$

The Laplacian of a scalar is found by using (2) and (3):

$$\nabla^2 V = \nabla \cdot \nabla V = \frac{1}{h_1 h_2 h_3}\left[\frac{\partial}{\partial u}\left(\frac{h_2 h_3}{h_1}\frac{\partial V}{\partial u}\right) + \frac{\partial}{\partial v}\left(\frac{h_3 h_1}{h_2}\frac{\partial V}{\partial v}\right) + \frac{\partial}{\partial w}\left(\frac{h_1 h_2}{h_3}\frac{\partial V}{\partial w}\right)\right] \tag{5}$$

Equations (2) to (5) may be used to find the divergence, gradient, curl, and Laplacian in any orthogonal coordinate system for which h_1, h_2, and h_3 are known.

Expressions for $\nabla \cdot \mathbf{D}$, ∇V, $\nabla \times \mathbf{H}$, and $\nabla^2 V$ are given in cartesian, circular cylindrical, and spherical coordinate systems inside the back cover.

A.3 VECTOR IDENTITIES

The vector identities listed below may be proved by expansion in cartesian (or general curvilinear) coordinates. The first two identities involve the scalar and vector triple products, the next three are concerned with operations on sums, the following three apply to operations when the argument is multiplied by a scalar function, the next three apply to operations on scalar or vector products, and the last four concern the second-order operations.

$$(\mathbf{A} \times \mathbf{B}) \cdot \mathbf{C} \equiv (\mathbf{B} \times \mathbf{C}) \cdot \mathbf{A} \equiv (\mathbf{C} \times \mathbf{A}) \cdot \mathbf{B} \tag{6}$$

$$\mathbf{A} \times (\mathbf{B} \times \mathbf{C}) \equiv (\mathbf{A} \cdot \mathbf{C})\mathbf{B} - (\mathbf{A} \cdot \mathbf{B})\mathbf{C} \tag{7}$$

$$\nabla \cdot (\mathbf{A} + \mathbf{B}) \equiv \nabla \cdot \mathbf{A} + \nabla \cdot \mathbf{B} \tag{8}$$

$$\nabla(V + W) \equiv \nabla V + \nabla W \tag{9}$$

$$\nabla \times (\mathbf{A} + \mathbf{B}) \equiv \nabla \times \mathbf{A} + \nabla \times \mathbf{B} \tag{10}$$

$$\nabla \cdot (V\mathbf{A}) \equiv \mathbf{A} \cdot \nabla V + V \nabla \cdot \mathbf{A} \tag{11}$$

$$\nabla(VW) \equiv V\,\nabla W + W\nabla V \tag{12}$$

$$\nabla \times (V\mathbf{A}) \equiv \nabla V \times \mathbf{A} + V\nabla \times \mathbf{A} \tag{13}$$

$$\nabla \cdot (\mathbf{A} \times \mathbf{B}) \equiv \mathbf{B} \cdot \nabla \times \mathbf{A} - \mathbf{A} \cdot \nabla \times \mathbf{B} \tag{14}$$

$$\nabla(\mathbf{A} \cdot \mathbf{B}) \equiv (\mathbf{A} \cdot \nabla)\mathbf{B} + (\mathbf{B} \cdot \nabla)\mathbf{A} + \mathbf{A} \times (\nabla \times \mathbf{B}) + \mathbf{B} \times (\nabla \times \mathbf{A}) \tag{15}$$

$$\nabla \times (\mathbf{A} \times \mathbf{B}) \equiv \mathbf{A}\nabla \cdot \mathbf{B} - \mathbf{B}\nabla \cdot \mathbf{A} + (\mathbf{B} \cdot \nabla)\mathbf{A} - (\mathbf{A} \cdot \nabla)\mathbf{B} \tag{16}$$

$$\nabla \cdot \nabla V \equiv \nabla^2 V \tag{17}$$

$$\nabla \cdot \nabla \times \mathbf{A} \equiv 0 \tag{18}$$

$$\nabla \times \nabla V \equiv 0 \tag{19}$$

$$\nabla \times \nabla \times \mathbf{A} \equiv \nabla(\nabla \cdot \mathbf{A}) - \nabla^2 \mathbf{A} \tag{20}$$

APPENDIX B

UNITS

We shall describe first the International System (abbreviated SI, for Système International d'Unités), which is used in this book and is now standard in electrical engineering and much of physics. It has also been officially adopted as the international system of units by many countries, including the United States.[1]

The fundamental unit of length is the meter, which was defined in the latter part of the 19th century as the distance between two marks on a certain platinum-iridium bar. The definition was improved in 1960 by relating the meter to the wavelength of the radiation emitted by the rare gas isotope krypton 86 under certain specified conditions. This so-called krypton meter was accurate to four parts per billion, a value leading to negligible uncertainties in constructing skyscrapers or building highways, but capable of causing an error greater than one meter in determining the distance to the moon. The meter was redefined in 1983

[1] The International System of Units was adopted by the Eleventh General Conference on Weights and Measures in Paris in 1960, and it was officially adopted for scientific usage by the National Bureau of Standards in 19[illegible]. It is a metric system which interestingly enough is the only system which has ever received specific sanction from Congress. This occurred first in 1866 and then again in 1975 with the Metric Conversion Act, which provides for "voluntary conversion" to the metric system. No specific time was specified, however, and we can assume that it will still be a few years before the bathroom scale reads mass in kilograms and Miss America is a 90–60–90.

in terms of the velocity of light. At that time the velocity of light was specified to be an auxiliary constant with an *exact* value of 299 792 458 meters per second. As a result, the latest definition of the meter is the distance light travels in a vacuum in 1/299 792 458 of a second. If greater accuracy is achieved in measuring c, that value will remain 299 792 458 m/s, but the length of the meter will change.

It is evident that our definition of the meter is expressed in terms of the "second," the fundamental unit of time. The second is defined as 9 192 631 770 periods of the transition frequency between the hyperfine levels $F = 4$, $m_F = 0$, and $F = 3$, $m_F = 0$ of the ground state ${}^2s_{1/2}$ of the atom of cesium 133, unperturbed by external fields. This definition of the second, complex though it may be, permits time to be measured with an accuracy better than one part in 10^{13}.

The standard mass of one kilogram is defined as the mass of an international standard in the form of a platinum-iridium cylinder at the International Bureau of Weights and Measures at Sèvres, France.

The unit of temperature is the kelvin, defined by placing the triple-point temperature of water at 273.16 kelvins.

A fifth unit is the candela, defined as the luminous intensity of an omnidirectional radiator at the freezing temperature of platinum (2042 K) having an area of 1/600 000 square meter and under a pressure of 101 325 newtons per square meter.

The last of the fundamental units is the ampere. Before explicitly defining the ampere, we must first define the newton. It is defined in terms of the other fundamental units from Newton's third law as the force required to produce an acceleration of one meter per second per second on a one-kilogram mass. We now may define the ampere as that constant current, flowing in opposite directions in two straight parallel conductors of infinite length and negligible cross section, separated one meter in vacuum, that produces a repulsive force of 2×10^{-7} newton per meter length between the two conductors. The force between the two parallel conductors is known to be

$$F = \mu_0 \frac{I^2}{2\pi d}$$

and thus

$$2 \times 10^{-7} = \mu_0 \frac{1}{2\pi}$$

or

$$\mu_0 = 4\pi 10^{-7} \qquad (\text{kg}\cdot\text{m/A}^2\cdot\text{s}^2\text{, or H/m})$$

We thus find that our definition of the ampere has been formulated in such a way as to assign an exact simple numerical value to the permeability of free space.

Returning to the International System, the units in which the other electric and magnetic quantities are measured are given in the body of the text at the time each quantity is defined, and all of them can be related to the basic units already defined. For example, our work with the plane wave in Chap. 11 shows that the velocity with which an electromagnetic wave propagates in free space is

$$c = \frac{1}{\sqrt{\mu_0 \epsilon_0}}$$

and thus

$$\epsilon_0 = \frac{1}{\mu_0 c^2} = \frac{1}{4\pi 10^{-7} c^2} = 8.854\ 187\ 817 \times 10^{-12}\ \text{F/m}$$

It is evident that the numerical value of ϵ_0 depends upon the defined value of the velocity of light in vacuum, 299 792 458 m/s.

The units are also given in Table B.1 for easy reference. They are listed in the same order that they are defined in the text.

Finally, other systems of units have been used in electricity and magnetism. In the electrostatic system of units (esu), Coulomb's law is written for free space,

$$F = \frac{Q_1 Q_2}{R^2} \qquad \text{(esu)}$$

The permittivity of free space is assigned the value of unity. The gram and centimeter are the fundamental units of mass and distance, and the esu system is therefore a cgs system. Units bearing the prefix stat- belong to the electrostatic system of units.

In a similar manner, the electromagnetic system of units (emu) is based on Coulomb's law for magnetic poles, and the permeability of free space is unity. The prefix ab- identifies emu units. When electric quantities are expressed in esu units, magnetic quantities in emu units, and both appear in the same equation (such as Maxwell's curl equations), the velocity of light appears explicitly. This follows from noting that in esu $\epsilon_0 = 1$, but $\mu_0 \epsilon_0 = 1/c^2$, and therefore $\mu_0 = 1/c^2$, and in emu $\mu_0 = 1$, and hence $\epsilon_0 = 1/c^2$. Thus, in this intermixed system known as the gaussian system of units,

$$\nabla \times \mathbf{H} = 4\pi \mathbf{J} + \frac{1}{c} \frac{\partial \mathbf{D}}{\partial t} \qquad \text{(gaussian)}$$

Other systems include the factor 4π explicitly in Coulomb's law, and it then does not appear in Maxwell's equations. When this is done, the system is said to be rationalized. Hence the gaussian system is an unrationalized cgs system (when rationalized it is known as the Heaviside-Lorentz system), and the International System we have used throughout this book is a rationized mks system.

Table B.2 gives the conversion factors between the more important units of the International System (or rationalized mks system) and the gaussian system, and several other assorted units.

Table B.3 lists the prefixes used with any of the SI units, their abbreviations, and the power of ten each represents. Those checked are widely used. Both the prefixes and their abbreviations are written without hyphens, and therefore 10^{-6} F = 1 microfarad = 1 μF = 1000 nanofarads = 1000 nF, and so forth.

TABLE B.1
Names and units of the electric and magnetic quantities in the International System (in the order they appear in the text)

Symbol	Name	Unit	Abbreviation
v	Velocity	meter/second	m/s
F	Force	newton	N
Q	Charge	coulomb	C
r, R	Distance	meter	m
ϵ_0, ϵ	Permittivity	farad/meter	F/m
E	Electric field intensity	volt/meter	V/m
ρ_v	Volume charge density	coulomb/meter^3	C/m^3
v	Volume	meter^3	m^3
ρ_L	Linear charge density	coulomb/meter	C/m
ρ_S	Surface charge density	coulomb/meter^2	C/m^2
Ψ	Electric flux	coulomb	C
D	Electric flux density	coulomb/meter^2	C/m^2
S	Area	meter^2	m^2
W	Work, energy	joule	J
L	Length	meter	m
V	Potential	volt	V
p	Dipole moment	coulomb-meter	C·m
I	Current	ampere	A
J	Current density	ampere/meter^2	A/m^2
μ_e, μ_h	Mobility	meter^2/volt-second	m^2/V·s
e	Electronic charge	coulomb	C
σ	Conductivity	mho/meter	℧/m
R	Resistance	ohm	Ω
P	Polarization	coulomb/meter^2	C/m^2
$\chi_{e,m}$	Susceptibility		
C	Capacitance	farad	F
R_s	Sheet resistance	ohm per square	Ω
H	Magnetic field intensity	ampere/meter	A/m
K	Surface current density	ampere/meter	A/m
B	Magnetic flux density	tesla (or weber/meter^2)	T (or Wb/m^2)
μ_0, μ	Permeability	henry/meter	H/m
Φ	Magnetic flux	weber	Wb
V_m	Magnetic scalar potential	ampere	A
A	Vector magnetic potential	weber/meter	Wb/m
T	Torque	newton-meter	N·m
m	Magnetic moment	ampere-meter^2	A·m^2
M	Magnetization	ampere/meter	A/m
$\mathscr{R}$	Reluctance	ampere-turn/weber	A·t/Wb
L	Inductance	henry	H
M	Mutual inductance	henry	H
ω	Radian frequency	radian/second	rad/s
c	Velocity of light	meter/second	m/s
λ	Wavelength	meter	m
η	Intrinsic impedance	ohm	Ω
γ	Propagation constant	complex neper/meter	m^{-1}
α	Attenuation constant	neper/meter	Np/m
β	Phase constant	radian/meter	rad/m

TABLE B.1 (continued)

Symbol	Name	Unit	Abbreviation
f	Frequency	hertz	Hz
$\mathscr{P}$	Poynting vector	watt/meter^2	W/m^2
P	Power	watt	W
δ	Skin depth	meter	m
Γ	Reflection coefficient		
s	Standing-wave ratio		
G	Conductance	mho	℧
Z	Impedance	ohm	Ω
Y	Admittance	mho	℧
Q	Quality factor		

TABLE B.2

Conversion of International to gaussian and other units (use $c = 2.997\,924\,58 \times 10^8$)

Quantity	1 mks unit	= gaussian units	= other units
d	1 m	10^2 cm	39.37 in
F	1 N	10^5 dyne	0.2248 lb$_f$
W	1 J	10^7 erg	0.7376 ft-lb$_f$
Q	1 C	$10c$ statC	0.1 abC
ρ_v	1 C/m^3	$10^{-5}c$ statC/cm^3	10^{-7} abC/cm^3
D	1 C/m^2	$4\pi 10^{-3}c$ (esu)	$4\pi 10^{-5}$ (emu)
E	1 V/m	$10^4/c$ statV/cm	10^6 abV/cm
V	1 V	$10^6/c$ statV	10^8 abV
I	1 A	0.1 abA	$10c$ statA
H	1 A/m	$4\pi 10^{-3}$ oersted	$0.4\pi c$ (esu)
V_m	1 A·t	0.4π gilbert	$40\pi c$ (esu)
B	1 T	10^4 gauss	$100/c$ (esu)
Φ	1 Wb	10^8 maxwell	$10^6/c$ (esu)
A	1 Wb/m	10^6 maxwell/cm	
R	1 Ω	10^9 abΩ	$10^5/c^2$ statΩ
L	1 H	10^9 abH	$10^5/c^2$ statH
C	1 F	$10^{-5}c^2$ statF	10^{-9} abF
σ	1 ℧/m	10^{-11} ab℧/cm	$10^{-7}c^2$ stat℧/cm
μ	1 H/m	$10^7/4\pi$ (emu)	$10^3/4\pi c^2$ (esu)
ϵ	1 F/m	$4\pi 10^{-7}c^2$ (esu)	$4\pi 10^{-11}$ (emu)

TABLE B.3

Standard prefixes used with SI units

Prefix	Abbrev.	Meaning	Prefix	Abbrev.	Meaning
atto-	a-	10^{-18}	deka-	da-	10^1
femto-	f-	10^{-15}	hecto-	h-	10^2
✓pico-	p-	10^{-12}	✓kilo-	k-	10^3
✓nano-	n-	10^{-9}	✓mega-	M-	10^6
✓micro-	μ-	10^{-6}	✓giga-	G-	10^9
✓milli-	m-	10^{-3}	tera-	T-	10^{12}
✓centi-	c-	10^{-2}	peta-	P-	10^{15}
deci-	d-	10^{-1}	exa-	E-	10^{18}

APPENDIX C

MATERIAL CONSTANTS

Table C.1 lists typical values of the relative permittivity ϵ_R or dielectric constant for common insulating and dielectric materials, along with representative values for the loss tangent. The values should only be considered representative for each material, and they apply to normal temperature and humidity conditions, and to very low audio frequencies. Most of them have been taken from "Reference Data for Radio Engineers,"[1] "The Standard Handbook for Electrical Engineers,"[2] and von Hippel,[3] and these volumes may be referred to for further information on these and other materials.

Table C.2 gives the conductivity for a number of metallic conductors, for a few insulating materials, and for several other materials of general interest. The values have been taken from the references listed previously, and they apply at zero frequency and at room temperature. The listing is in the order of decreasing conductivity.

Some representative values of the relative permeability for various diamagnetic, paramagnetic, ferrimagnetic, and ferromagnetic materials are listed in Table C.3. They have been extracted from the references listed above, and the data for

[1] See Suggested References for Chap. 11.
[2] See Suggested References for Chap. 5.
[3] See Suggested References for Chap. 11.

TABLE C.1
ϵ_R and $\sigma/\omega\epsilon$

Material	ϵ_R	$\sigma/\omega\epsilon$
Air	1.0005	
Alcohol, ethyl	25	0.1
Aluminum oxide	8.8	0.000 6
Amber	2.7	0.002
Bakelite	4.74	0.022
Barium titanate	1200	0.013
Carbon dioxide	1.001	
Ferrite (NiZn)	12.4	0.000 25
Germanium	16	
Glass	4–7	0.002
Ice	4.2	0.05
Mica	5.4	0.000 6
Neoprene	6.6	0.011
Nylon	3.5	0.02
Paper	3	0.008
Plexiglas	3.45	0.03
Polyethylene	2.26	0.000 2
Polypropylene	2.25	0.000 3
Polystyrene	2.56	0.000 05
Porcelain (dry process)	6	0.014
Pyranol	4.4	0.000 5
Pyrex glass	4	0.000 6
Quartz (fused)	3.8	0.000 75
Rubber	2.5–3	0.002
Silica or SiO_2 (fused)	3.8	0.000 75
Silicon	11.8	
Snow	3.3	0.5
Sodium chloride	5.9	0.000 1
Soil (dry)	2.8	0.05
Steatite	5.8	0.003
Styrofoam	1.03	0.000 1
Teflon	2.1	0.000 3
Titanium dioxide	100	0.001 5
Water (distilled)	80	0.04
Water (sea)		4
Water (dehydrated)	1	0
Wood (dry)	1.5–4	0.01

the ferromagnetic materials is only valid for very low magnetic flux densities. Maximum permeabilities may be an order of magnitude higher.

Values are given in Table C.4 for the charge and rest mass of an electron, the permittivity and permeability of free space, and the velocity of light.[4]

[4] Cohen, E. R., and B. N. Taylor: "The 1986 Adjustment of the Fundamental Physical Constants," Pergamon Press, Elmsford, NY, 1986.

TABLE C.2
σ

Material	σ, ℧/m	Material	σ, ℧/m
Silver	6.17×10^7	Graphite	7×10^4
Copper	5.80×10^7	Silicon	1200
Gold	4.10×10^7	Ferrite (typical)	100
Aluminum	3.82×10^7	Water (sea)	5
Tungsten	1.82×10^7	Limestone	10^{-2}
Zinc	1.67×10^7	Clay	5×10^{-3}
Brass	1.5×10^7	Water (fresh)	10^{-3}
Nickel	1.45×10^7	Water (distilled)	10^{-4}
Iron	1.03×10^7	Soil (sandy)	10^{-5}
Phosphor bronze	1×10^7	Granite	10^{-6}
Solder	0.7×10^7	Marble	10^{-8}
Carbon steel	0.6×10^7	Bakelite	10^{-9}
German silver	0.3×10^7	Porcelain (dry process)	10^{-10}
Manganin	0.227×10^7	Diamond	2×10^{-13}
Constantan	0.226×10^7	Polystyrene	10^{-16}
Germanium	0.22×10^7	Quartz	10^{-17}
Stainless steel	0.11×10^7		
Nichrome	0.1×10^7		

TABLE C.3
μ_R

Material	μ_R
Bismuth	0.999 998 6
Paraffin	0.999 999 42
Wood	0.999 999 5
Silver	0.999 999 81
Aluminum	1.000 000 65
Beryllium	1.000 000 79
Nickel chloride	1.000 04
Manganese sulfate	1.000 1
Nickel	50
Cast iron	60
Cobalt	60
Powdered iron	100
Machine steel	300
Ferrite (typical)	1000
Permalloy 45	2500
Transformer iron	3000
Silicon iron	3500
Iron (pure)	4000
Mumetal	20 000
Sendust	30 000
Supermalloy	100 000

TABLE C.4
Physical Constants

Quantity	Value
Electron charge	$e = (1.602\,177\,33 \pm 0.000\,000\,46) \times 10^{-19}$ C
Electron mass	$m = (9.109\,389\,7 \pm 0.000\,005\,4) \times 10^{-31}$ kg
Permittivity of free space	$\epsilon_0 = 8.854\,187\,817 \times 10^{-12}$ F/m
Permeability of free space	$\mu_0 = 4\pi 10^{-7}$ H/m
Velocity of light	$c = 2.997\,924\,58 \times 10^{8}$ m/s

APPENDIX

D

ANSWERS TO ODD-NUMBERED PROBLEMS

CHAPTER 1

1(*a*) $-0.845\mathbf{a}_x + 0.1690\mathbf{a}_y - 0.507\mathbf{a}_z$; (*b*) $-0.911\mathbf{a}_x + 0.391\mathbf{a}_y + 0.1302\mathbf{a}_z$; (*c*) $-0.793\mathbf{a}_x + 0.463\mathbf{a}_y - 0.396\mathbf{a}_z$; (*d*) 25.5. **3** (7,8.4,7.8). **5**(*a*) $-24\mathbf{a}_x - 4\mathbf{a}_y - 72\mathbf{a}_z$; (*b*) $-0.316\mathbf{a}_x - 0.0526\mathbf{a}_y - 0.947\mathbf{a}_z$; (*c*) $4x^4y^2 + 4(z-x)^2 + 9x^2y^2z^2 = 10^4$; (*d*) 2.04; (*e*) 7.17. **7**(*a*) $y = -x$ (plane), $z = 0, \pm 1, \pm 2, \ldots$ (planes); (*b*) $y = -x^2$ (parabolic cylinder); (*c*) $x^2 + y^2 = 10$ (circular cylinder). **9**(*a*) $0.0995\mathbf{a}_x - 0.995\mathbf{a}_z$; (*b*) 34.7°; (*c*) 36. **11**(*a*) $-0.688\mathbf{a}_x - 0.229\mathbf{a}_y - 0.688\mathbf{a}_z$; (*b*) 98.6°; (*c*) 21.4; (*d*) -0.651. **13**(*a*) 31.1; (*b*) $-0.835\mathbf{a}_x + 0.0642\mathbf{a}_y + 0.546\mathbf{a}_z$; (*c*) $-0.808\mathbf{a}_x - 1.077\mathbf{a}_y + 0.269\mathbf{a}_z$; (*d*) $-3.19\mathbf{a}_x + 3.08\mathbf{a}_y + 2.73\mathbf{a}_z$. **15**(*a*) $0.295\mathbf{a}_x + 0.494\mathbf{a}_y - 0.818\mathbf{a}_z$; (*b*) $0.674\mathbf{a}_x - 0.1740\mathbf{a}_y - 0.718\mathbf{a}_z$; (*c*) 31.9°. **17**(*a*) 20.3; (*b*) $\pm(0.0983\mathbf{a}_x - 0.836\mathbf{a}_y + 0.541\mathbf{a}_z)$; (*c*) $-0.851\mathbf{a}_x + 0.211\mathbf{a}_y + 0.480\mathbf{a}_z$. **19**(*a*) 5.01; (*b*) $-0.635\mathbf{a}_x - 0.489\mathbf{a}_y - 0.598\mathbf{a}_z$; (*c*) $-0.741\mathbf{a}_\rho + 0.306\mathbf{a}_\phi - 0.598\mathbf{a}_z$. **21**(*a*) $-2.13\mathbf{a}_x + 6.07\mathbf{a}_y + 3\mathbf{a}_z$; (*b*) $-2.35\mathbf{a}_\rho + 3.53\mathbf{a}_\phi + 4\mathbf{a}_z$; (*c*) 105.4. **23**(*a*) $-0.738\mathbf{a}_\rho + 0.510\mathbf{a}_\phi + 0.442\mathbf{a}_z$; (*b*) $-0.696\mathbf{a}_\rho - 0.565\mathbf{a}_\phi + 0.442\mathbf{a}_z$. **25**(*a*) $-\mathbf{a}_x + 4\mathbf{a}_y + 7\mathbf{a}_z$; (*b*) $-2.68\mathbf{a}_\rho + 3.13\mathbf{a}_\phi + 7\mathbf{a}_z$; (*c*) $-7.22\mathbf{a}_r - 2.03\mathbf{a}_\theta + 3.13\mathbf{a}_\phi$. **27**(*a*) $0.424\mathbf{a}_r + 0.424\mathbf{a}_\theta + 0.8\mathbf{a}_\phi$; (*b*) $0.424\mathbf{a}_x - 0.566\mathbf{a}_y - 0.707\mathbf{a}_z$. **29**(*a*) 14.91; (*b*) 3.86; (*c*) 36.8. **31**(*a*) 8.726 65; (*b*) 8.723 88.

CHAPTER 2

1(*a*) 16.67 nC; (*b*) 10.18°. **3**(*a*) 1835 N; (*b*) 60°. **5**(*a*) $5.62\mathbf{a}_y - 11.23\mathbf{a}_z$ V/m; (*b*) (0,2.07,−4.14). **7** 554, 225, 0 V/m. **9**(*a*) (0,0,±0.1414); (*b*) 17.30 kV/m. **11**(*a*) 0;

(*b*) 19.57 V/m; (*c*) 25.7 V/m. **13**(*a*) 320 C; (*b*) 926 C. **15**(*a*) 10 C/m^3; (*b*) 157.1 nC; (*c*) 1.678 mm. **17** $0.1425\mathbf{a}_x$ V/m. **19**(*a*) $-172.6\mathbf{a}_y + 86.3\mathbf{a}_z$ V/m; (*b*) -4.43 μC. **21**(*a*) $-119.8\mathbf{a}_z$ V/m; (*b*) $16.06\mathbf{a}_x - 7.65\mathbf{a}_z$ V/m. **23**(*a*) 158.5 V/m; (*b*) $0.962\mathbf{a}_x + 0.272\mathbf{a}_z$; (*c*) $-2.26\mathbf{a}_x$ μN/m. **25**(*a*) $-179.7\mathbf{a}_x - 360\mathbf{a}_y$ V/m; (*b*) (2.88,5.00,0). **27** $2.16 \times 10^{11}\mathbf{a}_z$ V/m. **29**(*a*) $3y^2 - x^2 = 23$, $z = -4$; (*b*) $0.976\mathbf{a}_x + 0.217\mathbf{a}_y$; (*c*) $0.217\mathbf{a}_x - 0.976\mathbf{a}_y$. **31**(*a*) $\cos 2y = 0.298e^{2x}$, $z = 0$; (*b*) $\pm(0.588\mathbf{a}_x - 0.809\mathbf{a}_y)$.

CHAPTER 3

1(*a*) penny: 5 nC, nickel: 0, dime: 0, can + lid: -5 nC; (*b*) penny: 5 nC, nickel: -1 nC, dime: -2 nC, can + lid: -2 nC. **3**(*a*) 113.9 nC; (*b*) 57.0 nC; (*c*) 55.0 nC. **5**(*a*) $49.5\mathbf{a}_z$ nC/m^2; (*b*) $4.96\mathbf{a}_x + 12.79\mathbf{a}_y + 38.1\mathbf{a}_z$ nC/m^2; (*c*) 8.70 μC. **7**(*a*) 403 nC; (*b*) $6.00\mathbf{a}_z$ nC/m^2. **9**(*a*) 16.76 nC; (*b*) $4r/3$ μC/m^2; (*c*) 18.75 cm. **11** $\rho < 1$ cm: 0, $1 < \rho < 3$ cm: $(4\cos 50\pi\rho + 200\pi\rho \sin 50\pi\rho - 2\pi)\mathbf{a}_\rho/(\pi^2\rho)$ pC/m^2, $\rho > 3$ cm: $-8\mathbf{a}_\rho/(\pi\rho)$ pC/m^2. **13**(*a*) 0, $13.33\mathbf{a}_\rho$, and $4.8\mathbf{a}_\rho$ nC/m^2; (*b*) -4 nC/m^2. **15** 63.4 μC/m^2 for each. **17**(*a*) $64a^3 + 32a^5$; (*b*) $64a^3$; (*c*) $\lim_{a\to 0} 32a^5/64a^3 = 0$. **19**(*a*) 0; (*b*) -9 nC; (*c*) $-(3.6y + 1.8)$ nC; (*d*) (*x*,2,2). **21**(*a*) 3; (*b*) 2; (*c*) 3; (*d*) 88.2. **23**(*a*) 8 C/m^2; (*b*) 0.64 C/m^2; (*c*) 1.287 C; (*d*) $10/r^2$ C/m^2. **25**(*a*) $12x^2$ C/m^3; (*b*) $12x^2$ C/m^3; (*c*) 32 C; (*d*) 32 C. **27**(*a*) 180 C/m^3; (*b*) 135 C/m^2; (*c*) 12.72 kC; (*d*) 12.72 kC. **29**(*a*) 0.3 C/m^3 and 0; (*b*) -0.0804 C. **31** 7.85 C. **33** 4.19.

CHAPTER 4

1(*a*) -160 pJ; (*b*) 240 pJ; (*c*) -480 pJ; (*d*) -560 pJ; (*e*) 226 pJ. **3**(*a*) 867 pJ; (*b*) -1533 pJ. **5**(*a*) 26 V; (*b*) 26 V. **7** 1530 J. **9**(*a*) 456 V; (*b*) 469 V; (*c*) 538 V; (*d*) 1463 V. **11** 1464 V. **13** 442 pJ. **15** 137.5 V. **17** 37.9 V. **19** $56.5 \ln [(0.01 + \sqrt{z^2 + 0.0001})/z]$. **21**(*a*) 882 V; (*b*) $-588\mathbf{a}_x - 1764\mathbf{a}_y - 252\mathbf{a}_z$ V/m; (*c*) 1876 V/m; (*d*) $-5.21\mathbf{a}_x - 15.62\mathbf{a}_y - 2.23\mathbf{a}_z$ nC/m^2; (*e*) 16.61 nC/m^2; (*f*) -17.67 nC/m^3; (*g*) $0.313\mathbf{a}_x + 0.940\mathbf{a}_y + 0.1343\mathbf{a}_z$. **23**(*a*) 80 V; (*b*) $-20\mathbf{a}_\rho + 46.2\mathbf{a}_\phi - 80\mathbf{a}_z$ V/m; (*c*) $-177.1\mathbf{a}_\rho + 409\mathbf{a}_\phi - 708\mathbf{a}_z$ pC/m^2; (*d*) -334 pC/m^3; (*e*) 94.5 V/m; (*f*) $0.212\mathbf{a}_\rho - 0.489\mathbf{a}_\phi + 0.846\mathbf{a}_z$. **25**(*a*) 150 V, $-50\mathbf{a}_r - 150\mathbf{a}_\phi$ V/m, $-0.443\mathbf{a}_r - 1.328\mathbf{a}_\phi$ nC/m^2, -0.885 nC/m^3; (*b*) -5.56 nC. **27**(*a*) 287.61 V; (*b*) 122.87 V/m; (*c*) 287.88 V; (*d*) 123.57 V/m. **29**(*a*) $8.99(10x - 3y + 5z - 13)/[(x-2)^2 + (y-4)^2 + (z-1)^2]^{1.5}$ V; (*b*) 3.69 V. **31**(*a*) 0.472 μJ; (*b*) 0.266 μJ. **33**(*a*) $5\mathbf{a}_r/4\pi r^2$ μC/m^2; (*b*) 2.81 J; (*c*) 4.45 pF. **35**(*a*) 25.0 kV/m; (*b*) 936 V; (*c*) 4.68 μJ.

CHAPTER 5

1(*a*) 169.1 kA; (*b*) 0; (*c*) 0. **3**(*a*) 70.7 A; (*b*) $9.33\mathbf{a}_r$ A/m^2; (*c*) 28.5°. **5**(*a*) $-200x^{1/3}\mathbf{a}_x$ V/m, $-1.771x^{1/3}\mathbf{a}_x$ nC/m^2, $-0.590x^{-2/3}$ nC/m^3; (*b*) -3.54 mA; (*c*) -3.54 mA. **7**(*a*) mass per second per square meter or mass rate of flow, mass per cubic meter or density; (*b*) -0.55 kg/$m^3\cdot$s. **9**(*a*) 2.04 m; (*b*) 83.4 V/m; (*c*) 5.89×10^7 A/m^2 RMS. **11**(*a*) 398 V, 159.2 Ω, $9.95\mathbf{a}_r/r^2$ V/m; (*b*) 320 V, 128.1 Ω,

198.9$\mathbf{a}_r/r$ V/m; (*c*) 800 W. **13**(*a*) 40 A; (*b*) $\mathbf{a}_r/8\pi r^2$ V/m; (*c*) 1.326 V; (*d*) 53.1 W. **15**(*a*) $r^2 = 2 \sin\theta \cos\phi$; (*b*) $215\mathbf{a}_r - 186.1\mathbf{a}_\theta + 124.1\mathbf{a}_\phi$ V/m; (*c*) ± 2.75 nC/m^2. **17**(*a*) $-10(x^3 + 5)\mathbf{a}_y$ V/m; (*b*) $V = 0$ for $y = 0$; (*c*) -1.240 nC. **19**(*a*) $\nabla \cdot \mathbf{D} = 0$; (*b*) $V = 0$ for $y = 0$; (*c*) $\mathbf{E} = -5000e^{-50x}\mathbf{a}_y$ V/m; (*d*) -0.885 nC; (*e*) 1.107 μJ. **21**(*a*) $63.4\mathbf{a}_x + 74.0\mathbf{a}_z$ V/m; (*b*) 275 V; (*c*) 7.96 nC/m^2. **23**(*a*) 403 V; (*b*) $0.223 = (z-1)/[x^2 + y^2 + (z-1)^2]^{1.5} + (z+1)/[x^2 + y^2 + (z+1)^2]^{1.5}$. **25**(*a*) 2.80 ℧/m, 0.357 Ω·m or 35.7 Ω·cm; (*b*) 892 Ω. **27** 0.005 21 ℧/m, 191.8 Ω·m or 1.918×10^4 Ω·cm. **29**(*a*) $4000\mathbf{a}_y$ V/m, $63.7\mathbf{a}_y$ nC/m^2, $28.3\mathbf{a}_y$ nC/m^2; (*b*) 0, 0, 0. **31**(*a*) $-13.29\mathbf{a}_x + 46.5\mathbf{a}_y + 19.93\mathbf{a}_z$ kV/m; (*b*) $-0.318\mathbf{a}_x + 1.112\mathbf{a}_y + 0.476\mathbf{a}_z$ μC/m^2; (*c*) 52.3 kV/m. **33**(*a*) 70 V/m; (*b*) $-30\mathbf{a}_x + 50\mathbf{a}_y$ V/m; (*c*) 58.3 V/m; (*d*) 91.1 V/m; (*e*) 39.8°; (*f*) 1.549 nC/m^2; (*g*) 2.07 nC/m^2; (*h*) $-1.062\mathbf{a}_x + 1.771\mathbf{a}_y + 1.549\mathbf{a}_z$ nC/m^2; (*i*) $-0.797\mathbf{a}_x + 1.328\mathbf{a}_y + 1.162\mathbf{a}_z$ nC/m^2; (*j*) 53.1°. **35**(*a*) $0.8\mathbf{a}_x + 2\mathbf{a}_y - 3\mathbf{a}_z$ nC/m^2; (*b*) $-3\mathbf{a}_z$ nC/m^2; (*c*) $0.8\mathbf{a}_x + 2\mathbf{a}_y$ nC/m^2; (*d*) $w_{E1} = 0.429$ μJ/m^3, $w_{E2} = 0.385$ μJ/m^3; (*e*) 144.3°; (*f*) 0.599; (*g*) 0.374. **37**(*a*) $0.743\mathbf{a}_x - 0.371\mathbf{a}_y + 0.557\mathbf{a}_z$; (*b*) $-0.345\mathbf{a}_x + 25.2\mathbf{a}_y + 32.2\mathbf{a}_z$ V/m. **39**(*a*) 100 pC; (*b*) 10 nC/m^2; (*c*) $1129/(1 + 500x)^2$ V/m; (*d*) 1.129 V; (*e*) 88.5 pF. **41** barium titanate. **43**(*a*) 3.41 nF/m^2; (*b*) 4.22 nF/m^2; (*c*) 4.92 nF/m^2; (*d*) 5.53 nF/m^2. **45**(a) 37.1 pF; (*b*) 34.9 pF. **47** 40.2, 3.01 nC/m, 12.37 m, 30.1 pF/m. **49**(*a*) 24.3 pF/m; (*b*) 1.464 cm.

CHAPTER 6

1 $C_{\text{exact}} = 47.83$ pF/m. **3** $C_{\text{exact}} = 78.70$ pF/m. **5** about 47 pF/m. **7** about 35 pF/m. **9** $V_x = 78$ V, $V_y = 26$ V. **11** 19 V. **13** 35 V. **15**(*a*) $2 < \rho < 3$ cm: $E_\rho = 0.315/\rho$ V/m and $J_\rho = 31.5/\rho$ A/m^2, $3 < \rho < 6$ cm: $E_\rho = 1.259/\rho$ V/m and $J_\rho = 31.5/\rho$ A/m^2; (*b*) 0.010 12 Ω; (*c*) 0.872 V. **17**(*a*) 3510 Ω per square; (*b*) 1.403 ℧/m. **19**(*a*) 2.81 nC; (*b*) 0.1268 A.

CHAPTER 7

1 $\nabla\epsilon = 0$, or $\partial\epsilon/\partial x = 0$, $\partial\epsilon/\partial y = 0$, and $\partial\epsilon/\partial z = 0$. **3**(*a*) -96 V; (*b*) 425 pC/m^3; (*c*) $96\mathbf{a}_x - 96\mathbf{a}_y + 32\mathbf{a}_z$ V/m; (*d*) $x^2yz + 12 = 0$; (*e*) any two of: $x^2 = 2y^2 + 2$, $x^2 = 2z^2 - 14$, $y^2 = z^2 - 8$; (*f*) no. **5**(*a*) $\nabla^2 V = 0$; (*b*) $A = 1250$ V and $B = 1.083$, or $A = -1250$ V and $B = 0.923$. **7**(*a*) $\nabla^2 V \neq 0$; (*b*) -4.37 nC. **9**(*a*) Yes, yes, yes, no; (*b*) no, no, no, no; (*c*) yes, no, no, no. **11**(*a*) 240 V; (*b*) $-10\mathbf{a}_x + 20\mathbf{a}_y - 50\mathbf{a}_z$ V/m. **13**(*a*) $3.16\mathbf{a}_\rho/\rho$ nC/m^2; (*b*) $1191[1 + (0.04/\rho)]\mathbf{a}_\rho$ V/m; (*c*) $-62.6 - 1191\rho - 47.7 \ln \rho$ V; (*d*) 198.8 pF/m. **15**(*a*) $-20\,000\phi + 3780$ V; (*b*) $20\,000\mathbf{a}_\phi/\rho$ V/m; (*c*) $177.1\mathbf{a}_\phi/\rho$ nC/m^2; (*d*) $177.1/\rho$ nC/m^2; (*e*) 84.78 nC; (*f*) $V = 28.7\phi + 194.9$ V, $\mathbf{E} = -28.7\mathbf{a}_\phi/\rho$ V/m, $\mathbf{D} = -254\mathbf{a}_\phi/\rho$ pC/m^2, $\rho_s = 254/\rho$ pC/m^2, $Q = 0.1216$ nC; (*g*) 84.90 nC, 471.7 pF. **17**(*a*) $(15/r) - 400$ V; (*b*) $15/r^2$; (*c*) -25.0 V; (*d*) 3.75 cm; (*e*) 8.34 nC; (*f*) 41.7 pF. **19**(*a*) $-4.19 \ln(\tan \frac{1}{2}\theta) - 2.24$ kV; (*b*) 1.458 kV; (*c*) $4.19\mathbf{a}_\theta/(r \sin\theta)$ kV/m; (*d*) $9.88\mathbf{a}_\theta$ kV/m; (*e*) 18.14 nC. **21**(*a*) $400/\sqrt{r}$ V; (*b*) $400/\sqrt{r}$ V. **23** 50.11 V. **25**(*a*) $200 \sinh(2\pi x/b) \sin(2\pi y/b)/\sinh(2\pi d/b)$ V (*b*) 28.88 V. **27**(*a*) Yes; (*b*) yes; (*c*) no; (*d*) yes; (*e*) no.

CHAPTER 8

1 0.796($\mathbf{a}_x + \mathbf{a}_y$) A/m. **3**(*a*) $-1.836\mathbf{a}_x + 1.224\mathbf{a}_y$ A/m; (*b*) $1.372\mathbf{a}_x - 0.549\mathbf{a}_z$ A/m. **5**(*a*) $0.5Ia^2/[a^2 + (z - z')^2]^{1.5}$ A/m; (*b*) $0.5K_0\{[z/\sqrt{z^2 + a^2}] - [(z - h)/\sqrt{(z - h)^2 + a^2}]\}$ A/m. **7** $0.709\mathbf{a}_x - 0.834\mathbf{a}_y + 0.376\mathbf{a}_z$ μA/m. **9** $2.95\mathbf{a}_z$ A/m. **11**(*a*) $2.05\mathbf{a}_x$ A/m; (*b*) $4.58\mathbf{a}_x$ A/m; (*c*) $5\mathbf{a}_x$ A/m. **13** 0, -0.6 A/m, -1 A/m, -0.4 A/m, 0. **15**(*a*) 0; (*b*) $35\mathbf{a}_\phi$ A/m; (*c*) $57.1\mathbf{a}_\phi$ A/m; (*d*) $30\mathbf{a}_\phi$ A/m; (*e*) $21\mathbf{a}_\phi$ A/m. **17**(*a*) $5 \times 10^6 \rho\mathbf{a}_\phi$ A/m; (*b*) $5\mathbf{a}_\phi/\rho$ A/m; (*c*) $3\mathbf{a}_\phi/\rho$ A/m; (*d*) $-750\mathbf{a}_z$ A/m. **19**(*a*) 80 A/m; (*b*) 104.1 A/m; (*c*) 40 A/m. **21**(*a*) $-b^3 - 4b^2$; (*b*) -0.041 A; (*c*) $(-2xy - 2y)\mathbf{a}_x + (z^2 + y^2)\mathbf{a}_y$; (*d*) -4 A/m^2; (*e*) $-4 \doteq -4.1$. **23**(*a*) 1.592×10^8 A/m^3; (*b*) $4.77 \times 10^9 \rho$ ℧/m. **25** 53.3, Stokes. **27**(*a*) $(4 - 3\rho)\mathbf{a}_z$ A/m^2; (*b*) 6.28 A; (*c*) 6.28 A, Stokes. **29**(*a*) -1959 A; (*b*) -1959 A. **31**(*a*) 0.213 μWb; (*b*) 2.85 μWb. **33**(*a*) $\rho = 7$ cm: $-429\mathbf{a}_z$ A/m, $z = 3$: $30\mathbf{a}_\rho/\rho$ A/m, $z = 0$: $-30\mathbf{a}_\rho/\rho$ A/m; (*b*) $30\mathbf{a}_\phi/\rho$ A/m; (*c*) 0.381 μWb. **35**(*a*) $40y - 8$ A; (*b*) $40\mu_0(x - 0.1)\mathbf{a}_z$ Wb/m. **37** $-30\phi + 18$ A. **39**(*a*) $x < -3$ m: $24\mathbf{a}_z$ A/m, $-3 < x < 3$: $-8x\mathbf{a}_z$ A/m, $x > 3$: $-24\mathbf{a}_z$ A/m; (*b*) $x < -3$: $24\mu_0\mathbf{a}_z$ T, $-3 < x < 3$: $-8\mu_0 x\mathbf{a}_z$ T, $x > 3$: $-24\mu_0\mathbf{a}_z$ T; (*c*) $24z - 120$ A; (*d*) $4\mu_0(4 - x^2)\mathbf{a}_y$ Wb/m. **41**(*a*) $-15\rho^{0.5}\mathbf{a}_\phi/\mu_0$ A/m; (*b*) $-22.5\rho^{-0.5}\mathbf{a}_z/\mu_0$ A/m^2; (*c*) $I = -30\pi/\mu_0$ A. **43** $A_z = (10^{-7}I/24)[(\rho^2/a^2) - 25 - 98 \ln(\rho/5a)]$ Wb/m.

CHAPTER 9

1(*a*) $P(0, 0.2 \text{ m}, 0.04 \text{ m})$; (*b*) $10^5\mathbf{a}_y + 4 \times 10^4\mathbf{a}_z$ m/s; (*c*) 1.04 μJ. **3**(*a*) $(9\mathbf{a}_x + 22\mathbf{a}_y + 8\mathbf{a}_z)10^4$ V/m; (*b*) 414 or -234 kV/m. **5**(*a*) $-7.2\mathbf{a}_y$ nN; (*b*) $-6.01\mathbf{a}_z$ nN; (*c*) $-5.14\mathbf{a}_y$ nN. **7**(*a*) $19.2\mathbf{a}_y$, 0, $-7.38\mathbf{a}_y$, 0 nN; (*b*) $11.82\mathbf{a}_y$ nN. **9** 0.0318 N/m. **11**(*a*) $\mathbf{F}_2 = \mu_0 I_1 I_2 \mathbf{a}_y/(2\pi d)$ N/m; (*b*) see end of Sec. 9.3. **13**(*a*) $30\mathbf{a}_x$ μN·m; (*b*) $30\mathbf{a}_x$ μN·m; (*c*) $2.27\mathbf{a}_x$ μN·m. **15** $9.6\mathbf{a}_x(l - \tan^{-1} l)$ μN·m. **17** (*a,b,c*) Proofs; (*d*) 3.42×10^{16} rad/s, 9.88×10^{-24} N·m, 9.88×10^{-24} A·m^2. **19**(*a*) $140\mathbf{a}_x$ A/m; (*b*) $0.7\mathbf{a}_x$ A/m; (*c*) $1819\mathbf{a}_z$ A/m; (*d*) 0, $60\mathbf{a}_\phi/\rho$ A/m, 0. **21**(*a*) 3.5; (*b*) 4.40 μH/m; (*c*) $11.37x\mathbf{a}_y$ kA/m; (*d*) $28.4x\mathbf{a}_y$ kA/m; (*e*) $11.37\mathbf{a}_z$ kA/m^2; (*f*) $28.4\mathbf{a}_z$ kA/m^2. **23** $z < 0$: 0; $0 < z < 2$: $-210\mathbf{a}_y$ A/m; $2 < z < 5$: $-100\mathbf{a}_y$ A/m; $z > 5$: 0. **25**(*a*) $5 < \rho < 15$ mm: $6\mathbf{a}_\phi/\pi\rho$ A/m, $6\mathbf{a}_\phi/\rho$ μT, $9\mathbf{a}_\phi/\pi\rho$ A/m; $15 < \rho < 20$ mm: $6\mathbf{a}_\phi/\pi\rho$ A/m, $2.4\mathbf{a}_\phi/\rho$ μT, 0; (*b*) $0 < \phi < \pi$: $1.091\mathbf{a}_\phi/\rho$ A/m, $3.43\mathbf{a}_\phi/\rho$ μT, $1.637\mathbf{a}_\phi/\rho$ A/m; $\pi < \phi < 2\pi$: $2.73\mathbf{a}_\phi/\rho$ A/m, $3.43\mathbf{a}_\phi/\rho$ μT, 0. **27** $4.71\mathbf{a}_x + 5.53\mathbf{a}_y - 1.816\mathbf{a}_z$ A/m. **29** 1.106 T, 1.048 T, 1.042 T. **31** 0.263 μWb. **33**(*a*) 0.628 J/m^3; (*b*) 1.257 J/m^3; (*c*) 1.257 J/m^3; (*d*) 2.51 J/m^3; (*e*) 1.026 J/m^3. **35**(*a*) $25\mathbf{a}_\phi/(\pi r \sin\theta)$ A/m; (*b*) 173.5 μJ. **37** 0.1388 μH. **39**(*a*) 1.228×10^{-3} m^3; (*b*) $\mu_R = 10$ for $3 < \rho < 10.33$ mm; (*c*) 2.61 μH/m. **41** $0.2Na \ln(1 + a/b)$ μH. **43**(*a*) Proof; (*b*) $2 \times 10^{-7}[0.25(a^4 - c^4) - c^2(a^2 - c^2) + c^4 \ln(a/c)]/(a^2 - c^2)^2$ H/m.

CHAPTER 10

1 $-2.33e^{-1453t}$ V. **3**(*a*) $-16 \sin 10^8 t$ V; (*b*) $15.93 \cos(10^8 t - 99.5°)$ V. **5**(*a*) -0.15 V; (*b*) -0.1414 V. **7**(*a*) $-1.44/(40t + 0.2)$ A; (*b*) $-0.0724/(0.010\,025 + 2t - t^2)$ A. **9** $142.1 \cos^2(6 \times 10^8 t + 78.5°)$ W. **11**(*a*) $(100/\rho)\cos(3 \times 10^8 t - z)\,\mathbf{a}_\rho$ V/m;

(*b*) $-(0.266/\rho)\sin(3\times10^8t-z)\,\mathbf{a}_\rho$ A/m^2. **13** $I_d=I_s=(4\pi\epsilon_R\epsilon_0\omega V_0\cos\omega t)/[(1/a)-(1/b)]$. **15** $8\times10^{-8}\cos(10^9t\mp50z)\,\mathbf{a}_x$ C/m^2, $\pm2\times10^{-5}\cos(10^9t\mp50z)\,\mathbf{a}_y$ T, $\pm1.6\cos(10^9t\mp50z)\,\mathbf{a}_y$ A/m, ±50 rad/m. **17**(*a*) 16.68 rad/m; (*b*) $-2210\mathbf{a}_x\sin10^9t$ A/m^2; (*c*) $9.83\cos(10^9t+42.2°)$ A. **19**(*a*) $\pm(-0.254\mathbf{a}_x+0.381\mathbf{a}_y+0.889\mathbf{a}_z)$; (*b*) $\pm3.49\cos10^6t$ nC/m^2. **21**(*a*) 1.884 Grad/s; (*b*) 4.43 nC; (*c*) -221 V$^{0.5}$/m$^{1.5}$. **23**(*a*) $\partial^2E_x/\partial z^2=\mu_0\epsilon_0\partial^2E_x/\partial t^2$; (*b*) $a=6.67\times10^{-7}$ A/m. **25** $-3\times10^5\mathbf{a}_y\sin3\times10^8t\sin z$ V/m, $796\mathbf{a}_x\cos3\times10^8t\cos z$ A/m.

CHAPTER 11

1 Proof. **3**(*a*) 250 rad/m; (*b*) 7.5×10^{10} rad/s; (*c*) 11.93 GHz; (*d*) 2.51 cm; (*e*) 377 Ω; (*f*) $0.531e^{-j250z+j30°}\mathbf{a}_y$ A/m; (*g*) 103.5 V/m. **5**(*a*) 9.42×10^8 rad/s; (*b*) 3.14 rad/m; (*c*) 2 m; (*d*) 3×10^8 m/s; (*e*) 377 Ω; (*f*) $(200\mathbf{a}_y+400\mathbf{a}_z)\cos(3\times10^8\pi t-\pi x)$ V/m; (*g*) $(-1.061\mathbf{a}_y+0.531\mathbf{a}_z)\cos(3\times10^8\pi t-\pi x)$ A/m. **7**(*a*) 2.41; (*b*) 1.659; (*c*) 0.0667 rad/m; (*d*) 94.2 rad/m; (*e*) 455 Ω. **9**(*a*) 2 rad/m, 3.14×10^8 rad/s, 1.571×10^8 m/s, 3.14 m, 3.64, 197.4 Ω; (*b*) $5\mathbf{a}_x$, $3.54\mathbf{a}_x-7.07\mathbf{a}_y$, $-10\mathbf{a}_y$, $-3.54\mathbf{a}_x-7.07\mathbf{a}_y$, $-5\mathbf{a}_x$ V/m. **11** ($\sigma=2.5$ m℧/m, $\epsilon_R=4$) 0.23 Np/m, 4.2 rad/m, 1.50 m, 1.5×10^8 m/s, 188 Ω. **13**(*a*) 0.226 Np/m; (*b*) 0.700 rad/m; (*c*) 8.97×10^7 m/s; (*d*) 8.97 m; (*e*) $429\underline{/17.86°}$ Ω; (*f*) $46.6e^{-0.226z}e^{j(-0.700z-17.86°)}\mathbf{a}_y$ mA/m; (*g*) $-4.59\mathbf{a}_x$ V/m. **15**(*a*) 0.999 cm, 0; (*b*) 0.994 cm, 0.496 dB/ft; (*c*) 0.589 cm, 1469 dB/ft. **17** P.F. $=[1+(\sigma/\omega\epsilon)^{-2}]^{-0.5}$, $Q=(\sigma/\omega\epsilon)^{-1}$. **19**(*a*) $106.2\mathbf{a}_z$, $0.531\mathbf{a}_z$, $104.1\mathbf{a}_z$, and $4.72\mathbf{a}_z$ W/m^2;
(*b*) $106.2\mathbf{a}_z$, $14.00\mathbf{a}_z$, $57.6\mathbf{a}_z$, and $85.6\mathbf{a}_z$ W/m^2. **21**(*a*) $120\pi[(\sin\theta)/r]\times\cos(3\times10^8t-r)\mathbf{a}_\theta$ V/m, $[(\sin\theta)/r]\cos(3\times10^8t-r)\mathbf{a}_\phi$ A/m;
(*b*) $120\pi[(\sin\theta)/r]^2\cos^2(3\times10^8t-r)\mathbf{a}_r$ W/m^2; $320\pi^2\cos^2(3\times10^8t-r)$ W.
23 $0.0287\mathbf{a}_z$ V/m, $[143\,400\rho-(41.4/\rho)]\mathbf{a}_\phi$ A/m, $[-4110\rho+(1.188/\rho)]\mathbf{a}_\rho$ W/m^2. **25**(*a*) 1.260 Ω/m; (*b*) 0.315 Ω/m; (*c*) 1.575 Ω/m. **27**(*a*) 4.56 Ω/m; (*b*) 0.0275 Np/m. **29** 1 GHz, 1.6×10^5 ℧/m. **31** 48.1 μm. **33**(*a*) $[1000e^{-j200x}+141.3e^{j(200x+106.1°)}]\mathbf{a}_y$ V/m; (b) $970e^{-j57.7x}e^{-j(208x-8.04°)}\mathbf{a}_z$ A/m; (*c*) $10.48e^{-57.7x}\times\cos(10^{10}t-208x-7.44°)\,\mathbf{a}_z$ A/m. **35**(*a*) 10^{-10} F/m, 4×10^{-6} H/m; (*b*) $\cos10^9t\cos20z\,\mathbf{a}_y$ A/m; (*c*) $-j200\sin20z\,\mathbf{a}_x$ V/m. **37**(*a*) 559 MHz; (*b*) 1.873. **39**(*a*) 2; (*b*) 4.5; (*c*) 2.93. **41**(*a*) 9; (*b*) 34.0; (*c*) 4.

CHAPTER 12

1(*a*) $0.1061+j2.00$ m^{-1}, 0.1061 Np/m, 2.00 rad/m, 2.50×10^8 m/s, 3.14 m, $125.3\underline{/-2.68°}$ Ω; (*b*) 21.7 m; (*c*) 0.785 m. **3**(*a*) 1.5 μH/m; (*b*) 16.67 pF/m; (*c*) $0.346\underline{/-140.3°}$, 2.06. **5**(*a*) 2.09×10^8 m/s, 20.9 m, 0.716 μH/m, 31.8 pF/m; (*b*) -1; (*c*) $j2120$ Ω. **7** $[10^{-7}I^2/4(c^2-b^2)]\{b^2-3c^2+[4c^4\ln(c/b)]/(c^2-b^2)\}$. **9**(*a*) $a=0.217$ cm, $b=0.5$ cm; (*b*) $a=0.700$ mm, $d=8.60$ mm; (*c*) $b=0.999$ cm, $d=0.530$ mm. **11**(*a*) 320 pF/m; (*b*) 160 μ℧/m; (*c*) 78.5 nH/m; (*d*) 0.1924 Ω/m; (*e*) 15.67 Ω. **13**(*a*) 1.161; (*b*) 1.000. **15**(*a*) $65.1e^{-j4.37°}$ V; (*b*) $75.7e^{-j36.4°}$ V. **17** $P_{25}=44.1$ W, $P_{100}=47.5$ W. **19**(*a*) 1.791 V; (*b*) 0.285 A. **21**(*a*) 0.324λ; (*b*) 4.00; (*c*) $75.4+j19.96$ Ω. **23**(*a*) 4.63 m; (*b*) $2.62+j0$; (*c*) 2.62; (*d*) 0.369 m. **25**(*a*) 2.16; (*b*) 19.20 cm;

(*c*) 27.0 cm; (*d*) 42.4 + *j*35.6 Ω. **27**(*a*) 2.62; (*b*) 0.5 − *j*0.5; (*c*) 1 − *j*1. **29**(*a*) 750 MHz; (*b*) 2.6; (*c*) 0.670 − *j*0.742. **31**(*a*) 63.6 − *j*23.0 Ω; (*b*) 33.2 − *j*9.51 Ω. **33**(*a*) d_1 = 8.05 cm, d = 12.65 cm; (*b*) d_1 = 1.950 cm, d = 1.649 cm. **35**(*a*) 0.758 m; (*b*) 20.8 pF.

CHAPTER 13

1(*a*) $(79.8/\rho)\cos(2\pi 10^8 t - 2.93z)$; (*b*) $(0.297/\rho)\cos(2\pi 10^8 t - 2.93z)$; (*c*) $100[\cos 2\pi 10^8 t - \cos(2\pi 10^8 t - 50.4°)]$ twice. **3** 164.9 MHz. **5** $-\oint_S (\mathbf{E} \times \mathbf{H}) \cdot d\mathbf{S} = \int_{vol} \mathbf{J} \cdot \mathbf{E}\, dv + \partial/\partial t \int_{vol} (\epsilon E^2/2 + \mu H^2/2)\, dv$ becomes $P_{in}(t) = V^2(t)/R + V(t)I_C(t) + V(t)I_L(t)$. **7**(*a*) 3 cos3*t* A for each; (*b*) *a*: −6 cos 3*t* V, all across *R*; *b*: −6 cos 3*t* V, uniformly distributed; *c*: 6 cos 3*t* V across source plus −6 cos 3*t* V across *R* gives zero. **9**(*a*) 419 MHz; (*b*) 26.6 pF, 5.43 nH, 16.67 kΩ; (*c*) 26.6 pF, 5.43 nH, 1.041 kΩ (neglects losses in short circuit); (*d*) 26.6 pF, 5.43 nH, 979 Ω. **11** 1.956 mW. **13** $(ad/8\sigma_c\delta)(H_{x0}^2 + H_{z0}^2)$. **15**(*a*) 94.8 MHz; (*b*) 25.1 pF; (*c*) 112.4 nH; (*d*) 160.9 kΩ; (*e*) 2403. **17**(*a*) $-0.283\mathbf{a}_x - 0.951\mathbf{a}_z$; (*b*) 0.258. **19**(*a*) $-j0.015\mathbf{a}_z e^{-j1000}$ V/m; (*b*) $-j0.015\mathbf{a}_y e^{-j1000}$ V/m; (*c*) $-j0.015(\mathbf{a}_y + \mathbf{a}_z)e^{-j1000}$ V/m; (*d*) $-12.40(\mathbf{a}_y + \mathbf{a}_z)$ mV/m; (*e*) 17.54 mV/m. **21**(*a*) 85.4 A; (*b*) 3.56 A. **23**(*a*) 0.2 V/m; (*b*) 0.000 236 + *j*0.2001 V/m; (*c*) 0.236×10^{-6} + *j*0.000 200 V/m.

INDEX

A (*see* Potential, vector magnetic)
Abbreviations and units, table of, 448–449
Absolute potential, 88
Acceleration, 266
  due to gravity, 179
Acceptor, 132, 201
Addition:
  graphical, 3
  vector, 3, 4
Adler, R. B., 155, 320
Admittance, 404, 406
Algebra:
  Boolean, 3
  scalar, 3
  vector, 3
Ampère, A. M., 217
Ampere, defined, 115, 446
Ampere-turn, 290
Ampère's circuital law, 218, 224–231, 241, 302, 320, 321
  point form of, 238, 255, 319, 325, 429
Ampère's law for the current element (*see* Biot-Savart law)
Ampère's work law (*see* Ampère's circuital law)
Amperian (bound) current, 283–286, 320
Analogy:
  electric-magnetic, 224, 243, 249, 320
  electrostatic-gravitational, 178–180
  electrostatic-hydraulic, 180
  uniform plane wave and transmission line, 382–386
Angle between vectors, 10–13, 18
Anisotropic material, 121, 136, 281, 286
Antenna, 433–436
  dipole, 433–435
  monopole, 435–436
  short, 433
Antenna patterns, 432
Antiferromagnetic material, 281–282
Area:
  of parallelogram, 13
  of vector surface, 10
Arithmetic, vector, 3
Arrow symbol, 3
Associative law, 3, 4
Atom, 119, 120, 278
Atomic nucleus, 119, 120, 278
Attenuation, 345, 348, 351, 373
Attenuation constant, 345, 348
Attwood, S. S., 110
Average power, 356, 361, 362, 368, 422
Axes, coordinate, 5, 6, 15, 19

B, defined, 243
Balmain, K. G., 259, 436, 437
Band:
  conduction, 119, 124, 131
  energy, 119–120
  forbidden, 120, 131
  valence, 119, 131
Bandwidth, 423
Barium titanate, 136
Baron, M. L., 181
Beaty, H. W., 155
Bewley, L. V., 317, 331
Biot, J. B., 217
Biot-Savart law, 217–224, 228, 229
Boast, W. B., 48, 175, 259
Boolean algebra, 3
Bound charge, 132–137, 140, 282
Bound (amperian) current, 283–286, 320
Boundary conditions:
  conductor-dielectric, 142–143, 162, 325–326, 384
  conductor-free space, 129, 325–326
  dielectric-dielectric, 138–142, 325–326
  magnetic, 287–289, 325–326
Branch cut, 249
Brown, R. G., 408

Cable, coaxial (*see* Coaxial transmission line)
Calculus, vector, 3
Candela, defined, 446
Capacitance:
  between coaxial cylinders, 147, 166, 196, 387–390
  between concentric spheres, 198
  between cone and plane, 200
  by current analogy, 177
  from curvilinear squares, 166
  between cylinder and plane, 152, 153
  defined, 144, 298
  distributed, 382, 387, 388, 391
  equivalent, of resonant cavity, 423
  between parallel cylinders, 155, 390, 391
  between parallel planes, 145, 146, 148–150, 392
  partial, 146
  of *pn* junction, 203
Capacitor:
  as circuit element, 416
  coaxial, 108, 147, 166, 196
  energy stored in, 146
  multiple dielectric, 148–150
  parallel-plate, 145–146, 148–150, 195
  spherical, 147–148, 198
Cartesian components, 6, 18, 21
Cartesian coordinate system, 4–14, 70, 85, 100, 190, 235, 441
  for expressing curl, 235
  for expressing divergence, 70
  for expressing gradient, 100
  Laplacian, 190
  transformations to other coordinate systems, 17–18, 21–22
Cavity:
  coaxial, 417–426
  resonant, 417–426
Cgs system of units, 447
Characteristic impedance, 385, 398
Charge, 27, 52, 115, 144
  bound, 132–137, 140, 282
  conservation of, 117
  on electron, 453
  forces on (*see* Force, on charge)
  free, 135, 140, 284–286
  magnetic, 244, 297, 323
  point, 27–30, 58, 88, 102, 190
    E field of, 30–32
    potential field of, 89–91
  test, 30, 81
  velocity of, 116, 120, 265, 266
Charge density:
  line, 37, 58, 93, 108, 190
    E field of, 39, 41
    potential field of, 88, 93
  surface, 43, 58, 93, 124, 138, 144, 190, 325
    E field of, 44, 195
    potential field of, 93, 195
  volume, 34, 58, 68, 93, 115, 135, 190
Chu, L. J., 155, 320
Circuit(s):
  distributed, 418
  electric, 94, 321, 382, 414–417
  magnetic, 289–298
  short, 316, 406
  substitution of, 317
Circuital law (*see* Ampère's circuital law)
Circular cylinder, 14
Circular harmonics, 215

Circulation, 235, 318
Closed line integral, 94, 233, 239, 313, 316
Closed path, 94, 225
Closed surface, 56–58, 241–242
Closed surface integral, 58, 74, 242
Coaxial cable (*see* Coaxial transmission line)
Coaxial capacitor, 147, 166, 196
  energy stored in, 108
Coaxial cavity, 417–426
Coaxial transmission line:
  E field of, 196
  H field of, 227–228
  R, *G*, *L*, and *C* of, 387–390
Coefficient:
  reflection, 366, 386, 398
  transmission, 366
Coercive force, 293
Cohen, E. R., 451
Collin, R. E., 76
Commutative law, 3, 10, 12
Complex frequency, 338
Complex scalar, 2
Component(s):
  cartesian, 6, 18, 21
  normal (*see* Normal component)
  tangential (*see* Tangential component)
  transformations of, 17–18, 21–22
  vector, 3–4, 6, 11
Component scalar, 8, 11
Component vector, 6, 11
Concentric-sphere experiment, 53
Concentric spheres, potential field of, 198
Conductance, 382, 387, 389, 391, 392
Conducting paper, 177
Conduction, 115, 119
  in metals, 115, 119–123, 144, 267
  skin effect, 357–363
Conduction band, 119, 124, 131
Conduction current, 121, 363
Conduction current density (*see* Current density, conduction)
  Conduction electron, 120
Conductivity, 121, 358
  defined, 121
  of lossy dielectric, 347
  magnetic analog for, 291
  of semiconductors, 131
  table of values, 451, 452
  variation with frequency, 352, 353
Conductor(s):
  filamentary, 115, 217, 302, 415
  metallic (*see* Conduction, in metals)
  moving, 312–318
  parallel, force between, 272–273
  perfect, 176, 325, 366, 370–372
Conductor-dielectric boundary conditions (*see* Boundary conditions, conductor-dielectric)
Conductor-free space boundary conditions, 129, 325–326
Cone(s), 19, 198–200
  potential field of, 199
Conservation of charge, 117
Conservative field, 91–95, 249
Constant:
  propagation, 345, 347, 358
  separation, 205
Constant field, 44
Continuity of current, 117
Continuity equation, 117, 118, 142, 218, 319
Convection current, 116
Convection current density, 116, 320, 323
Coordinate axes, 5, 6, 15, 19
Coordinate system:
  cartesian (*see* Cartesian coordinate system)
  cylindrical (*see* Cylindrical coordinate system)
  general curvilinear, 22, 440–443
  rectangular (*see* Cartesian coordinate system)
  right-handed, 5, 13, 16, 19
  spherical (*see* Spherical coordinate system)
  transformations between, 17–18, 21–22
Coordinates:
  curvilinear, 20, 440–443
  of point, 5
  polar, 14
Coplanar vectors, 3
Coulomb, Col. Charles, 27
Coulomb, 28, 115
  defined, 28
Coulomb's law, 27, 80, 218, 224, 267, 279
Cross product, 12–14
Crystalline lattice, 120
Crystalline solid, 120

Curie temperature, 281, 282
Curl, 231–238, 318, 329, 443
  defined, 234
Current (I), 115, 217, 290
  Amperian (bound), 283–286, 320
  conduction, 121, 363
    density of, 121, 175, 320, 323, 350, 357–363
  continuity of, 117
  convection, 116
  defined, 116
  displacement, 318–322, 359, 417
  filamentary, 115, 427
    H field of, 223
  line, **H** field of, 221, 223
  (*See also* Current density)
Current density, 115, 219, 290, 301
  conduction, 121, 175, 320, 323, 350, 357–363
  convection, 116, 320, 323
  defined, 116
  displacement, 319–322, 350, 358
  surface, 219, 326
Current element, differential, 217, 427
    force on, 267–270
Current loop, differential, 275–277
    torque on, 276
Current sheet, **H** field of, 229
Curvilinear coordinates, 22, 440–443
Curvilinear squares, 163–168, 176, 221, 249
Cylinder(s):
  capacitance of, to plane, 152
  circular, 14
  parallel, capacitance between, 155, 390, 391
Cylindrical coordinate system, 14–18, 70, 85, 100, 190, 235, 441
  for expressing curl, 235
  for expressing divergence, 70
  for expressing gradient, 85
  Laplacian, 190
  transformations between, 17–18

D, defined, 54
Degree, kelvin, 446
Dekker, A. J., 155, 212
∇, del operator, 72
Della Torre, E., 48
Density:
  current (*see* Current density)
  displacement, 54
  electric flux, 54, 125, 135–138, 163
  energy, 110, 297
    in electric field, 106–110, 354–355
    in magnetic field, 297, 354–355
  line charge (*see* Charge density, line)
  magnetic flux (*see* Magnetic flux density)
  power, 355, 356, 359
  surface charge (*see* Charge density, surface)
  volume charge (*see* Charge density, volume)
Depth:
  of penetration, 359
  skin, 359–363, 388, 392, 424
Determinant, 14, 234, 443
Diamagnetic material, 279, 286, 452
Dielectric, 53, 132–143, 175
  lossy, 347–353
  perfect, 138–142, 345, 366–369
Dielectric constant, 136
  table of values, 451
  (*See also* Permittivity)
Dielectric-dielectric boundary conditions, 138–142, 325–326
Dielectric hysteresis, 136
Dielectric interface, 138–142
Difference, potential, 87–90, 123, 144, 290
Differential current element, 217, 427
  force on, 267–270
Differential current loop, 275–277
  torque on, 276
Differential distance, 6, 17, 20, 85
Differential surface, 6, 17, 21, 58
Differential volume, 6, 17, 21
Dipole:
  E field of, 104
  electric, 102, 133, 276, 431
  magnetic, 276
  point, 104
  potential field of, 103, 105
Dipole antenna, 433–435
Dipole moment, 105, 133
  magnetic, 276–278
  per unit volume, 133
Directed distance, 8, 29
Direction of vector, 6, 41, 45
Displacement, 53
  directed distance, 8
Displacement current, 318–322, 359, 417

Displacement current density, 319–322, 350, 358
Displacement density, 54
Displacement flux, 53
Displacement flux density, 54
Distance:
differential, 6, 17, 20, 85
directed, 8, 29
Distributed circuit, 418
Distributive law, 4
Divergence, 68–70, 118, 162, 329, 442
defined, 69
Divergence theorem, 74–76
Division, vector by scalar, 4
Domain:
ferromagnetic, 281
frequency, 338
Donor, 132, 201
Doping, semiconductor, 132, 200
Dot operation, 72
Dot product, 10–12, 18, 21
Drift velocity, 120
Drill problems, directions for working, 9

E, defined, 30
Elastic membrane, 180
Electric circuit, 94, 321, 382, 414–417
Electric dipole, 102, 133, 276, 431
Electric field, energy density in, 106–110, 354–355
Electric field intensity, 30–34, 81, 95, 135–137, 266
defined, 30
of dipole, 104
of line charge, 39, 41, 86
magnitude of, 97
motional, 316
of *n* point charges, 33
of point charge, 30
of radial planes, 197
of sheet of charge, 44
of two sheets of charge, 44
Electric flux, 52–53, 57, 164
defined, 53
Electric flux density, 54, 125, 135–138, 163
Electric susceptibility, 136
Electrolytic trough, 176
Electromotive force (emf), 290, 312–318, 321
motional, 316

Electron, 28, 119, 132, 200, 453
charge on, 453
conduction, 120
free, 120
mass of, 28, 453
radius of, 28
Electron mobility, 120
Electron orbit, 119, 279–282
Electron spin, 279–282
Electrostatic fields, 94, 124, 129, 162, 290
Electrostatic potential, 87, 88, 246, 326
Element, volume, 6, 17, 21
Elephant, 13
Emf (electromotive force), 290, 312–318, 321
motional, 316
Emu system of units, 447, 449
Energy:
in gravitational field, 179
kinetic, 119
to move point charge, 81
potential (*see* Potential energy)
quantum of, 119
stored in capacitor, 146
stored in inductor, 300
thermal, 121, 131
(*See also* Work)
Energy band, 119–120
Energy density, 110, 297
in electric field, 106–110, 354–355
in magnetic field, 296–297, 354–355
Energy gap, 120, 132
Energy level, 120
Energy state, 120
English system of units, 28, 449
Equality of vectors, 4
Equipotential surface, 91, 97, 129, 144, 163
Esu system of units, 447, 449
Euler's identity, 337, 374, 383
External inductance, 303, 387–392

Fano, R. M., 155, 320
Farad, 28, 145
defined, 145
Faraday, M., 52, 311, 331
Faraday's law, 297, 312–318, 323, 324, 325, 327, 415
Ferrimagnetic material, 279, 281, 282, 285
Ferrite, 282, 285, 451

Ferroelectric material, 136
Ferromagnetic domain, 281
Ferromagnetic material, 280–282, 286, 292–295
Field(s), 2, 216
  conservative, 91–95, 249
  constant, 44
  electrostatic, 94, 124, 129, 162, 290
  flux-density type of, 53–54, 69
  force-type, 30, 54
  gravitational, 81, 89, 91, 95, 178–180
  inverse-cube, 105, 431
  inverse-distance, 41, 91, 93, 431
  inverse-square-law (*see* Inverse-square-law field)
  magnetic (*see* Magnetic field)
  nonconservative, 95
  potential (*see* Potential field)
  scalar, 2, 91, 99
  sketches of, 45–48
  vector, 2–4, 9
Field intensity:
  electric (*see* Electric field intensity)
  magnetic (*see* Magnetic field intensity)
Field map, 45, 163
Filamentary conductor, 115, 217, 302, 415
Filamentary current, 115, 223, 427
Fink, D. G., 155
Finney, R. L., 23, 76, 110
Fluid dynamics, 180, 235
Fluid-flow maps, 180
Flux:
  electric, 52–53, 57, 164
  fringing, 295
  leakage, 295
  magnetic (*see* Magnetic flux)
Flux density:
  displacement, 54
  electric, 54, 125, 135–138, 163
  magnetic (*see* Magnetic flux density)
  remnant, 293–294
Flux-density type of field, 53–54, 69
Flux line, 52–54
Flux linkage, 298, 318
Forbidden band, 120, 131
Force, 10
  on charge, 27–30, 81, 120, 266, 320, 324
  on closed circuit, 273
  coercive, 293
  on conductor, 269
  on differential current element, 267–273
  between differential current elements, 271–273
  Lorentz, 266, 324
  on magnetic material, 296–298
  moment of a, 274
  on moving charge, 266
  between parallel conductors, 272–273
Force-type field, 30, 54
Fourier integral, 344
Fourier series, 210, 344
Free charge, 135, 140, 284–286
Free electron, 120
Free space:
  permeability of, 243, 446, 453
  permittivity of, 28, 447, 453
Frequency, 353, 359–360
  complex, 338
Frequency domain, 338
Fringing flux, 295
Function, vector, 2, 9

Gap, energy, 120, 132
Gauss, K. F., 57
Gauss, defined, 243
Gauss' law, 56–58, 65, 80, 125, 325
  applications of, 60–68
  for the magnetic field, 244, 325
  point form of, 71, 137, 189
Gaussian surface, 58
Gaussian system of units, 447, 449
General curvilinear coordinate system, 22, 440–443
Gilbert, L., 27
Ginzton, E. L., 376
Gradient, 98, 162, 442
  defined, 98
  potential, 95, 246
Graphical addition, 3
Gravitational field, 81, 89, 91, 95, 178–180
Gravity, acceleration due to, 179
Grossman, S. I., 22
Ground, 89

H, defined, 217, 284
Hall effect, 267
Hall voltage, 267
Halliday, D., 320, 331

Harman, W. W., 318, 331
Harmonics, circular, 215
Hayt, W. H., Jr., 175, 181, 210, 212
Heaviside-Lorentz system of units, 447
Helmholtz, H. L. F. von, 340
Helmholtz equation, 340
Henry, J., 312
Henry, defined, 243, 299
Hertz, H. R., 340
Hole, 131, 132, 200
Homogeneous material, 144, 165, 247, 345
Horizontal pattern, 432
Hughes, W. L., 408
Hurd, F. K., 408
Hydraulics, 180, 235
Hysteresis:
  dielectric, 136
  magnetic, 281, 293

I (*see* Current)
Identities, vector, 443
Images, 102, 129–131
Impedance:
  characteristic, 385, 398
  input, 386, 394, 398, 401, 418
  intrinsic, 343, 346, 348–351, 361
    input, 374
  load, 386, 398
  normalized, 398
Incidence, normal, 364
Incident wave, 364, 366, 369, 386
Inductance, 298–304, 382, 387–392, 415
  of coaxial cable, 299, 387–390
  defined, 299
  external, 303, 387–392
  internal, 303, 387–392
  mutual, 303–304
  of planar transmission line, 392
  of solenoid, 304
  of toroid, 299
  of two-wire transmission line, 390–391
Input impedance, 386, 394, 398, 401, 418
Input intrinsic impedance, 374
Insulator, 120, 132, 451
  (*See also* Dielectric)
Integral:
  Fourier, 344
  line (*see* Line integral)
  surface, 58, 74
    closed, 58, 74, 242
  volume, 35, 74
Intensity:
  electric field (*see* Electric field intensity)
  magnetic field (*see* Magnetic field intensity)
Interface:
  dielectric, 138–142
  magnetic, 287–289
Internal inductance, 303, 387–392
International System of Units, 27, 445–449
International Telephone and Telegraph Co., Inc., 376
Intrinsic impedance, 343, 346, 348–351, 361
  input, 374
Intrinsic semiconductor, 131
Inverse-cube field, 105, 431
Inverse-distance field, 41, 91, 93, 431
Inverse-square-law field, 31, 41, 61, 91, 105, 431
Isotropic material, 121, 136, 281, 286
Iteration method, 169–175

J, defined, 115
Jordan, E. C., 259, 436, 437
Joule, 30
Junction, *pn*, 200

K, defined, 219
Kelvin, defined, 446
Kemmerly, J. E., 210, 212
Kilogram, defined, 446
Kirchhoff's voltage law, 95, 139, 291
Kraus, J. D., 76, 305
Krypton meter, 445

L, defined, 299
Laplace's equation, 169, 188–200, 204–212, 247
Laplacian, 190, 443
  defined, 190
  of a vector, 255–258
Latitude, 19
Lattice, crystalline, 120
Leakage flux, 295
Lenz, H. F. E., 312

Lenz's law, 312
Level, energy, 120
Light, velocity of, 330, 337, 341, 342, 446, 453
Line:
  flux, 52–54
  lossless, 393
Line charge:
  E field of, 39, 41, 86
  parallel, potential field of, 150–155
  potential field of, 88, 93
Line charge density (*see* Charge density, line)
Line current, **H** field of, 221, 223
Line integral, 82
  closed, 94, 233, 239, 313, 316
  path of, 82–89
Linearity, 30, 121, 136, 287
Linkage, flux, 298, 318
Load impedance, 386, 398
Longini, R. L., 155
Longitude, 19
Longo, C. L., 48
Lorentz force, 266, 324
Loss tangent, 350–353, 423, 451
Lossless line, 393
Lossy dielectric, 347–353

M, defined, 283
M, defined, 303
m, defined, 276
Macroscopic region, 122
Magnet, permanent, 217
Magnetic boundary conditions, 287–289, 325–326
Magnetic charge, 244, 297, 323
Magnetic circuit, 289–298
Magnetic dipole, 276
Magnetic dipole moment, 276–278
Magnetic field:
  energy density in, 297, 354–355
  Gauss' law for, 244, 325
Magnetic field intensity, 217, 284
  of coaxial cable, 226–228
  of current sheet, 229
  defined, 244, 325
  of finite line current, 223
  of line current, 221, 223
  of solenoid, 230–231
  of toroid, 231
Magnetic flux, 243–246, 291, 302, 312, 415–417
  defined, 244
Magnetic flux density, 10, 13, 244–246, 266, 290
  defined, 243
  remnant, 293, 294
Magnetic hysteresis, 281, 293
Magnetic imaging, nuclear, 279
Magnetic interface, 287–289
Magnetic materials, 278–282
  characteristics of, table, 282
Magnetic moment, 276–282
Magnetic pole, 323
Magnetic potential:
  scalar, 246–249, 290–291
    defined, 246
  vector, 249–253, 301, 326–330
    defined, 249–250
Magnetic susceptibility, 285
Magnetization, 282–289
Magnetization curve, 293–294
Magnetohydrodynamics, 266
Magnetomotive force (mmf), 290–295
Magnetostriction, 281
Magnitude, vector (*see* Vector magnitude)
Map:
  curvilinear square, 163–168, 176, 221, 249
  field, 45, 163
  fluid-flow, 180
Mass, 179, 446
Matching, 375, 404–407, 433
  stub, 404–407
Material(s):
  anisotropic, 121, 136, 281, 286
  antiferromagnetic, 281–282
  diamagnetic, 279, 286, 452
  ferrimagnetic, 279, 281, 282, 285
  ferroelectric, 136
  ferromagnetic, 280–282, 286, 292–295
  homogeneous, 144, 165, 247, 345
  isotropic, 121, 136, 281, 286
  magnetic, 278–282
  nonmagnetic, 286
  paramagnetic, 280, 282, 286
  superparamagnetic, 279, 282
  (*See also* Conduction, in metals; Dielectric; Semiconductor)
Matrix equation, 137, 286
Matsch, L. W., 155, 305
Maxwell, J. C., 94, 146, 155, 311

Maxwell's equations:
applications of, 336, 342, 353, 382, 384, 414, 427
non-time-varying, 71, 238, 244
time-varying, 314–326
Membrane, elastic, 180
Metallic conductor (*see* Conduction, in metals)
Meter, defined, 445–446
MHD (magnetohydrodynamic) generator, 266
Mho, defined, 121
Mks units, 445–449
Mmf (magnetomotive force), 290–295
Mobility, 120
Molecule:
nonpolar, 133
polar, 133
Moment:
dipole, 105, 133
per unit volume, 133
of a force, 274
magnetic dipole, 276–278
Monopole antenna, 435–436
Moore, A. D., 180, 181
Motional electric field intensity, 316
Motional emf, 316
Moving path, 312–318
Multiplication:
vector:
cross, 12–14
dot, 10–12, 17–18, 21
vector by scalar, 4
vector by vector, 4, 10, 12
Multipole, 106
Mutual inductance, 303–304

n-type semiconductor, 132, 200
Napier, J., 349
Nasar, S. Y., 260, 305
National Bureau of Standards, 445
Neper, 349
Newton, defined, 28, 446
Newton's law of gravity, 27
Newton's third law, 446
Noise, 35
Nonconservative field, 95
Nonlinearity, 136, 137, 290
Nonmagnetic material, 286
Nonpolar molecule, 133
Normal component:
at conductor boundary, 126
at dielectric boundary, 139
at magnetic boundary, 288
at perfect conductor, 325
Normal incidence, 364
Normal to surface, 10, 12, 44, 57, 98
Normalized impedance, 398
North Pole, 19
Notation, scalar and vector, 2
Nuclear spin, 279, 282
Nuclear magnetic imaging, 279
Nucleus, atomic, 119, 120, 278
Nussbaum, A., 331

Oersted, H. C., 312
Ohm, G. S., 121
Ohm:
defined, 123
per square, 178
Ohm's law, 123
point form of, 121, 132, 175, 290, 325, 416
Operation, dot, 72
Operator:
del, 72
scalar, 73
vector, 72
Orbit, electron, 119, 279–282
Orbital shell, 119–120, 279
Origin, 5, 12, 274
Owen, G. E., 318, 331

P, defined, 133
p, defined, 105
$\mathscr{P}$, defined, 355
p-type semiconductor, 132, 200
Paddle wheel, 235
Panofsky, W. K. H., 318, 331
Paper, conducting, 177
Parallel conductors, force between, 272–273
Parallel cylinders, capacitance between, 155, 390, 391
Parallel line charges, potential field of, 150–155
Parallel-plane transmission line, 392
Parallel-plate capacitor, 145–146, 148–150, 195
Parallelepiped, rectangular, 6
Parallelogram, area of, 13
Parallelogram law, 3
Paramagnetic material, 280, 282, 286
Parameters, transmission-line, 387–392
Paris, D. T., 408

Partial capacitance, 146
Particle trajectory, 180
Path:
  closed, 94, 225
  of integration, 82–89
  moving, 312–318
Patterns, antenna, 432
Paul, C. R., 260, 305
Penetration, depth of, 359
Perfect conductor, 176, 325, 366, 370–372
Perfect dielectric, 138–142, 345, 366–369
Permanent magnet, 217
Permeability, 282–286
  of free space, defined, 243, 446, 453
  relative, 285
    table of values, 452
Permittivity, 136–137, 352–353
  defined, 136–137
  of free space, 28, 447, 453
    defined, 447, 453
  relative, 136
    table of values, 451
Phase constant, 345, 351, 358–359
Phase shift, 345, 350, 366, 368
Phase velocity, 342, 385
Phasor, 2, 338, 383
Phasor vector, 339
Φ, defined, 244
Phillips, M., 318, 331
Pillbox, gaussian, 139
Planar transmission line, 392
Plane wave, uniform, 340–353, 357–364, 383
Plonsey, R., 76
Plonus, M. A., 76
pn junction, 200
Point, coordinates of, 4
Point charge, 27–30, 58, 88, 102, 190
  E field of, 31, 32
  potential field of, 89–91
Point charges:
  E field of, 33
  potential field of, 92
Poisson's equation, 189, 200–204
Polar coordinates, 14
Polar molecule, 133
Polarization, 133–138, 141, 324, 356
Pole, magnetic, 323
Polystyrene, 352
Potential:
  absolute, 88
  electrostatic, 87, 88, 246, 326
  retarded, 326–331, 428
  scalar magnetic, 246–249, 290–291
  vector magnetic, 249–253, 301, 326–330
  zero reference for, 89
Potential difference, 87–90, 123, 144, 290
  defined, 87
Potential energy, 95, 106–110, 119, 133, 296, 354
Potential field, 95
  of concentric spheres, 198
  of cones, 199
  of dipole, 103, 105
  of line charge, 88, 93
  of *n* point charges, 92
  of parallel line charges, 150–155
  of point charge, 89–91
  of radial planes, 197
Potential gradient, 95, 246
Potential reference, 89
Potentials, retarded, 326–331, 428
Poularikas, A. D., 213
Power, 353, 420
  average, 356, 361, 362, 368, 422
Power density, 355, 356, 359
Power series, 206
Poynting, J. H., 353
Poynting theorem, 354
Poynting vector, 355
Problems, drill, 9
Product:
  cross, 12–14
  dot, 10–12, 18, 21
  scalar, 10–12, 18, 21
  vector, 12–14
Product solution, 204
Projection:
  scalar, 12
  vector, 12, 21
Propagation, 330, 342, 347, 357, 430
Propagation constant, 345, 347, 358
Proton, 28
Pugh, E. M., 212
Pugh, E. W., 212
Pyranol 1467, 352

Q (*see* Charge; Quality factor)
Quality factor, 350, 422, 426
Quantum of energy, 119
Quantum theory, 119, 279

Radar, 349, 375
Radial planes:
  E field of, 197
  potential field of, 197

Radiation, 417, 427–436
Radiation resistance, 433–435
Radome, 375
Ramo, S., 155, 167, 212, 436
Rao, N. Narayana, 376
Rationalized mks units, 447
Rectangular coordinate system (*see* Cartesian coordinate system)
Rectangular parallelepiped, 6
Recurrence relationship, 206
Reference, potential, 89
Reflected wave, 364–369, 386
Reflection, 364–369, 386
Reflection coefficient, 366, 386, 398
Relative permeability, 285
  table of values, 452
Relative permittivity, 136
  table of values, 451
Relativity, 318
Relaxation method, 175
Relaxation time, 143
Reluctance, 291
Remnant flux density, 293–294
Resistance, 123, 176–178, 291, 363, 416
  defined, 123
  radiation, 433
  sheet, 177, 448
Resistivity, defined, 121
Resnick, R., 320, 331
Resonant cavity, 417–426
Retarded potentials, 326–331, 428
Right-hand rule, 13, 19, 313
Right-handed coordinate system, 5, 13, 16, 19
Right-handed screw, 5, 12, 13
Rochelle salt, 136
Rounding off, 171

Salvadori, M. G., 181
Savart, F., 217
Scalar, 2
  complex, 2
  component, 8, 11
Scalar algebra, 3
Scalar field, 2, 91, 99
Scalar magnetic potential, 246–249, 290–291
  defined, 246
Scalar operator, 73
Scalar product, 10–12, 18, 21
Scalar projection, 12
Scarborough, J. B., 175, 181
Schelkunoff, S. A., 48
Screw, right-handed, 5, 12, 13
Second, defined, 446
Seeley, S., 213
Self-inductance (*see* Inductance)
Semiconductor, 120, 131–132, 200
  intrinsic, 131
  n-type, 132, 200
  p-type, 132, 200
Semiconductor doping, 132, 200
Separation constant, 205
Series, 33, 206, 210, 344
Seshadri, S. R., 376, 408
Sharpe, R. A., 408
Sheet of charge, **E** field of, 44
Sheet resistance, 177, 448
Shell, orbital, 119–120, 279
Shielding, 64, 228
Short circuit, 316, 406
Silvester, P., 175, 181
Single-stub matching, 404–407
Sink, 69
Skilling, H. H., 76, 110, 235, 251, 260
Skin depth, 359–363, 366, 388, 392, 424
Skin effect, 357–363, 384
Smith, A. C., 155
Smith, P. H., 397
Smith chart, 397–407
Smythe, W. R., 213
Solenoid:
  H field of, 230
  inductance of, 304
Solid, crystalline, 120
Sonar, 349
Soroka, W. W., 181
Source, 69
Spangenberg, K. R., 180, 181
Sphere, 19, 59
Spherical capacitor, 147–148, 198
Spherical coordinate system, 19–22, 70, 85, 100, 190, 235, 441
  for expressing curl, 235
  for expressing divergence, 70
  for expressing gradient, 100
  Laplacian, 190
  transformations to cartesian, 21–22
Spiegel, M. R., 22
Spin:
  electron, 279–282
  nuclear, 279, 282
Standing wave, 367–368
Standing-wave ratio, 370–376, 386, 396–397, 404
Stokes' theorem, 238–241
Stratton, J. A., 441

Streamlines, 45–48, 52, 53, 163–168, 221, 249
Stub matching, 404–407
Substitution of circuits, 317
Subtraction of vectors, 4
Summation, 33, 37, 83
Superconductivity, 121
Superparamagnetic material, 279, 282
Superposition, 92
Surface:
closed, 56–58, 241–242
differential, 6, 17, 21, 58
equipotential, 91, 97, 129, 144, 163
gaussian, 58
normal to, 10, 12, 44, 57, 98
vector, 10, 57
Surface charge density (*see* Charge density, surface)
Surface current density, 219, 326
Surface integral, 58, 74, 242
Surfaces:
cartesian coordinate, 6
cylindrical coordinate, 17
spherical coordinate, 21
Susceptibility:
electric, 136
magnetic, 285
Swokowski, E. W., 23
Symbol:
arrow, 3
closed line integral, 94
closed surface integral, 58
unit vector, 6, 8
Symmetry, 37, 52, 61
System of units:
cgs, 447
emu, 447, 449
English, 28, 449
esu, 447, 449
gaussian, 447, 449
Heaviside-Lorentz, 447
International, 27, 445–449
prefixes for, 449
rationalized mks, 447
table, conversion, 449

Tangential component:
at conductor boundary, 125–126
at dielectric boundary, 138–139
at magnetic boundary, 288–289
at perfect conductor, 326
Tank, electrolytic, 176
Taylor, B. N. 451
Taylor's series, 66
Teflon, 138, 144, 352
TEM (transverse electromagnetic) wave, 343, 381, 382
Tesla, defined, 243
Thermal energy, 121, 131
Thomas, G. B., 23, 76, 110
Thomson, R. M., 155
Titanate, barium, 136
Toroid:
H field of, 231
inductance of, 299
Torque, 273–278
on differential current loop, 276
Trajectory, particle, 180
Transformations between coordinate systems, 17–18, 21–22
Transmission coefficient, 366
Transmission line:
coaxial (*see* Coaxial transmission line)
planar or parallel-plane, 392
two-wire, 390–391
Transmission-line chart, 397–407
Transmission-line parameters, 387–392
Transmission lines, 381–407, 418
Transmitted wave, 366, 369
Transverse electromagnetic (TEM) wave, 343, 381, 382
Traveling wave, 342, 370
Trough, electrolytic, 176
Two-wire transmission line, 390–391

Uncurling, 259
Uniform plane wave, 343, 346, 347, 357, 364, 383
Unit vector, 6–8, 12–13, 16, 19, 41, 44
Units, 445–449
abbreviations for, 448–449
conversion table for, 449
system of (*see* System of units)
table of, 448–449

V (*see* Potential, electrostatic; Potential, scalar magnetic)
Valence band, 119, 131
Van Duzer, T., 155, 167, 212, 436
Variables, relationships between, 17, 21
Vector(s), 2
angle between, 10–13, 18
component, 6, 11
coplanar, 3

Vector(s), (*Continued*)
direction of, 6, 41, 45
equality of, 4
Laplacian of a, 255–258
phasor, 339
Poynting, 355
projection of, 21
unit (*see* Unit vector)
Vector addition, 3, 4
Vector algebra, 3
Vector arithmetic, 3
Vector calculus, 3
Vector components, 3–4, 6, 11
Vector field, 2–4, 9
Vector function, 2, 9
Vector identities, 443
Vector magnetic potential, 249–253, 301, 326–330
Vector magnitude, 2, 4, 6, 8, 10–12, 41
Vector multiplication:
cross, 12–14
dot, 10–12, 17–18, 21
Vector notation, 3
Vector operator, 72
Vector product, 12–14
Vector projection, 12
Vector by scalar multiplication, 4
Vector subtraction, 4
Vector surface, 10, 57
Vector by vector multiplication, 4, 10, 12
Velocity:
of charge, 116, 120, 265, 266
drift, 120
of light, 330, 337, 341, 342, 446, 453
phase, 342, 385
of propagation, 330, 342, 346, 360, 430
Vertical pattern, 432
Volt, defined, 30, 87
Volume, differential, 6, 17, 21
Volume charge density (*see* Charge density, volume)
Volume element, 6, 17, 21
Volume integral, 35, 74
von Hippel, A. R., 353, 376, 450

Wave:
incident, 364, 366, 369, 386
reflected, 364–369, 386
standing, 367–368
TEM, 343, 381, 382
transmitted, 366, 369
traveling, 342, 370
uniform plane, 343, 346, 347, 357, 364, 383
Wave equation, 329, 339, 345
Wave motion, 337
Wavelength, 341, 346, 360, 417, 430
Weber, E., 176, 181, 213
Weber, defined, 243
Weeks, W. L., 408, 436
Wert, C. A., 155
Whinnery, J. R., 155, 167, 212, 436
Work, 10, 81, 84, 106
(*See also* Energy)

Zero reference for potential, 89

DIVERGENCE

CARTESIAN $$\nabla \cdot \mathbf{D} = \frac{\partial D_x}{\partial x} + \frac{\partial D_y}{\partial y} + \frac{\partial D_z}{\partial z}$$

CYLINDRICAL $$\nabla \cdot \mathbf{D} = \frac{1}{\rho}\frac{\partial}{\partial \rho}(\rho D_\rho) + \frac{1}{\rho}\frac{\partial D_\phi}{\partial \phi} + \frac{\partial D_z}{\partial z}$$

SPHERICAL $$\nabla \cdot \mathbf{D} = \frac{1}{r^2}\frac{\partial}{\partial r}(r^2 D_r) + \frac{1}{r \sin\theta}\frac{\partial}{\partial \theta}(D_\theta \sin\theta) + \frac{1}{r\sin\theta}\frac{\partial D_\phi}{\partial \phi}$$

GRADIENT

CARTESIAN $$\nabla V = \frac{\partial V}{\partial x}\mathbf{a}_x + \frac{\partial V}{\partial y}\mathbf{a}_y + \frac{\partial V}{\partial z}\mathbf{a}_z$$

CYLINDRICAL $$\nabla V = \frac{\partial V}{\partial \rho}\mathbf{a}_\rho + \frac{1}{\rho}\frac{\partial V}{\partial \phi}\mathbf{a}_\phi + \frac{\partial V}{\partial z}\mathbf{a}_z$$

SPHERICAL $$\nabla V = \frac{\partial V}{\partial r}\mathbf{a}_r + \frac{1}{r}\frac{\partial V}{\partial \theta}\mathbf{a}_\theta + \frac{1}{r\sin\theta}\frac{\partial V}{\partial \phi}\mathbf{a}_\phi$$